KU-722-976

RUBBER TECHNOLOGY

RUBBER TECHNOLOGY

THIRD EDITION

Edited by

MAURICE MORTON

Regents Professor Emeritus of Polymer Chemistry
The University of Akron
Akron, Ohio

Sponsored by the Rubber Division of the

AMERICAN CHEMICAL SOCIETY

VNR VAN NOSTRAND REINHOLD COMPANY
New York

Library of Congress Catalog Card Number 86-26791

ISBN 0-442-26422-4

Printed in the United States of America

Van Nostrand Reinhold Company Inc.
115 Fifth Avenue
New York, New York 10003

Van Nostrand Reinhold Company Limited
Molly Millars Lane
Wokingham, Berkshire RG11 2PY, England

Van Nostrand Reinhold
480 La Trobe Street
Melbourne, Victoria 3000, Australia

Macmillan of Canada
Division of Canada Publishing Corporation
164 Commander Boulevard
Agincourt, Ontario M1S 3C7, Canada

16 15 14 13 12 11 10 9 8 7 6 5 4 3 2 1

Library of Congress Cataloging-in-Publication Data

Rubber technology.

Includes bibliographical references and index.
1. Rubber industry and trade. I. Morton, Maurice.
TS1890.R86 1987 678′.2 86-26791
ISBN 0-442-26422-4

To my wife
Lilian
for her patience
and encouragement

PREFACE

About ten years after the publication of the Second Edition (1973), it became apparent that it was time for an up-date of this book. This was especially true in this case, since the subject matter has traditionally dealt mainly with the structure, properties, and technology of the various elastomers used in industry, and these are bound to undergo significant changes over the period of a decade.

In revising the contents of this volume, it was thought best to keep the original format. Hence the first five chapters discuss the same general subject matter as before. The chapters dealing with natural rubber and the synthetic elastomers are up-dated, and an entirely new chapter has been added on the thermoplastic elastomers, which have, of course, grown tremendously in importance. Another innovation is the addition of a new chapter, ''Miscellaneous Elastomers,'' to take care of ''old'' elastomers, e.g., polysulfides, which have decreased somewhat in importance, as well as to introduce some of the newly-developed synthetic rubbers which have not yet reached high production levels.

The editor wishes to express his sincere appreciation to all the contributors, without whose close cooperation this task would have been impossible. He would especially like to acknowledge the invaluable assistance of Dr. Howard Stephens in the planning of this book, and for his suggestion of suitable authors. He would also like to express his sincere thanks to Connie Morrison, of the Rubber Division, ACS, for her painstaking efforts in carrying out the large volume of secretarial work involved in this type of publication.

MAURICE MORTON

CONTENTS

1

INTRODUCTION TO POLYMER SCIENCE

MAURICE MORTON
Regents Professor Emeritus of Polymer Chemistry
The University of Akron
Akron, Ohio

Polymer science is concerned with the composition and properties of a large number of substances classed as ''polymers,'' which include rubbers, plastics, and fibers. The meaning of the term ''polymer'' will become clear during the course of this chapter, but it does involve some understanding of basic chemistry. The science of chemistry concerns itself with the composition of matter and with the changes that it undergoes. To the layman, the methods and processes of chemistry appear quite baffling and difficult to comprehend. Yet these are essentially based on a simple, logical development of knowledge about the substances which comprise our physical world. Since this book is not intended exclusively for readers trained in chemistry, the subject of polymer chemistry will be developed from the type of first principles comprehensible to anyone.

ATOMS AND MOLECULES

In this ''atomic'' age, it is certainly common knowledge that the atom is the basic unit of all matter. There are as many different kinds of atoms as there are elementary substances, i.e., elements, and these number about 100. It is the myriads of combinations of these atoms which make possible the hundreds of thousands of different substances comprising our world. Just as in the case of the 100-odd elements it is the atom which is the smallest unit, so too in the case of all other substances (*chemical compounds*), it is the particular *group of atoms*, or *molecule*, which is the smallest possible unit. If the molecule of a compound substance is broken up, the substance ceases to exist and may be reduced to the elementary substances which comprise it. It is the various possible combinations of element atoms to form different molecules which is the basis of chemistry.

Now, in this "nuclear" age, we hear much about "atomic fission" processes, in which atoms are split apart, with the accompaniment of considerable energy release. How then can an atom be split if it is supposedly a basic unit of matter? Well, we know today that atoms are not the indivisible particles they were originally believed to be. Instead, they, too, are made up of even smaller particles such as protons, electrons, neutrons, and others. The differences between atoms can be explained by the different numbers and arrangements of these subatomic particles.

The thing to remember, however, is that the different substances of our world make their appearance once atoms have been formed. This becomes obvious when we consider that all protons are the same, all neutrons are the same, and so on. There is no such thing as a "proton of iron" being different from a "proton of oxygen." However, an atom of iron is considerably different from an atom of oxygen. That is why the chemist is concerned mainly with the arrangement of atoms into molecules and considers the elemental atoms as the basic building blocks.

Valence

So far nothing has been said about the manner in which the subatomic particles are held together or what holds atoms together in molecules. Although this is not our immediate concern here, it should be understood that there are electromagnetic forces which hold all these particles together. Thus most of the subatomic particles carry either positive or negative electrical charges, which act as binding forces within the atoms as well as between different atoms. Because each atom contains a different number and arrangement of these charged particles, the forces acting between different atoms also are different. These forces, therefore, result in certain well-defined rules which govern the number and type of atoms which can combine with each other to form molecules. These rules are called the rules of *valence*, and refer to the "combining power" of the various atoms.

From the beginning of the nineteenth century, chemists have been carefully weighing, measuring, and speculating about the proportions in which the element substances combine to form compounds. Since they found that these proportions were always the same for the same substances, this was most easily explained by assuming combinations of fixed numbers of atoms of each element. This was the basis for Dalton's Atomic Theory, first proposed in 1803. In accordance with this hypothesis, an atom from the element hydrogen is never found to combine with more than one atom of any other kind. Hence, hydrogen has the lowest combining power, and is assigned a valence of 1. On the other hand, an atom of the element oxygen is found capable of combining with two atoms of hydrogen, hence can be said to have a valence of 2.

In this way, the kinds and numbers of atoms that make up a molecule of a given compound can be deduced and expressed as a *formula* for that compound. Thus the compound water, whose molecules are each found to contain 2 atoms of hydrogen and 1 atom of oxygen, is denoted by the formula H_2O, where the letters represent the kinds of atoms, and the numbers tell how many of each atom are present within one molecule. This formula could also be written as H—O—H to show how 1 oxygen atom is bonded to 2 hydrogen atoms. In the case of the well-known gas, carbon dioxide, the molecule would be represented as CO_2, where the carbon atom would show a valence of 4, since it is capable of combining with 2 oxygen atoms (each of which has a valence of 2). This formula could also be written as O=C=O, which is known as a *structural formula*, since it shows exactly how the carbon and oxygen atoms are bonded. Such structural formulas must be written in a way which will satisfy the valence of all atoms involved.

MOLECULES AND MACROMOLECULES

We have seen how the chemist is concerned with the structure of the molecules which comprise the substances in our world. At this point it might be of interest to consider the actual size of these molecules. It is now known that a molecule of water, for instance, measures about ten billionths of an inch, that is, one million such molecules could be laid side-by-side to make up the thickness of the type used on this page. Putting it another way, one drop of water contains 1,500,000,000,000,000,000,000 molecules! Most ordinary substances have molecules of this approximate size. Thus the molecule of ordinary sugar, which has the rather impressive chemical formula $C_{12}H_{22}O_{11}$, is still only a few times larger than the water molecule.

Such substances as water, sugar, ammonia, gasoline, baking soda, and so forth, have relatively simple molecules, whose size is about the same as indicated above. However, when chemists turned their attention to the chemical structure of some of the "materials" like wood, leather, rubber, and so forth, they encountered some rather baffling questions. For instance, *cellulose* is the basic fibrous material which comprises wood, paper, cotton, linen, and other fibers. Chemists knew long ago that cellulose has the formula $C_6H_{10}O_5$, as shown by chemical analysis, yet their measurements did not indicate that the molecule of cellulose existed in the form shown by the formula. Instead, they found evidence that the cellulose molecule was unbelievably large. Hence they did not really believe that it was a molecule, but considered the true molecule to be $C_6H_{10}O_5$, and that hundreds of such molecules were bunched together in aggregates.

Rubber, like cellulose, was also classed as a *colloid*, i.e., a substance which contains large aggregates of molecules. Over 150 years ago, the formula for rubber was established as C_5H_8, but here again the evidence indicated that the molecule was much larger than shown by the formula. It was only as recently as the 1920's that Staudinger advanced the revolutionary idea that the rubber molecule was indeed a giant molecule, or *macromolecule*. Today, it is generally accepted that the class of materials which form the useful fibers, plastics, and rubbers are all composed of such macromolecules. Hence the formula for cellulose, if it is to represent one molecule, should more correctly be written as $(C_6H_{10}O_5)_{2000}$, while rubber should be represented as $(C_5H_8)_{20,000}$.

Such molecules are, of course, many thousands times larger than the molecules of ordinary chemical substances, hence the name "macromolecules." However, it should be remembered that even these giant molecules are far from visible in the best microscopes available. Thus the cellulose or rubber molecule may have a diameter several hundred times larger than that of a water or sugar molecule, but its size is still only several millionths of an inch! Even so, however, their relatively enormous dimensions, in comparison with ordinary molecules, give them the unusual properties observed in plastics, rubbers, and fibers.

POLYMERS AND MONOMERS

The molecular formula for rubber, $(C_5H_8)_{20,000}$, at first glance appears formidable indeed. There can obviously be a vast number of possible arrangements of 100,000 carbon atoms and 160,000 hydrogen atoms! To deduce the exact way in which they are arranged seems an impossible task. Yet it is not as difficult as it would seem. Fortunately—for the chemists—these huge molecules are generally composed of a large number of simple repeating units attached to each other in long chains. Thus the strucutral formula for the molecule of natural rubber may be represented by the simple unit C_5H_8 multiplied many thousand times. It is actually shown thus:

$$\left[-CH_2-C(CH_3)=CH-CH_2- \right]_n$$

In the above structure, n represents a value of about 20,000, as stated previously. The arrangement of carbon and hydrogen atoms obeys the valence rules of 4 for carbon and 1 for hydrogen. The unattached bonds at each end of the unit are, of course, meant to show attachments to adjacent units.

Because these giant molecules consist of a large number of repeating units, they have been named *polymers*, from the Greek "poly" (many) and "meros" (parts). The repeating unit shown above would, therefore, be called the *monomer*. It is actually very similar to a simple compound, known as isoprene, a low-boiling liquid. In 1860, Greville Williams first isolated this compound as a decomposition product obtained when rubber was heated at elevated temperatures, out of contact with air. The relationship between the monomer and polymer structures, in this case, is illustrated below.

```
        monomer                          polymer

H          H H               ⎡ H          H  H   ⎤
|          | |               ⎢ |          |  |   ⎥
C══C────C=C                  ⎢—C────C══C—C—      ⎥
|   |        |               ⎢ |   |         |   ⎥
H H—C—H      H               ⎢ H H—C—H       H   ⎥
    |                        ⎢     |             ⎥
    H                        ⎣     H             ⎦n

     isoprene                polyisoprene (natural rubber)
```

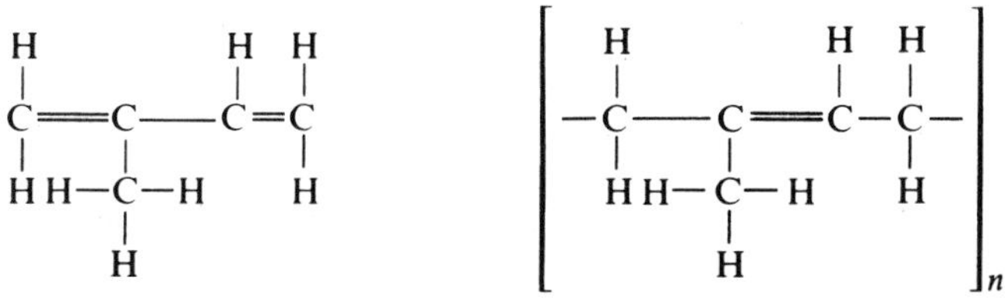

It can be seen here that the unit in the chain molecule corresponds exactly with the uncombined monomer molecule except for the necessary rearrangements of the bonds between the carbon atoms.

THE SYNTHESIS OF MACROMOLECULES

So far we have been discussing the macromolecules, or polymers, which occur ready-made in nature. Although the chemist may surmise, he cannot be entirely certain just how these large molecules are built up in nature. However, over the past half-century, chemists have become very proficient in producing synthetic polymers from a wide variety of possible monomers. It is from these processes that there have arisen all the new and varied synthetic rubbers that we know today.

In order to form a polymer molecule, it is, of course, necessary to find a way in which individual small molecules may join together in large numbers. This process is, for obvious reasons, called *polymerization*. There are a number of ways in which this process can be effected, but these can all be broadly classed under two main types of chemical reactions. One of these is known as *addition polymerization*, since it involves a simple addition of monomer molecules to each other, *without the loss of any atoms* from the original molecules. The other type is called *condensation polymerization*, since it involves a reaction between monomer molecules during the course of which a bond is established between the monomers, but some of the atoms present are lost in the form of a by-product compound. In this case, then, the polymer molecule is a *condensed* version of the original monomer molecules which reacted.

Addition Polymerization

The simplest example of addition polymerization is the formation of polyethylene from ethylene. This can be shown by means of the following molecular structures:

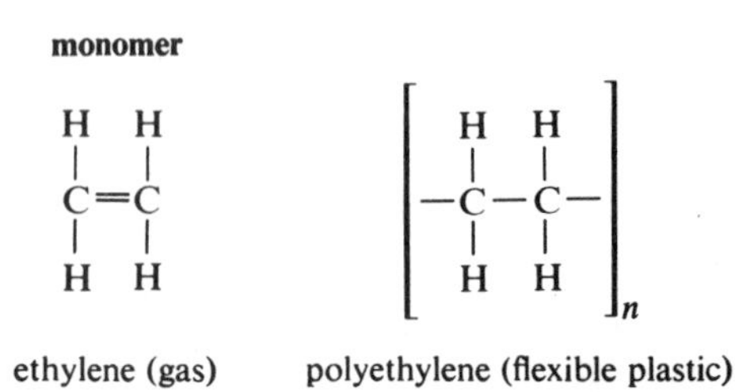

ethylene (gas) polyethylene (flexible plastic)

It can be noted that, in the ethylene molecule, after due regard is given to the requirements of valence, there is a "double bond" between the two carbon atoms. Furthermore, to cause polymerization, *one* of these two bonds must *open* and become available as *two bonds*, ready to unite with similarly available bonds from adjacent ethylene molecules. For this reason, ethylene is one of the compounds known as "unsaturated," due to the presence of this extra bond between the carbon atoms, a bond which can be "saturated" by bonding to *two* other atoms, carbon or otherwise. This makes it possible for an "unsaturated" molecule to form a long-chain polymer molecule, as shown.

The well-known line of "vinyl" and "acrylic" polymers in the plastics and fibers field are all based on this activity of the carbon—carbon double bond. As a matter of fact, the term "vinyl" is a generic chemical name for this bond. The following monomers are the starting materials for the better-known polymers in this field, as the names will indicate.

H H	H H	H Cl	H CH_3
| |	| |	| |	| |
C=C	C=C	C=C	C=C
| |	| |	| |	| |
H Cl	H C_6H_5	H Cl	H $COOCH_3$
vinyl chloride	styrene	vinylidene chloride	methyl methacrylate

A somewhat more complex form of addition polymerization occurs with the unsaturated monomers of the "diene" type. These differ from the vinyl monomers in possessing *two* double bonds. An important and typical diene is *butadiene*, which forms the basis of the most important synthetic rubber today. The structures of the monomer and polymer are shown below.

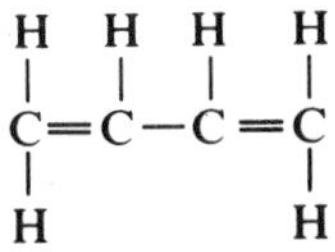

$$\begin{array}{cccc} H & H & H & H \\ | & | & | & | \\ C= & C- & C= & C \\ | & & & | \\ H & & & H \end{array} \qquad \left[\begin{array}{cccc} H & H & H & H \\ | & | & | & | \\ -C- & C= & C- & C- \\ | & & & | \\ H & & & H \end{array} \right]_n$$

Butadiene (gas) — Polybutadiene (a rubbery solid)

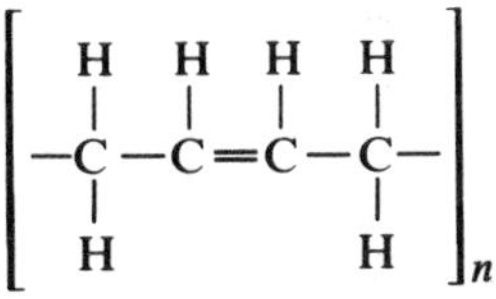

The similarity between butadiene and isoprene is immediately apparent. Isoprene is actually the same as butadiene, except for the presence of a "methyl" group (CH_3) on one of the carbon atoms, instead of a hydrogen atom. As a matter of fact, isoprene has the chemical name 2-methyl butadiene.

It can be seen at once that when a diene is polymerized, the polymer obtained differs in one important respect from the vinyl polymers, viz., it still contains a *double bond* in each unit of the chain. This is a natural outcome of the fact that the monomer has *two double bonds* at the outset. The residual unsaturation in these polymers plays a very important role in the process of vulcanization, which will be discussed later. It is quite obvious, however, from the foregoing discussion, that the presence of these double bonds in the polymer makes it possible for the latter to react further with agents, like sulfur, which can add to these available double bonds, and actually form crosslinks between adjacent chain molecules.

Condensation Polymerization

Polymers made by a condensation reaction are not as important in the rubber field as the addition polymers discussed above. However, there are one or two types which are of some importance. One of these is the class of polymers known as polysulfide polymers, and bearing the trade name of Thiokol. This was one of the first synthetic rubbers developed in the United States, as an oil- and solvent-resistant elastomer. Various grades of this material are prepared using variations in starting materials, but the following will illustrate the process:

$$\underset{\text{ethylene dichloride}}{Cl-CH_2-CH_2-Cl} + \underset{\text{sodium polysulfide}}{Na-S_x-Na} \longrightarrow \underset{\text{poly(ethylene polysulfide)}}{[-CH_2-CH_2-S_x-]_n} + \underset{\text{salt (sodium chloride)}}{NaCl}$$

where $x = 2 - 4$.

The above equation shows that this polymerization proceeds by means of the reaction between the sodium (Na) atoms and the chlorine (Cl) atoms, which combine to form salt (NaCl). This results in a bond being formed between the sulfur and carbon atoms, and a long-chain molecule is thus formed. Here the salt is the by-product of the condensation reaction and is not included in the polymer chain. In all such condensation reactions, provision must be made to remove such by-products in order to permit the reaction to proceed smoothly and to avoid contamination of the polymer. In the above case, the salt is conveniently removed by performing the reaction in the presence of water, which dissolves the salt but not the polymer.

THE PHYSICAL BEHAVIOR OF POLYMERS

Up to this point, we have discussed the basic principles involved in the synthesis and structure of polymers, and it would therefore be appropriate to consider in more detail the individual rubber polymers with which the technologist is concerned. However, before doing so it would seem highly desirable to consider the special features which make it possible for polymers to have their unique physical properties. We have seen how these macromolecules are composed of the same kinds of atoms as ordinary "chemicals," and are distinguished only by their enormous size. How does this difference in size convert a simple *chemical* into a strong, tough, durable *material*? That is indeed a question which scientists have been studying with a good deal of interest.

The most reasonable answer to the above question may be obtained by considering the two following aspects, viz., the type of forces which act on atoms and molecules in general and the structure and behavior of an individual polymer molecule.

Interatomic and Intermolecular Forces

We have already seen that there are electromagnetic forces operating between atoms, i.e., "valence" forces that bind atoms together. These forces, operating as they do at relatively short ranges, are very powerful indeed, so that a considerable amount of energy is required to break up a molecule of a compound. This is evident from the fact that high temperatures are usually necessary to decompose substances. The physical strength of interatomic bonds can best be exemplified by considering the basis for the strength of metals and minerals. It is generally agreed that such materials do not consist of molecules at all but contain a three-dimensional crystalline structure of atoms bonded to each other by the powerful valence forces mentioned above. Hence, to rupture the metal or mineral it is necessary to actually break these valence bonds, and this usually requires forces of many thousands of pounds per square inch.

On the other hand, a consideration of the forces operating *between molecules* soon shows that these are very much smaller than the interatomic forces. Hence it can be stated that those substances where atoms exist in small groupings, i.e., molecules, are either liquids or weak solids, since the intermolecular bonds are very weak. It is well-known that the "states" of matter depend on the balance between the intermolecular forces and the heat energy of the molecules, as indicated by temperature. Thus heat tends to make atoms or molecules move faster, i.e., gives them a greater kinetic energy, while the intermolecular forces pull them closer together and tend to restrain their motions. This is the distinction between gases, where the intermolecular forces are far too weak to overcome the kinetic energy of the molecules, and liquids or solids, where the opposite is the case. There is a further distinction: between liquids, which flow because these forces are too weak to hold the molecules firmly in place, and true solids, where such molecules (or atoms) are held in a rigid pattern, i.e., crystals.

Structure and Behavior of Macromolecular Chains

In this regard, the behavior of macromolecules is further influenced by their long-chain character. Although the chemical formulas shown thus far for these molecules indicate that they are very long chains, they do not show that the chains are *flexible* by virtue of the ability of the chain carbon atoms, held together by single bonds, to *rotate* around their axis. Hence each macromolecular chain can be expected to be capable of twisting into various convolutions. Depending on the temperature, therefore, it could be expected that these chains would be in a constant twisting motion and that they would thus be badly entangled with each other. In such a disordered state, if the intermolecular (interchain) forces are not very strong, the polymer could be considered to be a *liquid*, but a very *viscous* liquid because of the long-chain, entangled state of the molecules. In fact, such a material would be so viscous as to have the appearance of a solid, i.e., an *elastic* solid or a rubber.

The reason for the "rubbery state" is, of course, that any deformation will tend to "straighten out," or uncoil, the entangled mass of contorted chains, and these will tend to coil up again when the restraining force is released. Thus the elastic retractive force is really due to the violent contortions of the long, flexible chains.

It should be remembered that the long-chain character of polymer molecules does not make it impossible for them to form crystals, just as small molecules do. However, as in the case of simple compounds, this crystallization will occur only if the intermolecular (interchain) forces are strong enough to overcome the kinetic energy of the contorting chains. This can happen in two ways, i.e., either the chain atoms, or groups of atoms, can exert very powerful attractive forces, and/or the chain sections can "fit together" so well that they can come close enough for the interchain forces to take over. The only real distinction between the crystalline, solid state of simple compounds and polymers is that the crystallization is much less complete in the case of the latter. Thus, whereas sugar can be said to crystallize almost "perfectly," polyethylene is only a partly crystalline solid. This is because, as is obvious, it is very difficult, if not impossible, to form "perfect" crystals from a mass of entangled chains.

A large number of polymers exist as partly crystalline solids at normal temperatures, of which polyethylene is an excellent example. The reason this well-known, flexible plastic material is opaque is, in fact, due to the presence of very fine crystallites which refract and scatter light. These crystallites also "tie together" and restrict the motions of the polymer chains so that the latter lose their elastic character. Hence the material has only limited elasticity but retains its flexibility, as in the case of polyethylene and other flexible plastics.

The ability some macromolecular chains have of crystallizing under the right conditions plays a dominant role in the properties of two types of materials: *fibers* and *rubbers*. Thus, in the case of a fiber-forming polymer such as nylon, the polyamide chains can exert powerful attractive forces toward each other. Hence, when these chains are oriented, e.g., by cold drawing, the interchain forces are strong enough to cause the oriented chains to crystallize into elongated crystals containing bundles of rigid chains, i.e., *fibers*. These forces thus act to give the fibers very high strength, and the fiber "crystals" cannot be melted again even at temperatures high enough to cause chemical decomposition.

In the case of the rubbers, the macromolecular chains may or may not be capable of crystallizing on stretching, depending on their chemical structure and regularity. Thus natural rubber, which has a very regular chain structure, can undergo a high degree of crystallization on stretching and therefore becomes a "fiber" at high elongations. This results in a high tensile strength. However, the interchain forces which cause such strain-induced crystallization are not sufficiently powerful to maintain this fiber-like structure once the applied force is removed, so that the fiber-like crystals "melt" and the rubber chains retract to their normal configuration. This phenomenon of "temporary crystallization" therefore plays a very important role in controlling the mechanical properties of a rubber, especially strength, and these properties vary greatly depending on whether the polymer does or does not undergo crystallization on stretching.

It is important to note at this point that the ability of an elastomer to crystallize on stretching, advantageous for strength, also means that the elastomer will crystallize, at least partially, at some low temperature *without stretching*. This temperature is referred to as its *crystal melting point*, T_m, and is, of course, generally below room temperature; otherwise the rubber will crystallize on storage and harden considerably. As a matter of fact, this can sometime happen to natural rubber, whose T_m is just below normal ambient temperatures.

At lower temperatures, elastomers also exhibit another phenomenon, known as the "glass transition." This occurs regardless of whether or not the polymer is capable of crystallization, and results in a transformation of the elastomer into a rigid plastic. Thus the "glass-transition temperature," T_g, also known as the "glass point," of natural rubber is $-72°C$; that of SBR is about $-50°C$; while the synthetic polybutadienes exhibit T_g values as low as $-100°C$. Although the glassy "state" makes these materials into rigid solids, they are still not considered as being in the true solid state, as defined by scientific criteria, since they have not really crystallized. Instead, such "glasses" are defined as "supercooled liquids," just as a rubber is defined as a "liquid." The best-known example of such a "glass" is, of course, ordinary glass itself, which is a supercooled silicate. Other examples are the well-known glass-like plastics, such as polystyrene and poly(methyl methacrylate), which owe this property to the fact that their T_g is well *above room temperature*, i.e., about 100°C, above which they, too, become rubbery in character.

Finally, there is one more aspect of elastomers that needs careful consideration. Since these materials have been defined, in a scientific sense, as very viscous elastic *liquids*, it is not surprising that they can *flow*, especially as the temperature increases. This is due, of course, to the ability of the entangled long-chain molecules to *slip* past each other when under a distorting force. Hence it is necessary to "anchor" these elastic chains to each other in order to have the elastomer behave as a truly elastic material, i.e., to exhibit a high degree of elastic recovery and a minimum of "set." For this reason elastomers have to be *vulcanized* for optimum properties; this process consists simply of introducing *crosslinks* between the long-chain molecules to obtain a continuous *network* of flexible, elastic chains. Vulcanization is accomplished by means of various chemical reactions, depending on the chemical structure of the macromolecule, the most common process involving sulfur and its compounds, which works very well with the unsaturated elastomers such as natural rubber and the synthetic polydienes.

NATURAL RUBBER AND SYNTHETIC ELASTOMERS

On the basis of the foregoing general discussion of polymers, it should now be possible for the reader to examine the chemical structure of the better-known elastomers, and to understand the various methods used in the vulcanization of these materials. Hence the remainder of this chapter will be concerned with the structures of the more important elastomers together with their behavior on stretching and their method of vulcanization; the other elastomers are discussed in later chapters. Those elastomers which do not have the ability to crystallize on stretching exhibit inferior tensile strength, as might be expected. However, when mixed with "reinforcing" pigments, such as carbon black, these elastomers develop high strengths, equal to that of natural rubber. It is thought that the polymer chains actually form attachments to the surface of these reactive pigments so that the pigment particles act just like crystallites in increasing the tensile strength. All of this information is summarized in Table 1.1.

Natural Rubber (and Stereoisomerism)

It has already been noted that natural rubber has the chemical name of polyisoprene. It is also important to note that there is a special feature about its structure, that accounts for its special properties. This concerns the possible *isomers* that can occur in a polyisoprene chain as follows:

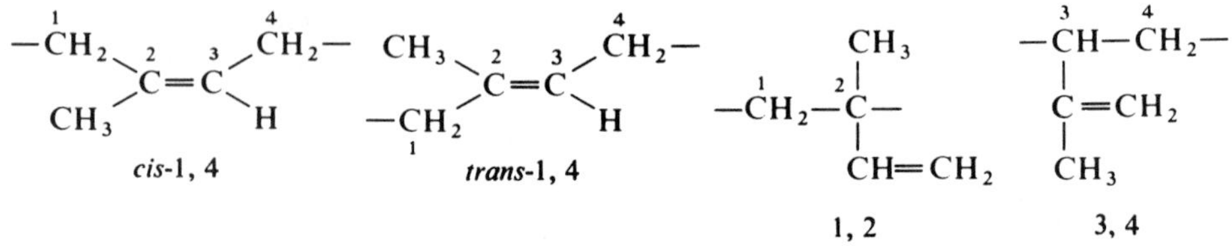

$$\begin{array}{c}-\overset{1}{C}H_2 \diagdown \quad \diagup \overset{4}{C}H_2- \\ \overset{2}{C}=\overset{3}{C} \\ CH_3 \diagup \quad \diagdown H\end{array}$$

cis-1, 4

$$\begin{array}{c}CH_3 \diagdown \quad \diagup \overset{4}{C}H_2- \\ \overset{2}{C}=\overset{3}{C} \\ -\underset{1}{C}H_2 \diagup \quad \diagdown H\end{array}$$

trans-1, 4

$$\begin{array}{c}CH_3 \\ | \\ -\overset{1}{C}H_2-\overset{2}{C}- \\ | \\ CH=CH_2\end{array}$$

1, 2

$$\begin{array}{c}-\overset{3}{C}H-\overset{4}{C}H_2- \\ | \\ C=CH_2 \\ | \\ CH_3\end{array}$$

3, 4

All of the above structures could, in theory, occur in a polyisoprene chain. The numbers refer to the particular carbon atoms in each unit which are attached to adjacent units. Thus a 1,4 structure means that carbon atoms 1 and 4 are joined in forming the chain. The terms *cis* and *trans* refer to the positions of the various carbon atoms with reference to the carbon–carbon double bond. Since a double bond is considered to prevent rotation of the attached atoms, it follows that other atoms or groups of atoms may occupy positions on *either side* of the double bond. Thus it can be seen that, in the *cis*-1,4 structure, carbon atoms 1 and 4 are *both* on the *same side* of the double bond, while, in the *trans*-1,4 structure, these two carbon atoms are on opposite sides of the double bond.

It turns out that natural rubber consists of polymer chains all having an almost perfect *cis*-1,4 structure; hence the true chemical name for this polymer is *cis*-1,4-polyisoprene. When the chain units in a macromolecule all consist of the *same isomer*, the polymer is said to be *stereoregular*. Because of this remarkable regularity, the natural rubber chains can attain a good regularity, especially when the rubber is stretched. Hence natural rubber crystallizes on stretching, resulting in high gum tensile strength.

Natural rubber is vulcanized with sulfur compounds that can crosslink the chains because of the presence of the reactive double bonds (unsaturation).

Synthetic "Natural Rubber" (Cis-1,4-polyisoprene)

Although the polymerization of isoprene dates back over 100 years, all attempts to synthesize the *cis*-1,4-polyisoprene structure were unsuccessful until the advent of the "stereospecific" catalysts during the decade of the 1950's. Prior to that, any of the synthetic polyisoprenes had a "mixed" chain structure, containing a random arrangement of the four isometric units. When *cis*-1,4-polyisoprene was finally synthesized, it was found to virtually duplicate the behavior and properties of natural rubber, e.g., crystallization on stretching and high gum tensile.

Polybutadiene

Stereospecific catalysts can also be used to polymerize butadiene to a high *cis*-1,4 structure, as shown below.

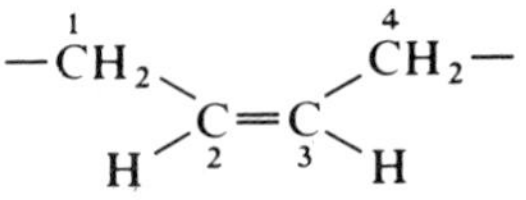

$$
\begin{array}{ccccc}
-\overset{1}{C}H_2 & & & & \overset{4}{C}H_2- \\
 & \diagdown & & \diagup & \\
 & & \underset{2}{C}=\underset{3}{C} & & \\
 & \diagup & & \diagdown & \\
H & & & & H
\end{array}
$$

This is an entirely new polymer which has been developed commercially during the past 25 years. Most of the polybutadienes produced today are of the *cis*-1,4 type, but some have a mixed chain structure. Although *cis*-1,4-polybutadiene, like its polyisoprene counterpart, is capable of crystallizing when stretched, it does not exhibit as high a gum tensile strength and is usually compounded with a reinforcing filler. Being an unsaturated elastomer, it is easily vulcanized with sulfur.

SBR Polymers

Styrene-butadiene rubbers (SBR) are the general-purpose synthetic rubbers today, and were originally produced by government-owned plants as GR–S. They are *copolymers*, i.e., polymer chains obtained by polymerizing a *mixture* of two monomers, butadiene and styrene, whose structures have been shown previously. The chains therefore contain random sequences of these two monomers, which gives them rubberlike behavior but renders them too irregular to crystallize on stretching. Hence these rubbers do not develop high tensile strengths without the aid of carbon black or other reinforcing pigments.

Nitrile Rubber

This, too, is a copolymer of two monomers, butadiene and acrylonitrile.

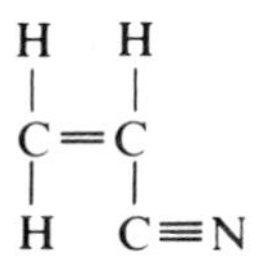

$$
\begin{array}{ccc}
H & & H \\
| & & | \\
C & = & C \\
| & & | \\
H & & C \equiv N
\end{array}
$$

acrylonitrile

It is prepared as a solvent-resistant rubber, the presence of the nitrile group ($C \equiv N$) on the polymer being responsible for this property. Like SBR, it also has an irregular chain structure and will not crystallize on stretching. Hence nitrile rubber requires a reinforcing pigment for high strength. Vulcanization is achieved by means of sulfur, as for SBR and natural rubber.

Butyl Rubber

This also is a copolymer, containing mostly isobutylene units with just a few percent of isoprene units. Hence, unlike the butadiene rubbers or natural rubber, this polymer contains only a few percent double bonds (due to the small proportion of isoprene).

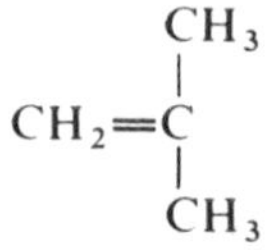

$$CH_2{=}\underset{\displaystyle CH_3}{\overset{\displaystyle CH_3}{\overset{|}{\underset{|}{C}}}}$$

isobutylene (gas)

This small extent of unsaturation is introduced to furnish the necessary sites for sulfur vulcanization, which is used for this rubber. The good regularity of the polymer chains makes it possible for this elastomer to crystallize on stretching, resulting in high gum tensile strength. However, because it is such a soft polymer, it is usually compounded with carbon black to increase the modulus.

Ethylene-propylene Rubbers

Like butyl rubber, the ethylene-propylene rubbers contain only a few percent double bonds, just enough for sulfur vulcanization, EPDM. They are copolymers of ethylene and propylene, containing in addition a few percent of a diene for unsaturation. These elastomers are also among the newer ones made possible by the advent of the stereospecific catalysts. Because the various chain units are randomly arranged in the chain, these elastomers do not crystallize on stretching and require a reinforcing filler to develop high strength.

Neoprene

This elastomer is essentially a polychloroprene, as shown. Chloroprene monomer is actually 2-chlorobutadiene, i.e., butadiene with a chlorine atom replacing one of the hydrogens. Since the polymer consists almost entirely of *trans*-1,4 units, as shown,

$$CH_2{=}\underset{\displaystyle Cl}{\underset{|}{C}}{-}CH{=}CH_2 \longrightarrow \begin{matrix} -CH_2 & & & & H \\ & \searrow & C{=}C & \nearrow & \\ Cl & \nearrow & & \searrow & CH_2- \end{matrix}$$

chloroprene (liquid) — trans-1, 4-polychloroprene

the chains are sufficiently regular in structure to crystallize on stretching. Hence neoprene exhibits high gum tensiles and is used in the pure gum form in many applications.

The vulcanization of neoprene is quite different from the elastomers considered so far. Unlike the others, it is not vulcanized by means of sulfur. Instead, use is made of the fact that the chlorine atoms on the chain can react to some extent with active metal oxides. Hence, zinc oxide or magnesium oxide are used to combine with some of the chlorine and interlink the polymer chains at those sites.

Polysulfide Elastomers

These rubbers, best known under the Thiokol trade name, have already been described as condensation polymers. These polymer chains do not crystallize on stretching; hence this material is used with fillers for reinforcement.

Here again vulcanization is not achieved by means of sulfur—which would not work, since this is not an unsaturated chain. Instead, use is made of the ability of the sulfur atoms to react with active metal oxides, like zinc oxide, which thus interlink the chains into a network.

Silicone Rubber

The silicone rubbers represent a completely different type of polymer structure from any of the others. This is because their chain structure does *not* involve a long chain of carbon atoms but a sequence of silicon and oxygen atoms, as shown:

polymer

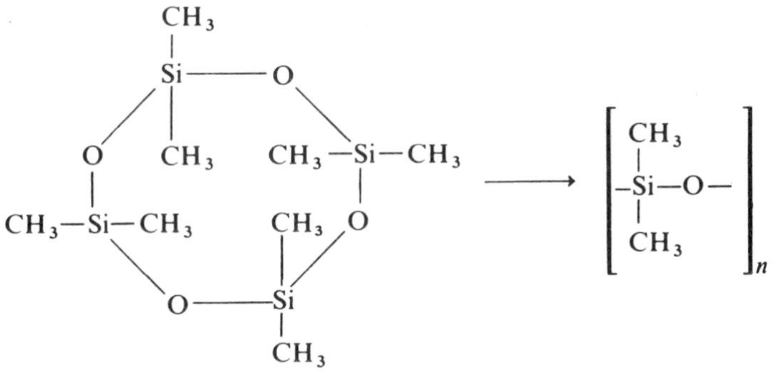

$$\text{cyclic } [Si(CH_3)_2-O]_4 \longrightarrow \left[-\overset{\displaystyle CH_3}{\underset{\displaystyle CH_3}{\overset{|}{\underset{|}{Si}}}}-O- \right]_n$$

cyclic siloxane polysiloxane

This siloxane structure results in a very flexible chain with extremely weak interchain forces. Hence the silicone rubbers are noted for showing little effect over a range of temperature. As might be expected, they show no tendency to crystallize on stretching and must be reinforced by a pigment, usually a fine silica powder.

Silicone rubber is customarily vulcanized by means of peroxides. These presumably are able to remove some of the hydrogen atoms from the methyl (CH_3) groups on the silicon atoms, thereby permitting the carbon atoms of two adjacent chains to couple and form crosslinks:

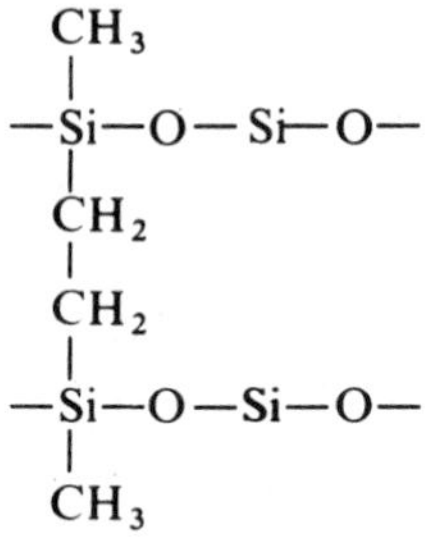

```
 CH3
  |
—Si—O—Si—O—
  |
 CH2
  |
 CH2
  |
—Si—O—Si—O—
  |
 CH3
```

Urethane Polymers

A novel modification in the polymerization field is represented by the urethane polymers. This is a novel method insofar as it involves a "chain extension" process rather than the usual polymerization reaction. In other words, these systems are able to make "big" macromolecules from "small" macromolecules, rather than from the monomer itself. Aside from being special processes for the preparation of polymers, these systems have special advantages, since the properties of the final polymer depend both on the *type* of original short-chain polymers used as well as on their *chain length*. Hence a wide variety of polymers can thus be synthesized, ranging from rigid to elastic types.

The short-chain polymers used are generally two types, i.e., polyethers and polyesters. These can be shown as follows:

$$\mathrm{HO[-R-O-]_{\mathit{n}}H}$$

polyether

$$\mathrm{HOR'O}\left[-\underset{\underset{\displaystyle O}{\|}}{C}-R-\underset{\underset{\displaystyle O}{\|}}{C}-O-R'-O-\right]_n \mathrm{H}$$

polyester

In the above formulas, the letter R or R′ represents a group of one or more carbon atoms. The value of n will vary from 10 to 50.

The chain extension reaction, whereby these short chains are linked, is accomplished by the use of a reactive agent, i.e., a diisocyanate. For this reaction, the above short-chain polymers must have terminal hydrozy (OH) groups, as shown, so that the following reaction can occur between these groups and the diisocyanate:

$$\mathrm{HO{-}P{-}OH} + \mathrm{O{=}C{=}N{-}R{-}N{=}C{=}O}$$

polyether or polyester — diisocyanate

$$\downarrow$$

$$\mathrm{HO}\left[\mathrm{P{-}O{-}\underset{\underset{\displaystyle O}{\|}}{C}{-}NH{-}R{-}NH{-}\underset{\underset{\displaystyle O}{\|}}{C}{-}O}\right]_x\mathrm{P{-}OH}$$

urethane polymer

Since the isocyanate group reacts vigorously with the hydroxy groups, a long-chain polymer results, with x having values up to 50 or 100.

In addition to this chain extension process, the diisocyanates are also capable of other reactions which can be utilized. Thus the diisocyanates can also react with some of the active hydrogen atoms attached to the polymer chain, leading to a crosslinking, i.e., vulcanization process. In this way, the urethane polymers lend themselves to a one-stage casting and curing process. Another reaction of the diisocyanates involves water, which reacts vigorously to form carbon dioxide gas, as follows:

$$\mathrm{OCN{-}R{-}NCO} + 2\,\mathrm{H_2O} \longrightarrow \mathrm{H_2N{-}R{-}NH_2} + \underset{\text{carbon dioxide}}{2\,\mathrm{CO_2}}$$

Hence, when a little water is mixed together with the short-chain polymer and diisocyanate, a process of simultaneous foaming, polymerization, and vulcanization results, leading to the rapid formation of foamed elastomers or plastics.

As elastomers, the urethane polymers show some outstanding physical properties, including high gum tensile strength.

Table 1.1. Elastomers and Their Characteristics.

Name	*Chemical Name*	*Structure*	*Vulcanization Agent*	*Stretching Crystallization*	*Gum Strength*
Natural rubber	*cis*-1,4-polyisoprene	$[-CH_2-\underset{\mid}{C}(CH_3)=CH-CH_2-]_n$	Sulfur	Good	Good
Polyisoprene (“synthetic natural”)	*cis*-1,4-polyisoprene	$[-CH_2-\underset{\mid}{C}(CH_3)=CH-CH_2-]_n$	Sulfur	Good	Good
Polybutadiene	Polybutadiene	$[-CH_2-CH=CH-CH_2-]_n$	Sulfur	Depends on structure	Poor to fair
SBR	Poly(butadiene-co-styrene)	$[(-CH_2-CH=CH=CH_2-)_5(-CH_2-\underset{\mid}{CH}(C_6H_5)-)]_n$	Sulfur	Poor	Poor
Nitrile	Poly(butadiene-co-acrylonitrile)	$[(-CH_2-CH=CH-CH_2-)_3(-CH_2-\underset{\mid}{CH}(CN)-)]_n$	Sulfur	Poor	Poor
Butyl	Poly(isobutylene-co-isoprene)	$[(-CH_2-C(CH_3)_2-)_{50}(-CH_2-\underset{\mid}{C}(CH_3)=CH-CH_2-)]_n$	Sulfur	Good	Good

EPR(EPDM)	Poly(ethylene-co-propylene-co-diene)	$[-(-CH_2-CH_2-)_{37}(-CH_2-\underset{\displaystyle CH_3}{CH}-)_{13}\text{diene}-]$	Peroxides (or sulfur)	Poor	Poor
Neoprene	Polychloroprene	$[-CH_2-\underset{\displaystyle Cl}{C}=CH-CH_2-]_n$	Mag. oxide or zinc oxide	Good	Good
Silicone	Polydimethylsiloxane	$[-\overset{\displaystyle CH_3}{\underset{\displaystyle CH_3}{Si}}-O-]_n$	Peroxides	Poor	Poor
Thiokol™	Polyalkylenesulfide	$[-CH_2-CH_2-S_{2-4}]_n$	Zinc oxide	Fair	Poor
Urethane	Polyester or polyether urethanes	$HO[-P-OCONHRNHCOO-]_nP-OH$	Diisocyanates	Depends on structure	Good

Note: The fluorocarbon rubbers have not been included since they are simply the partially fluorinated analogs of various synthetic elastomers (see Ch. 14).

2

THE COMPOUNDING AND VULCANIZATION OF RUBBER

HOWARD L. STEPHENS
Professor Emeritus of Polymer Science and Chemistry
The University of Akron
Akron, Ohio

In the rubber industry, the problem of selecting the basic raw materials for the preparation of a specific commercial product usually is assigned to the compounder. Traditionally, the compounder has been a trained chemist or chemical engineer. This background is necessary since some of the processes involve complicated chemical reactions, of which vulcanization is the most important. In addition, chemical analysis of the raw materials and of the completed products may be required. Thus this knowledge is necessary in order to select the proper test methods.

The compounder must be capable of describing these processes and the problems involved to the engineers, development chemists, and the sales-service personnel, as they are concerned with the production of a serviceable product at a reasonable cost.

COMPOUNDING RECIPES AND THEIR USE

In order to aid in the development of a rubber compound, the various ingredients to be used are compiled into a ''recipe''. Every recipe contains a number of components, each having a specific function either in the processing, vulcanization, or end use of the product. Two typical tire-tread recipes adapted from Vanderbilt's 1978 *Rubber Handbook* are given in Table 2.1.

In general, from the data given in Table 2.1, the following information can be obtained concerning compounding recipes:

1. All the ingredients used are normally given in amounts based on a total of 100 parts of the rubber or combination of rubbers (or masterbatches)

Table 2.1. Typical Passenger-tire Treads.

Ingredient	*phr*[a] *Radial*	*phr*[a] *Bias*	*Function*
Styrene-butadiene/oil masterbatch (1712)	82.5[b]	82.5[b]	Elastomer-extender oil masterbatch
Cis-polybutadiene (1252)	55[b]	55[b]	Elastomer-extender oil masterbatch
Carbon Black N-234 (ISAF-HS)	70	–	Reinforcing filler
Carbon Black N-339 (HAF-HS)	–	70	Reinforcing filler
Oil-soluble sulfonic acid	1	1	Processing aid
Stearic acid	2	2	Accelerator activator
Zinc oxide	3	3	Accelerator activator
Polymerized 1,2-dihydro-2,2,4-trimethylquinoline	2	2	Antioxidant
N-(1,3-dimethylbutyl)-N′- phenyl-*p*-phenylene-diamine	1	1	Antiozonant
Blended petroleum wax	3	3	Crack inhibitor and antiozonant
Sulfur	1.75	1.5	Vulcanizing agent
N-cyclohexyl-2-benzothiazole sulfenamide	1	1	Delayed-action accelerator
Total	222.25	222.0	
Specific gravity	1.13	1.13	

[a]Parts per 100 parts of rubber, by weight.
[b]Extended with 37.5 parts of highly aromatic oil.

used. This notation is generally listed as phr (parts per hundred of rubber). Thus, in comparisons of different recipes, the effects of varying any ingredient used are easily recognized when the physical properties or processing characteristics are compared.

2. Although the function of each component, as shown in this example, is never indicated in industrial or laboratory recipes, it is apparent that many different materials with specific purposes are used in every recipe.
3. In many recipes, the materials are listed in the general order that they are mixed into the rubber during processing. This method aids the compounder in setting up mixing schedules for processing various compounds and for the preparation of special masterbatches that may be used in many different products.
4. From the total amount of materials used (whether in grams, ounces, pounds, or other method of measurement), the cost of the total compound can be computed rather simply, as follows:

$$\text{Cost per pound} = \frac{\text{Total cost of all ingredients}}{\text{Total recipe weight}} \quad \text{or}$$

$$\text{Cost per volume-pound} = \underline{\text{Cost per pound}} \times \text{Specific gravity}$$

Generally, the cost is based on a volume-pound figure since this method produces a comparable value for all rubber vulcanizates, especially when the compounding ingredients are changed.

COMPONENTS OF THE RECIPE

Although the examples used in the above illustration are not typical of all recipes, in general, the materials utilized by the rubber compounder can be classified into nine major categories, which are defined as follows:

Elastomers: The basic component of all rubber compounds, it may be in the form of rubber alone, or "masterbatches" of rubber-oil, rubber-carbon black, or rubber-oil-carbon black, reclaimed rubber, or thermoplastic elastomers. Combinations or blends as given in the synthetic-tire-tread recipe are quite common. The elastomers are selected in order to obtain specific physical properties in the final product.
Processing Aids: Materials used to modify rubber during the mixing or processing steps, or to aid in a specific manner during extrusion, calendering, or molding operations.
Vulcanization Agents: With the exception of thermoplastic elastomers, these materials are necessary for vulcanization, since without the chemical crosslinking reactions involving these agents, no improvement in the physical properties of the rubber mixes can occur.
Accelerators: In combination with vulcanizing agents, these materials reduce the vulcanization time (cure time) by increasing the rate of vulcanization. In most cases, the physical properties of the products are also improved.
Accelerator Activators: These ingredients form chemical complexes with accelerators, and thus aid in obtaining the maximum benefits from an acceleration system by increasing vulcanization rates and improving the final product's properties.
Age-Resistors (Antidegradants): Antioxidants, antiozonants, and other materials used to reduce aging processes in vulcanizates. They function by slowing down the deterioration of rubber products. Deterioration occurs through reactions with materials that catalyze rubber failure, i.e., oxygen, ozone, light, heat, radiation, and so forth.
Fillers: These materials are used to reinforce or modify physical properties, impart certain processing properties, or reduce cost.
Softeners: Any material that can be added to rubber to aid mixing, promote greater elasticity, tack, or extend (or replace) a portion of the rubber hydrocarbon (without a loss in physical properties) can be classified as a softener.
Miscellaneous Ingredients: Materials that can be used for specific purposes but are not normally required in the majority of rubber compounds can be included

in this group. It includes retarders, colors, blowing aids, abrasives, dusting agents, odorants, homogenizing agents, and so forth.

All of the classes of components listed above will be discussed in greater detail in this and subsequent chapters.

PROCESSING METHODS

Both in the laboratory and factory, the most common methods for incorporating the compounding ingredients into the raw rubber involve the use of either a mill or a Banbury (internal) mixer. There are many sizes of each type, and some typical examples are given in Table 2.2 and shown in Figs. 2.1 through 2.4.

For the purpose of illustrating how these two methods of mixing are used, a typical laboratory recipe and mixing schedules which have been developed by the Committee D-11 on Rubber and Rubber-Like Materials of the American Society for Testing and Materials (ASTM) have been chosen. These with other standards and test methods are published annually in the ASTM *Book of Standards*; this particular part covers procedures adopted for rubber, carbon black, and gaskets; and all rubber technologists should become well acquainted with its contents.

The procedures are taken from "Carbon Black" in NR-Recipe and Evaluation Procedures, ASTM Designation D 3192-82. The recipe in Table 2.3 using natural rubber was chosen to show the use of mixing schedules during processing.

The batch size for a 6″ laboratory mill is four times the recipe; and for the BR Banbury (internal) mixer with a capacity of approximately 1200 cm^3 the batch size is equal to the volume of the mixer times the specific gravity of the stock. In this case it is approximately six times the recipe weight. The mixing cycles used are shown in Table 2.4.

FINISHING STEPS FOR BOTH PROCEDURES

Check the batch mass and record. If it differs from the theoretical value by more than 0.5%, reject the batch. From this stock, cut enough sample to allow testing of compound viscosity in accordance with Method D 1646, or curing characteristics in accordance with Method D 2084, or both if these are desired.

Open the mill and sheet-off to a 0.085 in. (2.2 mm) finished gage. Cool on a flat, dry metal surface.

To prevent absorption of moisture, condition the stock for 1 to 24 hours at a temperature of 73.4 ± 5.4°F (23 ± 3°C) in a closed container after cooling, unless the relative humidity is controlled at 35 ± 5% in accordance with Recommended Practice D 3182.

Table 2.2. Typical Size and Capacity of Mills and Banbury Mixers

2-roll Mills

Roll Size (in.)[a]	*Batch Size (lb)*[b]	*Motor (H.P.)*	*Weight (lb)*
6 × 13	1.25 to 2	7.5	6,200
8 × 16	2.5 to 4	10–15	8,000
10 × 20	5 to 8	15–20	10,000
12 × 24	10 to 18	30–40	13,800
14 × 30	20 to 30	40–50	18,000
16 × 42	30 to 50	70–75	24,000
18 × 48	45 to 70	75–100	30,000
22 × 60	75 to 125	125–150	39,500
24 × 72	125 to 200	150–200	50,000
26 × 84	150 to 250	150–200	65,000
28 × 84	175 to 300	200–250	70,000

Banbury Mixers

Size	*Capacity (lb)*[b]	*Motor (H.P.)*	*Weight (lb)*
Midget	0.67	7.5	3,300
BR	2.6	8.5–25	5,000
OOC	7	15–30	6,000
1	27	50–100	9,500
1A and 1D	27	50–100	12,900
3A and 3D	105	150–300	30,000
3A and 3D Unidrive	105	200–400	38,000
9	265	200–400	65,000
9D Unidrive	290	250–500	66,800
11 and 11D	370	300–600	90,000
11 and 11D Unidrive	370	400–800	106,000
27 and 27D Unidrive	930	1500	251,400

[a]Diameter × length.
[b]Based on 1.00 specific gravity stock.
Source: Farrel Company Bulletins 173E and 215.

From the above examples, it is apparent that the following information concerning mixing cycles and times can be extended to practical factory or laboratory processes:

1. There are definite temperature ranges for mixing rubbers, each specific rubber having an optimum temperature at which the desired dispersion of compounding ingredients in the mix is obtained.
2. Some rubbers require an initial breakdown period before the ingredients are added.

Fig. 2.1. 6″ × 13″ laboratory mill (front and rear). (Reprinted with the permission of the Farrel Company)

Fig. 2.2. 22″ × 60″ direct-coupled rubber mill. (Reprinted with the permission of the Farrel Company)

Fig. 2.3. 82 BR Laboratory Banbury mixer. (Reprinted with the permission of the Farrel Company)

Fig. 2.4. Size F620 Banbury mixer. (Reprinted with the permission of the Farrel Company)

3. A specific order of incorporation of the compounding ingredients is necessary.
4. The time of mixing for each step in the process is important.
5. A finishing step and control of the final temperature of the mix is necessary to prevent prevulcanization, and,
6. Banbury (internal) mixes require less time and handling and mix larger batches than the corresponding mill mixers.

Table 2.3. Natural-rubber Recipe[a]

	Mill Mixing	Internal Mixer
Natural rubber	100	100
Zinc oxide	5	5
Sulfur	2.5	2.5
Stearic acid	3	3
Benzothiazyl disulfide	0.6	0.6
Carbon black	50	50
Total	161.1	161.1
Specific gravity	1.13	1.13
Batch factor	4	6

[a]Standard reference materials available from sources are listed in the procedure.

Table 2.4. Mixing Cycles.

	Mill	Banbury
Temperature, °C(F)	70 ± 5°C (158 ± 9°F)	110–125°C (230–257)[a]
Mixing speed, slow roll	24 rpm	11 rpm
Roll ratio (slow to fast roll)	1 to 1.4	1 to 1.125

	Time (minutes)	
Mixing Steps	*Mill*	*Internal Mixer*
1. Set the mill opening at 0.055 in. (1.4 mm) and adjust and maintain roll temperature at 158 ± 9°F (70 ± 5°C)	0	–
Adjust the internal mixer temperature to achieve the discharge conditions in step 8. Close the discharge gate, start rotor, raise ram. Lower ram after each operation.	–	0
2. Add rubber and band on the front roll. Make two ¾-cuts from each side.	2.0	–
Add the rubber.	–	0.5
3. Set mill opening at 0.065 in. (1.65 mm). Add stearic acid. Make one ¾-cut from each side.	2.5	–
Add the benzothiazyl disulfide.	–	0.5
4. Add sulfur, accelerator, and zinc oxide. Make two ¾-cuts from each side.	2.0	–
Add stearic acid.	–	1.0

Table 2.4. (*Continued*)

	Time (minutes)	
Mixing Steps	*Mill*	*Internal Mixer*
5. Add all the black. When the portion of the carbon black that was added has dropped through to the mill pan and the bank is dry, make two $\frac{3}{4}$-cuts from each side. Open the mill to 0.075 in. (1.9 mm) and add the carbon black from the mill pan until all is incorporated. Make three $\frac{3}{4}$-cuts from each side.	7.5	–
Note 1: Do not cut any stock while free carbon black is evident in the bank or on the milling surface. Be certain to return any pigments that drop through the mill to the milling stock.		
Add the zinc oxide and one-half the carbon black.	–	1.5
6. Set the mill opening at 0.32 in. (0.80 mm) and pass the rolled batch endwise through the mill six times.	2.0	–
Add the remainder of the carbon black.	–	1.5
7. Open the mill to give a minimum stock thickness of 0.25 in. (6 mm) and pass the stock through the rolls four times, folding it back on itself each time.	1.0	
Add the sulfur. Clean the mixer throat and the top of the ram.	–	1.0
8. Dump at 7 minutes. Adjust internal mixer to give dump temperature between 230 to 257°F (110 to 125°C).	–	1.0
9. Set the mill opening at 0.32 in. (0.80 mm) and maintain the roll temperature at 158 ± 9°F (70 ± 5°C). Pass the rolled batch endwise through the mill six times.	–	2.0
10. Open the mill to give a minimum stock thickness of 0.25 in. (6 mm) and pass the stock through the rolls four times, folding it back onto itself each time.	–	1.0
Total Time	17.0	10.0

Although the Banbury mixer is capable of handling larger factory batches at a faster rate than mill mixing, there are other processing problems involved, to be discussed in detail in Chapter 4. It is only necessary here to point out that factory techniques require much stricter control tests than corresponding laboratory methods since large quantities of material are used. Hence labor, processing, and equipment costs are far more important in production than in compounding research, especially if a large factory batch must be scrapped due to improper mixing or compounding.

VULCANIZATION

After rubber compounds have been properly mixed and shaped into blanks for molding, or calendered, extruded, or fabricated into a composite item (as a tire), they must be vulcanized by one of many processes. During vulcanization, the following changes occur:

1. The long chains of the rubber molecules become crosslinked by reactions with the vulcanization agent to form three-dimensional structures. This reaction transforms the soft, weak plastic-like material into a strong elastic product.
2. The rubber loses its tackiness and becomes insoluble in solvents and is more resistant to deterioration normally caused by heat, light, and aging processes.

These changes generally occur with the use of the following vulcanization systems.

Vulcanization Systems

Sulfur Vulcanization. It is quite apparent from the data presented in Table 2.5 that the most common rubbers used are the general-purpose type, with the remaining types representing only about 19% of the total usage. Since these rubbers (general-purpose) contain unsaturation, vulcanization with sulfur is possible, and it is in general the most common vulcanizing agent used. With sulfur, crosslinks and cyclic structures of the following type are formed:

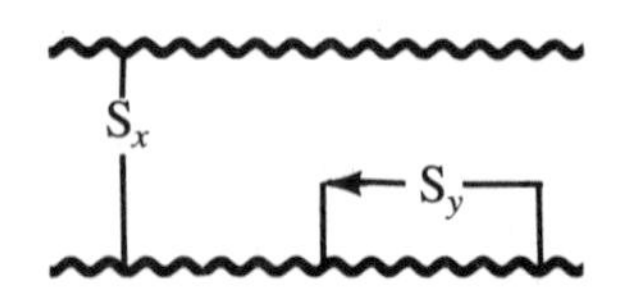

Generally, x in an efficient accelerated curing system is about 1 or 2, with little or no cyclic groups formed. In inefficient systems, x equals up to 8 and many

Table 2.5. Typical Commercial Rubbers.

Common Name	*ASTM Designation (D1418-81)*	*Consumption (U.S.) in Metric Tons (1983)*[a]	*Price*[b] *($/lb)*
	General Purpose		
Natural	NR	676,267	0.44
Polyisoprene	IR	*	0.72
Styrene-butadiene	SBR	887,005	0.66
Butyl	IIR	*	0.76
Ethylene-propylene	EPDM	141,490	1.01
Polybutadiene	BR	355,541	0.74
	Solvent Resistant		
Polysulfides (Thiokol)	T	*	1.83
Nitrile	NBR	57,239	1.10
Polychloroprene (Neoprene)	CR	85,096	1.28
Polyurethanes			
Polyester	AU	N.A.*	3.00
Polyether	EU	N.A.*	3.70
Epichlorohydrin	CO	*	2.24
Epichlorohydrin-ethylene oxide	ECO	*	2.19
	Heat Resistant		
Silicone	MQ	*	4.40
Chlorosulfonated-polyethylene (Hypalon)	CSM	*	1.32
Polyacrylates	ACM	*	2.66
Fluororubbers	CFM	*	15.10

[a]Natural Rubber News, April 1984.
[b]Average prices for most common grade, December 1984.
*Total for all rubbers, exluding urethanes 334,420.

Total consumption,[a]	synthetic rubbers	1,860,792
	natural rubber	676,267
	Total consumption	2,537,059

cyclic structures are formed. The total amount of sulfur combined in these networks is usually called the *coefficient of vulcanization* and is defined as parts of sulfur combined per 100 parts of rubber. For most rubbers, one crosslink for about each 200 monomer units in the chain is sufficient to produce a suitable vulcanized product (molecular weight between crosslinks equals ca. 8000 to 12,000).

It is these amounts of cyclic sulfur (y) and the excessive sulfur in the crosslinks (x) which contribute to the poor aging properties of the vulcanizates.

Sulfurless Vulcanization. Vulcanization effected without elemental sulfur, by the use of thiuram disulfide compounds (accelerators) or with selenium or tellurium, produces products that are more resistant to heat aging. With the thiuram disulfides, efficient crosslinks containing only 1 or 2 sulfur atoms are found, and in addition the accelerator fragments act as antioxidants.

Peroxide Vulcanization. The saturated rubbers cannot be crosslinked by sulfur and accelerators. Organic peroxides are necessary for the vulcanization of these rubbers. When the peroxides decompose, free radicals are formed on the polymer chains, and these chains can then combine to form crosslinks, as shown below:

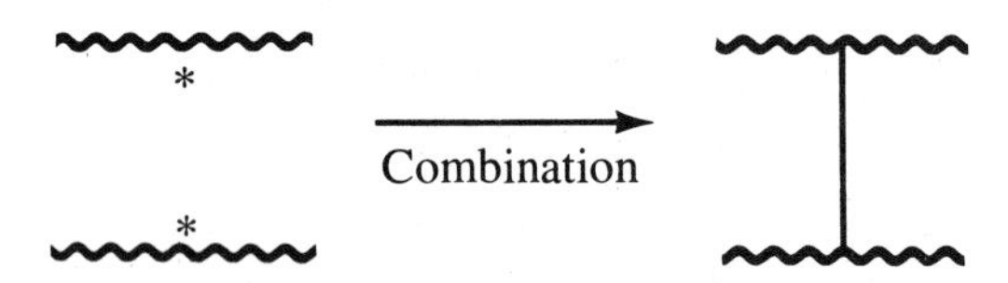

Crosslinks of this type only involve carbon-to-carbon bonds and are quite stable. They are also formed by gamma radiation and X-radiation.

Other Systems. Some elastomers can be vulcanized by the use of certain nonsulfur bifunctional compounds that form bridge-type crosslinks, for example, neoprene with metal oxides or butyl rubber with dinitrosobenzene. More detailed descriptions of the processes involved will be discussed in later chapters dealing specifically with each rubber.

Vulcanization Conditions

In vulcanization (curing) processes, consideration must be made for the differences in the thickness of the objects involved, the vulcanization temperature, and the thermal stability of the rubber compound.

Effect of Thickness. Rubbers are poor heat conductors, thus it is necessary to consider the heat conduction, heat capacity, geometry of the mold, heat-exchange system, and the curing characteristics of a particular compound when articles thicker than about $\frac{1}{4}$ inch are being vulcanized. This effect is best shown by immersing thermocouples at various depths in a rubber compound and measuring the time required to reach the vulcanization temperature as indicated by the press temperature. This effect is illustrated in Fig. 2.5.

Since the effects are complicated, generally an estimate of time required can be determined by adding an additional 5 minutes to the cure time for each $\frac{1}{4}$ inch of thickness. In exceptionally thick or complicated articles, the item may

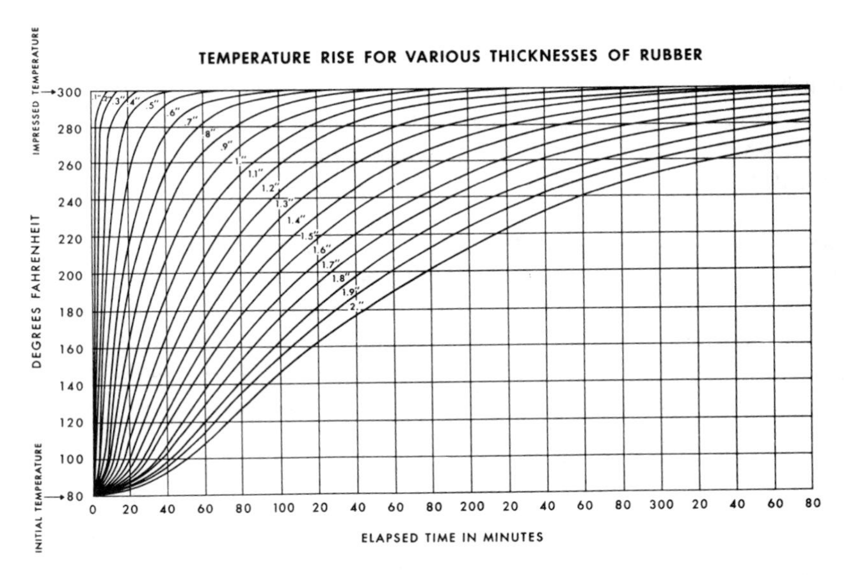

Fig. 2.5. Effect of thickness on temperature rise. (Reprinted with the permission of the E. I. du Pont de Nemours & Co., Inc.)

be built up using sections with different curing characteristics, or by controlling the rate at which the mold is heated or cooled.

Effect of Temperature. The vulcanization temperature must be chosen in order to produce a properly cured product having uniform physical properties in the shortest possible molding time. The *temperature coefficient of vulcanization* is a term used to identify the relationship that exists between different cure times at different temperatures.

With information of this type, optimum cure times at higher or lower temperatures can be estimated for many rubber compounds with known coefficients of vulcanization. For example, most rubber compounds have a coefficient of approximately 2 (1.8 to 2.2). This indicates that the cure time must be reduced by a factor of 2 for each 18°F (10°C) increase in cure temperature, or, if the temperature is reduced by 18°F, the cure time must be doubled.

Effect of Thermal Stability. Each type of rubber has a definite range of temperatures that may be used for vulcanization. These temperatures may vary somewhat but, to avoid deterioration, it is quite important not to exceed the maximum for each rubber type. The "thermal" effect is shown either by the appearance of the finished product or by its physical properties. Many of these effects will be discussed in the following chapters.

Vulcanization Techniques

Finished articles can be prepared by numerous vulcanization techniques; however, the specific methods used in most industries are usually based on producing suitable commercial goods utilizing standard techniques. The eight significant methods are briefly outlined:

Compression Molding. This method, or a modification of it, probably utilizes the most common type of mold used in the rubber industry. Essentially, it consists of placing a precut or shaped "slug" (or blank), or a composite item, into a two-piece mold that is closed. The pressure applied by the press forces the material to fit the shape of the mold, and the slight excess present flows out of the rim of the mold or through special vents. This excess is known as "mold flash." A simple mold of this type is shown in Fig. 2.6.

Tires are normally cured in a modification of the compression mold. A bladder, or airbag, forces and holds the "green" tire against the mold surface during vulcanization. This force reproduces the design of the tread, and heat (in the form of steam) is normally introduced into the bladder to aid the curing process. An excellent history on the development of tires and their vulcanization is given in reference 3. Fig. 2.7 shows a typical automatic tire-curing press.

Transfer Molding. As shown in Fig. 2.6, transfer molding involves the distribution of the uncured stock from one part of the mold (the pot) into the actual mold cavity. This process permits the molding of complicated shapes or the imbedding of inserts in many products; these procedures are difficult with the usual compression molds.

Although the molds are relatively more expensive than compression molds, the actual process permits shorter cure times through the use of higher temperatures and of better heat transfer, which is obtained due to the higher pressure applied to force the compound into the mold.

Injection Molding. In recent years, injection-molding processes, which are normally used for the production of plastics, have been developed so that rubber compounds can be molded and vulcanized by this method. A diagram of the process is shown in Fig. 2.6 and a commercial unit is illustrated in Fig. 2.8.

By careful temperature control of the feed stock, items can be vulcanized in less than several minutes (cure times are generally reported in seconds). This method can be completely controlled by programmed-feed, injection, and demolding cycles resulting in low rejection rates and lower finishing costs. The initial cost of both the molds and equipment has hindered the adoption of this type of molding.

Estimates have been made indicating the use of the injection method will

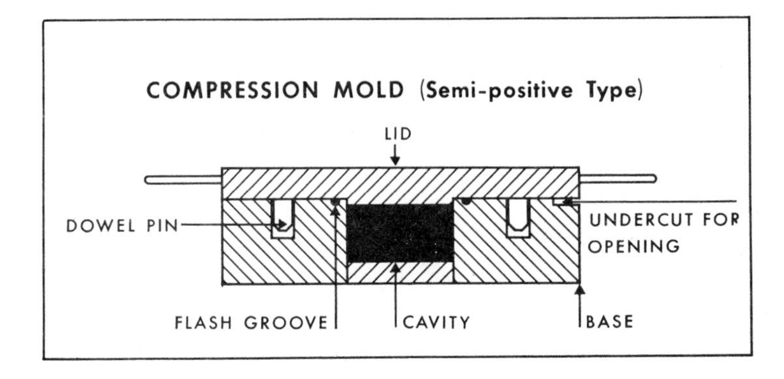

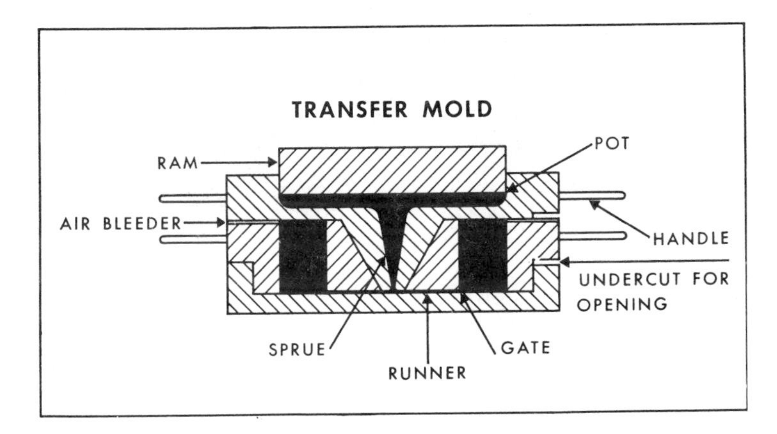

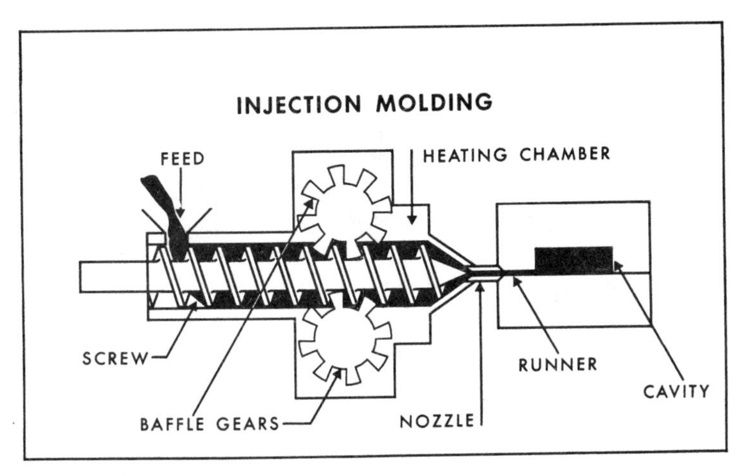

Fig. 2.6. Examples of compression, transfer and injection molding techniques. (Reprinted with the permission of the E. I. du Pont de Nemours & Co., Inc.)

Fig. 2.7a. Automatic tire-curing press showing curing blatter in place. (Reprinted with permission of McNeil Akron, Inc.)

expand from the present use of 5% to 25% (of the molding-under-compression type of vulcanization).

Open Cures. *Hot-air ovens* can be used to vulcanizate thin articles (balloons, and the like), items which have been preshaped (by extrusion) or by a combination of precuring in a mold followed by post-vulcanization in an oven. The last process is used to remove peroxide decomposition products from items cured with peroxides. The system is not too efficient due to the poor heat transfer of hot air, and longer cure times at lower temperatures are necessary to prevent the formation of porosity or deformation of the unvulcanized products.

Open steam can be used in closed containers called ''autoclaves'' (which resemble home pressure cookers); the process involves saturated steam under pressure, which acts as an inert gas, enabling better heat transfer; thus higher temperatures can be used and shorter cure times are possible, making this process more desirable than the air oven. Hose, cables, built-up footwear, and tires in pot-heater molds are cured by this method.

Water cures can be used for articles not affected by immersion. The method is useful for large items (for example, large containers or rubber-lined containers) and is especially useful for hard-rubber compositions. Direct contact

Fig. 2.7b. Automatic tire-curing press showing the removal of a cured tire. (Reprinted with permission of McNeil Akron, Inc.)

with water produces better heat transfer than do air or open steam; consequently, with this system less deformation and faster cures are obtained.

Lead sheathing can be used to cover soft, large extruded sections with a protective cover for vulcanization in steam. The process is used for garden and other hose; the lead sheath is usually applied immediately by hydraulic pressure or extrusion as the hose emerges from its extruder.

Continuous Vulcanization Processes (C.V.). Continuous vulcanization generally involves some form of heating in a manner such that the vulcanization step usually occurs immediately after the rubber article is formed. The process

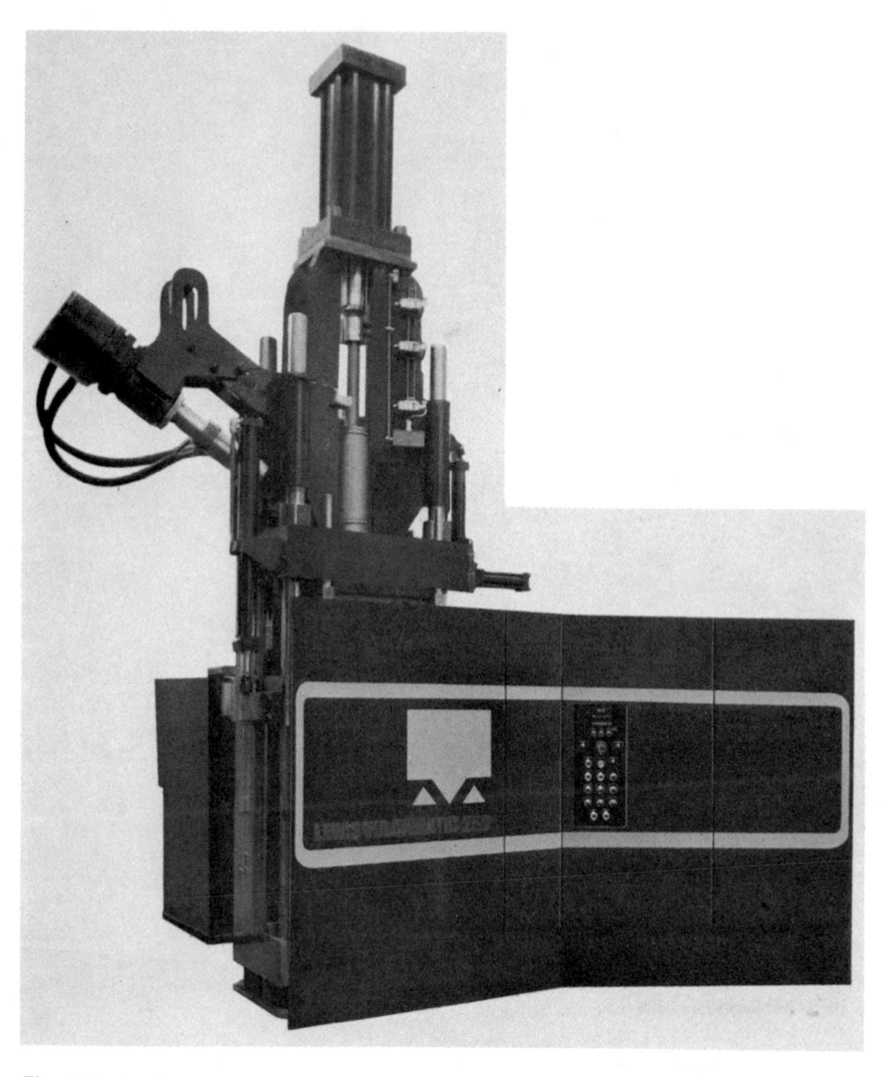

Fig. 2.8. Lewis Vulcamatic rubber-injection machine. (Reprinted with permission of McNeil Akron, Inc.)

is normally used for extruded goods, coated wiring, conveyor belts, and flooring.

Liquid curing methods (L.C.M.) involve suitable hot liquid baths in which extrusions may be vulcanized in a continuous process. Items can be cured rapidly at temperatures from 200° to 300°C; however, the compounds must be modified to prevent porosity, always a problem with any extrudate.

Suitable materials for the curing medium include bismuth-tin alloys; a eutectic mixture of potassium nitrate, sodium nitrite, and sodium nitrate; poly-glycols; and certain silicone fluids.

Fluidized beds consisting of small particles (glass beads) suspended in a stream of heated air are efficient vulcanization systems. They are normally used for continuous vulcanization of extrusions. Heat transfer is approximately 50 times greater than with hot air alone.

Continuous hot-air tunnels can be used in the same manner that hot air ovens are used. Thin articles, for example, dipped goods, can be vulcanized after dipping by passing an endless conveyor carrying the items through a series of heated tunnels (ovens) at a rate that will complete vulcanization by the time articles are stripped from the forms.

Steam tubes are used for the continuous vulcanization of sheathing. After extrusion of the rubber cover onto the wire, the cable passes into a jacketed stream tube containing steam under pressure. Depending on the thickness of the cover, the period of time in the tube can be controlled to complete vulcanization, or conversely, the length of the steam tube can be extended.

The Rotacure process can be used to vulcanize large conveyor belts and continuous flooring strips. The process involves the use of an endless steel band which presses the article against a large heated drum. Slow rotation of the drum permits vulcanization to occur after approximately 10 minutes contact time. Belt curing presses are also used; however, this system is not completely continuous, although long lengths of belting are made by this method.

Cold Vulcanization. Thin articles may be vulcanized by treatment with sulfur monochloride (S_2Cl_2) by dipping in a solution or exposure to its vapors. The process has been essentially replaced with ultra accelerators, which are capable of curing at room temperature.

High-energy Radiation. Systems using either gamma radiation from cobalt 60 or electron beams have been used for vulcanization. The electron-beam method has been used to cure both polyethylene and silicone rubbers, generally accomplished by passing the material through the beam on a conveyor.

Microwave Vulcanization. Ultrahigh frequency fields (UHF) developed by alternating electromagnetic circuits can be used to warm up or vulcanize articles with large or uneven cross sections. The process requires polar rubber mixtures since nonpolar materials will not absorb the energy produced. It is possible to warm articles up to 200°C within 30 seconds with some UHF systems. Continuous processes utilizing extruders and microwave heaters are available.

COMPOUNDING INGREDIENTS

Elastomers (Rubber and Rubber-like Materials)

Today's rubber compounder has many types of rubber and rubber-like materials available for use. In most instances, before choosing the proper rubber the compounder must first decide what particular use the end product will be subjected to. For this choice, rubbers are generally classified into three major classes (former ASTM classification):

General Purpose (R): For services where specific resistance to the action of petroleum-base fluids is not required.
Solvent Resistant (S): For services where specific resistance to the action of petroleum-base fluids is required.
Heat Resistant (T): For services where specific resistance to the effects of prolonged exposure to abnormal temperatures or compounded petroleum oils, or both, is required.

A listing showing the types in general usage is given in Table 2.5; included is the general ASTM code, current usage, and price data.

For each rubber, there are many types, depending upon the manufacturing process, grade, and polymer composition. The problem of selecting a specific type is somewhat simplified by the use of publications available from major suppliers, manufacturers, and associations.

Natural Rubber. The grades of natural rubber and the various types of each are listed in the "Green Book," which is the common name given to *The International Standards of Quality and Packing for Natural Rubber Grades*, published by The Rubber Manufacturers Association, Inc.*

In addition, The Rubber Research Institute of Malaysia has introduced innovations including "Standard Malaysian Rubber" (SMR). Information on these newer types of natural rubber is available from the Malaysian Rubber Bureau† and is discussed in detail in Chapter 6.

Synthetic Rubbers. It is quite apparent from the listing given in Table 2.5 that there are many types and compositions available when a synthetic rubber is considered for use. During the development of the synthetic rubber industry in World War II, a numbering system was used by the government for the identification of SBR rubbers. However, when the plants were sold to private companies, many of these numbering codes were changed. In 1960, the International Institute of Synthetic Rubber Producers Inc. (IISRP)‡ was formed in order to organize a systematic listing of all the available synthetic rubbers and

*1400 K St., N.W., Washington, DC 20005.
†1925 K St., N.W., Washington, DC 20006.
‡2077 South Gessner Road, Suite 133, Houston, TX 77063.

to promote the interests of all concerned in the manufacture and use of synthetic rubbers. As of 1985, 44 producers had memberships in the organization, and it currently publishes editions listing the procedures, the nomenclature developed for the classification of the polymers, and data on the various types of rubbers and latices actually available.

Information on styrene-butadiene, high-styrene resins, butadiene, isoprene, ethylene-propylene, butyl, chloroprene, and nitrile rubbers, and their latices is included in the tabulation. Since these rubbers will be discussed in detail in later chapters, no further classification will be made here.

Reclaimed Rubbers. Many rubbers can be reclaimed and utilized as a partial or complete replacement for new rubber in many articles. This usage represented about 8.6% of the total consumption of all rubber in the U.S. during 1969. Unfortunately, information is not available on present-day usage. A complete description of the reclaim processes and product uses will be given in Chapter 19.

Chemical Plasticizers (or Peptizing Agents)

Some rubbers, especially natural and high-viscosity synthetics, require an initial breakdown period during mixing to soften the material for processing or to increase building tack after compounding. This softening effect can be catalyzed by the addition of small amounts of chemical plasticizers (up to 2 phr) which help control the amount and speed of breakdown and aid in the dispersion of the other compounding ingredients. Since the process also can reduce the nerve and shrinkage of the compound, improved stock preparation (as in extrusion or calendering) and molding operations are achieved from the reduction of the molecular weight (chain length) of the rubber through oxidative chain scission during mastication.

For proper usage, the plasticizer should function only during the initial mixing period (nonpersistent) and its action should be stopped by the addition of carbon black, sulfur, or accelerators. Chemical plasticizers are normally used with natural and styrene-butadiene rubbers; typical examples are xylyl mercaptan (thioxylenols), oil-soluble sulfonic acids, zinc salt of pentachlorothiophenol, pentachlorothiophenol, 2-naphthlenethiol, and phenylhydrazine salts.

It is apparent from the above listing that the majority of the peptizers have active –SH groups that function as chain-terminating agents by reactions with the free radicals formed when the rubber chains rupture during mastication.

Vulcanization Agents

As stated earlier, these are the chemicals that are required to crosslink the rubber chains into the network that gives the desired physical properties in the final

Table 2.6. Comparison of Elemental Vulcanization Agents.

	Sulfur			
	Rhombic	*Amorphous*	*Selenium*	*Tellurium*
Atomic weight	32.06	32.06	78.96	127.61
Appearance	Yellow powder	Yellow powder	Metallic powder	Metallic powder
Specific gravity	2.07	1.92	4.80	6.24
M.P.°C	112.8–119	>110	217.4	449.8
Price $/lb	0.17	0.63	16.00	25.00

product. The type of crosslinking agent required will vary with the type of rubber used; however, they can usually be grouped in the following categories.

Sulfur and Related Elements. The most common agent used is sulfur, as it enters into reactions with the majority of the unsaturated rubbers to produce vulcanizates. In addition, two other elements in the same periodic family, namely selenium and tellurium, are capable of producing vulcanization.

Two forms of sulfur, rhombic and amorphous (or insoluble), are compared with selenium and tellurium in Table 2.6.

The rhombic form is normally used for vulcanization; it exists as a cyclic (ring) structure composed of eight atoms of sulfur, S_8. The amorphous form is actually polymeric in nature; it is a metastable high polymer with a molecular weight of 100,000 to 300,000. It is insoluble in most solvents and rubber, hence the name "insoluble sulfur." The amount of insolubility is usually determined by using carbon disulfide as the solvent. Because of this insolubility, amorphous sulfur is used to prevent "blooming" on uncured rubber surfaces where it is necessary to maintain "building tack." Insoluble sulfur must not be processed above 210–220°F (99–105°C) or it will revert to the rhombic form.

In general, about 1 to 3 phr of sulfur is used for most rubber products. Commercially, both forms of sulfur are available in forms that have been treated with small amounts of a material (carbon black, magnesium carbonate, and so forth) which produces free-flowing, noncaking powders. Oil-sulfur mixtures are used occasionally to improve dispersion. Masterbatches of sulfur with rubbers or rubber-like polymers are also used where processing safety and ease of dispersion are important.

Selenium and tellurium are used in place of sulfur where excellent heat resistance is required. They generally shorten cure time and improve some vulcanizate properties. Selenium is somewhat more active than tellurium.

Sulfur-bearing chemicals, accelerators, and similar compounds can be used as a source of sulfur for the vulcanization of natural and styrene-butadiene rub-

Table 2.7. Typical Compounds Used for "Low-sulfur" Vulcanization.

Compound	*Sulfur Content (%)*	*Price $/lb*
Tetramethylthiuram disulfide	13.3	1.47
Dipentamethylenethiuram hexasulfide	35.0	3.79
Dimorpholinyl disulfide	31.4	5.11
Dibutylxanthogen disulfide	21.4	11.18
Alkylphenol disulfide	22.3	1.23
Alkylphenol disulfide	30.5	1.50

bers in recipes that include very small amounts of elemental sulfur. Generally in these "low-sulfur" cures, less than 1 phr of sulfur is used in combination with 3 to 4 phr of the sulfur donor; in some cases, no elemental sulfur is included in the recipe.

The sulfur-bearing compounds used decompose at the vulcanization temperature and release radicals which combine with the chains to form crosslinks. With these systems, efficient crosslinking occurs as most of the sulfur is combined in crosslinks containing one or two sulfur atoms with little or no cyclic sulfur present. Consequently, this form of vulcanization produces products which resist aging processes at elevated temperatures much more effectively than those produced with normal curing systems. However, due to the large amounts of sulfur donors used, these systems are more expensive than normal sulfur cures and are only employed when necessary. Some typical compounds used in low-sulfur cures are shown in Table 2.7.

Nonsulfur Vulcanization. Most nonsulfur vulcanization agents belong to one of three groups: (1) metal oxides, (2) difunctional compounds, or (3) peroxides. Each will be discussed here separately; however, much more detail will be given in the chapters that follow.

Metal Oxides. Carboxylated nitrile, butadiene, and styrene-butadiene rubbers may be crosslinked by the reaction of zinc oxide with the carboxylated groups on the polymer chains. This involves the formation of zinc salts of the carboxylate groups. Other metal oxides are also capable of reacting in the same manner.

Polychloroprenes (neoprenes) are also vulcanized by reactions with metal oxides, zinc oxide being normally used. The reaction involves active chlorine atoms and is described in a later chapter. Chlorosulfonated polyethylene (Hy-

Table 2.8. Non-sulfur Vulcanization Compounds.

Compound	*phr Usage*	*Price $/lb*
	Metal Oxides	
Zinc oxide	5 (neoprene): 10 (Thiokol)	0.55
Litharge	25 (Hypalon)	0.53
Magnesia/pentaerythritol	4/3 (Hypalon)	1.02/0.75
	Difunctional Compounds	
Phenolic resins	12 (butyl)	0.98
p-Quinonedioxime	2 (butyl)	7.26
Hexamethylenediamine carbamate	<1.5 (fluororubber)	24.00
	Peroxides	
Dicumyl peroxide (40%)	2 (silicone), 5 (urethane)	4.13
2,5-bis(*t*-butylperoxy)-2,5-dimethylhexane	2 (polyethylene or EPDM)	5.66
Zinc peroxide	4 (Thiokol)	

palon) is also crosslinked in the same general way. Litharge (PbO), litharge/magnesia (MgO), and magnesia/pentaerythritol combinations are used.

In many of these systems, the metal oxides are used in combinations for the purpose of controlling the vulcanization rate and absorbing the chlorides formed.

Difunctional Compounds. Certain difunctional compounds form crosslinks with rubbers by reacting to bridge polymer chains into networks. Epoxy resins are used with nitrile, quinone dioximes with butyl, and diamines or dithio compounds with fluororubbers. Other examples are given later in the text.

Peroxides. Organic peroxides are used to vulcanize rubbers that are saturated or do not contain any reactive groups capable of forming crosslinks. This type of vulcanization agent does not enter into the polymer chains but produces radicals which form carbon-to-carbon linkages with adjacent polymer chains.

Typical examples of nonsulfur compounds are given in Table 2.8.

Accelerators

As stated earlier, the main reason for using accelerators is to aid in controlling the time and/or temperature required for vulcanization and thus to improve the properties of the vulcanizate.

Reduction in the time required for vulcanization is generally accomplished by changes in the amounts and/or types of accelerators used. This usage will be quite evident from the many examples given in the following chapters. However, compounders employ some common practices to establish suitable recipe changes without extensive research. They are as follows:

1. Single-accelerator systems (primary accelerators) that are of sufficient activity to produce satisfactory cures within specified times;
2. Combinations of two or more accelerators, consisting of the primary accelerator, which is used in the largest amount, and the secondary accelerator, which is used in smaller amounts (10 to 20% of the total) in order to activate the primary accelerator and to improve the properties of the vulcanizate. Combinations of this type usually produce a synergistic effect in that the final properties are somewhat better than those produced by either accelerator separately;
3. Delayed-action accelerators, not affected by processing temperatures (thus providing some protection against scorching) but producing satisfactory cures at ordinary vulcanization temperatures.

Classification. Although accelerators can be grouped by their acidic or basic nature, or by their activity in certain rubbers (as shown in Table 2.9 and Fig. 2.9) no single classification system is suitable since they behave differently in each rubber compound. Consequently, the chemical group classification given in Table 2.10 will be used to illustrate the types in general use. Reference should be made to which type is in use for the specific rubbers given later in the text.

The classification in Table 2.9 has normally been used to describe the activity of the various chemical types in current use; however, grouping by chemical type as shown in Table 2.10 is more important since each type may produce

Table 2.9. Relative Activity of Accelerators in Natural Rubber.[a]

	Relative Vulcanization	
Type	*Time at 284°F*	*Examples*
Slow	90 to 120 minutes	Aniline
Moderately fast	Ca. 60 min	Diphenylguanidine Hexamethylene tetramine
Fast	Ca. 30 min	Mercaptobenzothiazole Benzothiazyl disulfide
Ultra-accelerators	Several minutes	Thiurams Dithiocarbamates Xanthates

[a]Ref. 4.

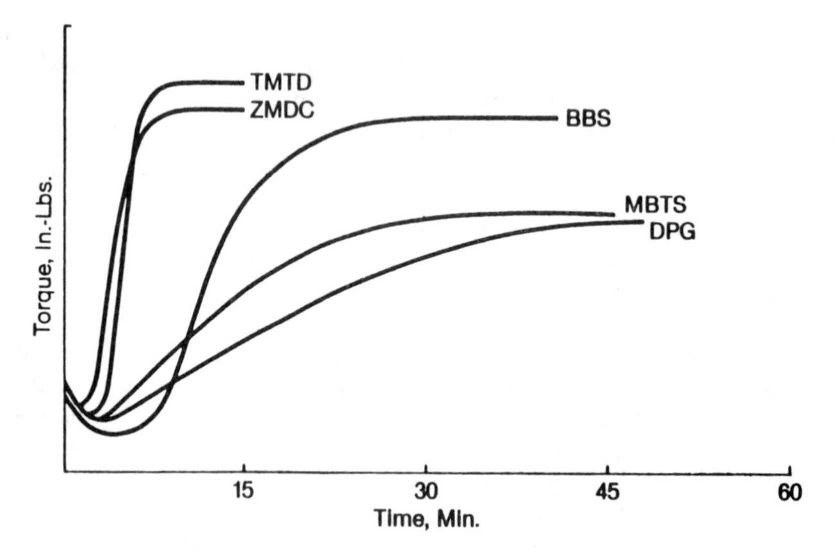

Fig. 2.9. An example of the effects of different accelerator systems on vulcanization rate. (Reprinted with the permission of the Monsanto Co.)

different crosslinks in the resulting vulcanizates. Reference 3 should be consulted for more detailed information.

Production. The production data for accelerators and related compounding ingredients are given in Table 2.11. From the information given, it is quite apparent that the thiazole derivatives are the most common type in use, representing 93% of all the cyclic accelerator compounds and 26.5% of all the rubber-processing chemicals produced in 1982.

Accelerator Activators.

These activators are used to increase the vulcanization rate by activating the accelerator so that it performs more effectively. It is believed that they react in some manner to form intermediate complexes with the accelerators. The complex thus formed is more effective in activating the sulfur present in the mixture, thus increasing the cure rate. Accelerator activators are grouped as follows:

Inorganic Compounds (mainly metal oxides), which include zinc oxide, hydrated lime, litharge, red lead, white lead, magnesium oxide, alkali carbonates, and hydroxides. Zinc oxide is the most common and it is generally used in combination with a fatty acid to form a rubber-soluble soap in the rubber matrix.

Zinc oxide is manufactured by two processes: the French, or indirect process;

Table 2.10. Chemical Classification of Accelerators.

Type	*Example*	*Price $/lb*	*Typical Use*
Aldehyde-amine reaction products	Butyraldehyde-aniline condensation product	2.60	Self-curing adhesives
Amines	Hexamethylene tetramine	0.60	Delayed action for NR
Guanidines	Diphenyl guanidine	2.33	Secondary accelerator
Thioureas	Ethylenethiourea	2.45	Fast curing for CR
Thiazoles	2-mercaptobenzo-thiazole	2.55	Fast curing, general-purpose, w/broad curing range
	Benzothiazole disulfide	1.57	Safe processing, general-purpose, moderate cure rate
Thiurams	Tetramethylthiuram disulfide	2.66	Safe, fast curing
Sulfenamides	N-cyclohexyl-2-benzothiazyl-sulfenamide	2.33	Safe processing, delayed action
Dithiocarbamates	Zinc dimethyldithiocarbamate	1.58	Fast, low temperature use
Xanthates	Dibutylxanthogen disulfide	11.18	General-purpose, low-temperature use
	Zinc isopropyl xanthate	2.34	Latex and adhesives room-temperature curing

Table 2.11. 1982 Production of Rubber-processing Chemicals.

	Production (1000 lbs)	*Unit Value*[a] *(per pound)*
Grand total	231,947	$1.72
Cyclic		
Total	207,740	1.79
Accelerators, activators and vulcanizing agents, total	65,985	1.82
Aldehyde-amine reaction products	280	–
Thiazole derivatives, total	61,480	1.66
2,2′-dithiobis (benzothiazole)	6,912	1.29
2-mercaptobenzothiazole	–	.88
2-mercaptobenzothiazole, zinc salt	890	1.38
N-oxydiethylene-2-benzothiazole-sulfenamide	–	1.99
All other thiazole derivatives	53,678	1.85
All other accelerators, activators, and vulcanizing agents[b,c]	4,225	3.04
Antioxidants, antiozonants, and stabilizers, total	132,006	1.72
Amino compounds, total	87,161	1.75
Substituted *p*-phenylenediamines	53,037	2.00
All other amino compounds[d]	34,124	1.47
Phenolic and phosphite compounds, total	44,845	1.67
Nonylphenyl phosphites, mixed	12,485	.80
Phenolic compounds:		
Phenol, alkylated	4,749	2.17
Phenol, styrenated, mixtures	629	1.13
All other phenolic and phosphite compounds	26,982	2.03
All other cyclic rubber-processing chemicals	9,749	2.19
Acyclic		
Total	24,207	1.21
Accelerators, activators, and vulcanizing agents, total	6,573	2.04
Dithiocarbamic acid derivatives, total[e]	5,009	2.04
Dibutyldithiocarbamic acid, nickel salt	306	–
Dibutyldithiocarbamic acid, sodium salt	201	–
Dimethyldithiocarbamic acid, zinc salt	1,334	1.48
All other dithiocarbamic acid derivatives	3,168	2.34
All other accelerators, activators, and vulcanizing agents[f]	1,564	2.06
All other acyclic rubber-processing chemicals[g]	17,634	.75

[a]Calculated from unrounded figures.
[b]Includes guanidines, dithiocarbamates, and other accelerators, activators, and vulcanizing agents.

Table 2.11. *(Continued)*

[c]Data on dithiocarbamates included in this table are for materials used chiefly in the processing of natural and synthetic rubber.
[d]Includes aldehyde- and acetone-amine reaction products.
[e]Includes blowing agents, peptizers, and other cyclic rubber-processing chemicals.
[f]Includes thiurams, xanthates, sulfides, and other accelerators, activators, and vulcanizing agents.
[g]Includes blowing agents, polymerization regulators, shortstops, and other acyclic rubber-processing chemicals.

Source: Synthetic Organic Chemicals: Investigation 332-135, Publication 1422, United States International Trade Commission, Washington, DC 20436.

and the American, or direct process. In the French process, zinc ore and coal are heated to produce zinc metal, which is vaporized and oxidized to oxide. This two-stage process is more effective than the American process, in which zinc ore and coal are heated in air to form the oxide. Impurities from the impure zinc ore and coal are lead, cadmium, iron, and sulfur compounds. Typical ASTM specifications for zinc oxides are shown in Table 2.12.

The majority of metal oxides are used in coated or treated forms in order to disperse more readily in the rubber mixtures. From 2 to 5 phr are normal usages.

Organic Acids. These are normally used in combination with metal oxides; they are generally high-molecular-weight monobasic acids or mixtures of the following types: stearic, oleic, lauric, palmitic, and myristic acids, and hydrogenated oils from palm, castor, fish, and linseed oils.

The usage of each particular type depends on the accelerator used and the amounts of other compounding ingredients present. Normally from 1 to 3 phr are used.

Alkaline Substances. These will increase the pH of a rubber compound, and in most instances the cure rate. As a rule of thumb, in the majority of recipes, any material that makes the compound more basic will increase the

Table 2.12. ASTM Specifications for Zinc Oxide.

	American	*French*	*Secondary Type*
Zinc oxide, min. (%)	99.0	99.5	99.0
Total sulfur, max. (%)	0.15	0.02	0.02
Moisture and other volatile matter, max. (%) 105°C loss	0.25	0.25	0.25
Total impurities, max. (%)	2.0	1.0	-
Coarse particles, residue on 44 micron sieve (No. 325), max. (%)	0.10	0.05	0.10
Lead (%)	0.10	0.002	0.10
Cadmium (%)	0.05	0.002	0.05
Surface area (m^2/g)	3.5	4.5	2.5

Table 2.13. Typical Accelerator Activators.

Type	*Sp. Gr.*	*Price $/lb*
Metal Oxides		
Zinc oxide, lead-free (American process)	5.6	0.465
Zinc oxide, lead-free (French process)	5.6	0.48
Red lead (98% Pb_3O_4)	9.0	0.30
Magnesium oxide	3.38	1.02
Litharge	9.5	0.50
Organic Acids		
Lauric acid	0.87	0.71
Stearic acid	1.02	0.26
Oleic acid	0.89	0.46
Amines		
Diethanolamine	1.09	1.18
Triethanolamine	1.12	0.60

Note: The specific gravity and price vary with the purity of the substances, and in the case of the oxides, whether or not the particles have been surface treated to improve dispersion, or are supplied in the dispersed form in a wax-like base.

cure rate since acidic materials tend to retard the effect of accelerators. Typical examples of these ingredients include ammonia, amines, salts of amines with weak acids, and reclaim rubbers made by the alkali process.

The phr usage would depend only on how the above materials are to be used in any recipe. Typical examples are given in Table 2.13.

Age Resistors (Antidegradants)

All rubbers are sufficiently affected by natural or accelerated aging processes so that it is necessary to add materials that are capable of retarding this type of deterioration.

The loss in physical properties, associated with aging processes, is normally caused by chain scission, crosslinking, or some form of chemical alteration of the polymer chains. Consequently, the age-resistors used must be capable of reacting with the agents causing aging (ozone, oxygen, prooxidants, heat, light, weather, and radiation) to prevent or slow the polymer breakdown, to improve the aging qualities, and to extend the service life of the product involved. Examples of how oxidation affects the various elastomers are shown in Table 2.14.

Chemical Protectants. There are three general types of compounds used for their protective qualities:

Secondary amines	$R_2N{-}H$
Phenolics	$R(OH)_x$
Phosphites	$(RO)_3P$

In general, the amines tend to discolor ("staining") and are used only where color is not important. The phenolics are "nonstaining" and are used mainly in light-colored goods, for which color retention is important. Phosphites are mainly used as stabilizers for SBR.

Physical Protectants. Products used in installations where little or no movement is involved can be protected with waxy materials which migrate ("bloom") to the surface of the rubber part and form a protective coating that shields the part from the effects of oxygen, ozone, and so forth.

Table 2.14. Effects Observed from Attack of Oxygen upon Elastomers[a]

ASTM D 1418 Designation	*Chemical Designation*	*Effect of Oxidation*	*Maximum Operating Temperature*
NR	Natural polyisoprene	Softens	212°F (100°C)
IR	Synthetic polyisoprene	Softens	212°F (100°C)
CR	Chloroprene	Usually hardens	225°F (107°C)
SBR	Styrene-butadiene	Hardens	250°F (121°C)
NBR	Acrylonitrile-butadiene	Hardens	275°F (135°C)
IIR	Isobutylene-isoprene	Softens	212°F (100°C)
CIIR	Chloro-isobutylene-isoprene	Hardens	–
BR	Polybutadiene	Hardens	250°F (121°C)
T	Polysulfide	Hardens	180°F (82°C)
EPM	Ethylene propylene	Hardens	300°F (149°C)
EPDM	Ethylene propylene terpolymer	Hardens	300°F (149°C)
CSM	Chlorosulfonated polyethylene	Hardens	250°F (121°C)
VMQ	Methyl-vinyl siloxane		
PVMQ	Phenyl-methyl-vinyl siloxane		
FPQ	Trifluoropropyl siloxane	Hardens	450°F (232°C)
AU	Polyurethane diisocyanate	Hardens	212°F (100°C)
FPM	Fluorinated hydrocarbon	Hardens	450°F (232°C)
ACM	Polyacrylate	Hardens	325°F (163°C)

[a]Adapted from *Handbook on Antioxidants and Antiozonants*, Goodyear Tire & Rubber Co., Akron, OH 44316.

Classification. The phr usage of any age-resistor would depend on the type of service to which the rubber part is subjected. However, normal usage is about 2–3 phr. Table 2.15 gives a classification of the types used, and production data are given in Table 2.11. It is important to note that the amines represent over 65% of the total production.

Many examples of the use of age-resistors are given in the following chapters, and test methods for determining the effectiveness of any particular component in aging studies are given in Chapter 5.

Softeners (Physical Plasticizers)

Physical plasticizers of this type do not react chemically with the rubbers involved, but function by modifying the physical characteristics of either the compounded rubber or the finished vulcanizate. In all cases, whether the softener is used as a processing aid (usually 2 to 10 phr) or it is used to alter the finished product (up to 100 phr), it must be completely compatible with the rubber and the other compounding ingredients used in the recipe. Incompatibility will result in poor processing characteristics or "bleeding" in the final product, or both.

Drogin has rated the various classes of materials used as physical plasticizers according to the effects they produce. This rating is shown in Table 2.16. In addition, the breakdown in Table 2.17 gives uses for materials of this type. The most general use is probably for the preparation of oil-extended rubbers of the SBR type with the naphthenic, paraffinic, and aromatic oils.

The reader should refer to the chapters dealing with each rubber in order to determine what material (including amounts) is in general use. It is quite typical for many of these materials to act as dual purpose ingredients, i.e., processing aids can also increase elongation, reduce hardness, improve tack, and so forth, depending on the amount and type used and the rubber involved.

Miscellaneous Ingredients

Materials of this type are used whenever some particular effect or property is desired in a vulcanizate. Some typical uses follow (examples of each are given in Table 2.18):

Abrasives: Erasers, grinding and polishing wheels require some type of abrasive for proper usage. Mineral ingredients such as ground silica and pumice are suitable for this purpose.

Blowing Agents: Some type of gas-generating chemical is necessary for preparing blown sponge and microporous rubber. Suitable agents must be capable of releasing gas during the vulcanization period. Azo compounds and carbonates are suitable gas-releasing chemicals.

Table 2.15. Examples of Age-resistors.

Chemical Type	*Example*	*Use*	*S.G.*	*Price $/lb*
	Antioxidants			
Hindered phenol	Styrenated phenol	Nonstaining	1.08 (liq.)	1.15
Hindered bis-phenol	2,2′-methylene-bis- (4 methyl-6-*t*-butyl- phenol)	Nonstaining	1.08	3.60
Amino-phenol	2,6′-di-*t*-butyl-α-dimethylamino-p-cresol	Nonstaining	0.97	2.18
Hydroquinone	2,5-di-*t*-amyl hydroquinone	Nonstaining	1.26	4.23
Phosphite	Tri (mixed mono- and di-nonylphenyl) phosphite	Nonstaining	0.99 (liq.)	1.02
Diphenylamine	Octylated diphenyl-amine	Semistaining	0.99 (liq.)	1.68
Naphthylamines	Phenyl-β-naphthyl-amine	Staining	1.24	3.16
Alkyldiamine	N,N′-diphenyl-ethylene diamine	Staining	1.14	1.30
Aldehyde-amine condensation product	Acetone-diphenyl-amine reaction product	Staining	1.10	1.69
Quinoline	Polymerized 2,2,4-trimethyl-1,2-dihydroquinoline	Staining	1.08	1.68
Phenylenediamine	N,N′-diphenyl-*p*-phenylene diamine	Staining, and flex-crack-resistant	1.28	4.25
	Antiozonants			
Dialkyl-phenylene diamine	N,N′-bis(1-methyl-heptyl)-*p*-phenylene-diamine		0.90 (liq.)	2.85
Alkyl-aryl-phenylene-diamine	N-isopropyl-N′-phenyl-*p*-phenylene-diamine		1.17	3.27
Carbamate	Nickel dibutyldithio-carbamate		1.26	4.42
Physical Type				
Waxes	Blended petroleum waxes		0.90	0.82
	Microcrystalline waxes		0.90	0.73

Table 2.16. Examples of Properties Obtained from Physical Plasticizers.[a]

	Properties	*Code*
	Fatty Acids	
Cotton seed	1	1. Improved tubing
Rincinoleic	1	2. Better tack
Lauric	1	3. Increased plasticity
	Vegetable Oils	
Gelled oils (sulfonated)	1, 6, 12, 13	4. Low modulus
Solid soya	4	5. Increased tensile
Tall oil	3, 4, 5, 13, 16	6. Improved elongation
Soya polyester	13	7. Softer cured stocks
	Petroleum Products	
Unsaturated	1	8. Harder cured stocks
Mineral oils	3, 4, 6, 7, 9, 11	9. Higher rebound
Unsaturated asphalt	3	10. Better tear
Certain asphalts	7, 10, 11	11. Low hysteresis
	Coal-tar Products	
Coal-tar pitch	1	12. High hysteresis
Soft cumars-tars	3	13. Improved flex life
Soft-coal tar	5, 6	14. Improved processing
Cumar resins	5, 11	15. Improved dispersion
	Pine Products	
Crude gum turpentine	2, 4, 5, 12, 13	16. Mold lubricant
Rosin oil	2, 5, 6	17. Flame retardant
Rosin	2, 8, 12	
Pine tar	3, 4, 5, 6, 7	
Dipentene	6, 13	
Certain rosins	13	
Rosin ester	2, 14, 15	
	Esters	
Dicapryl phthalate	3	
Butyl cuminate	9	
Dibutyl phthalate	3, 7, 9	
Butyl lactate	10	
Glycerol chlorobenzoate	10	
Chlorodibutyl carbonate	13	
Methyl ricinoleate	2	
Butyl oleate	3, 7, 13	
Dibutyl sebacate	3, 7, 13, 14	
Dioctyl phthalate	3, 7, 13	
Methyl oleate	1, 3, 7	
Tricresyl phosphate	2, 7, 17	

Table 2.16. (*Continued*)

	Properties	*Code*
	Resins	
Coumarone-indene	2, 5, 7, 15	
Phenol-formaldehyde	2, 3	
Shellac	8	
	Miscellaneous	
Amines	6	
Wool grease	7	
Pitches	8, 12	
Diphenyl oxide	9	
Benzoic acid	10	
Benzyl polysulfide	10	
Waxes	11	
Fatty acids	11	

[a]Compiled from a lecture given by Dr. I. Drogin, N.Y. Rubber Group, Elastomer Technology Course, Oct. 17, 1955, and the 1984 Blue Book listings.

Table 2.17. Examples of Typical Physical Plasticizers.

Use	*Example*	*S.G.*	*Cost $/lb*
Extenders	Extender oils		
	Naphthenic	0.92	1.66 gal
	Paraffinic	0.90	1.86 gal
	Aromatic	0.98	1.33 gal
	Mineral rubber	1.04	0.19
Processing aids	Castor oil	0.96	0.96
	Low-molecular-weight polyethylene	0.91	0.94
	Tall oil	0.94	0.31
Reduced-stock hardness	Mineral oil	0.93	3.51 gal
	Vulcanized vegetable oil (Factice)	1.04	1.08
Tackifiers	Coumarone-indene resins	1.04	0.65
	Ester gum	1.10	0.90
	Oil-soluble phenolic resin	1.01	0.69

Table 2.18. Typical Miscellaneous Ingredients.

Type	*Example*	*S.G.*	*Price $/lb*
Abrasive	Pumice (ground)	2.35	0.30
Blowing agent	Azodicarbonamide	1.63	2.49
Colorants	Titanium dioxide	4.20	0.75
	Ultramarine blue	2.35	1.30
Flame retarder	Antimony oxide	5.20	1.80
Homogenizing agent	Mixture of aliphatic-naphthenic-aromatic resins	1.05	0.60
Internal lubricant	Low-molecular-weight Polyethylene	0.92	0.94
Odorant	Essential oil blend	0.94	6.15
Retarder	Salicyclic acid	1.37	1.34

Colorants: Materials used for coloring nonblack goods utilize either inorganic pigments or organic dyes. They must be stable, color-fast, and reasonably priced.
Flame Retardants: Chlorinated hydrocarbons, phosphate and antimony compounds may be added to reduce flamability.
Homogenizing Agents: Resin mixtures used to aid in blending two or more elastomers. They reduce processing time and give more uniform blends.
Internal Lubricants: Certain amines, amides, fluoro-carbons and waxy materials based on low-molecular-weight polyethylene act as internal lubricants providing good mold release and fidelity.
Odorants: Aromatic compounds are capable of screening out or masking odors from rubber compounds. These components are normally used for wearing apparel and drug sundries. Some are effective as germicides.
Retarders: These ingredients should reduce accelerator activity during processing and storage. Their purpose is to prevent scorch during processing and prevulcanization during storage. They should either decompose or not interfere with the accelerator during normal curing at elevated temperatures. In general, these materials are organic acids that function by lowering the pH of the mixture, thus retarding vulcanization.

The use of retarders should be avoided if possible by the proper selection of accelerator-sulfur combinations and careful control of processing conditions. Careful temperature control and proper storage (controlled temperature and humidity) are important, especially during the summer months when factory temperatures are abnormally high.

Fillers

Fillers have been used since the discovery of rubber vulcanization to color, reinforce, extend, and cheapen compounds. Two major classes are used: carbon blacks and nonblack fillers.

The carbon blacks were developed specifically for use in elastomers and are currently named by the process by which they are produced: thermal, furnace, and acetylene blacks. Channel blacks, the first type of blacks to be utilized, are no longer produced because the furnace process can be modified to produce blacks with the same characteristics as the channel and thermal types.

The nonblack fillers are not as well-classified as the blacks. They include clays (semireinforcing), calcium carbonate (extending fillers), precipitated silica (reinforcing) and titanium dioxide (pigmenting filler). Many other materials are used in elastomer compounds but only those with relatively small particle sizes are reinforcing to nearly the same extent as carbon blacks. The majority are used to dilute or reduce cost of compounds, or perhaps to impart a special property.

The selection of a particular filler for a compound will depend on the processing characteristics and physical properties desired, the cost, and, above all, the final performance of the product. For example, many fillers may reduce the cost and give good tensile properties, but the performance given by the vulcanizate may not match those of a higher-cost, quality-controlled filler.

The compounder has to start with the best possible materials and then adjust the compound to meet the product specifications and cost. These modifications and other characteristics of fillers are discussed in detail in Chapter 3.

SUMMARY

Many ingredients are used to prepare rubber products. In general, each may be classified according to its specific use; however, many are capable of functioning in more than one manner in certain rubber compounds. Typical of this versatility is zinc oxide, which may function as an accelerator-activator, vulcanizing agent, filler, or colorant depending on which it is selected for.

Today one of the problems in selecting any ingredient is whether or not OSHA will list the material as a possible carcinogen. In the past 10 years, many standard-usage materials have been banned, and in some cases suitable replacements are difficult to find.

There are many references available giving listings of compounding ingredients. *Materials and Compounding Ingredients for Rubber and Plastics*, prepared by *Rubber World*, New York, was used in compiling the examples given in this chapter.

REFERENCES

1. *The Vanderbilt Rubber Handbook*, G. G. Windspear, ed., R. T. Vanderbilt Co., 1978.
2. *1983 Book of ASTM Standards*, Pt. 9.1: "Rubber; Carbon Black; Gaskets," American Society for Testing and Materials, 1916 Race St., Philadelphia, PA 19103.

3. G. Alliger and I. J. Sjothun, *Vulcanization of Elastomers*, Reinhold, New York, 1964.
4. J. LeBras, *Rubber: Fundamentals of its Science and Technology*, Chemical Publishing Co., New York, 1957.

ADDITIONAL TEXTS

H. J. Stern, *Rubber: Natural and Synthetic*, 2nd Ed., Maclaren and Sons, London, 1967.

Rubber Technology and Manufacture, 2d ed., C. M. Blow and C. Hepburn, eds., Butterworth Scientific, London, 1982.

W. Hofmann, *Vulcanization and Vulcanization Agents*, Palmerton, New York, 1967.

3

FILLERS

Part I

Carbon Black

J. T. BYERS
Ashland Chemical Co.
Akron, Ohio

INTRODUCTION

Although the phenomenon of carbon black reinforcement of rubber was discovered about 80 years ago, the nature of the mechanism is still a subject of considerable debate. Multiple theories are presented in the literature.[1-18] There is evidence that both physical and chemical attachments are capable of giving reinforcement effects, such as altering the modulus or tensile strength of the rubber phase. A reinforcing carbon black produces a partial immobilization of the chain segments of the polymer. The modification of an elastomer by carbon black reinforcement and vulcanization generates a unique three-dimensional visco-elastic network that transforms the soft elastomer into a strong, elastic product.

Available carbon blacks have a range of physical and chemical attributes (particle size, surface area, structure, surface activity) and can change elastomer properties in different ways and to different degrees. Carbon black loading is a critical factor in achieving the degree of elastomer modification desired to give a rubber compound properties suitable for a given end-use application. A rubber compounder must select the right polymer, the right carbon black grade/loading level combination, and the right vulcanization system to give a rubber compound with the desired properties at the lowest possible cost.

The first carbon blacks used in rubber were made using gas as the source of carbon as well as of heat. The channel black process produced small particle-size blacks that were rather acidic. These blacks worked quite well in natural rubber—the only polymer available at the time—to give compounds with good abrasion resistance. The acidic nature of the channel blacks also contributed to delaying scorch of the fast-curing natural rubber. Other gas-furnace processes were developed to produce larger particle size, semi-reinforcing blacks. The channel and early gas-furnace processes were not very efficient and polluted the air for miles downwind of the manufacturing plant. Increasing gas prices and air-pollution pressures stimulated the search for a cleaner, lower-cost process for producing carbon blacks. Today thermal blacks are the only gas-furnace blacks that are still produced in the United States and account for less than 5% of the rubber-grade carbon black consumed. The large particle-size, low-reinforcement thermal blacks can be used at high loadings to give benefits in some rubber compounding applications.

The timing was perfect for the introduction of the oil-furnace process in 1943. This new process used a low-grade petroleum feedstock for its source of carbon, was more efficient, offered greater manufacturing flexibility, and did not pollute the air. The oil-furnace blacks were alkaline and higher in structure compared with the channel blacks. Structure is the degree of interlinkage between carbon particles. These characteristics well suited the GR-S synthetic rubber (SBR) that was entering the scene at about the same time. Synthetic rubber lacked the inherent strength of natural rubber and vitally needed carbon black reinforcement. The early SBR polymers were hard to process and slow curing. The higher structure of the oil-furnace blacks contributed to improved processing and their alkaline nature resulted in a faster cure rate.[19] The faster curing tendencies were a disadvantage in natural-rubber compounds, but developments in delayed-action acceleration permitted successful usage. The oil-furnace process rapidly displaced the channel and gas-furnace processes for economic and environmental reasons. Because about 97% of the rubber-grade carbon blacks used in the U.S. today are oil-furnace blacks, these will be given primary emphasis. Besides the oil-furnace and thermal blacks, several other types are also available in limited quantities: lamp blacks, acetylene blacks, and channel blacks (very limited). In addition, a product made from bituminous coal, one made from rice hulls, and others obtained by the pyrolysis of tires are also available. Each of these can differ substantially in chemical, physical, morphological, and compound-performance properties.[20]

MANUFACTURE AND MORPHOLOGY

A schematic diagram of a typical oil-furnace reactor is shown in Fig. 3.1. The reactor itself is a refractory lined tube and may be either horizontal or vertical. Feedstock oil, natural gas or other fuel, and air (all except the natural gas are preheated) are injected into the combustion zone at specific rates for the carbon

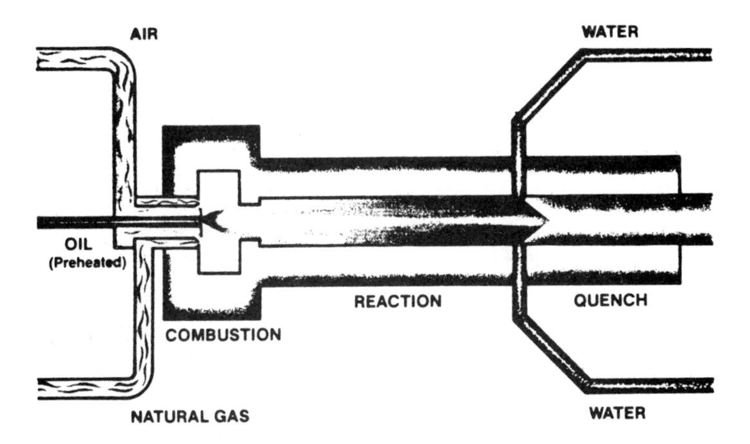

Fig. 3.1. Generalized carbon black reactor. (Source: Ashland Chemical Co., Carbon Black Div.)

black grade produced. The burning gas generates a very hot, turbulent atmosphere for cracking the feedstock oil. The primary source for feedstock (about 90% of U.S. feedstock) is refinery heavy-bottom oils. Minor sources include ethylene cracker tars and coal tars produced during coke production. The heavier aromatic oils and tars give better yields than lighter oils containing more hydrogen. The reactions by which the aromatic feedstock is converted to elemental carbon are complex and not well-understood. Spherical carbonaceous particles are formed which become increasingly viscous. The collision of particles still in a liquid-like state produces aggregates of spherical particles fused together in random grape-cluster configurations as illustrated in Fig. 3.2. The term *structure* is used to describe the degree of aggregation of the particles. A low-structure black may have an average of 30 particles per aggregate, whereas a high-structure black may average (wide distribution) up to 200 particles per aggregate.[11,12] Particle size is defined as the arithmetic mean diameter (AMD) of the individual particles that make up the aggregate. A sample of any oil-furnace carbon black grade will have a distribution of particle sizes and number of particles per aggregate.

Downstream of the combustion or cracking zone, a water quench is used to rapidly reduce the temperature and terminate the reaction. The "smoke" that exits the reactor is a mixture of carbon black aggregates, combustion gases, and moist air. As the smoke stream is cooled further, the heat is utilized to preheat the feedstock and air, and to generate steam for use elsewhere in the plant. The fluffy black and the combustion gases are separated by filtration, and the combustion gases are used as a heat source later in the process.

At this stage, the primary properties of particle size, surface area, and primary (or persistent) structure have already been established. The aggregates

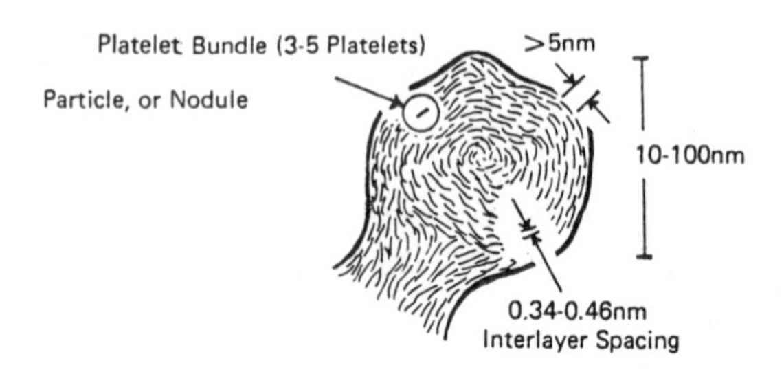

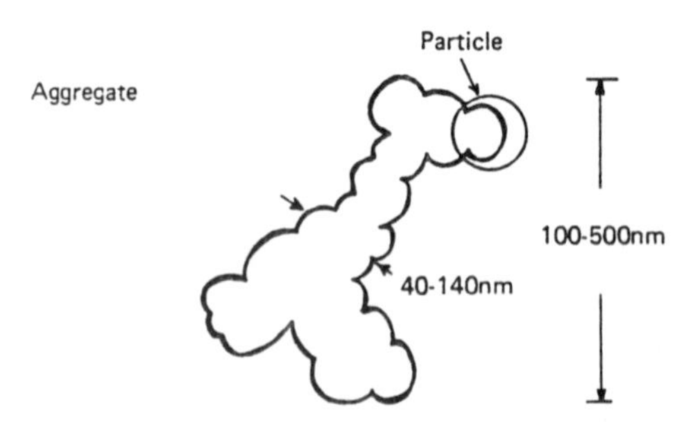

Fig. 3.2. Carbon black particle and aggregate. (Source: Ashland Chemical Co., Carbon Black Div.)

have a marked tendency to cluster together with other aggregates to form agglomerates during subsequent processing operations. This agglomeration is called *secondary* or *transient structure* because the bonds between the aggregates are low-energy bonds (easily broken) as opposed to the fusion linkages between particles that make up the primary structure. The majority of the secondary structure is broken during the mixing and dispersion of carbon black into a rubber compound, providing that mixing is adequate to properly disperse the black.

Various combinations of the primary black properties are obtained by process changes, such as feedstock flow rate, gas rate, air rate, and quench position. Particle size is principally controlled by feedstock rates and temperatures. A higher combustion ratio increases the temperature and reduces the particle size as well as the yield. Within limits, surface area can be controlled independently of either particle size or structure by changes in reactor dimensions and temperatures. Structure is more dependent on the manner of feedstock injection into the reactor. Structure can also be lowered with additives, usually small quantities of alkali metal salts.

The carbon black from the reactor is separated from the combustion gases by bag filters. The collected black is known as ''loose black'' and has a bulk density of 20 to 50 kg/m^3. The remaining combustion gas stream is called *tailgas*. The loose black and any extraneous material (such as particles of refractory or coke) are pulverized to 325 mesh sieve size or less prior to pelleting.

In the U.S., the wet-pelleting process is the only one used. In this process, a rotating, pin-studded shaft mixes the loose black with water, which may contain a binding agent such as molasses, to produce small round beads or pellets. The formation and quality of pellets, as well as the level of secondary structure, are highly dependent on pelletizer shaft speeds and feed rates of black and water. The pelleting process increases the bulk density to 300–500 kg/m^3 for ease of handling and shipping.

The wet pellets are fed into a rotary drier heated by combustion of the tailgas. The steam that is generated is removed by an exhaust fan and can be partially replaced with air. Oxygen can then react with the carbon black. This surface oxidation influences the chemical properties of the carbon black and, in turn, the cure rate and properties of rubber vulcanizates.

The pelleted black is screened for uniformity and passed over magnetic separators to remove any metallic contamination that may have gotten into the product stream as it progressed through the process equipment. The finished product is conveyed to the bulk-storage tank for packaging and shipment.

Furnace blacks are broadly classified into two groups—*reinforcing* (hard, tread) and *semireinforcing* (soft, carcass). The reinforcing blacks have the smaller particle sizes and lower yields, hence higher prices than those of the semireinforcing blacks.

PHYSICAL-CHEMICAL PROPERTIES

Particle Size

The only direct method for measuring particle size and particle size distribution is provided by electron micrographs, such as shown in Fig. 3.3. The average diameter of the ultimate particles of rubber-grade furnace blacks ranges from about 19 to 95 nm (nanometers) as determined by direct measurement of micrographs. Many particles must be measured to obtain a representative average. Their definition is often vague and the method is extremely time-consuming. Therefore, other tests are often used to estimate particle size. The tint-strength test (ASTM D3265) is an example. In this procedure, a carbon black sample is mixed with zinc oxide and a soybean oil epoxide to produce a black or grey paste. This paste is then spread to produce a suitable surface for measuring the reflectance of the mixture with a photoelectric reflectance meter. This reflectance is then compared to the reflectance of paste containing the Industry Tint Reference Black (ITRB) prepared in the same manner. The tint test is affected by the structure as well as by the particle size of the black. For a given particle

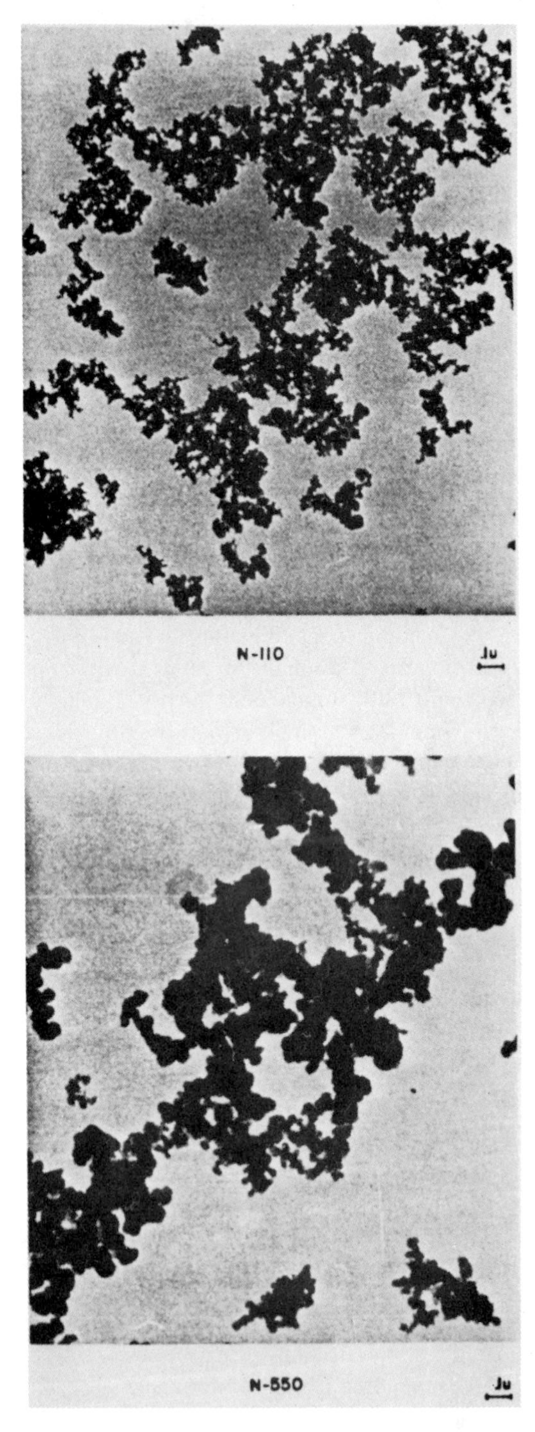

Fig. 3.3. Micrographs of carbon blacks. The unit of measure (bar) below each micrograph is 0.1 micron. (Source: Ashland Chemical Co., Carbon Black Div.)

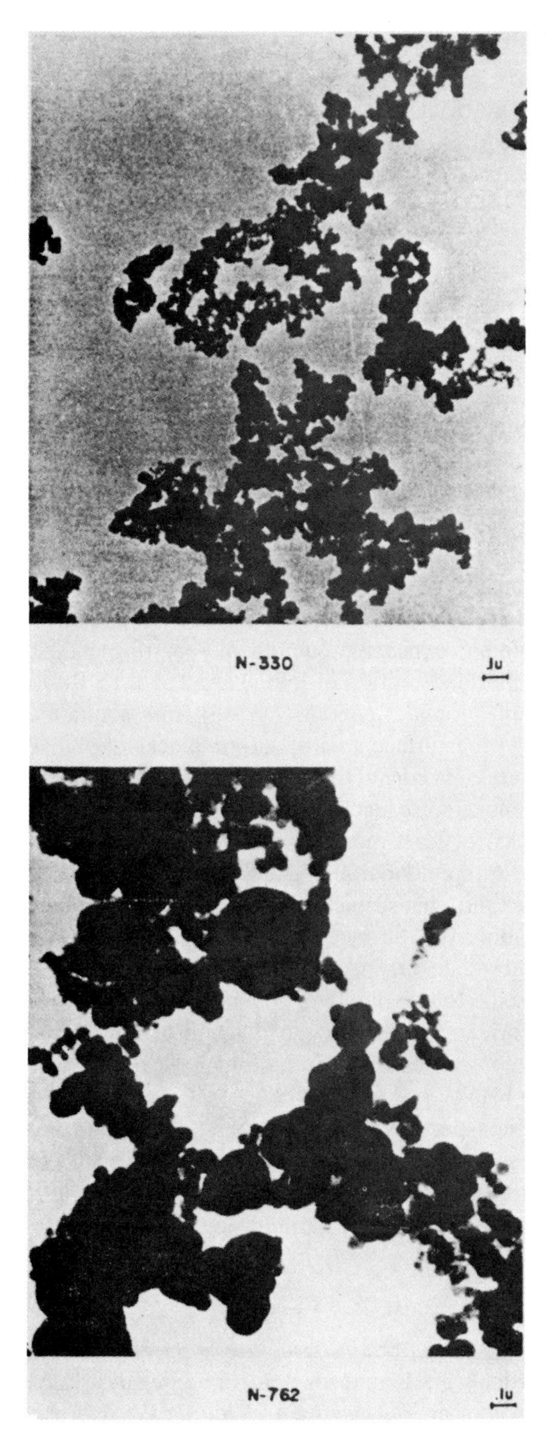

Fig. 3.3. (*Continued*)

size, higher-structure blacks have lower tinting strengths. An average particle size can be estimated from statistical equations that relate tint strength and structure to particle size as measured from electron micrographs.[21]

Surface Area

The surface area of a carbon black is very important because it defines how much surface is available for interactions with other materials present in a rubber compound. Generally, a small particle-size black will have a high surface area, but the texture or nature of the surface can also influence the surface area.

A procedure to determine the surface area of particulate matter was developed by Brunauer, Emmett, and Teller in 1938.[22] The BET method is based on the adsorption of a gas, usually nitrogen, on the surface. A simpler one-point nitrogen adsorption technique (ASTM D3037) has become the standard technique for measuring carbon black surface areas. The surface area of carbon blacks can also be calculated from measurements made on electron micrographs. However, this method has the problems that are associated with electron microscopy: time-consuming, expensive equipment requiring skilled operators, and rather subjective. Both adsorption of iodine (ASTM D1510) and CTAB (cetyltrimethylammonium bromide) (ASTM D3765) from solution are also used to obtain a measure of the surface area of carbon blacks.

Some high-surface-area conductive blacks and specialty color blacks are made porous through controlled oxidation. Generally it is believed that the standard rubber-grade blacks, at least those blacks with nitrogen surface areas of less than about 130 m^2/g, are nonporous. The CTAB surface area is influenced less by porosity than the nitrogen surface area, because the CTAB molecule is larger than the nitrogen molecule. Surface area measurements by CTAB and by nitrogen surface area correlate very well for the oil-furnace blacks commonly used in rubber compounds. Iodine numbers obtained on devolatilized blacks (1 hour at 837°C) also correlate well with the other measurements; but without the devolatilization step, the results may be influenced by residual oils present on the surface of the black.

Generally, the nonspecialty furnace blacks give a good inverse correlation between nitrogen surface area and particle-size measurements from electron micrographs. Because the nitrogen surface area can be determined easily and accurately, it is often used in lieu of particle-size measurements.

Structure

As noted earlier, the term *primary structure* is used to describe the degree to which carbon particles are bound together in aggregates. *Secondary structure* is the agglomeration of aggregates due to Van der Waals forces and is formed during the collection and pelletizing of the carbon black. Aggregation contributes nearly all of the structure-related effects observed in rubber compounds, while agglomeration is easily destroyed during rubber-mixing and has only minor effects on compounded rubber properties.

Because nonspherical particles pack differently from spheres, the volume of void space between aggregates per unit weight of carbon black increases with the number of particles per aggregate. Structure measurements are based on measuring these voids by filling them with a liquid, as in the DBP (dibutylphthalate) absorption test (ASTM D2414), or by measuring the compressibility, as in the void-volume test. The DBP test results are influenced by the amount of work done on the black in the pelletizer. In general, increasing the speed in the pelletizer will lower DBP absorption. The compressed DBP test (ASTM D3493) and the void-volume test, described below, give results that are influenced less by the amount of pelletizer work.[20] Thus these tests are felt to be more indicative of the carbon black primary structure.

In the void-volume test, a measured quantity of black is compressed at a constant pressure. The specific volume of the compressed pellet is calculated from its dimensions and then the volume occupied by the carbon (as determined by helium density) is subtracted to obtain the volume of voids.

Measurements obtained from electron micrographs have also been used to characterize aggregate morphology in terms of shape, size, and bulkiness. A procedure to characterize the aggregate in terms of length, width, and shape categories is described in ASTM D3849. Carbon blacks show a wide distribution of the number of particles per aggregate and a distribution of particle sizes from one aggregate to another. For practical purposes, this method of morphological characterization has several associated problems, such as the difficulties of obtaining a micrograph of a representative sample and dispersing the sample to its primary aggregate, cost of the instrument, and the time required. A three-dimensional approach to characterize aggregates via stereo micrographs has also been tried.[24]

Chemical Properties and Surface Activity

The chemical nature of a carbon black surface is variable. There is evidence for the presence on the surface of at least four oxygen-containing groups: carboxyl, phenol, quinone, and lactone. The surfaces can also differ in adsorptive capacity and in the distribution of sites of high energy.

Elastomers of a polar nature, such as neoprene or nitrile rubber, will interact more strongly with filler surfaces having dipoles such as OH and COOH groups or chlorine atoms. However, carbon blacks with chemically modified surfaces have not demonstrated any noticeable improvement in reinforcement in the general-purpose hydrocarbon rubbers. Chemical surface groups do affect the rate of cure with many vulcanizing systems. However, the physical adsorption activity of the filler surface is of much greater overall importance for the mechanical properties of the general-purpose rubbers than is its chemical nature.[25]

There are no standard procedures for the quantitative estimation of surface chemistry and surface activity of carbon blacks. However, a relative measure of the surface oxides can be obtained from the pH of carbon black slurries (ASTM D1512) and the weight loss on heating dry carbon black at 950°C. ''Bound-rubber'' measurements are sometimes used to obtain a measure of the surface activity, but these are affected by loadings, the molecular weight distribution of the rubber, and the thermal history of the rubber sample.

It is generally accepted that the oxygen content influences the cure rate of sulfur-cured stocks. In general, increasing the oxygen content gives a longer scorch period, a slower rate of cure, and a lower modulus at optimum cure. The amount of oxidation that occurs during the pellet drying operation thus can affect the cure rate and modulus of rubber compounds.

Carbon blacks are typically evaluated in standard test recipes, such as ASTM Natural Rubber (D3192) or SBR (D3191), and compared with a standard control black. This serves indirectly to indicate the surface activity of the black.

Electrical and Thermal Conductivity

Carbon blacks are generally electrically conductive because of the highly conjugated bonding scheme present in the crystalline regions. Increasing the structure and surface area both contribute to increased electrical conductivity. The surface oxygen groups can also play a significant role in determining electrical properties, and the groups' removal by heating in an inert atmosphere can lead to a significant decrease in electrical conductivity. Acetylene black is used to obtain highly conductive systems in carbon-zinc cells, dry cells, batteries, and so forth. Acetylene black not only has high conductivity but also has a high absorptive capacity for electrolytes and excellent fluid retention—properties required for electrolytic cells.

In thermal conductivity, the effects of the elastomer overshadow the effects of the carbon black. Specialty blacks for high electrical conductivity are usually high in structure and may be porous. These blacks do not necessarily give the highest thermal conductivity, perhaps due to their porous nature that provides a measure of heat insulation.

Pellet Quality

Pellets are required for economical transportation and for adequate flow in bulk-shipping containers and conveying equipment. These pellets must have sufficient strength to resist physical breakdown during transportation and plant handling yet be weak enough to break up and disperse during rubber mixing.

The pellet hardness test (ASTM D3313) involves applying an increasing force to an individual pellet until it fractures. The force at fracture varies with pellet size, so the common industry practice is to use the −12/+14 mesh sieve fraction. Pellet hardness requirements are governed by the nature of the compound in which the pellets will be used. Compounds that use low-viscosity polymers usually require soft pellets.

The percent fines and attrition (ASTM D1508) are considered as indicators of bulk-handling characteristics. High fines may also signal the possibility of mixing difficulties. Mass strength (ASTM D1937) reflects carbon black flowability in bulk shipments.

Pellet size distribution (ASTM D1511) is sometimes important. A high concentration of large pellets is generally believed to be undesirable because of their tendency to fragment rather than to crumble and hence to give poor dispersion.

Purity

The purity of carbon blacks is generally judged from toluene discoloration (ASTM D1618), ash (D1506), grit residue (D1514), and sulfur content (D1619) tests. Toluene discoloration gives a rough estimate of the amount of extractable material present. The ash arises mostly from salts in quench and pelleting water with some contribution from non-hydrocarbon impurities in the feedstock. *Grit* or *sieve residue* is material sufficiently large to remain on a U.S. 35- or 325-mesh screen. Primary sources of grit are coke formed in the reactor, pieces of refractory that have eroded from the reactor, and metal scale from the process equipment.

Sulfur in furnace blacks comes from the feedstock and varies with the sulfur content of the feedstock. Most of the sulfur is chemically bound and appears to be distributed throughout the carbon black aggregate. Carbon black with sulfur levels up to 1.5 percent do not appear to influence cure rates.

The oxygen present is chemically combined on the surface of the black, while hydrogen is distributed throughout the particle. The sulfur content depends mostly on the sulfur content of the feedstock and ranges up to 2%. Ash content is governed mostly by the impurities in the water used for quenching and pelletizing. Carbon blacks also contain small amounts of materials that can be extracted with solvents such as benzene or toluene. These extractables are predominantly fused ring aromatics with no side chain.

CLASSIFICATION AND NOMENCLATURE

A standard classification system for carbon blacks used in rubber is described in ASTM D1765. The grade identification consists of a letter followed by three digits. The letter is either "N" for normal cure rate or "S" for slow-curing

Table 3.1. Current vs. Previous Nomenclature.

ASTM D1765	*Old Name*
N110	SAF (super-abrasion furnace)
N220	ISAF (intermediate super-abrasion furnace)
N330	HAF (high-abrasion furnace)
N358	SPF (super-processing furnace)
N660	GPF (general-purpose furnace)
N762	SRF (semireinforcing furnace)

blacks (such as oxidized blacks). The first digit following the letter indicates the particle-size range—lower numbers for smaller-particle-size blacks. The last two digits are arbitrarily assigned by ASTM, and do not have any descriptive meaning. Prior to adoption of the ASTM classification, letter names were used to indicate the particle-size range and the structure level. As new grades were introduced, the old nomenclature became very cumbersome. Some examples, along with the current ASTM identification are listed in Table 3.1.

A table of typical properties for rubber-grade carbon blacks is also given in ASTM D1765. The values listed are averages of typical values from several manufacturers, and may not necessarily be an accurate representation of that grade from any single manufacturer. Some of the grades listed in the D1765 table are used in low-volume applications and may be of limited availability.

Several other classification and identification systems for carbon blacks have been proposed to replace the current system, but the adoption of a new system is not anticipated soon.

MIXING AND DISPERSION

Carbon black must be adequately dispersed in order to obtain the maximum benefits of carbon black reinforcement. The initial step in mixing the carbon black into a polymer involves the breakup of the carbon black pellets. A soft pellet (low pellet-hardness) is required for adequate dispersion if low shear forces are present during the carbon-black-addition step of the mixing operation. A soft polymer (or oil-extended polymer) may be squeezed into a relatively hard pellet without breaking it, while a stiffer polymer will crush the same pellet into fragments. As mixing continues, the pellet fragments break up and the polymer penetrates the voids between the carbon black aggregates. When all voids are filled with rubber, the black is considered "incorporated," but may still exist as concentrated microagglomerates of black held together by the rubber vehicle. Additional mixing may be required to reach the desired degree of black dispersion throughout the compound mixture.

Small-particle-size (high surface area) carbon blacks are difficult to incorporate. High-structure blacks contain more void volume to be penetrated by the rubber and thus take longer to incorporate; but once they are incorporated, they disperse more rapidly than lower-structure blacks.[25]

In a simple mixture of polymer and carbon black, the second-power peak of a mixing energy profile roughly identifies the point of incorporation. Very soon thereafter, the viscosity of the mixture starts to decrease due to polymer degradation while the black is being further dispersed. If mixing is halted at the second power peak, the compound modulus may have already achieved its maximum level. However, additional mixing is required to achieve maximum tensile strength.[26]

CARBON BLACK PROPERTIES VS. COMPOUND PROPERTIES

In general, high-surface-area and high-structure carbon blacks are associated with increased reinforcement in a rubber compound. The particle size of carbon black affects the abrasion resistance, heat build-up (or resilience), tensile strength, and tear strength of the rubber compound. Carbon black structure has more effect on the modulus, hardness, and extrusion-die swell.

Four carbon blacks can be used to demonstrate the effects of varying surface area, structure, and black loadings on various compound properties. These carbon blacks have the physical-chemical properties listed in Table 3.2.

The effects of structure differences can be observed by comparing N339 with N356 *and* N650 with N660. The blacks of each pair have nearly equivalent nitrogen surface areas, but also have large differences in structure as judged from DBP absorption and void-volume data. The effects of a large change in surface area can be observed by comparing the two HAF blacks (N300 series) with two GPF blacks (N600 series).

Three compound recipes based on different polymers enable the observation of changes in carbon black effects from one polymer to another. The three recipes are listed in Table 3.3. Nine to eleven loading levels of each black were mixed in each recipe and the resulting compounds were tested. Some of these data are plotted against the black loading in Figures 4 through 12 to demonstrate the effect of black-loading variations as well as the effects of surface area and structure.

Table 3.2. Carbon Black Properties.

	Nitrogen Surface Area m^2/g	*DBP Absorpt. cc/100 g*	*Void Volume cc/100 g*	*Tint Strength %ITRB*
N339	98.9	118.4	69.2	108.8
N356	100.1	160.2	77.7	103.2
N650	38.8	123.7	57.4	52.7
N660	39.3	89.7	44.5	59.2

Table 3.3. Formulations.

EPDM Compound	
EPDM	100.0 phr
Naphthenic oil	12.0
Zinc oxide	5.0
Stearic acid	1.0
Processing aid	2.0
Sulfur	1.5
MBT	0.5
TMTD	3.0
Carbon black	Varies
SBR Compound	
SBR-1500	100.0 phr
Aromatic oil	5.0
Zinc oxide	3.0
Stearic acid	1.5
Sulfur	1.75
CBS	0.85
DPG	0.28
Carbon black	Varies
NR Compound	
Natural rubber	100.0 phr
Highly aromatic oil	15.0
Zinc oxide	5.0
Stearic acid	2.5
Antioxidant	2.0
Antiozonant	2.0
Sulfur	1.5
CBS	1.6
Carbon black	Varies

Compound Property Group 1

Compound viscosity, modulus, and hardness tests measure the degree of stiffening that carbon black contributes to the polymer-carbon black composite. High structure and an increase in the amount of carbon black surface available for attachment to the polymer both contribute to make the composite more viscous and less elastic. The amount of carbon black surface that is available is a function of the black's surface area and loading. The structure contribution also increases as the black loading increases. Thus these compound properties are dependent on black loading, structure, and surface area, and increase as any of the three carbon black factors are increased.

Fig. 3.4, a plot of Mooney viscosity in the SBR compound recipe; Fig. 3.5,

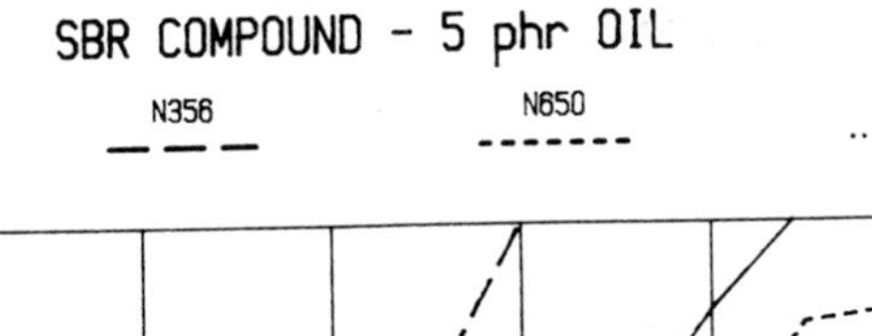

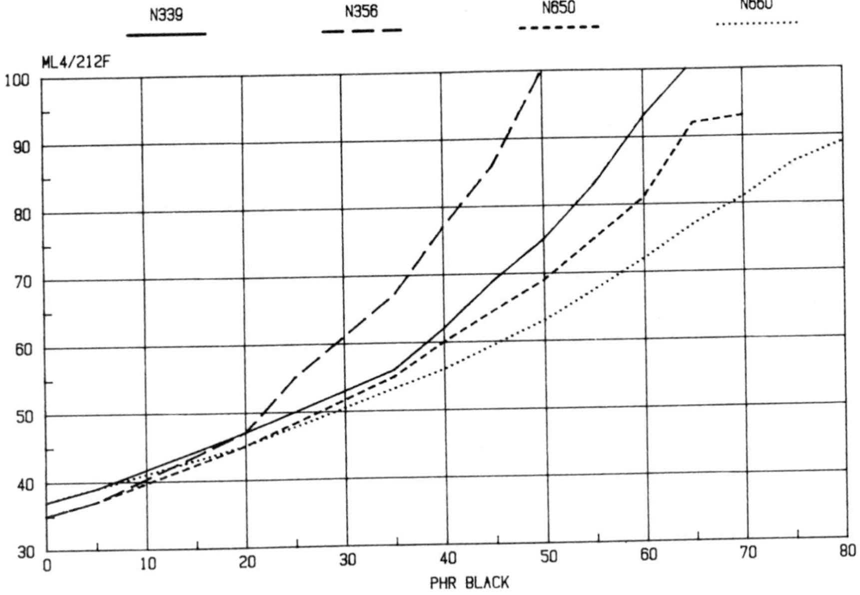

Fig. 3.4. Mooney viscosity. (Source: Ashland Chemical Co., Carbon Black Div.)

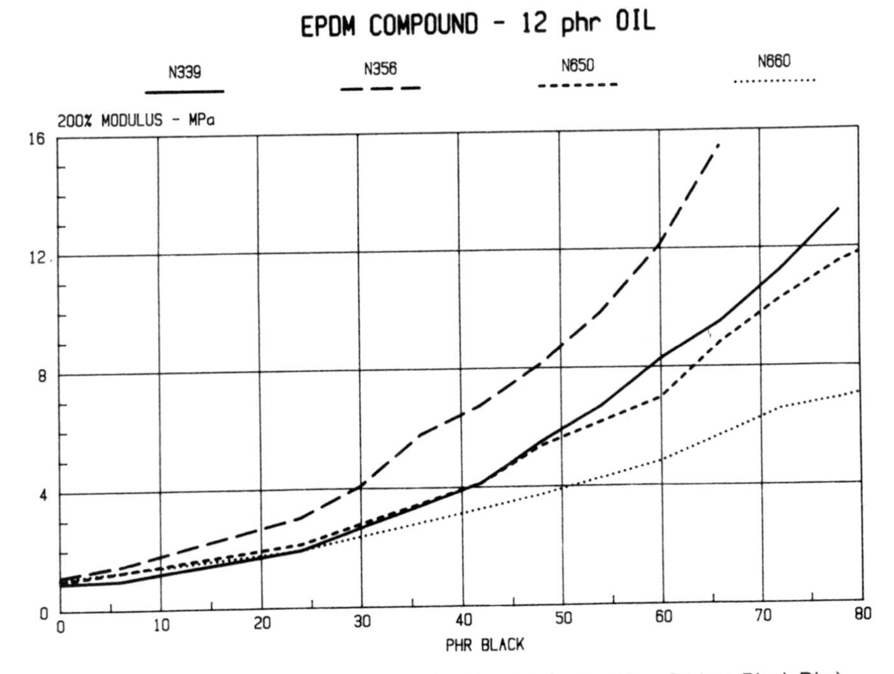

Fig. 3.5. Modulus at 200% elongation. (Ashland Chemical Co., Carbon Black Div.)

a plot of 200% modulus in the EPDM compound; and Fig. 3.6, a plot of hardness in the NR compound, all show increases for the four blacks as the loading is increased. Each graph is also typical of the data for the other two recipes. The plot of die-swell data (diameter of the extrudate) shown in Fig. 3.7 demonstrates the reduction in elasticity that accompanies the increase in viscosity and also illustrates that black loading, structure, and surface area all have an effect.

Compound Property Group 2

Abrasion resistance, tear strength, and tensile strength are destructive tests that measure resistance to failure under several different types of stress. These strength-related properties are enhanced by higher carbon black surface area and increased black loading up to a limiting value that is dependent on the packing characteristics (morphology) of the carbon black aggregates. Increasing the surface area or increasing the black loading serves to make additional carbon black surface available for rubber attachments.

As the black loading is increased, eventually a level is reached where the carbon aggregates are no longer adequately separated by polymer and the com-

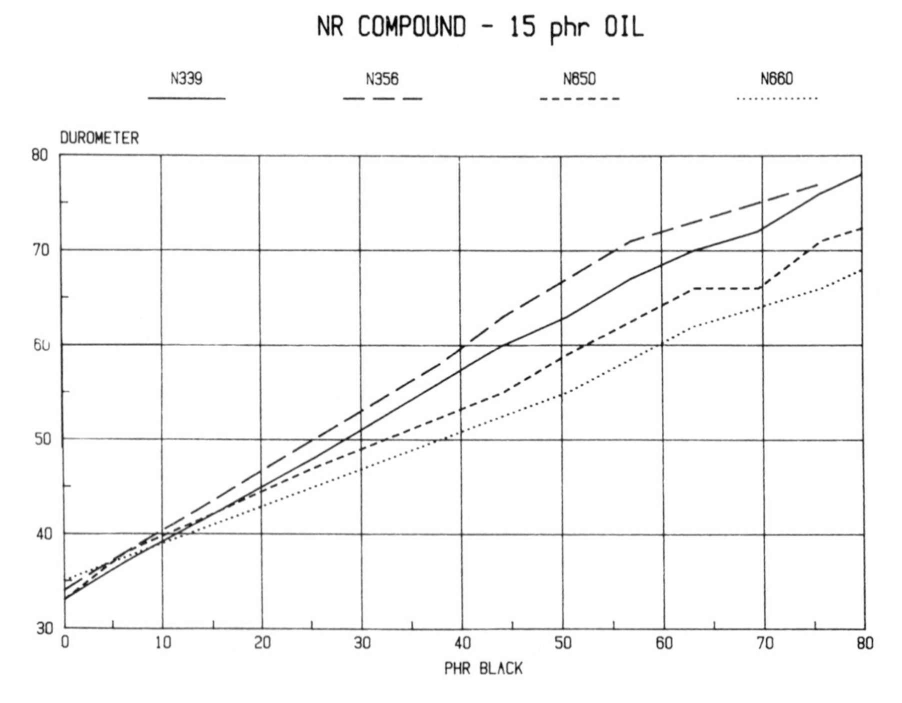

Fig. 3.6. Shore A hardness. (Source: Ashland Chemical Co., Carbon Black Div.)

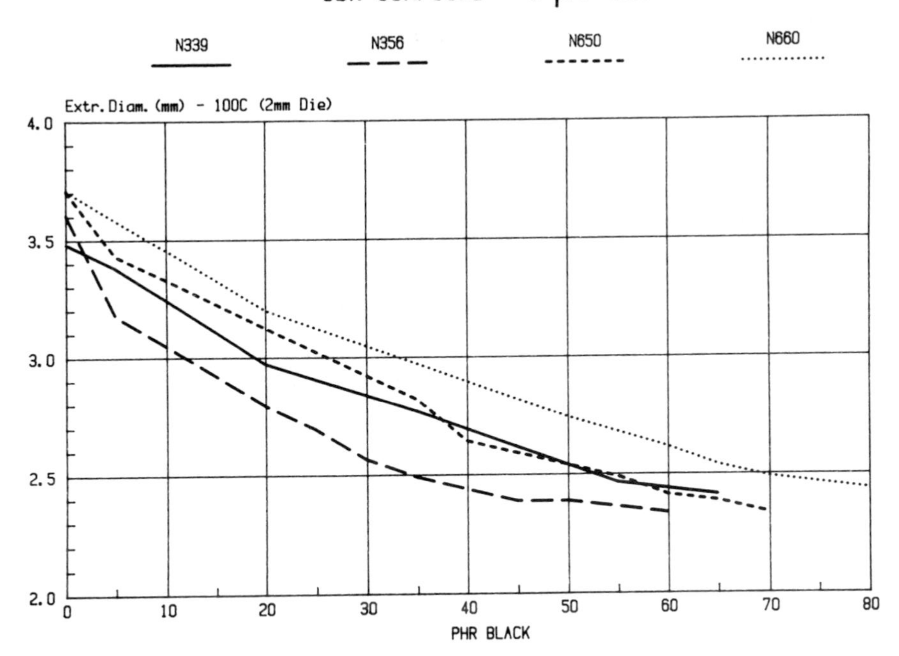

Fig. 3.7. Die swell (extrudate diameter). (Source: Ashland Chemical Co., Carbon Black Div.)

posite starts to weaken. The reduction in strength may be due to agglomeration of the carbon black aggregates to form a domain that acts like a foreign body, or may be simply the result of physical contact between adjacent aggregates. The packing characteristics of the carbon black that determine the upper loading limit are functions of the particle size (or surface area) and the structure of the carbon black.

Figs. 3.8A, 3.8B, and 3.8C illustrate that the surface area and loading (the latter is affected by structure) are the most important factors for abrasion resistance as measured with a laboratory angle-abrasion test. The lower-surface-area GPF blacks (N650 and N660) contributed a rather small improvement in abrasion resistance, regardless of the loading level, in all three compound recipes. The abrasion resistance is considerably better for the compounds containing the HAF blacks (N339 and N356), but is very dependent on the black loading and influenced by structure. The compounds containing the higher-structure N356 black reach maximum abrasion resistance at lower loadings than the compounds containing N339, but N339 ultimately gives a higher abrasion resistance in two out of the three recipes. Comparing the two blacks at equal loading would favor

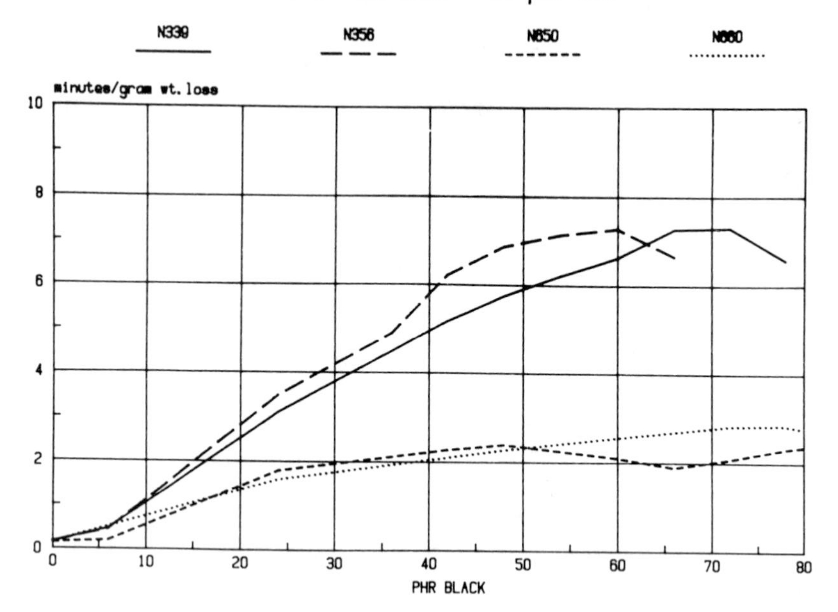

Fig. 3.8a. Abrasion resistance (angle ABR). (Source: Ashland Chemical Co., Carbon Black Div.)

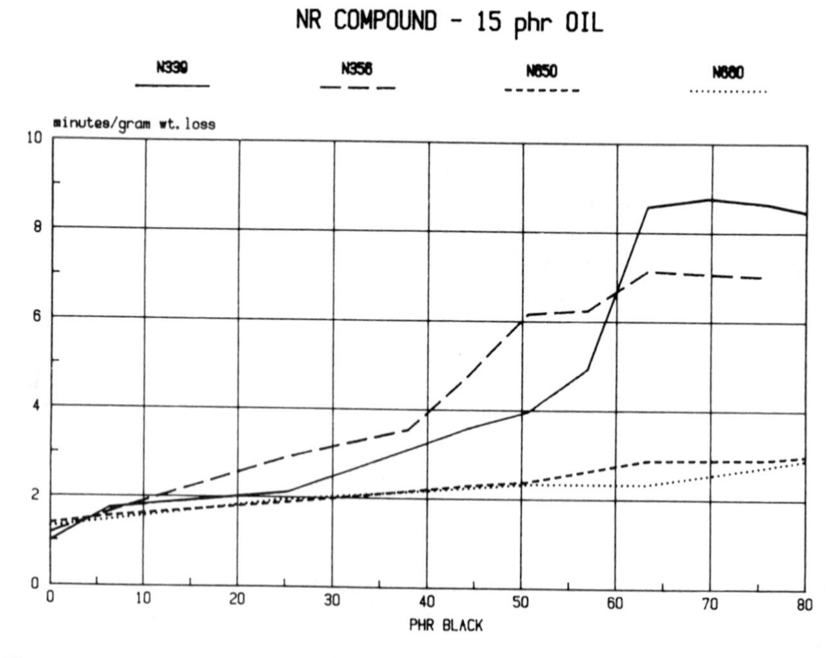

Fig. 3.8b. Abrasion resistance (angle ABR). (Source: Ashland Chemical Co., Carbon Black Div.)

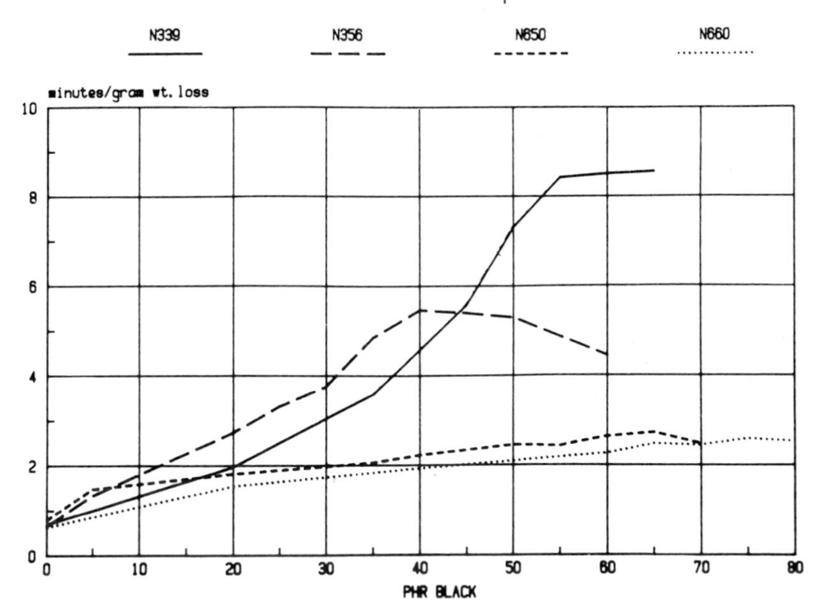

Fig. 3.8c. Abrasion resistance (angle ABR). (Source: Ashland Chemical Co., Carbon Black Div.)

N356 if the loading chosen for the comparison is relatively low, but would favor N339 when compared at a higher loading.

The low oil level in the SBR compound recipe is reflected in the shift of its curves toward the left (lower black loadings) compared to the EPDM and NR compounds.

Tear-strength data from the NR and EPDM compounds are shown in Fig. 3.9A and Fig. 3.9B, respectively. The greater tear strength obtained from natural rubber becomes obvious in comparing the two graphs. The data for N339 and N356 in the NR recipe (Fig. 3.9A) demonstrate the expected relationship as the black loading is increased, i.e., strength increasing to a peak, then declining. An effect of structure is also indicated by the shift of the N356 curve to the left (lower limiting value for loading because of the effect of higher structure on packing). The data for the GPF blacks show a general increase in tear strength as the black loading is increased. The EPDM compound data charted in Fig. 3.9B show rather small effects from differences in the carbon blacks, but the curve for the N356 black again appears to be shifted to the left compared to the N339 curve.

Fig. 3.10A illustrates the tensile strength relationships in the EPDM compound recipe (the SBR compound data gave a similar plot). The unfilled com-

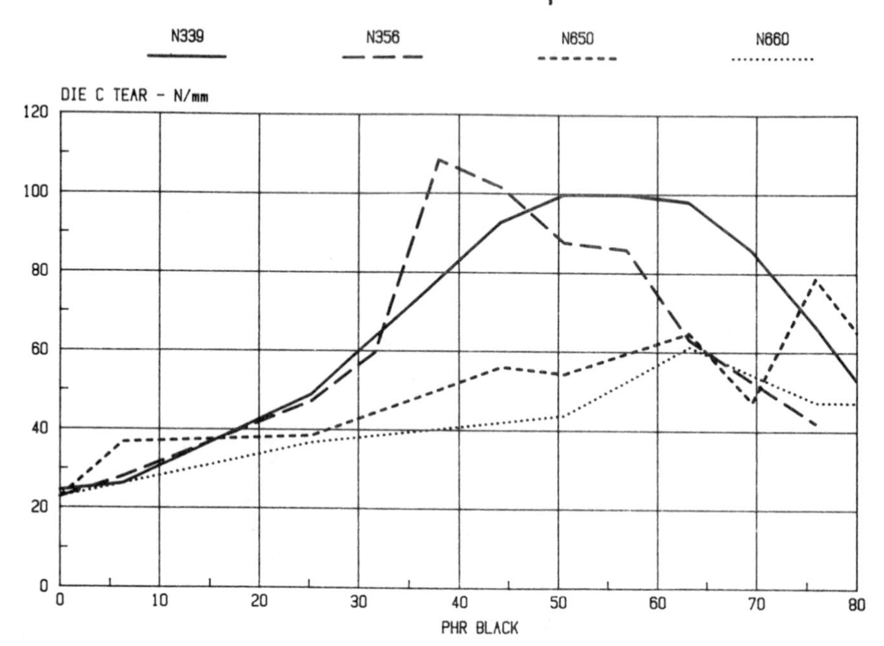

Fig. 3.9a. Tear strength (die C). (Source: Ashland Chemical Co., Carbon Black Div.)

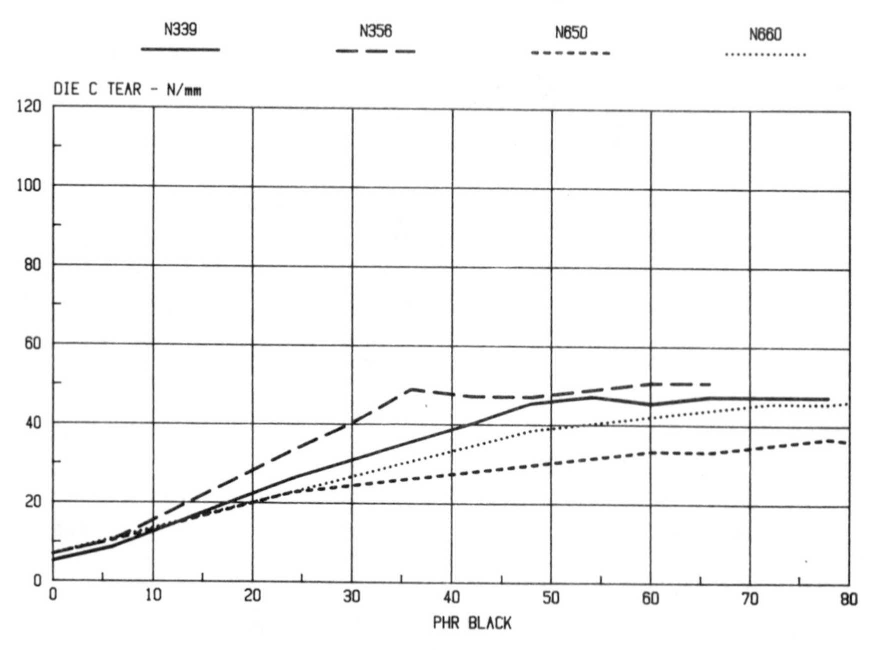

Fig. 3.9b. Tear strength (die C). (Source: Ashland Chemical Co., Carbon Black Div.)

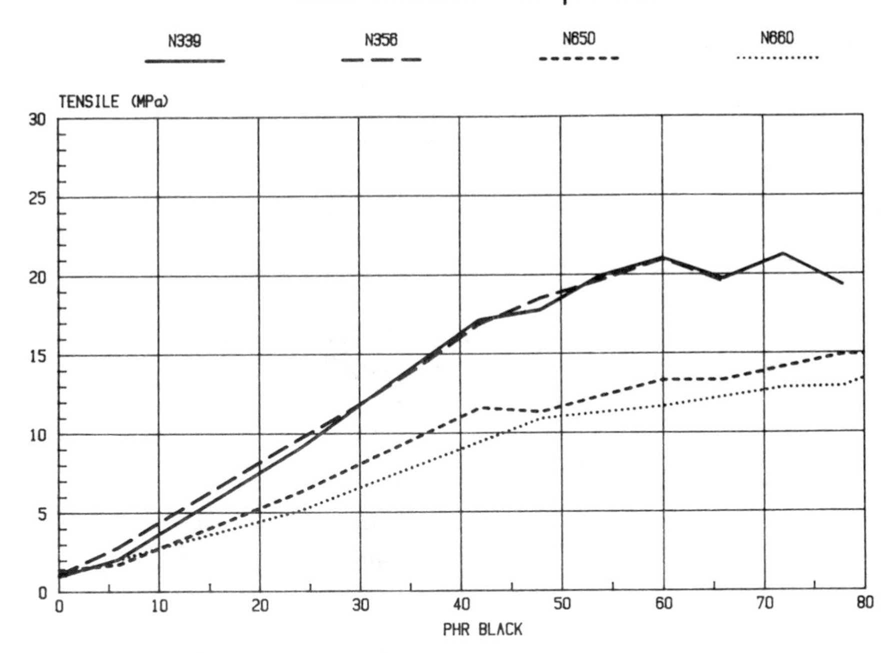

Fig. 3.10a. Tensile strength. (Source: Ashland Chemical Co., Carbon Black Div.)

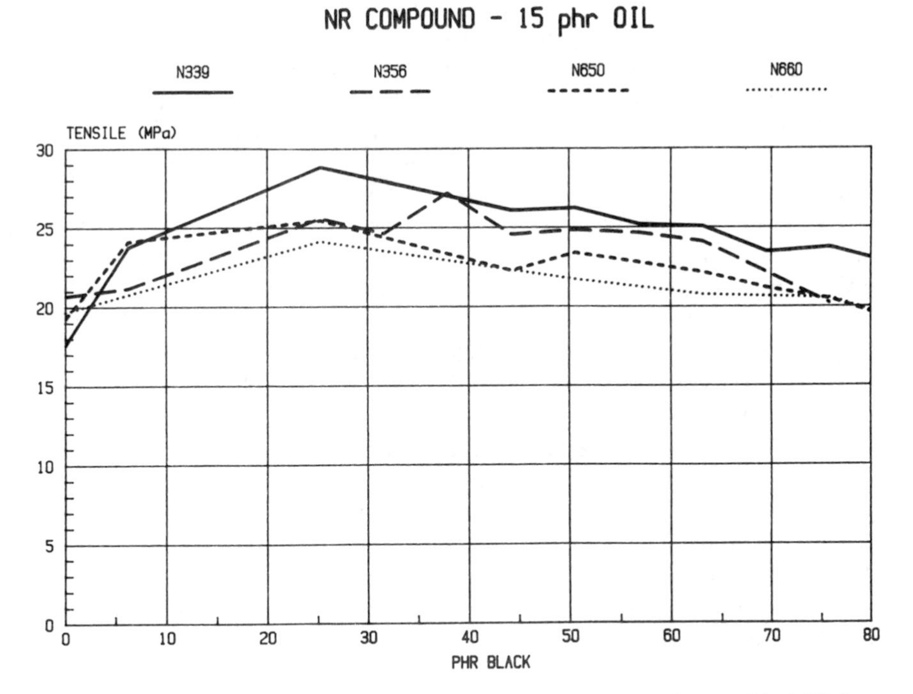

Fig. 3.10b. Tensile strength. (Source: Ashland Chemical Co., Carbon Black Div.)

pound has very little tensile strength. The tensile strength increases dramatically as a carbon black is added until a maximum tensile strength is attained. As the loading is increased further, the tensile strength either remains constant or starts to decrease. The higher-surface-area HAF blacks give improved tensile strength compared to the GPF blacks, but the differences in structure do not show any significant effect on the relationship between tensile strength and black loading.

In contrast, the NR compound data shown in Fig. 3.10B illustrate the inherent higher tensile strength of unfilled natural rubber because of its crystallizing ability, but show less change as a result of carbon black addition or the differences in black surface area. The tensile strength reaches its maximum at relatively low carbon black loadings (20–40 phr) and shows a decreasing tendency as the black loading is increased.

Compound Property Group 3

The third group of tests measure the energy absorption or hysteresis characteristics. This group includes tests such as rebound or resilience, heat build-up or running temperature, and dynamic tests of various types that can separate the elastic (E') and loss (E'') modulus contributions.

The principal theories or mechanisms which have been proposed for the effect of carbon black on the dynamic properties of rubber have been reviewed by Medalia[12] and are closely related to the theories proposed for the reinforcement mechanism. In a compound with normal loadings of carbon black that is at rest or being cycled at very low strain, the black aggregates are believed to be associated into agglomerates and a network (called "interaggregate association" by Medalia) via Van der Waals forces. When a large stress is applied or the sample is subjected to cycles of high strain, the secondary network of aggregates breaks up. Cycles of intermediate strain are believed to cause some breakdown and reformation of these secondary networks, resulting in energy losses or hysteresis. The lost energy is reflected as a temperature rise. Rebound tests give a simple estimate of energy loss under constant energy-input conditions and have been shown to correlate with "tangent delta" (tan δ) or (E''/E'). Several authors[27,28] have demonstrated that rebound, resilience, and tangent delta are affected by carbon black loadings and surface area, but not by structure.

Fig. 3.11 is typical of the running temperature relationships observed for all three recipes of Table 3.3, except that NR gives generally lower running temperatures than either SBR or EPDM. Structure differences have no significant effect on the running temperature. The effects due to surface area and loading differences are quite apparent.

The rebound data shown in Fig. 3.12 leads to the same conclusion regarding the effects of surface area and loading level on compound properties related to hysteresis. Carbon black structure appears to have little influence on running

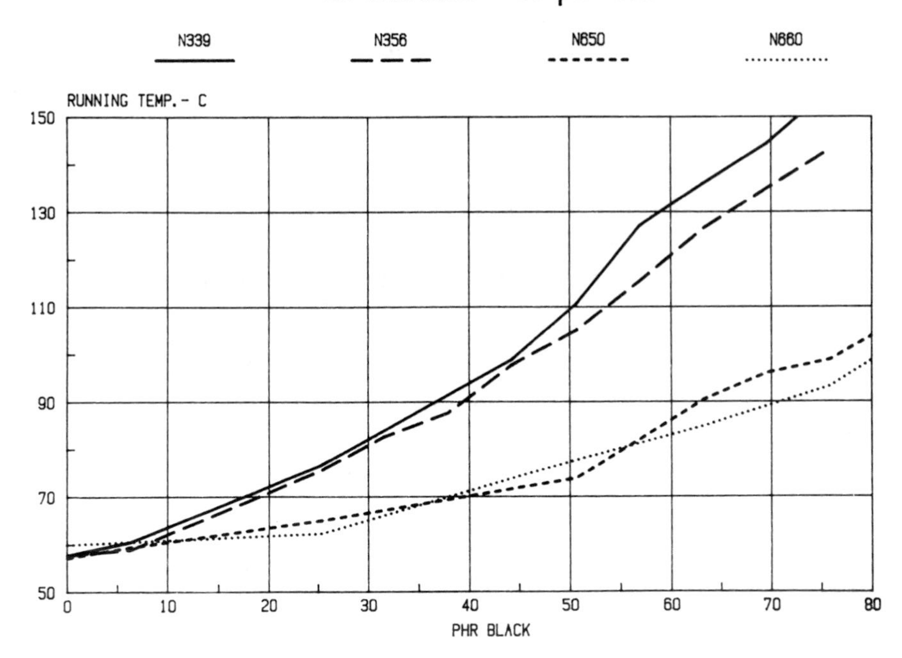

Fig. 3.11. Running temperature (Flexometer). (Source: Ashland Chemical Co., Carbon Black Div.)

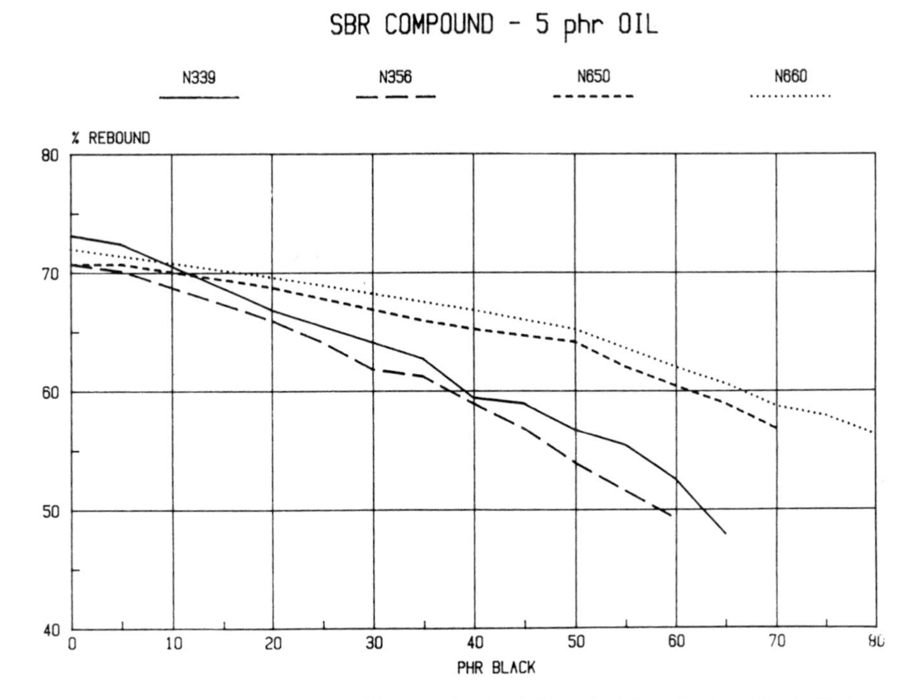

Fig. 3.12. Pendulum rebound. (Source: Ashland Chemical Co., Carbon Black Div.)

temperature or rebound. However, in all three recipes there is a trend toward slightly lower rebound for the higher-structure blacks similar to that shown in Fig. 3.12.

General trends in compound properties that result from changes in carbon-black surface area, structure, and loading are summarized in Table 3.4.

Carbon black properties and loading must be considered together to arrive at

Table 3.4. Compound Property General Trends.

Property	*Increasing Surface Area*	*Increasing Structure*	*Increasing Loading*
	Stress/strain		
Tensile strength	+	Nil	P
300% modulus	+	+	+
Hardness	+	+	+
Elongation	Nil	−	−
	Dynamic		
Heat build-up	+	Nil	+
Rebound	−	Nil	−
Flex life	−	−	P
	Processing		
Viscosity	+	+	+
Scorch time	Nil	−	−
Extrusion rate	−	−	P
Die swell	−	−	−
Processing temperature	+	+	+
Mixing ease/dispersion	−	+	P
	Cure		
Time to 90% cure	Nil	Nil	−
Minimum torque	+	+	+
Maximum torque	+	+	+
	Other properties		
Abrasion resistance	+	*	P
Tear Strength	+	*	P
Compression set	Nil	−	−
Traction/Skid resistance	+	*	P
Solvent resistance	Nil	Nil	+
Electrical conductivity	+	+	+

(+) Increase in numerical value. (−) Decrease in numerical value.
(Nil) Negligible influence by changes in property.
(P) Numerical value increases to a peak value and then begins to decline.
(*) Depends on the Loading.

the right combination for the desired compound properties. Compounding to equal hardness or equal modulus is commonly used in rubber compounding, and is accomplished by adjusting the loading level of blacks that are being compared. A somewhat similar technique of compounding to equal dynamic viscosity (reinforcement factor) has also been described.[29,30] Generally, carbon blacks compared at equal hardness or modulus show smaller differences in many of the compound properties, but abrasion-resistance and tensile-strength differences are usually still evident.

Another compounding approach is to switch to a more reinforcing carbon black, with no adjustment in black level, and to increase the oil or plasticizer level to maintain the modulus or hardness.[31] This approach usually offers an advantage of reduced compound cost. The effect of the additional oil or plasticizer as well as the effect from the carbon black difference enters into the overall changes in the properties of the compound. For example, ''polymer dilution'' tends to reduce abrasion resistance and tensile strength and may negate the improvement in these properties gained from the more reinforcing black.

High loadings of fillers and plasticizer oils are desirable in some applications where abrasion resistance, tensile strength, and hysteresis are not particularly critical. These highly loaded compounds are usually less resilient, but may give low compression-set and low volume-swell properties, provided an adequate state of cure is obtained. A low-reinforcement carbon black such as N762, N754, or N990 (thermal black) is used in many highly loaded compounds. Combinations of the semireinforcing blacks with lower-cost fillers such as clays, calcium carbonate, and so forth, are also widely used in such applications.

FUTURE GRADE AVAILABILITY

In recent years, the trend has been to reduce the number of carbon black grades available. Some new grades continue to be introduced, but a greater number of low-volume grades have been discontinued by most carbon black manufacturers in an attempt to improve efficiency and economics. Grade availability is largely dependent on the volume of usage, and thus is strongly influenced by the tire industry. Approximately 64% of U.S. carbon black production is used in tires, 10% in various automotive rubber parts (hoses, belts, seals, and so on), 20% in industrial rubber goods of various types, and about 6% in nonrubber applications such as inks, paints, paper, and plastics.

Current high-volume blacks that should continue to be readily available include:

Reinforcing Grades
N220, N299, N326, N330, N339, N351, N375
Semireinforcing Grades
N550, N650, N660, N754, N762, N774

The three N700 series (SRF) blacks are used primarily in nontire compounds for both automotive and general industrial rubber parts.

Low-volume grades with a more questionable future availability include:

Reinforcing Grades
N110, N121, N231, N234, N347, N356, N358
Semireinforcing Grades
N539, N630, N683, N787

Several other grades such as N472 and N293 should continue to be available for special electrically-conductive compound applications.

N356 and N358 are both very-high-structure HAF blacks competing for a rather limited volume of usage; N358 currently has the lead in volume. In many cases, the low-volume grades are produced by fewer carbon black manufacturers than are the more popular grades. A full range of carbon black surface areas will likely continue to be available in future years, but the selection of structure levels may become more limited.

REFERENCES

1. G. Kraus, *Reinforcement of Elastomers*, Interscience, John Wiley & Sons, New York (1965).
2. J. Winkler, A. M. Gessler, W. M. Hess, A. I. Medalia, "Reinforcement of Elastomers with Carbon Black," Educat. Lecture Course, Rubber Division, A.C.S. (1976).
3. J. B. Donnet and A. Voet, *Carbon Black—Physics, Chemistry, and Elastomer Reinforcement*, Marcel Dekker, New York (1976).
4. A. F. Blanchard, *J. Polym. Sci.*, **8,** Part (1), 813, 835 (1970).
5. E. M. Dannenberg, *Trans. Inst. Rubber Ind.* **42,** T26 (1966).
6. E. M. Dannenberg, *Rubber Chem. Tech.* **48,** 410 (1975).
7. A. M. Gessler, *Rubber Age* **101,** 54 (1964).
8. A. M. Gessler, *Rubber Chem. Tech.* **42,** 858 (1969).
9. G. Kraus, *Rubber Chem. Tech.* **51,** 297 (1978).
10. G. C. McDonald and W. M. Hess, *Rubber Chem. Tech.* **50,** 842 (1977).
11. A. I. Medalia, *Rubber Chem. Tech.* **45,** 1171 (1972).
12. A. I. Medalia, *Rubber Chem. Tech.* **51,** 437 (1978).
13. L. Mullins and N. Tobin, *J. Appl. Polym. Sci.* **9,** 2993 (1965).
14. A. R. Payne, *Rubber Plast. Age*, August 1961, p. 963.
15. Z. Rigby, *Rubber Chem. Tech.* **55,** 1180 (1982).
16. M. L. Studebaker, *Rubber Chem. Tech.* **30,** 1401 (1957).
17. J. D. Ulmer, W. M. Hess, and V. E. Chirico, *Rubber Chem. Tech.* **47,** 729 (1974).
18. A. J. Voet, *J. Polym. Sci.* **15,** Part D, 327, (1980).
19. E. M. Dannenberg, *Rubber Plast. News* **14** (2), 32 (Aug. 13, 1984).
20. A. C. Patel and W. A. Brown, "Carbon Black," Rubber Division, ACS Lecture Series (1984). "Compounding, Processing and Testing of Elastomers," Rubber Division Education Course.
21. N. L. Smith, *Rubber World* **162** (2), 51(adv.) (1970).
22. S. Brunauer, P. H. Emmett, and E. Teller, *J. Am. Chem. Soc.* **60,** 309 (1938).
23. A. C. Patel, *Elastomerics* **110** (9), 69(adv.) (1980).

24. H. N. Mercer, A. H. Boyer, P. L. Brusky and M. L. Deviney, *Rubber Chem. Tech* **49,** 1068 (1976).
25. B. B. Boonstra in *Rubber Technology*, 2nd ed., M. Morton, ed., Van Nostrand Reinhold, New York (1973).
26. S. Wolff, *Rubber World* **190** (3), 28 (1984).
27. G. Kraus and J. Janzen, *Kaut. Gummi. Kunst.* **28,** 253 (1975).
28. J. M. Caruthers, R. E. Cohen, and A. I. Medalia, *Rubber Chem. Tech.* **49,** 1076 (1976).
29. A. C. Patel and J. T. Byers, *Elastomerics* **114** (2), 29 (1982).
30. J. T. Byers and A. C. Patel, *Rubber World* **188** (3), 21 (1983).
31. J. T. Byers, *Rubber World* **189** (3), 26 (1983).

Part II
Nonblack Fillers

M. P. WAGNER
PPG Industries, Inc.
Barberton, Ohio

INTRODUCTION

Nonblack fillers have always been an important part of the rubber industry. They represented the principal fillers initially, but later took their place alongside carbon black to fill an important role in rubber products.

The early nonblack fillers were mainly naturally occurring minerals or by-products of manufacturing, such as clay, whiting, barytes, zinc oxide, zinc sulfide, blanc fixe, mica, asbestos, kieselguhr, magnesium carbonate, iron oxide, litharge, and the like. They were added to natural rubber to reduce tack, increase hardness, improve durability and reduce cost. The tire industry used zinc oxide, iron oxide, and clay as its chief reinforcing fillers in the early part of this century.

The rapid growth of the automobile industry between 1920 and 1950 put greater demands on tires. These greater demands required new elastomers and new reinforcing fillers that essentially displaced the nonblack fillers in this segment of the rubber industry. Clays and whitings retained their importance in many rubber products where they were used as economical fillers to modify processing and performance of natural and synthetic rubbers.

The need for more reinforcing nonblack fillers in many rubber applications led to the introduction between 1940 and 1960 of calcium carbonates, calcium silicates, hydrated silicas, and fumed silicas. These nonblack fillers were characterized by very small particle size—substantially smaller than the natural products and similar to the reinforcing carbon blacks. They achieved the highest reinforcement attained with nonblack fillers and they were accepted readily for products that benefited from their unique combinations of properties.

During the past 25 years, further refinements in hydrated and fumed silicas have produced a number of grades for specific types of performance. A special technology for compounding nonblack fillers has developed, helping to fully utilize the silicas. Means to increase the surface activity of nonblack fillers, and the development of special elastomers, have further improved the reinforcing properties of many of these, especially the clays and silicas.

Today, the principal nonblack fillers are clays, calcium carbonates (both natural and precipitated), and silicas (both hydrated and fumed). Their consumption in the rubber industry is nearly 600,000 tons per year. Clays are used to the greatest extend (54%), followed by calcium carbonates (27%) and silicas (15%).

A broad spectrum of performance properties—from nonreinforcing, but economical, to highly reinforcing—exists in these nonblack fillers. Properties are related ultimately to the average particle size of the filler which ranges from 10 microns (whitings) down to 0.01 microns (fumed silicas), a span of 1,000-fold. This variety of characteristics provides rubber products that can be maximized readily for economy and performance.

As mentioned, a wide variety of nonblack fillers for rubber exists. Some of these are listed in Tables 3.5 and 3.6, along with three important properties useful for compounders. These properties are particle size, specific gravity (in rubber), and price.

The major nonblack fillers (Table 3.5) include calcium carbonates (natural and precipitated), clays, precipitated silicas and silicates, and fumed silicas. These represent fillers with average particle sizes ranging from greater than 10

Table 3.5. Comparison of Principal Nonblack Filler Properties.

	Ultimate Particle Size Microns	*Effective Specific Gravity*	*Nominal Price ¢/lb*	*Relative Volume Price*
	Natural Calcium Carbonates			
Dry ground	10–20	2.72	2	5.4
Water ground	1–10	2.70	5	13.5
	Precipitated Calcium Carbonates			
Fine	0.1–1.0	2.70	10	27
Ultra fine	<0.1	2.70	20	54
	Clays			
Dry ground	1–10	2.6	3	7.8
Water washed	0.5–5	2.6	9	23
Calcined	0.5–5	2.6	14	36
	Silicas/Silicates			
Precipitated	0.02–0.1	2.0	45	90
Fumed (anhydrous)	0.01–0.02	2.0	210	410

Table 3.6. Property Comparison of Miscellaneous Nonblack Fillers.

	Particle Size Range, Microns	*Specific Gravity*	*Nominal Price ¢/lb*	*Relative Volume Price*
Talcs	0.5–10	2.72	8	22
Wollastonite	5–50	2.9	7	20
Magnesium carbonate	10–50	2.22	80	178
Alumina trihydrate	2–10	2.42	14	34
Barytes	10–50	4.45	12	53

microns to about 0.01 micron (10 nm). The reinforcement or rubber, dependent on particle size, also has a wide range, with natural calcium carbonates providing virtually none and the silicas providing very high reinforcement. The term "reinforcement" applies to an improvement in mechanical durability, usually related to tensile strength, tear strength, and abrasion resistance.

Specific gravity and prices are two additional properties that have an important impact on compounding. The specific gravity of a filler affects the molded product's final weight, which can be important for application specifications that include the specific gravity (e.g., handballs). However, high specific gravity adversely affects the price of the product by requiring more weight of compound in a fixed-volume mold. The mathematical product of specific gravity and price per ton gives the relative volume price contributed by the filler. In an extreme example, barytes (Table 3.6) is lower in ton-price than ultrafine calcium carbonate, but they contribute equally to the unit cost of the product.

The miscellaneous nonblack fillers (Table 3.6) find minor usage in the rubber industry. They provide certain properties useful in rubber compounding. Talcs are used mainly for dusting rubber, in mold lubricants, and as semireinforcing fillers. Wollastonite, because of its needle-like (acicular) shape, potentially can replace asbestos in some rubber products. Magnesium carbonate has a small demand in resilient or transparent rubber products, but its escalating price has reduced its use in rubber drastically. Alumina trihydrate is used mainly to impart flame-resistant properties to rubber. Barytes is used to adjust the specific gravity of rubber compounds or to provide X-ray-shielding rubber products. None of these miscellaneous fillers is considered primarily for mechanical properties improvement.

In the following sections, each of the principal nonblack fillers will be discussed separately. Special attention will be given to the requirements for processing and vulcanization demanded by each type. Also, the development of specific properties associated with each filler will be considered. Surface modification and coupling agents—important new technology—will be reviewed, as well. Finally, an application section will cover some examples of current product formulations based on nonblack fillers.

CALCIUM CARBONATES

Manufacture and Grades

Natural calcium carbonates are derived from serveral sources, including high calcium or dolomitic limestone, or marble. While limestone deposits occur throughout the world, only the white and bright sources are mined for fillers.

The limestone is either dry- or wet-ground and is classified to various size grades. The finer grinds are surface-treated with calcium stearate for faster incorporation and improved dispersion. The standard grades are 1–4 wet ground and 5–8 dry ground. Of the eight grades of ground limestone, the rubber industry uses mainly grade 2 (wet ground) for maximum brightness and grade 5 (dry ground) for general purposes (Table 3.7). Grade 2 also is supplied as a stearate-treated version for improved dispersion in rubber.

Precipitated calcium carbonates are manufactured by the formation of calcium carbonate from milk of lime via the use of carbon dioxide. The particle size and shape can be controlled by process variables, and the rubber-grade products are characterized by high purity and brightness. The ultrafine grades (0.03 to 0.1 microns average particle size) are prepared for rubber products, and surface-treated grades (1–2% calcium stearate) often are used for easier incorporation and improved dispersion.

Characteristics

High loadings of ground calcium carbonate fillers can be incorporated into most rubbers with a resultant, relatively slow increase in Mooney viscosity. Loadings to 250 phr are feasible with Mooney viscosity below 80. Precipitated calcium carbonates, on the other hand, increase viscosity more rapidly. This especially is true with the ultrafine grades. Coated grades provide shorter incorporation times and reduce the Mooney viscosity of the final batch.

Scorch and cure rates, like viscosity, are not major problems with calcium carbonates. Conventional accelerators, based on thiazoles, sulfenamides, thiurams, guanidines or their combinations can be used depending on the requirements for scorch time and cure time. No activators normally are required. In fact, due to the alkaline pH of calcium carbonates (Table 3.8), slight activation

Table 3.7. Common Grades of Ground Limestone Used in Rubber.

		Particle Size, Microns		
Grade	*Type*	*Average*	*Maximum*	*G.E. Brightness*
2	Wet ground	3	15	108
5	Dry ground	7	44	65

Table 3.8. Comparison of Filler and Rubber Properties With Various Calcium Carbonates.

Carbonate Type	*Natural*				*Precipitated*	
Average diameter, microns	10	4	2	.45	0.1	.035
pH	7.1	7.5	7.8	7.8	9.0	9.2
Properties in Natural Rubber (100 phr)						
Modulus, MPa	8.5	10.0	11.4	13.2	14.1	15.2
Tensile strength, MPa	13.1	14.5	16.2	17.7	19.3	21.6
Crescent tear, kN/m	19	21	25	30	38	99

of cure usually is observed. Cure activation becomes more pronounced with increased loading and smaller average particle size, possibly requiring a reduction in accelerator content to maintain suitable scorch and cure times.

Modulus, tensile strength, hardness, tear strength and abrasion resistance are directly dependent on loading, and are inversely related to average particle size. At a constant loading (100 phr) in natural rubber, this relationship to particle size is evident for modulus, tensile strength and crescent tear strength (Table 3.8). The low tensile strength obtained with the coarsest calcium carbonates indicates the appropriateness of the term "nonreinforcing" filler.

Modulus and tensile strength are changed very little with up to 250 phr of the coarsest (ground) particle size. Both properties, however, increase as particle size decreases. Tensile strength reaches a maximum at a loading of 125 to 175 phr, and this optimum loading decreases with decreasing particle size, consistent with the behavior of most fillers. Hardness increases with loading, but no significant effect related to particle size results. Tear strength, like tensile strength, passes through a maximum, and this maximum increases with decreasing particle size.

One grade of precipitated calcium carbonate has been modified with a low molecular weight, carboxyl-terminated polybutadiene. Called Fortimax 423 (ICI, Inc),[1] this product provides improved performance to SBR when compared to the untreated calcium carbonate (Table 3.9). Modulus, tensile strength and tear strength are increased, while abrasion and compression set are decreased. The surface treatment makes the performance of precipitated calcium carbonate more like that of MT carbon black, emphasizing the need for appropriate filler surface chemistry to obtain improved rubber properties.

CLAYS

Manufacture and Grades

Clays represent the largest volume (overall quantity) nonblack filler used in rubber. They are second to carbon blacks in this respect. They owe this popu-

Table 3.9. Functionalized Calcium Carbonate in SBR.[a,1]

Filler (40 Volumes)	*Ultrafine Calcium Carbonate* Untreated[b]	*Ultrafine Calcium Carbonate* Treated[c]	MT-Black
300% modulus, MPa	1.7	4.5	7.4
Tensile strength, MPa	6.1	7.6	8.0
Elongation, %	580	504	366
Hardness, IRHD	61	67	67
Tear strength, kN/m	22	30.5	29
Abrasion (DIN), mm^3	600	444	309
Compression Set, % (25% strain for 24 hrs at 70°C)	30	33	26

[a]Formulation: SBR-1502, 100; zinc oxide, 5; stearic acid, 1; sulfur, 2.5; Vulcafor 9 (MBT/DPG), 2; Permanex WSP (phenolic antioxidant), 1.5; filler (40 volumes).
[b]Winnofil, Imperial Chemical Industries, Ltd., Mond. Div.
[c]Fortimax 423, Imperial Chemical Industries, Ltd., Mond. Div.

larity to a combination of low cost, low-to-moderate reinforcement, and processing benefits (especially extrusion and calendering).

There are many types of clays, but kaolins are used almost exclusively in rubber. The major U.S. deposits for this purpose occur in South Carolina and eastern Georgia.

After these high-purity (>90% kaolin) deposits have been mined, the ore is dried, pulverized, and refined by either air- or water-separation processes to remove impurities (mainly quartz and mica), and to produce clay in the desired particle-size range.

Through a combination of mineral-source and particle-size classification (using either "air-floating" or "water-washing" methods), relatively coarse and fine particle size products are produced. These are termed, respectively, soft and hard clays—terms derived from the relative rubber hardness produced.

Each of the four types of clays are included in Table 3.10 to emphasize the effect of the process on quality. The weight-percent of fine particle (i.e., less than 2 microns) is larger for hard clays than for soft clays. Generally, particles of less than two microns constitute more than 80% of hard-clay weight.

Water-washing is more efficient for particle size fractionation than is air-floating. Water-washed clays also are virtually free of foreign minerals such as quartz and mica. If air-floated clays are used, such impurities can be source of problems with abrasiveness and die plate-out (coating of die) in the extrusion of rubber.

Some representative clays and their respective rubber properties also are given in Table 3.10. The differences in particle size of hard and soft clays produce the characteristic changes in tensile strength and ASTM D624 Die-C tear. Air-floated and water-fractionated clays show typical differences in pH, brightness, 90% cure times, and relative compound cost.

Table 3.10. Comparison of Various Types of Clay.[a,4]

	Air Floated		*Water Fractionated*	
	Soft	*Hard*	*Soft*	*Hard*
	Clay Properties			
Percent less than 2μ	50	85	30	90
Avg. particle size, μ	1.5	0.46	1.5	0.5
G.E. brightness	80.5	82.5	83.0	87.0
pH	4.5	4.7	4.6	7.0
Residue, % on 325 mesh	0.17	0.20	0.15	0.005
	Rubber Properties			
90% cure, min./160°C	6.0	6.2	7.4	7.2
300% modulus, MPa	0.41	1.9	0.79	0.79
Tensile strength, MPa	4.0	5.5	4.5	5.6
Elongation, %	650	610	630	670
Hardness	40	61	47	45
Die C tear, kN/m	5.8	16	7.9	9.8

[a]Formulation: SBR-1715, 100; stearic acid, 1; zinc oxide, 5; Neville LX-1065, 15; Rubar, 5; Helizone, 6; Circolight Oil, 25; Titanox RA, 30; MBTS, 4.5; TMTD, 0.5; sulfur, 2.2; clay, 200.

Clays can be modified in a variety of ways to produce improvements in certain properties. One of the earliest modifcations was calcining, which removes the bound water in clay by high-temperature treatment. The resulting product produces excellent dielectric properties in wire and cable coatings. The calcination process contributes substantially to overall product cost, and also reduces the reinforcement of the clay.

Several commercial clays have been treated with silane coupling agents to improve their performance in rubber. A clay treated with a mercapto-containing silane coupling agent is compared with untreated clay in Table 3.11. Since the silane coupling agent provides a means to bond the clay particles to the rubber, increased modulus and tensile strength are obtained. Also, heat build-up is reduced.

Rubber Compounding

Clays can be formulated to rather high loadings in most elastomers, with soft clays allowing somewhat higher loadings than hard clays. Viscosity builds moderately with loading, but processible formulations with 150–200 phr are feasible. Process aids, such as coumarone-indene resin, frequently are used to improve dispersion and obtain maximum tensile strength.

Durometer hardness is the usual starting point for a guide in selecting clay loading. A rule-of-thumb in many elastomers is to use 5–6 phr of hard clay and

Table 3.11. Silane-Treated Clay.[a]

	A	*B*
Water-fractionated hard clay, phr	70	-
Silane-treated clay,[b] phr	-	70
Scorch (T-5), 132°C	>30	28
90% cure, min./150°C	29.5	29.0
Cure: 45 min. at 150°C		
300% modulus, MPa	1.3	2.1
Tensile strength, MPa	3.1	4.6
Elongation, %	730	660
Hardness	42	42
Goodrich heat build-up, °C	19	13

[a]Formulation: SBR-1778, 87.3; cis-BR, 35; filler, as shown; Circosol 596, 10; stearic acid, 2; paraffin wax, 2; MBTS, 1; DPG, 0.8; sulfur, 2; zinc oxide, 1.5
[b]Nucap 200L, J. M. Huber Corp.

7–8 phr of soft clay for each point in hardness above the base hardness of the unfilled compound. When clay is used to replace other fillers, it is recommended that the replacement be based on equal volumes. Refinements in the formulation to accommodate processing and hardness can be made by the addition of process oil.

The main factor to consider in adding clay to most formulations is its reduction of cure rate. This reduction will require the addition of an activator (e.g., triethanolamine, diethylene glycol, or polyethylene glycol) and/or an increase in acceleration. A combination of thiazole or sulfenamide with thuiram offers a useful accelerator system for many sulfur-cured elastomers.

A typical formulation with SBR shows that hard clay, at equal-volume loadings, produces properties similar to those of N-990 (MT) black (Table 3.12).

Table 3.12. Hard Clay in SBR (50 Volumes).[a]

	Hard Clay 130 phr	*N-990 Black 100 phr*	*N-650 Black (20 phr) Hard Clay (104 phr)*
ML 1 + 4 (100°C)	72	57	71
300% modulus, MPa	1.9	5.3	7.0
Tensile strength, MPa	14.2	10.0	18.6
Elongation, %	590	610	620
Hardness	70	70	78
Compression set (B),[b] %	60	28	45
Die C tear, kN/m	16	14	17.9
Abrasion index	27	23	27

[a]Formulation: SBR-1502, 100; zinc oxide, 5; stearic acid, 1; TMTD, 0.1; morpholinobenzothiazole sulfenamide, 1.5; sulfur, 3; triethanolamine, 3 (clay-loaded stocks only); filler (50 volumes), as shown.
[b]70 hours at 100°C.

Alone, hard clay matches the tensile strength, Die C tear, and abrasion index of N-990 black, but it is deficient in modulus and compression set. The combination of hard clay (40 volumes) and N-650 black (10 volumes) increases modulus and tensile strength while reducing compression set and maintaining lower compound cost.

SILICAS

Manufacture and Grades

Precipitated Silicas. Precipitated silicas offer the highest reinforcement of the economical nonblack fillers. They are noted for their unique combination of tear strength, adhesion, abrasion resistance, age resistance, color and economics in many applications. A comprehensive review of precipitated silicas and silicates in rubber was published in 1976.[2]

These fillers are manufactured by the controlled precipitation from sodium silicate with acid or alkaline earth metal salt. The ultimate particle size is controlled closely by the conditions of precipitation. In this way, a variety of fillers, ranging from highly reinforcing to semireinforcing are produced.

The properties of three products from one manufacturer (Table 3.13) illustrate the range of available silicas. Silicas are produced by twelve manufacturers throughout the world, and their products have been tabulated.[2]

The reinforcement of the silicas is determined by the ultimate particle size, even though the particles are aggregated to some extent in the rubber. Since particle-size analysis is a laborious procedure, specific surface area customarily is used to designate particle size.

Silicas may vary in pH (measured in 5% aqueous slurry) from slightly acidic to slightly alkaline. Since they are hydrated silicas, they adsorb ambient moisture. Typical amounts range from 3 to 9 percent. This adsorbed moisture plays an important role in compounding with silicas, as will be discussed later.

Soluble salts are formed during the precipitation, most of which are washed

Table 3.13. Representative Silica Properties.

Filler Type[a]	*HS-200*[b]	*HS-500*[c]	*HS-700*[d]
Particle size, nm	20	40	80
Surface area, m^2/g	150	60	35
pH	7	7.5	8.5
Loss at 105°C, %	6	6	7
Soluble salts, %	1.2	1.2	1.2

[a]See Reference 2 for classification system.
[b]Hi-Sil® 233, PPG Industries, Inc.
[c]Hi-Sil® 532 EP, PPG Industries, Inc.
[d]Silene® 732 D, PPG Industries, Inc.

from the product. For economic reasons, a small amount (1–2%) of this salt remains in commercial silicas. This residue has no significant effect on rubber performance in most applications.

Fumed Silicas. The smallest-particle-size silicas are manufactured in a high-temperature reaction between silicon tetrachloride and water vapor. They are variously labeled "fumed," "pyrogenic," or merely "anhydrous" silicas. These silicas are characterized by extremely fine particle size (7-15 nm), high purity (99+% silica), and low adsorbed water (<1%). Because of the manufacturing process, a high price is another characteristic (cf. Table 3.5).

High-quality silicone rubber products are virtually the only rubber application of fumed silicas. Fumed silicas are the only fillers combining high reinforcement, high-temperature stability, electrical properties, and the transparency for which silicone rubber products are noted. Compounding silicas in silicone rubber will be discussed later.

Compounding with Silicas

Processing and Vulcanization. The use of precipitated silicas in rubber requires special considerations not encountered to the same degree with other rubber fillers. Adding silicas to rubber tends to build viscosity more rapidly than most fillers. This is true especially of silicas with high surface area. In addition, the most frequently used accelerator systems are severly deactivated by silicas. As a result, both the optimum cure and the state of cure are reduced substantially. The magnitude of both of these effects increases with the total available surface (i.e., a combination of loading and specific surface area) of the silica.

Viscosity and cure rate also are dependent on the adsorbed moisture present on the silica during compounding. This moisture behaves as a psuedo-plasticizer and cure activator. However, this moisture can vary with ambient humidity and with rubber processing conditions. Specific additives and attention to mixing procedures are recommended to compensate for these effects.

Viscosity. Because small particle size contributes to increased viscosity, the primary emphasis on improving viscosity has been with the more reinforcing silicas. Zinc oxide interaction with silica during mixing affects viscosity and other properties. The early addition of zinc oxide and silica results in lower Mooney viscosity, greater extrusion swell and lower modulus. Withholding zinc oxide until after the silica has been dispersed, either later in the first-stage mix or even in the second-stage mix, causes a significant increase in Mooney viscosity.

Certain activators used with silica-filled rubber produce a viscosity reduction disproportionately large compared to the quantity added. Diethylene glycol

(DEG) or triethanolamine (TEA) at 2 phr can reduce Mooney viscosity by 30% or more while polyethylene glycol (PEG) is only half as effective.

Mooney viscosity also can be lowered by using hydrocarbon process oils. These behave similarly for all fillers. Other plasticizers can be used to advantage with silica-loaded elastomers.

Plasticizers of vegetable origin appear to be advantageous in natural rubber. Tall oil and hydrogenated resin are typical examples. On the basis of viscosity reduction, 5 phr of tall oil is equivalent to 30 phr of naphthenic oil. Tall oil generally provides better properties as well.

In SBR, aromatic resins have special advantages with silica fillers. Coumarone–indene resins, for example, aid in the incorporation rate and dispersion of silica fillers. While their effect on Mooney viscosity is comparable to that of process oils, the retention of properties using aromatic resins is superior. Tensile strength, tear strength, abrasion resistance, and extrusion smoothness generally are superior with the addition of aromatic resin (at 10–20 phr) as a process aid.

The viscosity of silica-filled elastomers is affected by other additives. Some of these have synergistic effects with silicas. In addition to those already discussed, also included should be accelerators, antidegradents, and coupling agents. These will be discussed in subsequent sections; effects on viscosity, when significant, will be noted.

Curing Systems. The systematic development of a sound technology of silica compounding must be based on a thorough understanding of cure behavior and the use of suitable cure systems. The surface chemistry of precipitated silicas differs profoundly from that of other rubber fillers, leading to a unique set of compounding variations.

Sulfur-cured rubbers containing precipitated silica must be modified in several ways to obtain optimum performance. Using low amounts of thiazole or sulfenamide accelerators alone is inadequate. Even the addition of a secondary accelerator (e.g., guanidine or thuiram types) in usual amounts generally is insufficient, though helpful, in obtaining practical cure cycles.

In many sulfur-cured elastomers (NR, SBR, BR, NBR), the use of thiazole or sulfenamide primary acceleration at a level of 0.75 to 1.5 phr with guanidine at 0.75 to 1.5 phr, or thiuram at 0.1 to 0.5 phr as a secondary accelerator, forms a basis for developing a suitable cure system. The specific amounts depend on the loading and type of silica.

With loadings of fine-particle silicas above about 20 phr, diethylene glycol (DEG) or polyethylene glycol (PEG 4000) reduces the accelerator requirements. This usually is an economic advantage, but glycols serve a more important function by buffering the effects of variable moisture.

Efficient vulcanization (EV) systems rely on sulfur donors, or accelerators with available sulfur, and little or no elemental sulfur. Such systems can be used in most sulfur-curable elastomers, but especially in NR, IR, SBR, BR, and NBR. The advantages derived for silica-filled rubbers include less sensitivity to cure retardation, better optimization of properties, and enhanced aging resistance.

Many accelerator combinations are available for the development of practical cure systems with silica-filled elastomers. The specific combinations depend on many factors, including scorch and cure rate needed, loading of silica, type of silica, and influence of other additives. Some general guidelines include the desirability of:

1. Combinations of two or more accelerators, at least one from the thiazoles or sulfenamides, the second from the guanidines, thiurams, or dithiocarbamates.
2. Adding glycol activators (TEA, DEG, or PEG) to lessen accelerator demand and buffer moisture variations.
3. Sulfur donor (or EV) cure systems that provide efficient crosslinking and benefits in property optimization and heat-aging.

Vulcanizate Properties. The primary goal of compounding is to develop formulations having suitable performance. Laboratory properties usually serve as a guide to achieving this goal. These properties often are influenced by cure state as well as the filler used. The following discussion assumes that an appropriate cure state has been achieved.

Silicas and silicates are known for the relatively low modulus imparted to rubber, compared to that expected from carbon black of comparable particle size (Table 3.14). Despite this characteristic, tensile strengths produced are similar for the two types of fillers. As with carbon blacks, tensile strength increases with decreasing particle size (increasing BET surface area). Accompanying the lower modulus and comparable tensile strength are higher elongation at break and lower resistance to abrasion. This last property also increases with increasing surface area.

High tear strength is one of the outstanding properties obtained with silica, and high tear strength also increases with increasing surface area. Rubber reinforced with small-particle silicas tears by an irregular "saw-tooth" process, often called *knotty tear*. High tear strength and knotty character are extremely susceptible to change in cure state. Consequently, optimization of tear strength depends on careful adjustment of the curing system (See reference 3 for more details).

Aside from cure state and silica particle size, reinforcement is enhanced with coupling agents. The most effective coupling agent is mercaptopropyltrimethoxysilane. This silane provides strong adhesion between the silica and the rubber, resulting in a substantial increase in reinforcement.

Table 3.14. Silica and Silicates in NR[a]

BET, surface area, m^2/g:	40	70	150
90% ODR cure at 144°C Min.	10	17	19
Viscosity, ML 1 + 4 (100°C)	45	50	70
Original stress-strain			
hardness	50	56	63
300% modulus, MPa	2.8	3.8	2.4
Tensile strength, MPa	22	23	26
Elongation, %	710	650	760
Trouser tear, kN/m	8.9	16[b]	26[b]
PICO abrasion index	53	70	73
Compression set, % (70 hours at 100°C)	45	54	67
Resilience, %	82	76	66
Heat build-up, °C (Goodrich)	16	17	32

[a]Formulation: NR 50; IR 50; Antiox 1.5; stearic acid, 2; aromatic resin–1.5; PEG-3; zinc oxide, 3; Sulfur-3; MBS, 1.5; TMTM, 0.1.
[b]Knotty Tear

The effect is illustrated with SBR containing 60 phr of reinforcing silica (Table 3.15). Mercaptosilane at only 2% of the silica weight increases modulus nearly threefold and increases the abrasion index, while reducing heat build-up. Similar results are obtained in most sulfur-cured elastomers.

In using silane coupling agents, it is essential to allow the silica, rubber, and coupling agent to mix thoroughly before adding other compounding ingredients. It is particularly critical that zinc oxide not be present during this phase of the mixing. Adding zinc oxide late in the first-stage mixing cycle, or even in the second-stage cycle, avoids interference with the coupling action. Other com-

Table 3.15. Silane Coupling Agent in Silica-filled SBR.[a]

	A	*B*
Mercaptosilane,[b] phr	0	1.2
Mooney scorch, T_5 (132°C), min.	>30	>30
90% ODR cure, min., at 150°C	30	39
300% modulus, MPa	4.7	12.0
Tensile strength, MPa	20.0	23.4
Elongation, %	640	450
Hardness	72	66
Goodrich heat buildup, °C	46	28
Pico abrasion index, %	71	119

[a]Formulation: SBR-1502, 100; Hi-Sil 210, 60; mercaptosilane, as shown; aromatic oil, 10; stearic acid 1; antioxidant, 1; CBS, 1.5 (A), 1.0 (B); DOTG, 1.5 (A), 1.2 (B); TMTD, 0.2 (A only); sulfur 2.5; zinc oxide (on the mill), 4.
[b]Silane A-189, Union Carbide Corp.

Table 3.16. Filler Comparison in Silicone Rubber.[a,4]

	Loading	*Hardness*	*Strength, MPa*	*Elongation, %*
Ground quartz	200	58	1.31	160
Diatomaceous earth	100	65	1.59	160
Precipitated silicas				
BET: 45 m^2/g	60	44	2.79	330
BET: 60 m^2/g	60	45	3.86	350
BET: 150 m^2/g	60	67	6.72	260
Fume silica				
BET: 250 m^2/g	40	44	8.28	625

[a]Formulation: VMQ silicone, 100; proprietary ''pacifier,'' as required for processing; 2,5-bis(*t*-butylperoxy)-2,5-dimethyl hexane, 0.5. Cure: 10 min at 171°C.

pounding ingredients appear to have less effect. It is recommended, however, that only those ingredients essential for processibility (e.g., plasticizers) be present during the early mixing cycle.

Silicone-rubber Compounding. Silicone rubber is supplied in a variety of compositions ranging from modified gum bases to complete formulations for specific applications. Three types of silica can be used as fillers depending on cost and performance. Fumed silicas are used for highest strength and best electrical properties. Pure sources of ground quartz are used to reduce cost, but with a severe strength penalty. Precipitated silicas offer moderately good tensile strength, tear strength, water absorption, and electrical insulating properties at an intermediate cost. Special grades of precipitated silica provide silicone rubber with reduced water absorption and increased electrical resistance.

The various silica fillers are compared in a silicone rubber formulation in Table 3.16. With the use of a vinyl methyl silicone gum (VMQ) and a proprietary ''pacifying'' agent (or process aid) of unspecified amount, the different fillers are at near-optimum loading. Reinforcement, as measured by tensile strength, is very low with extender fillers such as diatomaceous earth and ground quartz. Their high loading and low cost permit economical silicone compounds to be formulated, however.

The highest reinforcement occurs with high-surface-area precipitated or fumed silicas. Although not shown, water-volume swell after 24 hours immersion is less than 1% with fumed silica and 5–10% with the precipitated silicas.

APPLICATIONS OF NONBLACK FILLERS IN THE RUBBER INDUSTRY

Nonblack fillers are used in many rubber compounds to provide suitable properties at a competitive price. In general, the natural calcium carbonates and soft clays tend to be used in less demanding applications in which high strength and

Table 3.17. Low-Cost Mat Compound.

	PHR	%
SBR-1715 (50 phr oil)	150	20.4
Soft clay	220	29.8
Hard clay	125	16.9
Silene® 732D	50	6.8
Whiting	50	6.8
Other	142.5	19.3
Total	737.5	100.0
ML 1 + 4 (100°C)	58	
90% cure, min./160°C	6	
Tensile strength, MPa	5.4	
300% modulus, MPa	3.5	
Elongation, %	460	
Hardness	60	

durability are not required. These also result in less costly compounds. Perhaps one extreme example is given in Table 3.17, in which less than 15% rubber and about 60% fillers are combined to provide a serviceable floor mat for automobiles. Despite the high filler loading, the product has moderate tensile strength, good elasticity and low viscosity. Because of the high loading of clays and whiting, it also has a low cost and high density.

Shoe soles represent the one class of rubber compounds in which nonblack fillers dominate. This is a result of the combinations of requirements, which include nonmarking or light-colored features, flexibility, hardness and abrasion resistance.

Two qualities of soft-shoe soles are given in Table 3.18. Quality in a shoe sole is usually gauged by the NBS Abrasion Index (ASTM D1630). In these formulations, hard clay and whiting provide a low NBS Abrasion Index of 30, while the use of precipitated silica (Hi-Sil® 233) provides a high index of 65. The use of the silica with small particle size provided increased abrasion resistance.

Three qualities of a hard work sole are given in Table 3.19. High-styrene resin is used to provide increased hardness after vulcanization while maintaining lower viscosity while processing the rubber compound. The increased abrasion index is obtained by replacing silica and clay in formulation A with all silica in formulation B. Additional resistance to abrasion also is obtained by replacing part of the SBR-1502 in formulation B with butadiene rubber in formulation C.

Note that commonly measured properties, such as modulus, tensile strength, and hardness, do not reflect the change in abrasion resistance. Of course, this rule applies to many service-related failure properties, such as abrasion resistance and fatigue resistance.

One of the largest commercial uses of precipitated silica has been in treads

Table 3.18. Nonblack Fillers in Deck and Tennis Shoe Soles.

	A	*B*
Cis-1,4-polyisoprene	80	-
SBR-1009	20	-
SBR-1778	-	100
Antioxidant	2	1
Zinc oxide	5	2
Stearic acid	1	1
Hard clay	100	-
Whiting	40	-
Hi-Sil® 233	-	60
Naphthenic oil	10	10
Coumarone indene resin	5	-
Wax	.75	-
PEG 4000	-	2.00
MBTS	1.50	0.75
DPG	0.75	1.50
Sulfur	2.75	2.00
Total	268.75	180.25
Cure, min./155°C	6	12
Tensile strength, MPa	10.2	11.7
300% modulus, MPa	5.2	2.8
Elongation, %	445	640
Hardness (A)	55	60
NBS abrasion index	30	65
Specific gravity	1.43	1.19

of off-the-road tires, as for earthmovers and mining vehicles. A typical formulation is shown in Table 3.20. The use of 15–25 phr of precipitated silica, such as Hi-Sil® 210 silica along with a highly reinforcing carbon black, produces excellent resistance to cutting and chipping while maintaining high abrasion resistance.

Many and varied rubber applications depend on nonblack fillers for required performance with the most economical formulation. The initial incentive to choose a particular nonblack filler is end-use performance. For this, such properties as abrasion resistance, tear resistance, flex resistance, adhesion, age resistance, temperature resistance, color, processibility, and finally cost, should be considered.

A VIEW OF FILLER REINFORCEMENT

The reinforcement of rubber with filler particles has been shown to be derived mainly from average particle size and coupling of filler to rubber. Particle shape and filler structure play a very minor role in this phenomenon.

Table 3.19. Nonblack Fillers in Nuclear Soling.

	A	*B*	*C*
SBR-1502	100	100	50
Cis-1,4-polybutadiene	–	–	50
High-styrene resin	35	35	35
Antioxidant	1	1.25	1
Zinc oxide	5	5	5
Stearic acid	2	2.5	2
Hi-Sil® 233	25	65	65
Hard clay	75	–	–
Coumarone indene resin	10	10	10
Naphthenic oil	5	5	5
PEG 4000	2	2.5	2.5
Wax	0.5	0.5	1
Salicylic acid	–	1.00	–
MBTS	1.75	2.00	2.00
DPG	1.25	1.00	1.00
Sulfur	3.00	2.50	2.50
Total	266.50	233.25	232.00
Cure, min./150°C	12	15	15
Tensile strength, MPa	15.0	20.0	14.7
300% modulus, MPa	7.8	6.0	6.7
Elongation, %	485	600	575
Hardness, (A)	93	93	93
NBS abrasion index	36	67	81
Specific gravity	1.31	1.18	1.18

The particle size of the filler may be considered the physical contribution to reinforcement, while surface activity provides the chemical contribution. In the absence of strong coupling bonds, the polymer is physically absorbed on the surface of the filler, which may result in a reduced mobility of the rubber molecules near the surface of the filler. Such a region may play a role in increased tear strength with fine-particle silicas. The tear propagation through this region of polymer would result in greater energy absorption and thus higher inherent tear resistance.

In the presence of strong coupling bonds, the detachment of rubber from the surface of the filler requires higher energy. Properties that depend on such a process, including modulus, tensile strength, and abrasion resistance, are substantially improved. At the same time, the rubber molecules near the surface of the filler become an integral part of the network and their mobility is increased. Heat build-up and tear strength are correspondingly decreased because they depend on molecular mobility.

The relative effects of particle size and surface activity can be summed up by a comparison of these two properties on tire-abrasion resistance (Table 3.21).

Table 3.20. Silica in OTR Tire Tread.

Natural rubber	100	
Hi-Sil® 210	20	
N-285 black	35	
Zinc oxide	5	
Antioxidant	2	
Tall oil	3	
Aromatic resin	3	
Microcrystalline wax	1	
Stearic acid	2	
PEG 4000	0.5	
TBBS	2	
Sulfur	2.5	
	177.4	
ML 1 + 4 (100°C)	62	
MS-138°C (T5), min.	5	
90% ODR cure, min./144°C	9.5	
	Original	*Aged 5 Days/100°C*
Hardness (A)	62	77
300% modulus, MPa	12	-
Tensile strength, MPa	23	17
Elongation, %	500	290
Die C tear, kN/m	79	49
Trouser tear, kN/m		
(Cure: 40 min./144°C)		
Room temp.	28	-
70°C	22	-

Three nonblack fillers, with and without a silane coupling agent, are compared with carbon black. The three nonblack fillers include clay (500 nm), HS-400 silica (40 nm), and HS-200 silica (22 nm). Tire treadwear varies linearly with the logarithm of the filler particle size in the tread. The coupling agent results in a greater slope, or sensitivity to particle size.

Table 3.21. Effect of Filler Particle Size and Coupling on Reinforcement.

	Treadwear Index[a]		
	Hard Clay	*Hi-Sil® 532EP*	*Hi-Sil® 233*
Particle Size, nm:	*500*	*40*	*20*
Without coupling	22	58	66
With coupling	36	76	99

[a]N-285 Carbon Black = 100.

In this particular case, the physical contribution to reinforcement is approximately 55% and the chemical contribution is 45%. Both parameters (particle size and surface activity) are highly important for full reinforcement.

REFERENCES

1. Fortimax, a white reinforcing filler for rubber, Imperial Chemical Industries, Ltd., Mond. Division.
2. M. P. Wagner, *Rubber Chem. Technol.* **49,** 703–774 (1976).
3. N. L. Hewitt, *Rubber World* **186**(3), 24 (1982).
4. A. Barbour, ''Clays,'' paper presented at Educational Symposium on ''Compounding with Nonblack Fillers,'' Rubber Division ACS Meeting, Cleveland, OH, Oct. 1979.

ABBREVIATIONS

BMDC	Bismuth dimethyl dithiocarbamate
DEG	Diethylene glycol
DETU	Diethyl thiourea
DMBPPD	N-(1,3-dimethyl butyl)-N′-phenyl-*p*-phenylenediamine
DPG	Diphenyl guanidine
DPTH	Dipentamethylene thiuram hexasulfide
DPTM	Dipentamethylene tetrasulfide
DTDM	Dithiodimorpholine
ETU	Ethylene thiourea
HMT (hexa)	Hexamethylenetetramine
MBS	Morpholinothio benzothiazole
MBT	Mercaptobenzothiazole
MDB	Morpholinodithio benzothiazole
MPPPD	N-methyl-2-pentyl-N′-phenyl-*p*-phenylenediamine
ODPA	Octylated diphenylamine
PEG 4000	Polyethylene glycol
PPDC	Piperidinium pentamethylene dithiocarbamate
TBBS	N-*t*-butyl benzothiazole sulfenamide
TB	Ammonia-formaldehyde-ethyl chloride (trimene base)
TBTD	Tetrabutyl thiuram disulfide
TMTD	Tetramethyl thiuram disulfide
TMTM	Tetramethyl thiuram monosulfide
TMQ	Polytrimethyl dihydroquinoline
TMETD	Tetramethyl thiuram disulfide/tetraethyl thiuram disulfide
ZBDC	Zinc dibutyl dithiocarbamate
ZMDC	Zinc dimethyl dithiocarbamate

4

PROCESSING AND VULCANIZATION TESTS

A. B. SULLIVAN AND R. W. WISE
Monsanto Polymer Products Co.
Akron, Ohio

INTRODUCTION

In this chapter we will direct out attention to the test procedures and instruments used to measure two of the most critical properties of rubber and rubber compounds, *processibility* and *vulcanization*. These characteristics are vitally important because they define the operating window available for converting uncured rubber compound into a usable product. Previous chapters have outlined the fundamental principles involved in the rubber manufacturing process; however, for the sake of clarity some of the basics of the technology will be reviewed here.

The rubber industry, maybe to a greater extent than other industries, has developed a terminology of its own. Thus, the meaning of some terms is not always self-evident. This is probably not surprising when one considers the apparent paradox under which the industry labors. In general, it takes a tough, sometimes elastic raw material and spends large amounts of energy converting this to a soft, pliable "green stock," incorporating additives (fillers, oils, chemicals, and so forth) along the way. This operation, broadly referred to as "compounding," involves "breaking-down" the rubber and combining additives by "mixing," "milling," or other forms of "mastication" to prepare a "stock."

At this point, the stock can be formed (shaped) by the application of force, and since it is predominantly plastic, it will retain the shape imposed upon it. This can be accomplished by squeezing the stock between rolls (calendering), pushing it through an orifice having the desired shape (tubing or extruding), or by confining it under pressure in a mold or cavity of the required dimensions.

The ease with which the aforementioned operations can be accomplished is defined by ASTM as the *processibility of the stock*. The relative performance of a compound in these steps is dependent upon its plasticity/elasticity relationship which in turn depends upon the temperature, the force, and the rate at which that force is applied. It follows that these last three parameters are very important in evaluating processibility.

After the stock has been formed to the desired shape, the compound or green stock needs to be converted to the strong, elastic material required for the end use. This is accomplished by the process of "curing," more commonly referred to as *vulcanization*, which is the chemical bonding (crosslinking) of the rubber chains, usually via the action of sulfur and accelerator under pressure at elevated temperature. During this process, the stock changes from an essentially plastic intermediate to a predominantly elastic final product. As a result, its inherent resistance to deformation increases as does its strength, resilience (bounciness), and toughness. Vulcanization tests are used to measure the performance of a compound during this process. Since curing is basically a chemical reaction, temperature is the most important variable in testing vulcanization behavior.

Some other terms need clarification before the discussion of test methods and apparatus is presented.

Scorch: "Scorch" is premature vulcanization, i.e., the stock becomes crosslinked during processing. It reduces the plastic properties of the compound, thus interefering with the forming operation and ruining the final product. Scorch is the result of the combined effects of time and temperature (heat history) on a compounded stock. The "scorch time" generally defines the time to onset of vulcanization at a particular temperature and thus represents the time available for processing.

Rate of Cure: The "rate of cure" (crosslinking rate) is the rate at which stiffness (modulus) develops after the scorch point. During this period, the compound changes from a soft plastic to a tough elastic material required for its end use. This effect results from the introduction of crosslinks connecting the long polymer chains. As more crosslinks are introduced, the network becomes tighter and the stiffness or modulus of the compound increases.

Cure Time: Cure time is the time required during the vulcanization step for the required amount of crosslinking to occur, yielding the desired level of properties. Its two components are *scorch time* and *crosslink time*, the latter being controlled by the *rate of cure*.

State of Cure: In general, "state of cure" is a term used to indicate the level of a particular property of the rubber with respect to the ultimate value of that property in the fully vulcanized stock. For example, as crosslinking proceeds, the modulus of a compound increases indicating higher states of cure. The so-called optimum state of cure is the amount of crosslinking that results in the maximum modulus. However, one must keep in mind that all properties imparted by vulcanization do not reach their optimum values at the same level of cure.

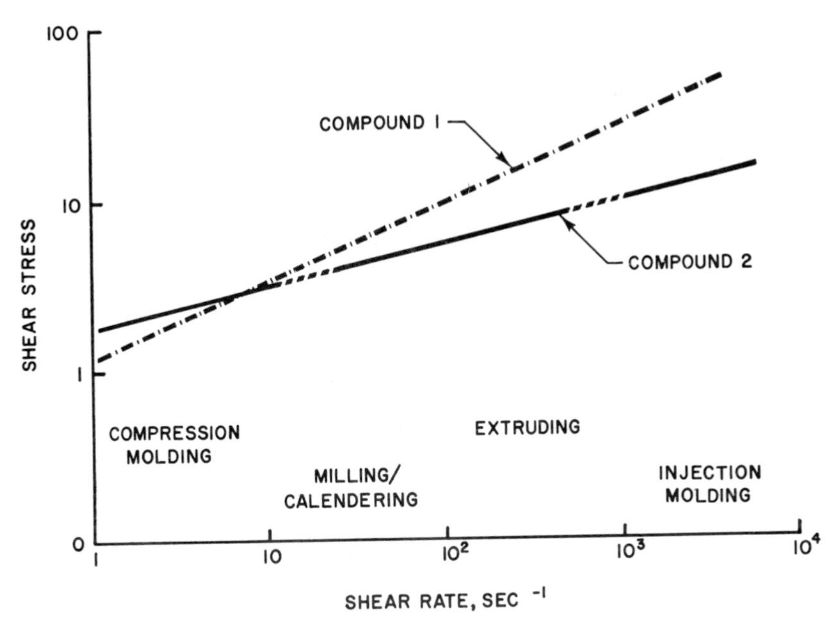

Fig. 4.1. Effect of shear rate on shear stress.

Overcure: A cure time which is longer than optimum results in an "overcure." There are generally two possible results of overcure depending on the type of rubber and the curing system. The stock may continue to harden with a loss in tensile strength and elongation, or it may soften with a loss in modulus, tensile, and elongation. The latter effect is commonly called "reversion" and it is most noticeable in natural-rubber compounds.

Shear Rate: This is the rate at which the dimensions of a unit of material are changed by the application of an external force. It is, of course, a function of the rate of application of the force as well as the properties of the material (see Fig. 4.1).

The test procedures needed to best define the various steps involved in the conversion of raw rubber to a usable product are examined in the remainder of this chapter. The narrative is subdivided into tests used to characterize successive steps in the process. The first major section is devoted to measurement of processibility, i.e., flow and scorch characteristics. Following that will be a section covering tests used to define the behavior of the processed compound during vulcanization.

PROCESSIBILITY

In a 1981 review on processibility testing, Norman and Johnson[1] pointed out that although many processibility tests have been devised, they are generally empirical and measure an obscure range of effects. Yet these measured properties have been found to correlate over a limited range with factory operations and thus are quite useful. In this regard, two of the most significant properties of a rubber stock are its viscosity (or plasticity) and the time to the onset of crosslinking (scorch). The former establishes the amount of energy required to mix and form the rubber, while the latter defines the time available for these operations.

Plasticity Tests

The tests used to define plasticity can be roughly classified on the basis of shear rate. This seems to be a logical basis for dividing the discussion since the various rubber processes—mixing, extrusion, calendering, molding—each involves different levels of shear and shear rate. This is graphically illustrated in Fig. 4.1, which shows the influence of shear rate on shear stress for two compounds. Although the two compounds have the same flow characteristics at the low shear rates encountered in compression molding, they exhibit markedly different flow behavior at the high shear rates encountered during extrusion. Therefore, tests to ascertain the processing characteristics of a particular compound should ideally expose the test sample to shear rates similar to those encountered in the actual process step. This is not always possible, since the shear rate achieved by most test instruments is lower than the rate of the process being simulated. As a result, it is frequently advisable to test at temperatures below the process temperature since lowering the test temperature is equivalent to raising the shear rate.[1]

Compression Instruments. These low-shear-rate ($<0.1\ s^{-1}$) instruments were the first developed, and in general, they utilize the compression of a cylindrical test piece between two parallel surfaces. The instruments are referred to as Williams[2] or Modified Williams[3] Plastometers depending on the size of the parallel surfaces with respect to the sample dimensions. In the widely used Williams Parallel Plate Plastometer, a preheated 2 cm^3 rubber pellet is placed between two parallel plates and a 5-kg load is applied for a standard time period, commonly 3 to 10 minutes. The thickness of the sample after the prescribed time period is recorded in mm and multiplied by 100 to yield the "plasticity number." The "recovery" value represents the increase in thickness that occurs after a prescribed period of time following removal of the load. These values relate to the flow and elastic properties, respectively, of the test specimen. They are useful in predicting processibility characteristics such as ease of forming and extrusion.

It should be noted that by this method plasticity is represented by the final height of the specimen; thus stocks that give a high degree of plasticity (more

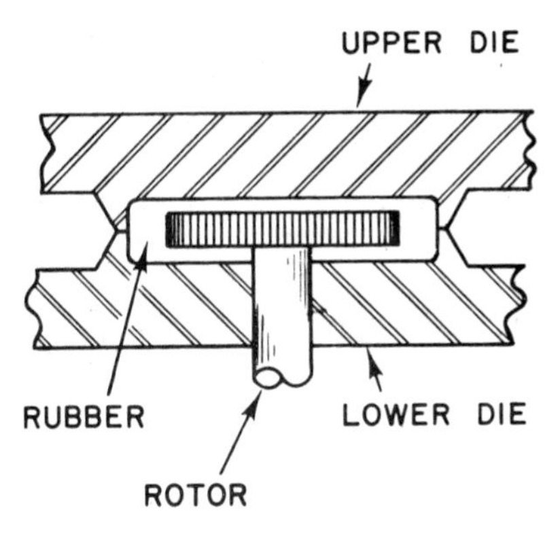

Fig. 4.2. Mooney chamber and rotor.

flow) during the test are indicated by a low plasticity number. As this is an apparent contradiction in terms, the results should always be referred to as the plasticity number. This method is described in detail in ASTM Standard D 926-83.[4]

Rotary Shear Instruments. A number of Plastometers of this type were developed during the rapid growth of the rubber industry in the 1920–1940 period. However, the shearing disk viscometer developed by the late Melvin Mooney[5] has become the "standard" of the rubber industry. In this instrument a flat, serrated disc rotates in a mass of rubber contained in a grooved cavity under pressure, as shown in Fig. 4.2. The torque required to rotate the disc at 2 rpm at a fixed temperature (usually 100°C) is defined as the *Mooney viscosity.**

The Mooney viscometer, which is widely used as a laboratory control instrument, operates at an average shear rate of about 2 s^{-1}. This is an order of magnitude higher than compression plastometers but still within the range of compression molding operation. (The modern instrument and its operation is described in detail in ASTM Standard D 1646-81.)[6] This procedure involves placing a square of sample on either side of the rotor and filling the cavity by pneumatically lowering the top platen. The platens are electrically heated and controlled at a preset temperature. A torque transducer measures the torque required to turn the rotor at 2 rpm. During viscosity measurements, the sample is allowed to warm up for 1 minute after the platens are closed. The motor is then started and the torque is recorded after the prescribed time period.

Fig. 4.3 shows typical viscosity vs. time curves for two rubbers. The ASTM

*As pointed out by Norman and Johnson[1], one should use the expression *Mooney torque*; however the misapplication of the term *viscosity* has become standardized.

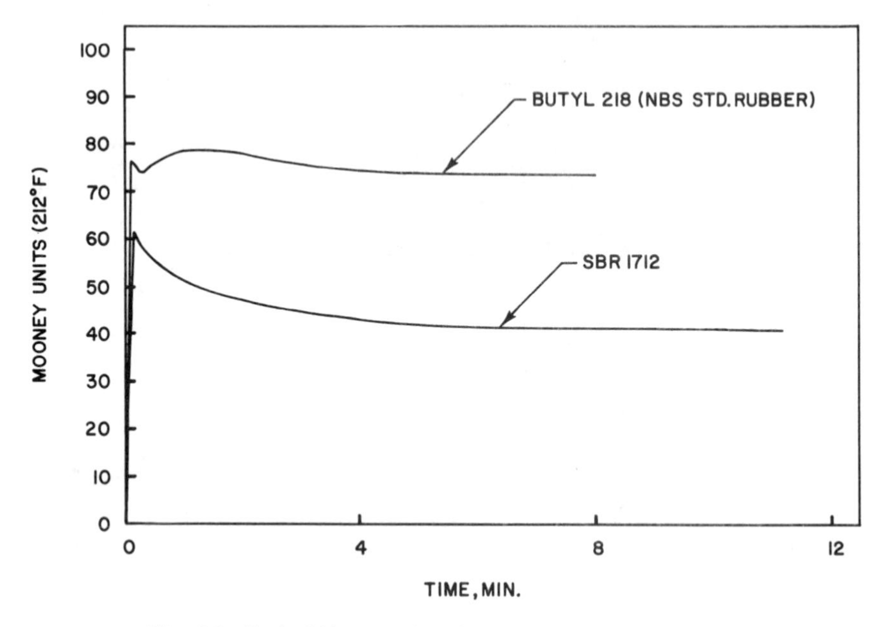

Fig. 4.3. Typical Mooney viscosity curves for different rubbers.

Standard gives standard test conditions for raw rubber viscosity measurements. Results of a typical test are normally reported as follows:

$$50\text{-ML } 1 + 4\ (100°\text{C})$$

where 50-M is the Mooney viscosity number, L indicates the large rotor, 1 is the time in minutes that the specimen is preheated, 4 is the time in minutes after starting the motor at which the reading is taken, and 100°C is the temperature of the test.

There are two standard rotors for the Mooney viscosity test, a large one for general use and a smaller one for very stiff materials. The ratio of viscosity results with the two rotors (L/S) is about 1.8, although this is somewhat dependent upon the type of rubber.

The Mooney viscometer is also widely used to determine the scorch characteristics of fully compounded rubber. Consideration of that aspect will be included in the section on scorch.

Mixing Tests. As indicated earlier, the ideal processibility test should expose the sample to shear rates similar to those encountered in the actual process. Ideally, this can be achieved only by using the factory equipment as its own processibility tester. Although this seems to be an absurd concept, the work input to a factory mixer can be used with considerable success to define in-process properties. Van Buskirk et al.[7,8] quantified high-shear mixing using a "work-per-unit-volume," unit-work parameter. With this approach, these and other authors[9] were able to illustrate a power-law relationship between work input and Mooney viscosity independent of the size or speed of the mixer. In fact, Van Buskirk found that passenger tire tread mixes have the same viscosity when mixed in a Brabender Plastograph (model PL-750) or Banbury mixers B, 1A, 11, and 27 if the same unit work is performed in all cases. This represents a scale-up factor of over 8000.

With this link between laboratory and factory mixers being firmly established, one can use small-scale mixing tests to define processibility parameters. The work input, which is the integral over time of the mixing power curve, can be conveniently measured on lab as well as factory machines, using a power integrator.[10] This instrument electronically corrects for nonproductive work, i.e., work required to run the empty mixer.

These "mixing-in-miniature" tests can be carried out on various instrumented mixers such as the Brabender Plastograph (Haake Rheomix), Brabender Plasti-Corder, Haake Rheocord, Hampden/RAPRA Variable Torque Rheometer, and so forth, as well as lab-size Banbury mixers. However, certain precautions are necessary when utilizing the "unit-work" concept. In addition to the obvious need to use identical raw materials (same lot of rubber) in lab simulations, operationally it is essential that the relative fill factor be constant (about 70%), the same loading procedure be followed, and the time/temperature profile be matched as closely as possible with the factory operation.

Extrusion Testing. Application of the work-input concept discussed above indicates that only about 20% of the total unit-work performed during a typical factory process is expended in the mixing phase.[7] This leaves about 80% of the unit-work for the very-high-shear-rate (100–1000 s^{-1}) extrusion steps. This breakdown highlights the influence of the extrusion component on total processibility, thus underscoring the importance of extrusion testing.

Historically the Garvey Die Test Described in ASTM D 2230-83[11] has been extensively used to test extrudability of unvulcanized compounds. This test involves the use of a variable-speed screw-type lab extruder such as the Brabender Plasti-Corder to extrude through a specially shaped die having a combination of flat surfaces, sharp corners, and thin sections. After the test conditions have been optimized with a standard compound, extrudate is then prepared with the test sample, and it is rated visually for smoothness, sharp corners, and integrity of thin sections. Although this test is only qualitative, it has been shown to be reproducible and correlatable with factory experience.

A more quantitative measure of extrusion characteristics can be obtained from the Pliskin Die Swell Tester (DST).[12] This semi-automated test involves a calculation of the die swell (ratio of extrudate cross-section to die cross-section) by measuring the velocity of the extrudate downstream from the die in a constant-volume extrusion. The proportionality between die swell and velocity assumes that the die swell resulting from the compound's elastic memory occurs almost instantaneously after the compound exits the die (this is not true for all rubber compounds, however).

Using this test, Van Buskirk[7] found a unique relationship between die swell and the unit-work parameter, i.e., the maximum in the die swell vs. unit-work curve is an indicator of the minimum work required to obtain low levels of undispersed black. More recently Tokita[13] has used the DST to simulate a factory extrusion process. His simulation gives good scale-up information provided that the operating temperature, the residence time in the die, and the die length to diameter ratio (L/D) are the same.

Capillary rheometers have been used for some years to measure basic rheological properties of liquids and polymer melts, but little work was done on elastomeric systems until the 1960's. It was not until commercial instruments became available in the 1970's that capillary rheology was commonly used to model extrusion processes on compounded rubber.

Where there once was a lack of commercial instrumentation, there is now an abundance. A variety of rheometers are available operating in two basic modes, i.e., measuring output rate with constant pressure on extrusion piston, or measuring extrusion pressure at fixed piston rate. The most versatile instrument according to Norman and Johnson[1] is the Monsanto Processibility Tester (MPT),[14,15] a constant-rate instrument that also includes a laser device for directly measuring extrudate die swell. The MPT has been utilized for measurement of processibility parameters on raw polymers as well as on compounded stock. Extrusion processes have been modeled using MPT data.[16] The model successfully predicts extruder stock temperatures, output rates, head pressures, and die swell.

The ideal testing situation would be to continuously monitor processibility characteristics "on line" so that changes can be detected and corrected in real time. Theoretically this could be done with the newly developed Rheometrics On-line Rheometer (ROR), an instrument designed for continuous measurement of viscoelastic characteristics at variable frequencies corresponding to shear rates of 0.1 to 500 s^{-1}.[17] The ROR operates on a side stream that is continuously pumped through a chamber consisting of two concentric cylinders. An oscillatory strain is imposed on the outer cylinder by a DC motor and the resulting stress (torque) experienced by the inner cylinder is monitored with a torque transducer. Although reported uses of the ROR to date have been restricted to thermoplastics, it could also be used on vulcanizable materials prior to scorch.

Relaxation Testing. The last type of processibility test to be considered in this section brings us full circle since it is a very-low-shear-rate test and, in fact, is a type of compression plasticity test. Flow occurs when compounded rubber is subjected to shear stress; however, when the stress is removed the flow decays with time. The characteristic time for a particular compound is called its *relaxation time*. It is a result of the "elastic memory" characteristic of viscoelastic materials. (Die swell, previously discussed, is at least in part a result of this behavior.)

Moghe has developed the Dynamic Stress Relaxometer DSR[18] in which a sample is forced into a conical cavity of fixed volume containing a rotor assembly attached to an air cylinder through a lever and arm. After a preheat period, the rotor is instantaneously rotated through a small angle and then held stationary. The torque, measured on a load cell attached to the arm, is monitored throughout both the strain and the relaxation period (1–2 s). The parameters measured are the torque, the time for the torque to decay to a preselected fraction of the peak torque, and the integral of the torque/time curve over a chosen time period. Of course, the instrument is capable of operating at preselected temperatures. This instrument has been used in the factory for testing the acceptability of a stock at various stages of processing.[18] This is done by determining the "integral torque" values at a fixed set of conditions for each batch being processed, and accepting or rejecting the batch based on control limits established with standard mixes. This approaches on-line testing since the test can be run very quickly.

A similar instrument is the RAPRA/Monsanto Stress Relaxation Processibility Tester (SRPT).[19] The SRPT is a plate-type compression-relaxation instrument, in which a cylindrical pellet of fixed volume is subjected to an initial fast compression between two parallel plates after a preheat period. The stress relaxation is then monitored with time via a pressure transducer in the upper platen. With the SRPT, the peak stress and the time for the stress to decay to a fixed fraction are the calculated relaxation parameters. This instrument can be used to evaluate stock processibility in a manner similar to the DSR. It has also been shown in factory trials to give results that correlate somewhat with die swell.[19]

Scorch

Scorch, which was previously defined as premature vulcanization, is really a vulcanization property and as such will be covered in that section. However, since it is extremely important in defining processibility limits, it will be briefly considered here. It is obvious that the viscosity of a fully compounded stock held at elevated temperature will increase (albeit nonlinearly) with time as a result of crosslinking. Thus, continuous measurement of viscosity at processing temperatures will indicate the time available for further processing. In this regard, the Mooney viscosity test described in ASTM Standard D 1646[6] is almost universally used to determine the scorch characteristics of compounded rubber.

From the chart of Mooney units vs. time as shown in Fig. 4.4, the time required for the compound to scorch is determined. The most common measure of scorch is the time to a 5-point rise above the minimum as shown in the figure. Normally the test is run at temperatures encountered during processing, i.e., between 120 and 135°C. The values normally taken from the Mooney cure curve are:

MV = Minimum viscosity

t_5 = Time to scorch at $MV + 5$ units

t_{35} = Time to cure at $MV + 35$ units

Δt_L = Cure index $= t_{35} - t_5$

Scorch, or premature vulcanization, may occur during the processing of a compound due to accumulated effects of heat and time. Therefore the time required for the compound to scorch will decrease as it moves through each stage of the process. Thus, samples taken from the same batch at different stages in the process will have progressively shorter scorch times. The value for a batch

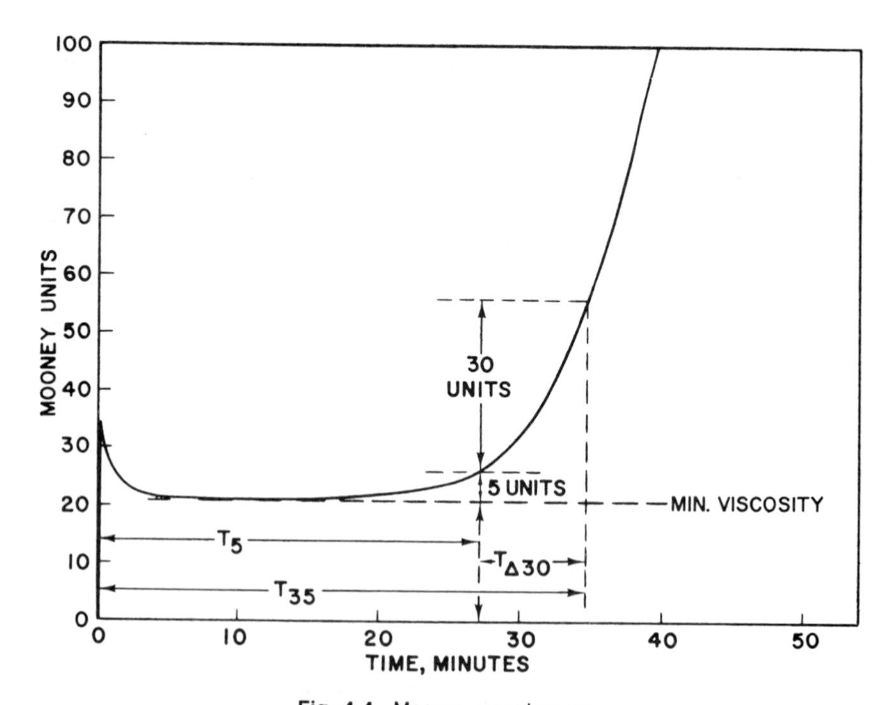

Fig. 4.4. Mooney scorch curve.

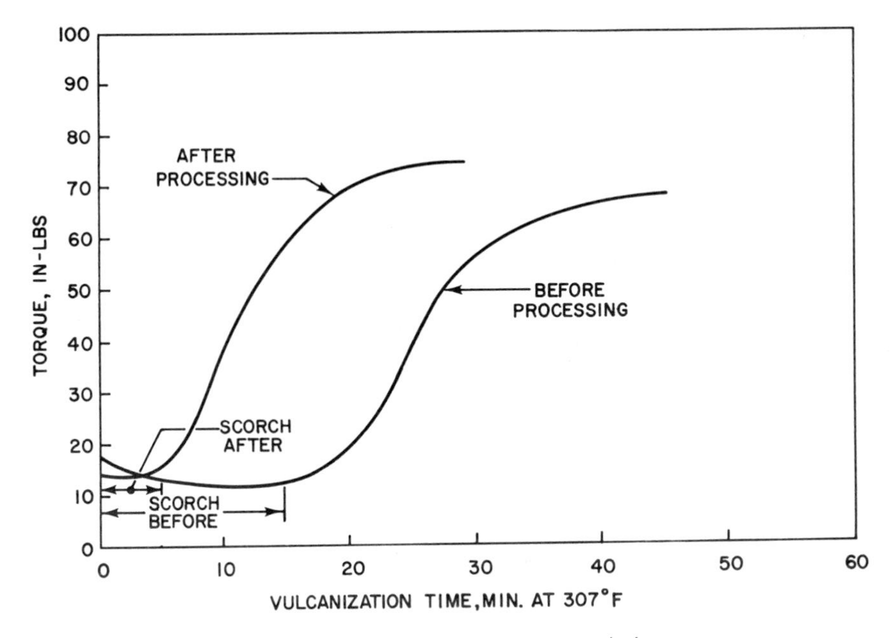

Fig. 4.5. Effect of heat history on scorch time.

at a particular point in the process is called its "residual scorch time." Fig. 4.5 illustrates this effect of heat history on scorch time. (The cure curves presented in Fig. 4.5 were obtained with a curemeter, which will be described in the following section on vulcanization.) Any good factory stock will have a scorch time slightly longer than the equivalent of the maximum heat history it may accumulate during processing.

VULCANIZATION

Assuming that the aforementioned stock is suitable for conversion to the final product (it still has residual scorch time), it is placed in a mold or otherwise subjected to heat to accomplish the vulcanization step. As stated in the introduction, this process involves chemically crosslinking the individual polymer chains to obtain an elastic network that will exhibit the desired "rubbery" properties. The performance of the stock in this phase of the process is assessed with vulcanization tests. Although the process is basically chemical in nature, the tests generally used are based upon the physical changes which occur in the rubber.

These changes generally occur in three stages: (1) an induction period, (2) a curing or crosslinking stage, and (3) a reversion or overcure stage. The location

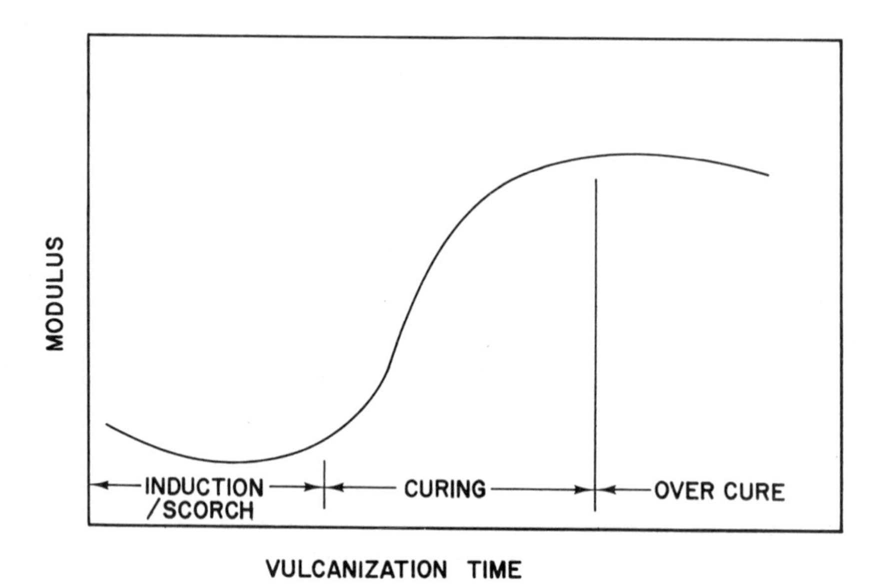

Fig. 4.6. Steps in the vulcanization process.

of these three stages in the vulcanization cycle is shown in Fig. 4.6. The induction period represents the time at vulcanization temperature during which no measurable crosslinking occurs. This is roughly analogous to the Mooney t_5 discussed in the previous section. Following the induction period, crosslinking proceeds at a rate dependent on the temperature, the type of rubber, and the curative system. As the curatives and crosslink sites become depleted, the crosslinking reactions slow until an optimum stiffness or modulus is achieved. This represents full cure. Further heating may result in a very slow increase in stiffness (''marching modulus'') or stock softening (''reversion''), depending on the type of rubber. These changes are generally refered to as *overcure*.

In the development or production of a rubber compound, the technologist strives to balance these three stages. For example, sufficient scorch time is mandatory to get the stock through the process to avoid scrap, yet not with so much residual scorch time that the cure time will be excessively long, thus tieing up expensive equipment. A fast cure rate is desired for the same reason, however it should not be so fast that insufficient mold flow occurs. Finally, one wants to achieve maximum utilization of the curing system by removing the product from the mold at the optimum time while avoiding the adverse effects of overcure. This balance can be best achieved through careful vulcanization testing.

Vulcanization Tests

Step-cure Method. The classical method for profiling vulcanization involves vulcanizing a series of sheets at increasing time intervals and then measuring the stress-strain properties of each sample and plotting the results as a function of vulcanization time. Typically, sheets are cured at the desired temperature for increasing time periods (5- to 10-minute intervals), dumbbell specimens are died from each sheet, and the physical properties, such as 300% modulus and tensile strength at break, are determined according to procedures described in ASTM D 412.[20] Fig. 4.7 shows typical data on two natural rubber stocks using different accelerators.

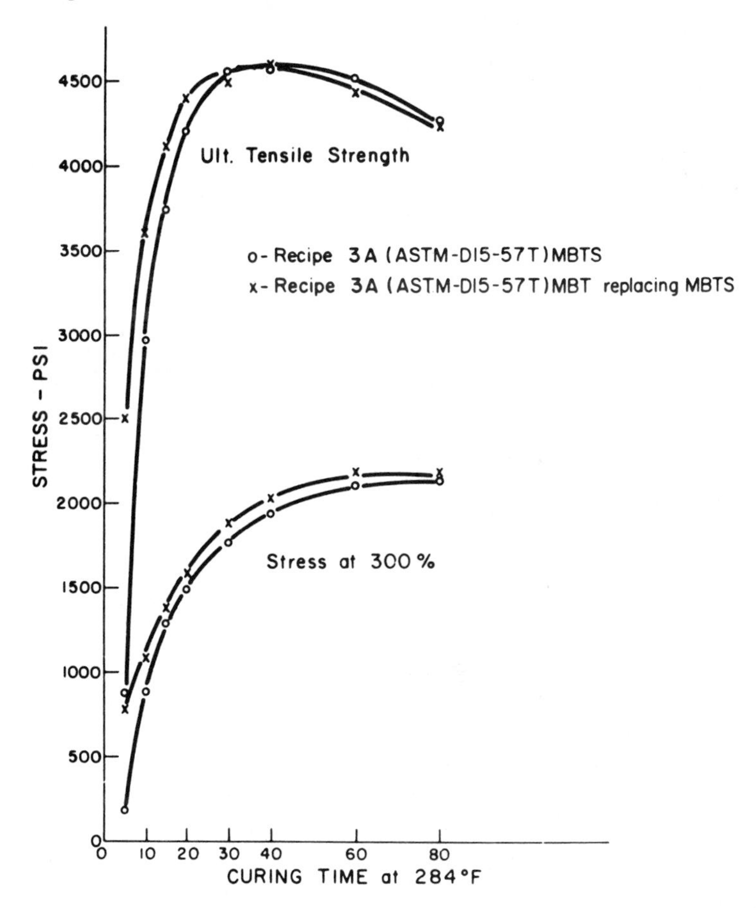

Fig. 4.7. Comparison of cure curves for natural rubber tread type stocks accelerated with MBTS and MBT.

This method is satisfactory for discerning large differences, but the inherent testing errors limit its applicability for detecting smaller variations in curing behavior. Of course, the test is inadequate for obtaining data during the induction period since useable tensile specimens cannot be prepared prior to cross-linking. However, when this test is coupled with the Mooney scorch test, a reasonable approximation of the cure curve is possible.

One advantage of the "step-cure method" is that it allows flexibility in choosing optimum vulcanization time depending upon the tensile property of greatest interest. (Note that in Fig. 4.7 maximum tensile strength occurs at a shorter vulcanization time than does maximum modulus.) This flexibility in choice is significant because the required vulcanization time often depends upon which property of the rubber is the most critical for the intended application.

However, in most real-world situations the compounder must choose a cure time that will give satisfactory values for the most critical performance property without reducing other properties below acceptable limits. For example, crack-growth resistance decreases as the cure becomes tighter while resilience (ability of the rubber to release stored energy) improves. Thus the optimum state of cure in this hypothetical case depends on whether the product being made is a rubber boot or a rubber ball. In the final analysis, the optimum state of cure is often a compromise. It is referred to as the "technical cure" since it gives the best balance of properties that can be technically obtained.

A modification of this test, generally referred to as the "rapid modulus test," is widely used in the industry for production control. A single sample taken from a production batch is vulcanized at a high temperature and its tensile modulus measured. Temperatures as high as 190°C are used to speed vulcanization and reduce the test time as much as possible. Any modulus value outside predetermined acceptance limits indicates that the batch is defective.

Continuous-measurement Method. In a paper published in 1957, Peter and Heidemann[21] describe an instrument for obtaining continuous measurements of rubber stiffness during cure. This instrument, and a later adaptation[22] was variously referred to as the Agfa, Bayer, or Bayer-Frank Vulkameter. It involved continuous measurement of the low-frequency shear modulus of a heated test specimen throughout the vulcanization cycle. The sample arrangement is shown schematically in Fig. 4.8. A paddle located between two test pieces, held by heated platens, was oscillated at a fixed frequency and amplitude. The force required to move the paddle was measured as a function of time. Using this instrument, Peter and Heidemann were able to obtain vulcanization curves for a number of rubber compounds that agreed quite well with data they obtained by the step-cure method. They also showed that the early part of the curve had a shape similar to the Mooney scorch curve when run at the same temperature.

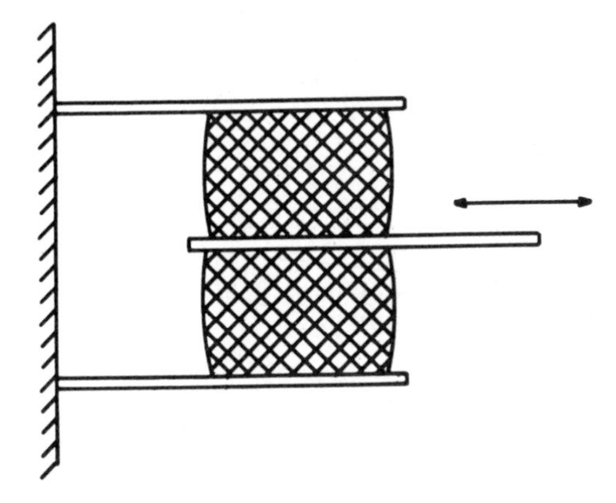

Fig. 4.8. Vulkameter and curemeter apparatus and test specimen.

The Wallace-Shawbury Curometer, reported in 1958,[23,24] was based on similar principles. In both the Vulkameter and the Curometer, the test specimen was cured at atmospheric pressure. As a result, sample porosity or dimensional changes (flow) often occurred during the test and the data did not agree with press cures under pressure. In an attempt to resolve this problem, the oscillating disk rheometer (ODR) was introduced in 1962.[25] This instrument, which got its inspiration from the Mooney viscometer, is depicted schematically in Fig. 4.9. The sample is held under pressure between heated platens surrounding a biconical rotor while the rotor is oscillated through a small arc with a motor-driven eccentric. As shown in the schematic, the mechanism includes transducers to measure simultaneously the displacement and corresponding torque required to shear the rubber specimen at the selected frequency. This instrument was originally designed to operate at either low frequency (1 cpm) for simple cure monitoring or at higher frequency (15 cps) for continuous measurement of dynamic properties during the entire cure process; an example of the latter is shown in Fig. 4.10.

Although less sophisticated versions were available, this particular configuration was sold as the LHS Rheometer (low/high speed). In 1967, an improved table model became available which gained wide acceptance for routine batch quality control. As a result, curemetering progressed from the compound-development laboratory to production control.

A number of other curemeters were developed through the 1970's designed to correct deficiencies in the ODR. It is beyond the scope of this chapter to discuss the details of these efforts; however, we will mention two other instruments that have achieved some acceptance and briefly describe their design. (Some of the more recent developments in curemeters will be discussed in a later section.)

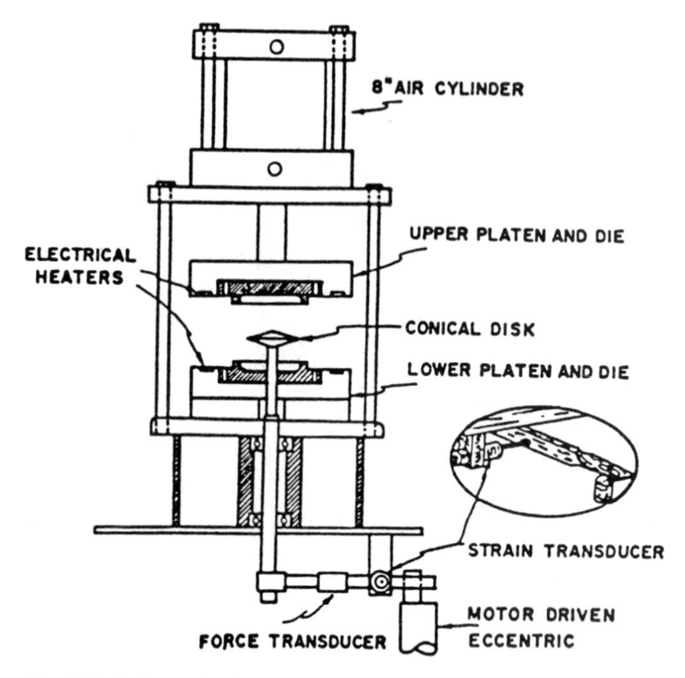

Fig. 4.9. Schematic diagram of oscillating disk rheometer. (Ref. 25)

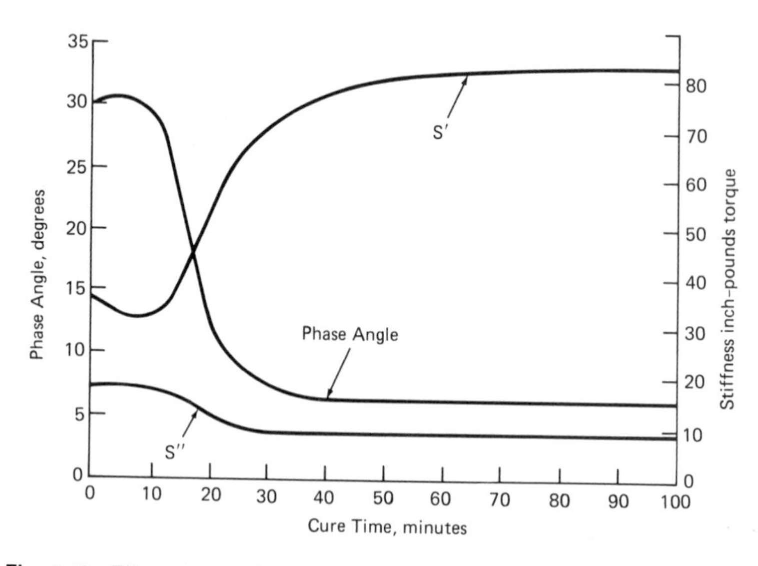

Fig. 4.10. Effect of cure of dynamic properties, SBR tread stock at 144°C. (Ref. 25)

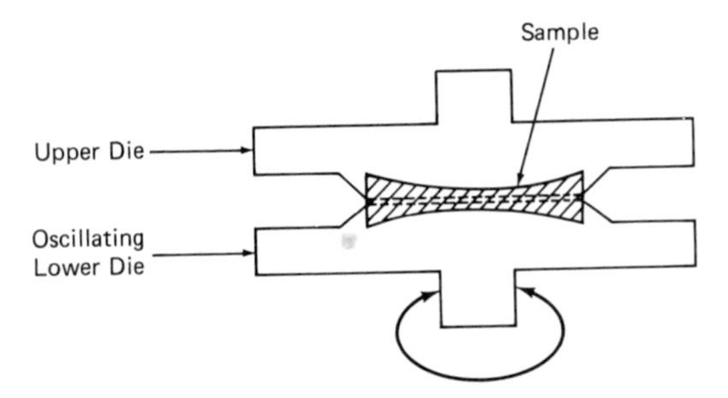

Fig. 4.11. Rotorless curemeter with uniform shear across directly heated dies. (Dotted lines represent gap between upper and lower dies. The lower die is oscillated while the transducer in the upper die monitors the force transmitted through the rubber sample. (Ref. 38)

The JSR Curelastometer, first reported in 1966[26], is a rotorless, quasi-closed chamber instrument that eliminates some of the perceived problems associated with the rotor in the ODR (frictional torque, slow thermal response, and difficult sample removal). The 1980 version of this instrument, known as the JSR Curelastometer III[27], is available commercially but has not gained wide acceptance.

A second rotorless rheometer, described by Gottfert in 1976[28] and known as the Gottfert Elastograph, uses two directly heated opposed biconical dies that are designed to achieve a constant shear gradient over the entire sample chamber (Fig. 4.11). The lower die is oscillated and the torque transducer on the upper die senses the force being transmitted through the rubber. This arrangement allows a fast thermal response as well as a constant shear gradient over the entire chamber area; however, since sample extrusion is involved in the cavity seal, high cavity pressure is not assured. This instrument is used extensively in Germany but not elsewhere.

Uses of Curemeters. It was more than a coincidence that the early publications on the chemistry of delayed action sulfur vulcanization appeared at about the same time that curemeters were being developed. There is a bit of irony in the fact that the comprehensive work by Monsanto's Campbell and Wise[29] on the characterization of the fate of accelerators and sulfur during the vulcanization process was done with the aid of a Bayer Vulkameter.[21,22] It is rather obvious that the value of this instrument in rapidly monitoring cure in support of their research provided incentive to the Monsanto workers for the development of the ODR.[25] (The coincidence of the latter instrument's development and the parallel development of new rubber chemical products illustrates the autocatalytic nature of technology-driven research programs. For those interested in this scenario, it is described in a 1983 Chemtech paper by A. Y. Coran, "The Art of Vulcanization."[30]

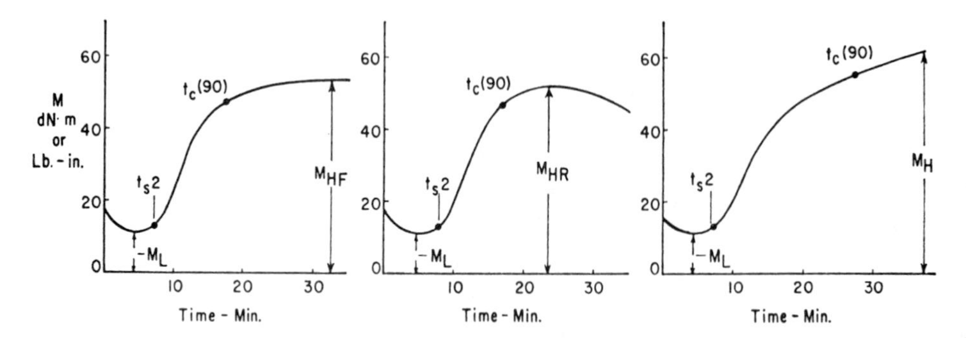

Fig. 4.12. Types of cure curve. (Ref. 31)

The value of curemeters was so obvious that their use was rapidly adopted in many rubber laboratories and subsequently in production control facilities. The driving force was, of course, the speed and accuracy with which the scorch and vulcanization characteristics of a compounded stock could be assessed. In response to this situation, ASTM adopted a tentative standard for the oscillating disc rheometer in 1968 and an approved standard in 1975.[31] Fig. 4.12 shows three different types of cure curves obtained on the ODR with the ASTM parameters noted. The ASTM definitions of the indicated values are as follows:

M_L = Minimum torque

M_{HF} = Maximum torque where curve plateaus

M_{HR} = Maximum torque of reverting curve

M_H = Highest torque attained during specified period of time when no plateau or maximum torque is obtained

t_S2 = Minutes to 2 units rise above M_L

t_X = Minutes to x% of maximum torque

This figure illustrates the types of cure curves that are obtained with different types of rubber. For example, some synthetic rubber compounds attain a constant or equilibrium torque level (M_{HF}) while most natural-rubber stocks exhibit reversion (M_{HR}). The "marching modulus" characteristic noted in the curve to the right is observed with some curing systems. Thus the cure curve gives a fingerprint of the scorch and vulcanization characteristics of the compound and, as a result, is very useful in categorizing rubber formulations.

The examples cited above show how gross changes in rubber stock composition are reflected in the rheometer curve. However, in practical rubber-compound development, the technologist is more frequently interested in the subtle changes that occur with compounding adjustments. Curemeter tests are ideally suited for this job. The effect of compound changes on viscosity and scorch can be determined from the early portion of the curve, while the effect on cure rate and relative modulus is seen in the crosslinking and plateau regions. In this regard, Fig. 4.13 indicates how relatively minor adjustments in the concentration of a compounding ingredient can be detected.

The ability of the curemeter to quickly detect minor compound changes and thus provide input for production control was noted as early as 1968.[32] Since that time it has become widely accepted as a production control test. For example, by operating at test temperatures in the 175 to 200°C range, a test can be completed in approximately 5 minutes, which coincides closely with a typical Banbury mixing cycle. The procedure usually followed is to establish specification limits at several points along the cure curve, as shown in Fig. 4.14. Then, a cure curve is obtained on each batch of production stock prior to its use in fabricating the final product and this curve is compared against the specification limits. Batches that yield curves falling outside these limits are rejected, thus minimizing unproductive processing.

Curemeters are also used to aid in selecting the proper cure time for laboratory specimens that require physical property tests. Since good agreement exists between the cure curve obtained with a curemeter and tensile modulus measurements made on step cures (see Fig. 4.15), a widely used laboratory practice is to determine the optimum cure time directly from the cure curve and prepare only a single specimen for stress-strain testing. With this approach, the curemeter can be used for functional evaluation of raw materials such as polymers, fillers, accelerators, and so forth, with considerable time savings.

Effect of Temperature on Vulcanization Rates. Since vulcanization is simply a chemical reaction in a polymer matrix, the rate of this process is significantly affected by temperature. The temperature sensitivity is governed by the well-known Arrhenius equation, which states that the logarithm of a reaction rate varies linearly with $(I/T)°K^{-1}$, i.e.:

$$\log k \propto I/T$$

Thus if one plots the log of a reaction rate parameter, such as optimum cure time vs. the reciprocal of the cure temperature expressed in °K, for a particular rubber stock cured at various temperatures, a straight line should be obtained as illustrated in Fig. 4.16. The slope of this line gives the activation energy, E_a, for the reactions involved, thus allowing the compounder to estimate the cure time at different temperatures from the cure time at the reference temperature. A rule-of-thumb for vulcanization as for many other chemical reactions

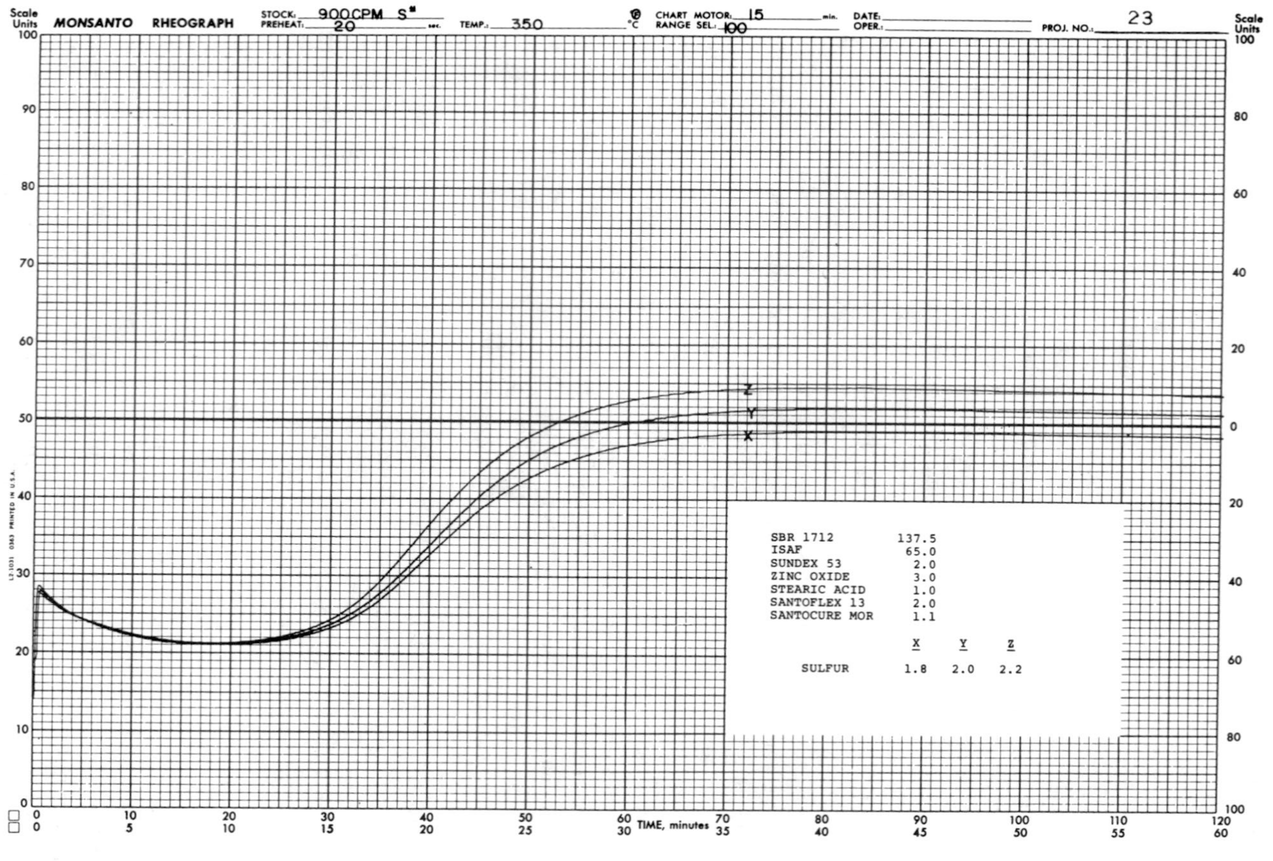

Fig. 4.13. Sensitivity of curemeter cure curves to minor changes in sulfur content. (Monsanto Technical Bulletin O/R C-3)

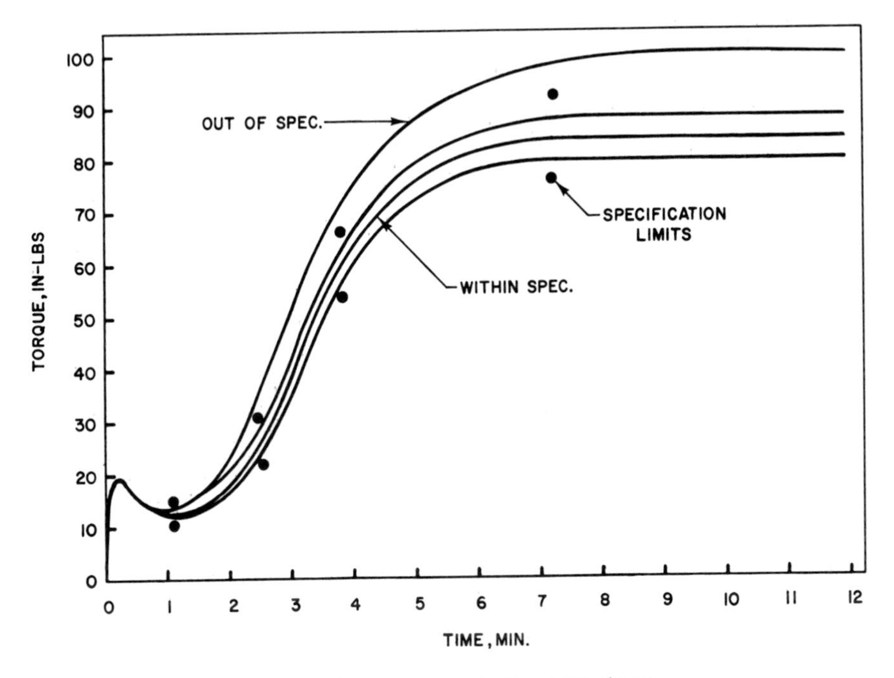

Fig. 4.14. Curemeter production control test.

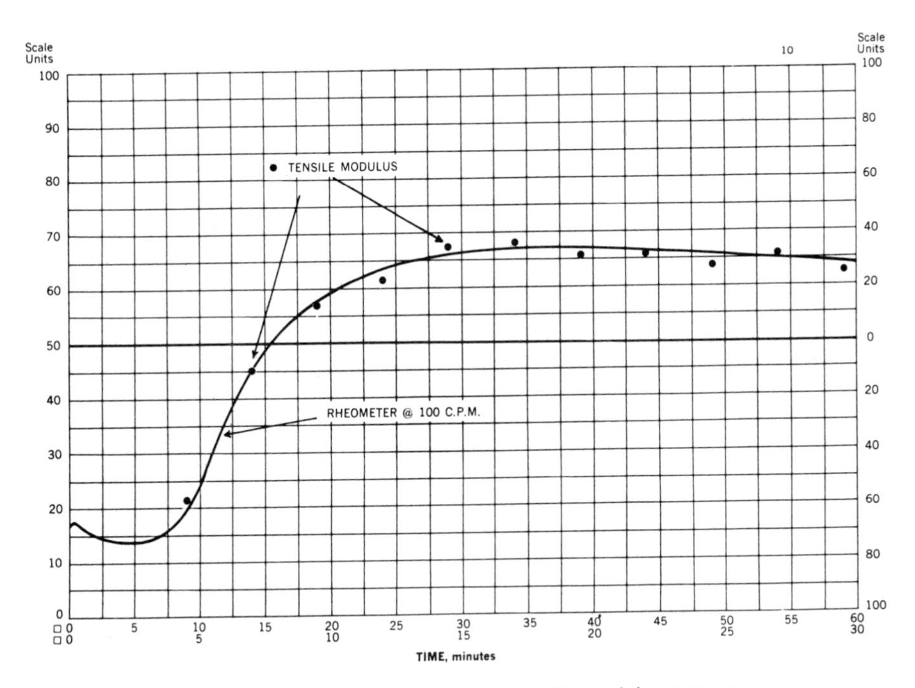

Fig. 4.15. Comparison of curemeter and tensile modulus cure curves.

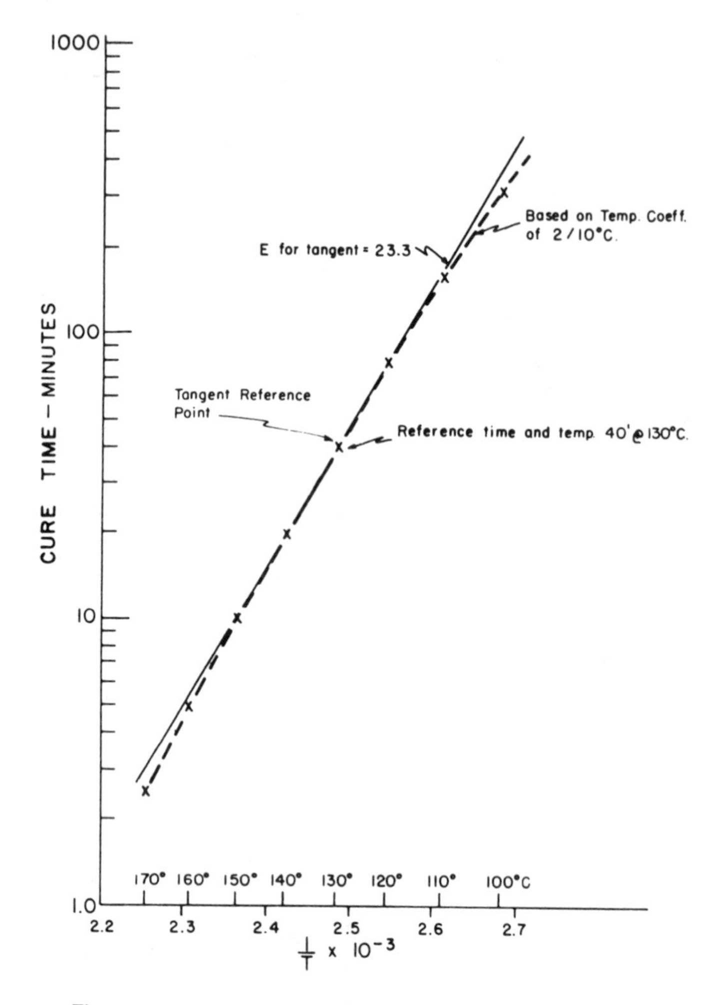

Fig. 4.16. Error involved in use of a temperature coefficient.

is that the rate doubles for every 10°C increase in temperature. The errors introduced by this approximation are also illustrated in the figure. However, with the use of a curemeter, the compounder can determine the optimum cure time at a variety of temperatures, construct an Arrhenius plot, and calculate the optimum cure time at any mold temperature without resorting to the approximation.

This, of course, assumes that curing is done under isothermal conditions. However, in practice this is not the case since tire sections are relatively thick and rubber is a very poor heat conductor. To illustrate, a typical time/temperature profile during cure is shown in Fig. 4.17 for different regions within a

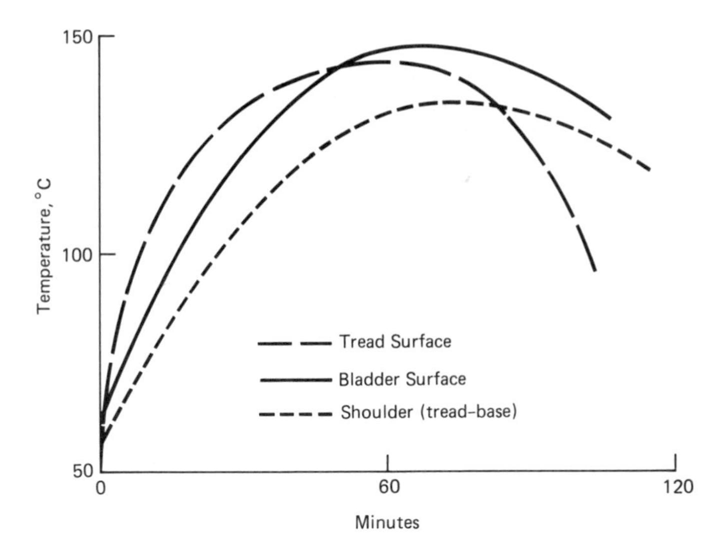

Fig. 4.17. Typical time/temperature curves for truck tires. (Ref. 34)

truck tire. These data were obtained by positioning thermocouples within the uncured tire. Note that the tread surface reaches curing temperature much before the internal regions and conversely cools down much faster after removal from the mold. In practice, tires are generally removed from the mold when the shoulder region has reached a relatively low state of cure (~25–35%). Vulcanization is completed during a postcure period utilizing the residual heat. It thus becomes necessary to design the curing system for each component of the tire to best take advantage of this temperature gradient and maximize productivity while achieving optimum end properties.

Past practice has been to determine the time/temperature profile for each tire component during cure with precisely positioned thermocouples. Lengthy calculations are then employed, using isothermal curing characteristics measured on a curemeter, to estimate equivalent states of cure under the actual time/temperature conditions experienced in the mold. Although these estimates represent a good approximation, a test is needed to directly duplicate factory cure cycles in the laboratory. This was achieved by utilizing program temperature cureometry in the Cure-Simulator.[33] This instrument is basically an oscillating disk rheometer with directly heated low-mass dies, which enable the temperature to be programmed by the operator. After establishing the time/temperature profile for the actual tire cure, the operator programs the instrument to follow the profile during the rheometer run.

A Cure-Simulator cure curve is shown in Fig. 4.18 for the center of a typical truck tread formulation.[34] From this curve it is possible to determine the standard curing parameters in the same way it is done with an isothermal rheograph.

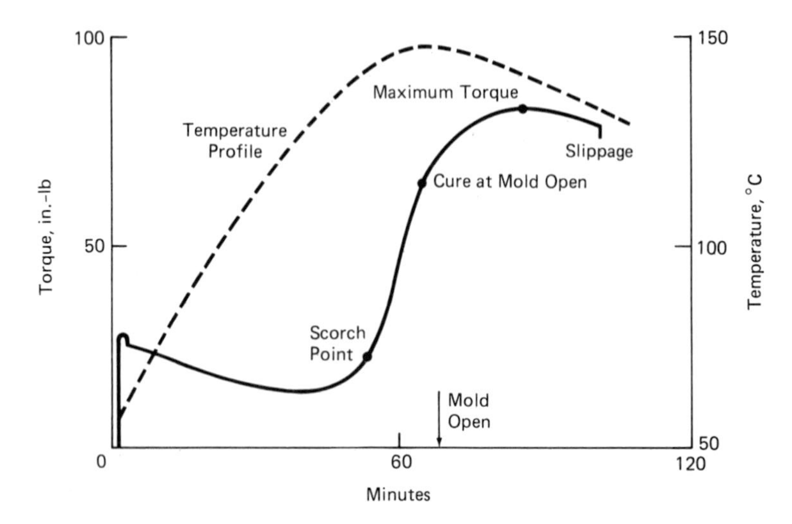

Fig. 4.18. Interpretation of cure simulator curve for truck tire. (Ref. 34)

In this hypothetical case, the projected mold opening would occur at 63 minutes corresponding to t_{70}, i.e., 70% cure. Although slighly undercured at this point, the tread would achieve full cure during the post-cure period.

Fig. 4.19 illustrates the very dramatic difference between isothermal and program temperature cures. In this example, the Cure-Simulator curve represents the rate of vulcanization at the center of a rubber article in a 170°C mold. It is

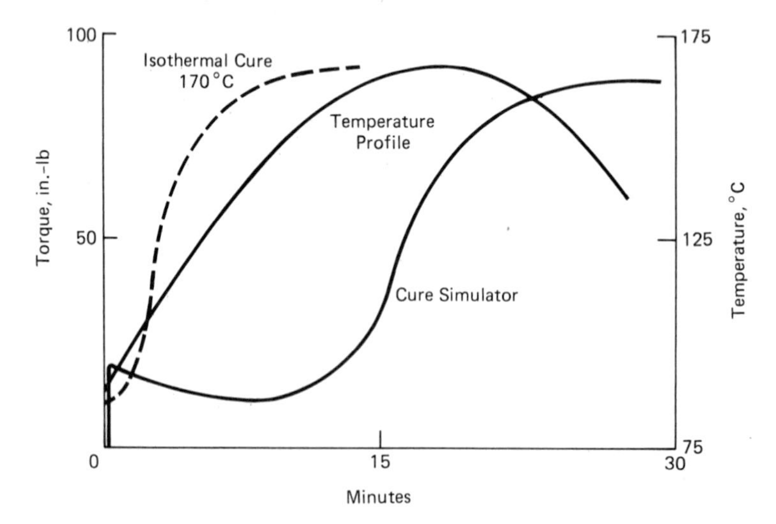

Fig. 4.19. Comparison of isothermal and cure simulator curves for passenger tire. (Ref. 34)

obvious that the isothermal rheogram very much underestimates the cycle time for this product. Of course, the surface of the item would see a temperature profile approaching the isothermal condition and its cure curve would be intermediate between the two extremes shown.

New Developments in Curemeters. After the introduction of curemeters in the late 1950's and early 1960's, there was a rapid development in vulcanization technology.[30] Some of these developments led to the improvements in test instrumentation discussed earlier. Two topics that seem to have particular significance to the rubber technologist will be briefly discussed in this section.

The typical rheometer output gives the familiar S-shaped cure curve when modulus (extent of cure) is plotted against time. Although this presentation is adequate for most purposes, added kinetic information about curative behavior can be obtained when "cure rate," in addition to "extent of cure," is plotted against time. This was demonstrated by Hartel in a 1978 paper, "Differential Vulcametrics."[35] In Hartel's terminology, the standard modulus-vs.-time curve is called the "network isotherm," i.e.:

$$y = f(t)$$

and the first derivative curve is referred to as the "network speed," i.e.:

$$y' = dy/dt = f'(t)$$

This latter function is determined with commercial curemeters by on-line electronic differentiation so that both functions (y and y') can be recorded vs. time. An example of this presentation for four curing systems is shown in Fig. 4.20. Although a detailed discussion of the value of this presentation is beyond the scope of this chapter, Kramer in a 1980 paper illustrated how the y' curve detects subtle differences between cure systems.[36]

The bimodal character of the y' curve in the rheogram depicted in Fig. 4.21 results from the overlapping crosslinking reactions that occur with this mixed accelerator system. This effect is not recognizable in the normal curve shown below. Thus this mode of curemeter presentation provides added sensitivity for detecting subtle differences between compounds, and as a result, can be very useful in quality control as well as in compound development and research.

Although isothermicity of the commercial curemeters has been assumed in the earlier discussions, a number of workers have shown that this is not actually the case. In recent publications, Hands and Horsfall compared various curemeters with data obtained on a truly isothermal minipress via the step-cure procedure.[37] Samples 0.5 mm thick were cured with a heat-up time of just over 1 s. In all cases, the existing curemeters tended to overestimate the cure time,

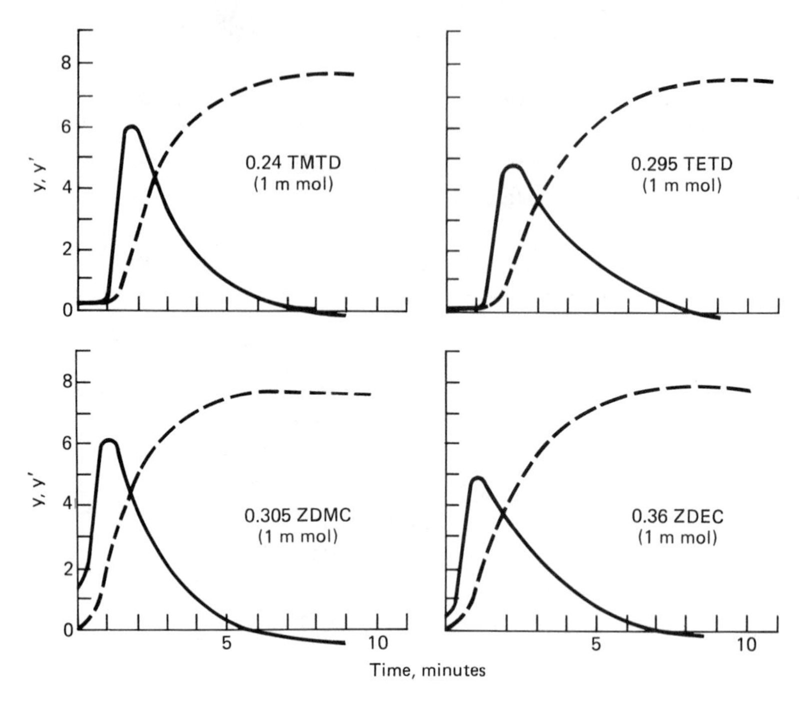

Fig. 4.20. Network isotherms y and network speeds y' for various accelerators with equimolar dosage. (Ref. 35)

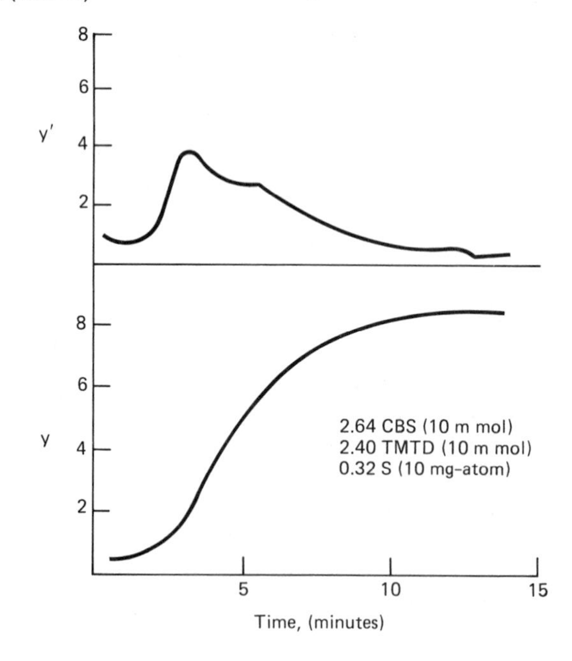

Fig. 4.21. Reaction rate curves for overlapping crosslinking reactions. (Ref. 36)

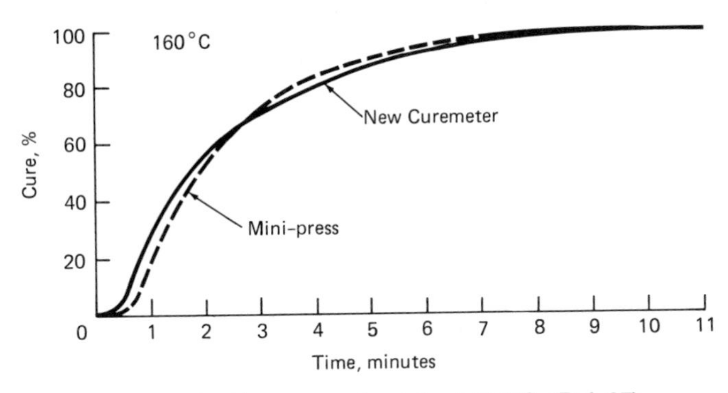

Fig. 4.22. New curemeter data at 160°C. (Ref. 37)

especially at higher temperatures. This overestimation was generally attributed to the thermal lag of the sample in the standard curemeters. Following up on this observation, these authors reported on a new rotorless curemeter (RAPRA #1) which approaches the isothermal character of the mini press. The comparison for a standard natural-rubber formulation is shown in Fig. 4.22.

Acting on this same concern, Monsanto has recently introduced a pressurized-cavity, rotorless curemeter that more nearly achieves isothermicity.[38] This instrument, called the Moving Die Rheometer (MDR), uses concentric cylinders on the die (see Fig. 4.23) to keep the sample cross-section thin and to improve the thermal response as well as increase signal output. Comparative results on the MDR and ODR are shown in Fig. 4.24. The shorter scorch time with the MDR is a result of this faster thermal response. Also, the elimination of the rotor in this instrument has made automatic sample loading and unloading possible. This feature together with automated data acquisition makes this instrument particularly useful for high-volume quality-control testing.

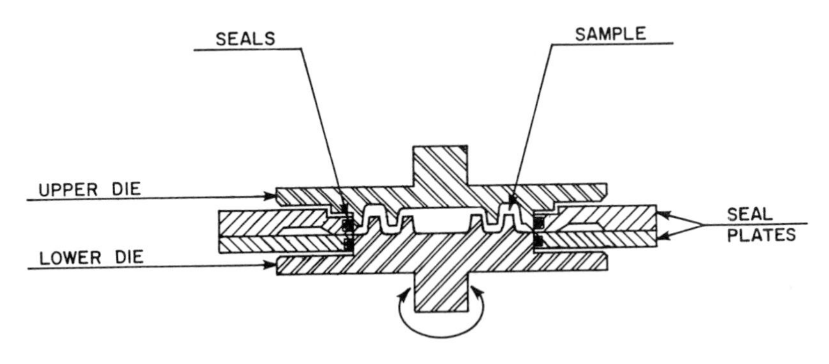

Fig. 4.23. Pressurized rotorless curemeter with concentric cylinders on the dies to increase signal compared to previous rotorless systems without increasing the size of the sample. (Ref. 38)

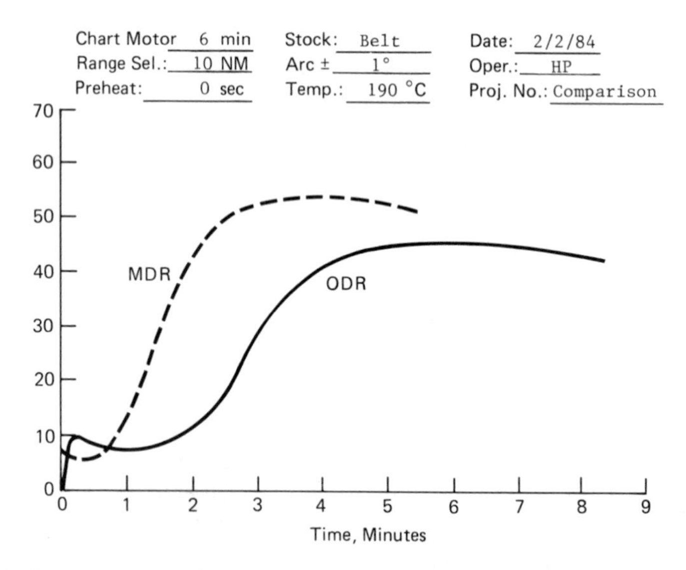

Fig. 4.24. Cure curve from ODR and MDR using the same stock and test conditions. (Ref. 38)

REFERENCES

1. R. H. Norman and P. S. Johnson, *Rubber Chem. Technol.* **54,** 493 (1981).
2. I. Williams, *Ind. Eng. Chem.* **16,** 362 (1924).
3. A. van Rossem and H. van der Meyden, *India Rubber J.* **76,** 360 (1928).
4. ASTM D 926-83, "Rubber Property—Plasticity and Recovery (Parallel Plate Method)."
5. M. Mooney, *Ind. Eng. Chem., Anal. Ed.* **6,** 147 (1934).
6. ASTM D 1646-81, "Rubber—Viscosity and Vulcanization Characteristics (Mooney Viscometer)."
7. P. R. Van Buskirk, S. B. Turetzky, and P. F. Gunberg, *Rubber Chem. Technol.* **48,** 577 (1975).
8. S. B. Turetzky, P. R. Van Buskirk, and P. F. Gunberg, *Rubber Chem. Technol.* **49,** 1 (1976).
9. G. E. O'Connor and J. B. Putman, *Rubber Chem. Technol.* **51,** 799 (1978).
10. S. W. Newell, J. P. Porter, and H. L. Jacobs, Paper No. 32, presented at a meeting of the Rubber Division, ACS, Cleveland, OH, May 6–9, 1975. Abstract in *Rubber Chem. Technol.* **48,** 1099 (1975).
11. ASTM D 2230-83, "Rubber Property—Extrudability of Unvulcanized Compounds,"
12. I. Pliskin, *Rubber Chem. Technol.* **46,** 1218 (1973).
13. N. Tokita, *Rubber Chem. Technol.* **54,** 439 (1981).
14. R. I. Barker, G. L. Hanna, and E. R. Rogers, Paper No. 70, presented at a meeting of the Rubber Division, ACS, San Francisco, Oct. 4–8, 1976; Paper No. 30, presented at International Rubber Conference, Institution of the Rubber Industry, Brighton, U.K., May 1977.
15. J. C. Warner and J. A. Jerdonek, *Eur. Rubber J.* **162,** 11 (Jul. 1980).
16. J. A. Sezna, *Elastomerics* **116,** (8), 26 (Aug. 1984).
17. R. D. Orwoll, *Advances in Polymer Technology* **3,** 23 (1983).
18. S. R. Moghe, *Rubber Chem. Technol.* **49,** 247 (1976).

19. J. P. Berry, R. W. Sambrook, and J. Beesley, *Plast. Rubber Process.* **2,** 97 (1977); J. P. Berry and R. W. Sambrook, Paper No. 28, presented at International Rubber Conference, Institution of Rubber Industry, Brighton, U.K., May 1977.
20. ASTM D 412-83, ''Rubber Properties in Tension.''
21. J. Peter and W. Heidemann, *Kautschuk und Gummi* **10,** 168 (1957); transl. *Rubber Chem. Technol.* **31,** 105 (1958).
22. J. Peter and W. Heidemann, *Kautschuk und Gummi* **11,** 159 (1958).
23. A. R. Payne, *Rubber J. Int. Plast.* **134,** 915 (1958).
24. A. R. More, S. H. Morrell, and A. R. Payne, *Rubber J. Int. Plast.* **136,** 858 (1959).
25. G. E. Decker, R. W. Wise, and D. Guerry, Jr., *Rubber World* **147,** 68 (1962). *Rubber Chem. Technol.* **36,** 451 (1963).
26. Y. Yoshida, presented at the 13th annual meeting of IISRP, Kyoto, 1972. M. Umeno, *Nippon Gomu Kyokaishi* (Japan) **46,** 369 (1973).
27. C. Tosaki, K. Ito, and K. Ninomiya, Paper No. 38, presented at a meeting of the Rubber Division, ACS, Denver, Oct. 1984.
28. O. Gottfert, *Kautschuk und Gummi* **29,** 261 (1976).
29. R. H. Campbell and R. W. Wise, *Rubber Chem. Technol.* **37,** 635, 650 (1964).
30. A. Y. Coran, *Chemtech* **13,** 106 (1983).
31. ASTM D 2084-81, ''Rubber Property—Vulcanization Characteristics Using Oscillating Disk Cure Meter.''
32. R. W. Jones, *Rubber Age* **100,** 52 (1968).
33. W. R. Deason and R. W. Wise, Paper No. 19, presented at a meeting of the Rubber Division, ACS, Los Angeles, May 1969. Abstract in *Rubber Chem. Technol.* **42,** 1481 (1969).
34. E. R. Rodger, ''Determination of Cure Cycles from Laboratory Tests,'' presented at a meeting of the Akron Rubber Group, Jan. 1977.
35. V. Hartel, *Kautschuk und Gummi* **31,** 415 (1978).
36. H. Kramer, *Plastics and Rubber: Processing*, Mar. 1980, p. 21.
37. D. Hands and F. Horsfall, *Kautschuk und Gummi* **33,** 440 (1980); Paper No. 11, presented at a meeting of the Rubber Division, ACS, Houston, Oct. 1983; D. Hands, R. H. Norman, and P. Stevens, ''A New Cure Meter,'' presented at the PRI Rubber Conference 84, Birmingham, U.K., Mar. 1984.
38. H. A. Pawlowski and A. L. Perry, ''A New Automatic Curemeter,'' presented at the RPI Rubber Conference 84, Birmingham, U.K., Mar. 1984.

5

PHYSICAL TESTING OF VULCANIZATES

F. S. CONANT
Standards Testing Laboratories, Inc.
Massillon, Ohio

INTRODUCTION

Scope

The purpose of this chapter is to introduce the subject of physical testing of rubber in terms of its rationale, methods, trends, and limitations. As the name implies, *physical testing* involves measurement and evaluation of *physical properties*. This could include a wide diversity of characteristics, but, by experience, certain categories of tests have emerged as being satisfactory adjuncts to economical development, production, and acceptance of rubber products.

Brief discussions of the physical properties of rubber are included to demonstrate the necessity for certain tests, to aid in selecting the most appropriate test in a given situation, and to explain the meaning of the test results. References are included for those who wish to pursue this aspect of rubber technology.

The methods chosen for discussion are mostly those that have been standardized by the American Society for Testing and Materials (ASTM).[1] This does not imply that these are necessarily superior to others but that they are widely used and representative. References are given to ASTM standards, where these exist, for detailed requirements on apparatus and method and to literature discussions on nonstandardized tests. Operating instructions for a particular apparatus are normally provided by the manufacturer.

Much more detailed treatments of the subject have been given in the excellent book on physical testing of rubber by Brown[2] and the more general treatment of all polymeric solids.[3] A review by Kainradl[4] approaches the subject from the viewpoint of the rationale for tests.

Reasons for Physical Testing of Vulcanizates

Tests used for other solid materials are seldom directly applicable to rubbers because of their many unique properties. Stresses, for example, are very dependent on time and temperature and are nonlinear with strain. Because of the ultra-high deformability and low modulus of rubbers, their test equipment must be accurate at higher displacements and lower forces than are needed for most engineering materials. Good reproducibility of results can be obtained only from thoroughly developed methods, which is one of the reasons for the many standardized tests that exist. Rubbers may also have unique Poisson's ratio, frictional properties, Gough-Joule effect, and very-high-energy storage. Hysteresis is also high and there is no metallic-type yield point. Many of these differences from other engineering materials relate directly to the nature of the force of retraction from a deformation, which is mostly due to entropy in rubbers and to internal energy in more rigid materials.

Nature of Tests

Rubber and rubber products are tested for initial and continued performance of the unique functions that are enabled by their viscoelastic properties. Specimens are, therefore, almost always deformed either statically or dynamically or both by the test procedure.

Static or low-deformation-rate tests for the common stress-strain properties of modulus, tensile strength, and elongation are usually made only for quality control since rubber products are seldom elongated in service. Low-rate tests for creep, stress relaxation, and set, however, reflect concern for limitations of service because of the time-dependent properties that are peculiar to rubbers. Examples are motor mounts, gasket seals, and bridge bearings. The amount of static deformation also affects such other properties as gas permeability, friction, electrical characteristics, and crystallization that are important, for example, for tire innerliners, tire treads, antistatic applications, and for some rubbers when used under arctic conditions. Flexibility rather than continuous deformation is involved directly or indirectly in such applications as coatings, diaphragms, and wearing apparel. Hardness is largely a function of modulus.

Elastomers are usually deformed dynamically in service, thereby transforming mechanical into heat energy. Deformation is thus resisted by both elastic and damping forces. The principal purpose of dynamic mechanical tests is to evaluate these forces and the attendant loss of useful energy. A bewildering number of tests for storage, loss, and combination properties has been developed from viewpoints of material property, engineering property, and technological property. This relation of physical test to the property involved has been treated in detail by Freakley and Payne[5] and by Hepburn and Reynolds.[6]

Test results are usually quite dependent on both the test method and the chosen conditions within that method. For example, the statement that a vulcanizate has a modulus of 5.0 MPa (725 psi) is meaningless unless the state and rate of deformation is given along with the test temperature. Furthermore, the value must be identified as Young's modulus or rigidity modulus (in shear), as equilibrium or transient, as tangent or secant, whether based on original or attained cross-section, and whether determined on the first or subsequent deformations. These details, which will be explained in later sections, provide some of the rationale for standardized tests.

A rubber laboratory may perform any or all of the following functions:

1. Quality control of incoming material.
2. Quality control of in-process material.
3. Quality control of finished goods.
4. Factory technical service.
5. Customer technical service.
6. Preparation of samples for sales department.
7. Development compounding.
8. Research.
9. Advertising assistance.
10. Product performance testing.

These are often condensed into three principal categories: compliance with specifications, quality control, and research and development.

Specifications are requirements, usually physical rather than chemical, imposed on a material or an end product. They dictate the tests to be made and acceptable test results. Specifications may be made by a customer to ensure a uniform product of adequate quality or by the manufacturer to maintain processibility. Duplicate tests are sometimes made by the final inspection department of a producer and by the acceptance laboratory of a customer, each to ensure that stated specifications have been met. Fortunately, the current trend in specification writing is to require only those properties which have a demonstrable value to further processing or to end use rather than the formerly common "general quality indices," such as tensile strength or compression set, whether or not these qualities were actually necessary. The system for classifying rubbers by "line call-out" given in ASTM D2000 is useful in specifications for rubbers to be used in automotive applications and elsewhere.

Control tests are made by the manufacturer, at any stage in the fabrication process, for the purpose of maintaining processibility or quality of a finished product. Such tests do not have as much need for standardization as do specification tests that may require that two laboratories get the same results. Cooperation between laboratory and production is essential for efficient utilization of control testing results. Unless some use is made of the data, it is completely nonproductive. Testing for processibility is covered in some detail in Chapter 4.

In both specification and control testing, the subject either passes or fails; these are often called "go-no-go" tests. In research and development tests, on the other hand, much more information is desired; single-point data are no longer sufficient. Results are often obtained in a series of situations in which a component or a test condition is varied and the results are plotted to show trends. Special instruments devised for such tests often serve as the basis for later development of standardized tests.

In summary, physical testing is always done for economic reasons. Unfortunately, however, the programs and evaluations are not always carried out in the most efficient manner. Specific instances of inefficiency are pointed out in some of the following sections.

GENERAL CONSIDERATIONS

Standardization

Standardization of a test means that many of its users have agreed on all of the critical requirements and that these have been written up by a standardizing group. Furthermore, the method will have been tried in an interlaboratory program so that its repeatability within a laboratory, and reproducibility between laboratories, are known. Each developed country has a national body with overall responsibility for such standardization. In the U.S., this is the American National Standards Institute (ANSI), which coordinates the various groups that actually establish standards.

Committee D-11 of ASTM has the principal responsibility for standards that affect producer-consumer relationships in the American rubber industry. These are published in Vol. 09.01 and 09.02 of the Annual Book of ASTM standards. Other American standards include Federal Test Method Standard 601,[7] military and other federal standards,[8] motor vehicles standards (U.S. Dept. of Transportation), and methods written into specifications such as Federal Specifications 501 and *Society for Automotive Engineers (SAE) Handbook*. Specifications for rubber products are written by many other groups such as Underwriters Laboratories, AMS Aeronautical Board, National Hospital Association, and International Electrotechnical Commission. The predominant international standards group is the International Organization for Standardization (ISO), of which Technical Committee 45 is concerned with rubber products.[9] Activities of this latter group have included many physical tests on rubber performed in a standard manner worldwide.

Standards are not static documents that are perpetuated without change. Those in ASTM, for example, are continuously updated as new instrumentation, new materials, and new knowledge become available. Much cooperative research usually precedes the writing of new standards or substantive revision of existing ones. Standards include specifications, guides, test methods, classifications, definitions, practices, and related materials such as proposals. A *specification* gives minimum requirements for materials or products, so it may refer to several test methods. A *guide* is a series of options or instructions that do not recommend a specific course of action. A *practice* is a definitive procedure for performing one or more specific operations or functions that do not produce a test result. Discussions of each of these and other explanatory material on standards are given in the Foreword of each volume of the annual *Book of ASTM Standards*.

References to an ASTM method should include the date of its latest revision unless the reference is quite general. For example, ASTM D1415-83, "Standard Test Method for Rubber Property—International Hardness," is the write-up as revised and adopted in 1983. It differs in some respect from the previous version, D1415-81.

Standardization of rubber testing has also been extended beyond the document stage toward monitoring of two important determiners of laboratory precision—standard materials and reference testing programs.

Reference or "standard" compounds serve three valuable functions in physical testing of vulcanizates: quality control of raw materials, controls against which experimental compounds are compared, and a system check on bias of test equipment and operating procedures. Such standard compounds require standard compounding materials. Since many of these, especially polymers, may degrade with aging, sources and verification of materials that are consistent in critical properties are necessary. A task group of ASTM Subcommittee D-11.20 supervises maintenance of such industry reference materials (IRMs) as natural rubber, SBR 1500, zinc oxide, sulfur, stearic acid, accelerators, carbon blacks, and age resistors, mostly as listed in ASTM D3182. Detailed procedures for preparing standard compounds and test pieces from different basic polymers are given in ASTM D3182 through D3192 plus D3403 and D3568.

A "Collaborative Reference Program" to monitor testing proficiency has been established and is technically supported by the Rubber Divison of The American Chemical Society.[10] Programs currently available are tensile properties (ASTM D412), hardness (ASTM D1415 or D2240), Mooney Viscosity (ASTM D1646), and vulcanization characteristics (D2084). Participants receive uniform test samples four times per year. They perform the test and return results for analysis. They then receive a summary report that shows test averages and standard deviations from all participants.

The functions served by industry reference materials and the collaborative reference program is explained more fully in the following section.

Validity, Accuracy, and Precision

A test is *valid* if the results are actually a measure of the desired property. If a valid wear test for tire tread rubber existed, for example, there would be little need for expensive fleet testing for wear resistance. Even though an apparatus and method may be *valid*, however, the *accuracy* can be impaired by either instrumental or procedural defects. An oven aging test, for example, is inaccurate if the temperature measurement is incorrect. A test is *precise* if results are closely reproducible. A low temperature brittle point test, for example, may be both valid and accurate if results on many specimens are averaged, yet have a wide scatter of results, and hence be imprecise, on certain types of materials. Precision of results within a given laboratory is "repeatability"; that between laboratories is "reproducibility."

Problems of validity, accuracy, and precision each exist in varying degrees in all physical tests on rubber. Juve[11] has pointed out many weaknesses in rubber testing and concluded that much of it is nonproductive. Examples of nonvalid testing include unnecessary specifications by the consumer. High tensile strength may be specified routinely when, in fact, it is totally unnecessary to the proposed use. In this case, it is not a valid test. Another example cited is use of the "peanut" or "angle" specimen for tear testing. Since results are unduly influenced by stock modulus, they do not truly represent tear resistance. While tests may be valid in a limited application, such as the simple hysteresis tests (rebound, Yerzley oscillograph), they may become invalid because of faulty *interpretation* of data. "The inadequacy of the simple device lies in the acceptance of the data as an unqualified measure of hysteresis, regardless of the service conditions for which the stock is intended"[11]. The validity of a method for a particular application should be indicated in its scope. A further caution is that tests should be chosen which can be interpreted. Some tests measure such complex combinations of basic factors that it is nearly impossible to understand the results.

Interlaboratory tests on rubber usually give a wide scatter of data. Invariably, some laboratories report results which are so far from the mean of the others that some instrumental or procedural problem is indicated. In other words, the data are *inaccurate* to a degree unsuspected by personnel of that laboratory. Surprisingly often, the problem is temperature measurement or calibration of force-measuring equipment. These problems can be avoided by testing standard compounds periodically or by continual reciprocal testing with other laboratories.[10]

Repeatability or reproducibility problems may be caused by variability in sample material or preparation, improper sampling, inadequate control of test conditions, or instrument problems such as friction or electronic drift. Certainly the precision of a test should be known, either from previous records or by testing many specimens, so that confidence levels of test results can be established statistically.

ASTM D3488 offers guidelines for measuring and using the precision of ASTM test methods, while D3040 gives directions for preparing precision statements. The statistical content of each includes important concepts such as confidence level, bias, least significant difference, sampling, analysis of variable (ANOVA), and test design.

Specimen Preparation

Most physical tests on vulcanizates are performed on specimens prepared from submitted samples. Some variations in properties must be expected even on supposedly identical samples because of nonuniformities in raw materials and processing techniques. For this reason, a *control* stock is often included in each group of experimental stocks. Dispersion of test results on the control must then be considered in determining a degree of confidence in evaluations of results on the experimental stock.

Some standard methods for stock mixing, sample curing, and specimen forming are given in ASTM methods D3182 through D3568. Many properties of rubber important to good physical testing practice are implicit in these directions. For example, different aging periods after mixing, after remilling, and after curing are required for compounds based on different polymers. In general, no tests should be made until at least 16 hours after vulcanizing a sample because significant post-vulcanization changes in structure of the material occur during this period. Of course, such delays are intolerable in control tests, so the change in properties between actual testing time and 16 hours after curing should be ascertained and taken into consideration.

Some test results are influenced by flow of rubber in the mold, so a good general practice is to use no specimens cut from material near a mold edge. Also the *grain* in rubber caused by the action of the mill in the final pass may affect certain properties. Tensile tests, for example, are always made along the grain, i.e., parallel to direction of passage through the mill.

Specimens are sometimes required from a finished product rather than from laboratory prepared samples. In such cases, they are often buffed to obtain uniform thickness, to remove fabric or fabric impressions, or to remove glaze or skin coats from the rubber. Abrasive grinding wheels of about 30 grit, 5-inch diameter, running at 2500 to 3500 rpm are satisfactory for this purpose. The grinder should have a slow feed so that very little material is removed at each cut, to avoid both high heat and coarse abrasive pattern at the surface.

Standard Temperature

ASTM D1349 gives a list of standard test temperatures which agrees with the ISO list. An advantage of using only the standard temperatures is that many unnecessary temperature changes in environmental chambers are eliminated. A standard humidity is also provided although it is doubtful that humidity has much effect on physical properties of rubber. Properties of fabrics or rubber-fabric composites, however, are very sensitive to moisture.

Break-in

For some tests a rubber specimen should be *broken in* before recording any results. This means that it should be deformed several times to the same extent that it is in the recorded test. This usually gives a softer material, but one on which results are reproducible. Upon resting, the specimen properties gradually return toward the initial values. The rebound test is an example of one that requires a break-in.

Fundamental Constants of Rubbers

Many physical properties of rubber are of the basic type, that is, they are independent of the measuring instrument, e.g., thermal conductivity, specific heat, thermal expansion, density, refractive, index, and so forth. Typical values for such properties have been collected by Wood.[12] Indices, such as abrasion resistance, are based on instrument constants. Some, such as tear resistance, are highly dependent on type of specimen.

STRESS-STRAIN TESTS

Stress-strain Terminology

Physical testing of rubber often involves application of a force to a specimen and measurement of the resultant deformation or, conversely, application of a deformation and measurement of the required force. Two common modes of deformation, tensile and shear, are illustrated in Fig. 5.1. Since we are interested in material property, test results should be expressed in a manner that is independent of specimen geometry. For this purpose, the concepts of *stress* and *strain* are used. Stress is the force-per-unit cross-sectional area, i.e., F/A for either tensile or shear deformations. Strain is the deformation per unit original length ($\Delta L/L_0$) in tensile tests or deformation per unit distance between the contacting surfaces (S/D) in shear tests.

Stress is usually expressed in pounds per square inch (psi) in traditional units or pascals (Pa) in SI units. Strain is often expressed as a percentage rather than a ratio. Since it is obtained from the ratio of two lengths, it is dimensionless. An elongation of 300%, for example, means that the specimen has been stretched to four times its original length.

The term *extension ratio* is often used in rubber technology. This is the length of the specimen at a given point in the test divided by its original length. An undeformed specimen, thus, has a strain of zero and an extension ratio of unity. At a strain of 1.0, the extension ratio is 2.0.

In the common parlance of rubber technology, the stress required for a given elongation is used to represent the material stiffness. This quantity is called the modulus. A 300% modulus, for example, means the stress required to produce a 300% elongation. In mechanical engineering usage, however, the term *modulus* is defined as the ratio of stress to strain. If this ratio is a constant, the material is said to obey Hooke's law and the constant is called *Young's modulus*. In practice, the term *Young's modulus* is often used to represent the ratio of stress to strain even in situations where it may vary with change in elonga-

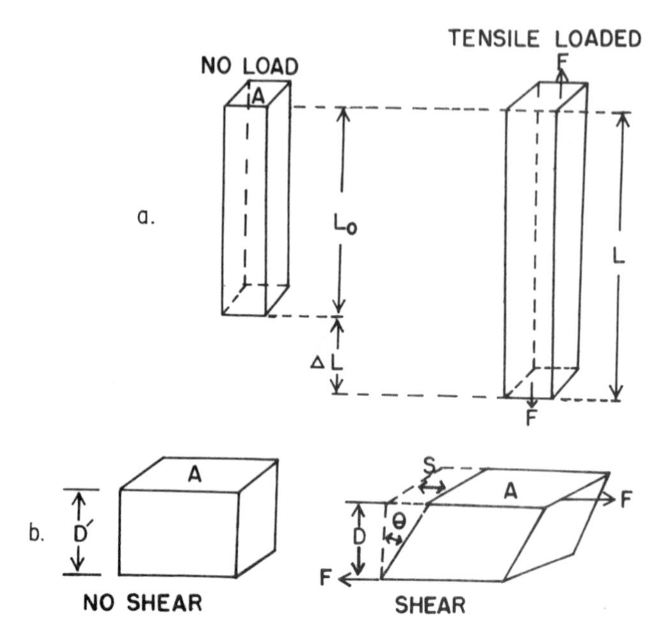

Fig. 5.1. a. Tensile stretching of a bar. b. Shear of a rectangular block. (Ref. 13)

tion. In the terms used in Fig. 5.1a, Young's modulus E is given by:

$$E = \frac{F/A}{\Delta L/L_0}$$

The same equation applies when the bar is decreased in length by a compressive force. Compressive tests, at low deflection, give moduli equal to or slightly greater than do tensile tests at comparable deflection.

Shearing of a rectangular block is illustrated in Fig. 5.1b. The rigidity or shear modulus G is defined as the ratio of shearing stress to shearing strain:

$$G = \frac{F/A}{S/D}$$

Shear is also involved in torsional deformation of beams but calculation of modulus from torsional tests involves a shape factor which, for rectangular beams, depends upon the ratio of thickness to width.[14]

A third type of modulus is the *bulk modulus B*. It is defined as the ratio of the hydrostatic pressure to the volume strain:

$$B = \frac{\text{Hydrostatic pressure}}{\text{Volume change per unit volume}}$$

Rubber is highly incompressible, so its bulk modulus is much higher than its Young's modulus.

When a material is stretched, its cross-sectional dimensions decrease. *Poisson's ratio*, ν, is the constant relating these changes in dimension:

$$\nu = \frac{\text{Change in width per unit of width}}{\text{Change in length per unit of length}}$$

If the volume of a material remains constant when it is stretched, Poisson's ratio is 0.50. This value is approached for liquids, and unfilled rubbers.

The indices discussed are interrelated by the following equations, which are strictly valid only in the small-strain regime:

$$E = 2G(1 + \nu) = 3B(1 - 2\nu)$$

Young's modulus is seen to be three times the shear modulus if Poisson's ratio is 0.50. The factor may actually be as high as four, however, for highly filled rubbers.[2]

Tensile Tests

The stress-strain test in tension, including ultimate tensile and elongation, is probably still the most widely used test in the rubber industry. Among the purposes for such tests are to determine effects of liquid immersion[15] on aging, to ensure that all compounding ingredients have been added in the proper proportions, to determine rate of cure and optimum cure for experimental polymers and compounds, for specification purposes, and to obtain an overall quality check on the compound. Proportionately fewer tensile tests are run now than in previous years for two principal reasons: the cure meters, ASTM D2084 for example, have been given a large share of the tests for state of cure; and more emphasis is being given to testing for the properties desired in a particular stock than to tests for general quality. High tensile strength is seldom required in service and, by itself, does not guarantee the level of any other property. However, since a single test can yield modulus at specified elongations, ultimate elongation, and ultimate tensile strength in well-standardized tests and with short testing time, tensile tests are far from outdated.

Standard methods for tensile testing of vulcanized rubbers are given in ASTM D412. Four shapes of test specimen are permitted: dumbbells and rectangles died from flat sheets, rings died or cut from flat sheets, and rings cut from tubular specimens.

In the U.S., the most common tensile specimen is the dumbbell, so-named for its shape, which has tabbed ends for gripping in the test machine and tapering to a central constricted section of uniform width. *Bench marks*, often one inch apart, may be stamped on the constricted section to facilitate manual or optical[16] following of the elongation during test. Six different shapes of dies for *clicking out* dumbbell specimens are currently permitted by ASTM D412. Advantages of the dumbbell specimen include its easy preparation, breakage in a predetermined area (usually), means for following the elongation, and the immense accumulated background and specifications for such data. Many styles of grips are available so that specimens of widely varying properties may be held properly. These include air clamps that maintain constant clamping pressure. The straight-sided rectangular specimen is least preferred for ultimate strength and elongation tests because of its tendency to break in the clamps. It may be required, however, because of the size or shape of the sample, e.g., from tubing or electrical insulation. Also the clamp separation can be used as a fair measure of the elongation.

Two sizes of ASTM rings and two sizes of ISO rings are died out from flat sheets. For testing, the ring is held by lubricated rollers in the machine so that the stress is equalized. Advantages of this specimen include the possibility of using clamp separation as a good measure of elongation and distribution of stress over a greater length than in the dumbbell so as to include more possible defects. This method for measuring elongation becomes important when autographic recorders are to be used and in tests at high speed or at other-than-normal room temperature or atmosphere. A disadvantage, especially for low-elongation tests, or high-modulus compounds, is that the inside of a ring cut from a flat sheet is stretched more than the outside.

Three types of rings have been used to reduce the disparity of stress across the width of a ring specimen cut from a flat sheet. The disparity arises if the width is greater than the thickness so that the ring has to straighten to in-test configuration.

1. A ring like the cut ring type 1 in ASTM D412 in which the width is no greater than the material thickness, so that a cut edge rather than a molded edge rides on the pins in a tensile test.
2. Oval rings, such as those used by Myers and Wenrick,[17] which are cut to approximate the shape attained during tensile testing.
3. Rings cut from tubing rather than a flat sheet.

A comparison of results from dumbbell (or tensile bar), ring, and oval[18] showed agreement for values of stress and strain above 100% strain but below failure. Each type of specimen had important limitations, however. The ring or oval is well adapted to computerized data collection.

Tensile stress is calculated as the ratio of observed force to the cross-sectional area of the unstretched specimen. Elongation for straight and dumbbell specimens is given by:

$$\text{Elongation, percent} = \frac{L - L_0}{L_0} \times 100$$

where L = observed distance between bench marks on the stretched specimen and L_0 = original distance between bench marks.

In order that stress can be read directly in megapascals or pounds per square inch, many tensile machines have provision for adjustments to compensate for the varying thickness (gauge) of different specimens. In the formerly common pendulum type machines, this was done mechanically by positioning a weight along the calibrated pendulum arm. The load-cell instruments have electronic compensators.

Tensile stress at rupture is usually higher for a specimen having a small cross-sectional area than for one having a larger cross-sectional area and higher for a specimen having a short test section than for one having a long section.[10] The reason is the same for both cases. Rubber in tension normally fails at a flaw, which may be caused by the die used to cut the specimen, by a region of poor dispersion, by porosity, by inclusion of foreign matter, or by any of a number of other reasons. Obviously, an increase in the amount of rubber being strained increases the chance of a flaw being in the strained area. For this reason, comparison of results from different types of specimens should be made with caution.

The standard speed for tension-testing of rubber is at a machine jaw or spindle separation rate of 500 ± 50 mm (20 ± 2 in.) per minute. A standard speed is necessary because all tensile properties vary with change in elongation rate. Of course, if a vulcanizate is to be used at a high deformation rate, it should be tested at a high rate. As the rate is increased, the modulus increases, the ultimate elongation decreases, and the tensile strength may either increase or decrease. The rate of change of each of these properties with change in rate of elongation depends upon both the test temperature and the glass transition temperature (T_g) of the material; the effect becomes less as the test temperature is moved farther above T_g. The effect of rate of elongation is related to stress relaxation in the specimen. There is time for more stress to relax during elongation at low rates than at high rates. As the glass transition temperature is approached, stress relaxation becomes progressively slower so the test piece registers higher stress (at a given rate of elongation) than it would at higher temperatures. This point is discussed in more detail in the section "Glass Transition Temperature."

An illustrative stress–strain diagram for rubber is given in the first curve of Fig. 5.2. The small hump in the curve near the origin is characteristic of most rubber compounds. The slope at the origin, if accurately measured, is directly related to the hardness of the material. Stress, in this case, is load per unit original cross-sectional area. If the attained cross-sectional area were used, the indicated stress would be much higher. Rubber does not have a yield point in

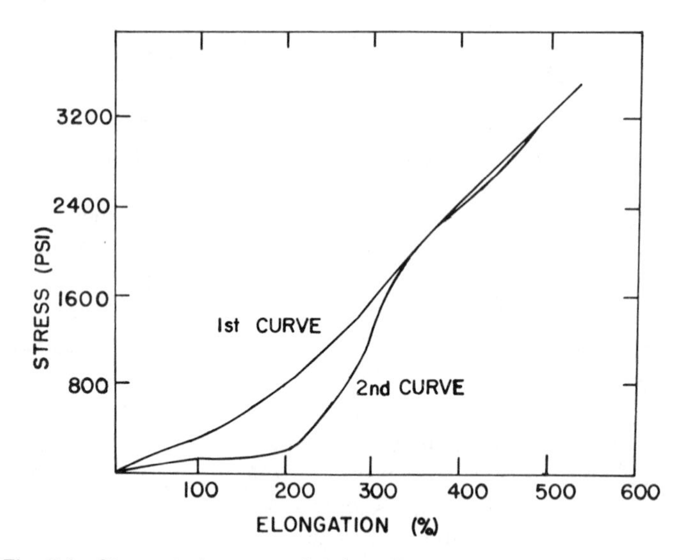

Fig. 5.2. Stress-strain curves showing effect of prestretching. (Ref. 12)

the conventional sense as associated with metals, so that its stress continues to increase as the strain increases until rupture. If the material crystallizes under strain, as exemplified by gum natural rubber, the slope of the stress–strain curve becomes very high near the breaking point.

The slope of a curve, such as that in Fig. 5.2, is a modulus in the engineering sense, since it is the ratio of stress to strain. Because the slope is not constant, however, two kinds of moduli should be distinguished. The slope of a straight line from the origin to a point on the curve at a particular elongation may be called a *secant modulus* at that elongation, whereas the slope of a straight line tangent to the curve at any point may be called a *tangent modulus* at that point.

Within a given rubber system, the tensile strength predicts tire-tread wear to some degree; the area between the stress–strain curve and the strain axis is even more indicative. The ultimate elongation values, both original and after aging, are indications of fatigue life. The compound with the higher elongation at failure will almost always give the better fatigue resistance. The rate and extent of the upsweep at high elongation is an indication of orientation or crystallization effect.[20]

Although tensile data, like all physical testing data, are usually shown as plotted through exact points, they should, in reality, be plotted as bands or belts to allow for error and statistical variation. This implies that in obtaining these values we should always use random and multiple sampling. Distribution of tensile data is nearly normal at room temperature, only slightly skewed in the direction of low values.[21] This means that a standard deviation is a good measure of variation. Since it is impractical, however, to test enough specimens of each sample to obtain a reliable standard deviation, recourse may be had to average standard deviations for the type of material and type of test being used. The ASTM method specifies that the middle value obtained from tests on three specimens should be used. This eliminates any effect of abnormally low values associated with defects but gives no information on precision of test or reproducibility of samples. The compounder should have some knowledge of these factors in order to assess a degree of confidence in the results.

The stress–strain curve is almost always obtained on a specimen which has not been stretched previously. This is a unique curve which cannot be repeated on subsequent stretchings because of material changes that are only partially reversible. These changes are more evident in compounds containing a reinforcing filler than in gum compounds. Fig. 5.2 shows an original stress–strain curve and a second curve obtained by prestretching a duplicate specimen three times to 300% elongation. Up to 300% elongation, the second curve shows lower stress, but beyond that elongation, it almost coincides with the first curve. Neither the ultimate tensile strength nor elongation is much affected. This behavior has been called the *Mullins effect*, after an early investigator, or *stress softening*. Many of the bonds that are broken in the first stretch will reform if the specimen is rested, more rapidly if it is heated.

Tensile strength, of natural rubber at least, drops abruptly at a critical temperature somewhere between 40°C and 130°C. This has been related to the size of naturally occurring flaws.[21]

EVALUATION OF RATE AND STATE OF CURE

Tensile Methods

After an *induction time* at curing temperatures, a vulcanizable compound starts to stiffen because of crosslinking. The crosslinking proceeds at a rate determined by the stock composition and the temperature. Every physical property of the stock is affected as the cure progresses, but at different rates, as is illustrated by the tensile properties in Fig. 5.3. The average *rate* of cure might be considered as the inverse of the time to reach a full cure after the induction period. A fast cure is desirable, but this must be balanced against safety from *scorching* (premature stiffening from accumulation of heat history).

The *optimum cure* for a rubber product is the one which optimizes the properties desired in that product. This has traditionally been judged by use of tensile strength, modulus, and ultimate elongation data to estimate a best *technical* cure. Standard methods for performing these tests are given in ASTM D412, but interpretation of the results to establish optimum cure is not so well standardized.[22]

In a series of cures, the time that gives the maximum tensile strength has often been taken as optimum. On this basis, the best cure for the SBR compound

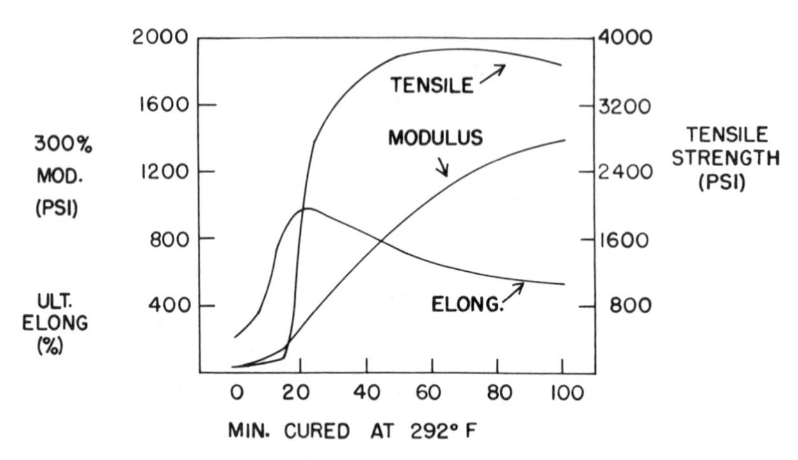

Fig. 5.3. Effect of state of cure on tensile strength, modulus, and elongation of a butadiene-styrene compound tested at 25°C (77°F). (Ref. 13)

in Fig. 5.3 might be chosen at about 65 minutes. Many prefer to select a somewhat lower time since product aging in effect causes an increase in cure, so about 45 minutes might be chosen.

The time at which a modulus curve bends sharply away from the modulus axis may also be selected as optimum, especially for compounds such as those based on SBR in which the modulus does not reach a maximum but continues to increase as the cure increases. For the example in Fig. 5.3, a cure time of about 70 minutes might be selected by use of this index. Another method[23] is to draw a straight line from the origin to a point on the modulus curve at a long cure time, such as 100 minutes in the figure. A second line is then drawn parallel to the first and tangent to the modulus curve. The time indicated at the point of tangency, again about 70 minutes for the example, is taken as optimum. As in most tensile testing, first-stretch modulus is normally used. In reinforced compounds, this contains contributions from filler-polymer interactions—a fugitive structure which breaks down on repeated stretching (Mullins effect). Perhaps, then, modulus measurements for determining state of cure should be made on specimens which have been stretched repeatedly beyond the elongation used in the modulus test.

When the modulus shows either a peak or a plateau at its maximum value, the cure time required to reach 90% of that value is often taken as optimum. If the modulus continues to increase over the normal range of cure times, however, the 90% index cannot be read directly. In such a case, it has been suggested that another series of cures be made at a temperature high enough to get a modulus curve showing a peak or a plateau. The time for 90% of full modulus on this curve can be translated to time at the lower temperature by the approximate rule that curing time is doubled by a 10°C (18°F) temperature decrease.

Ultimate elongation is usually considered along with tensile strength and modulus in selecting an optimum cure time. For most compounds, the breaking elongation reaches a well-defined maximum at quite low curing times, about 25 minutes in the illustration. If it drops off sharply beyond the peak, this may be a strong factor in selecting an optimum cure since a vulcanizate which breaks at low elongation is usually undesirable.

Properties other than tensile strength, modulus, and elongation can also be used to determine optimum cure. In fact, the property of greatest consequence in the proposed application should be used, since different optimum cures will generally be obtained when based on different properties.

Cure Meters

Instead of curing each test compound at separate ranges of temperatures and making separate tensile tests, many laboratories use a cure meter, of which several types have been developed.[24] Modulus change is monitored during the cure in these instruments. Many properties are obtainable from the data; the most common being minimum and maximum stiffness, scorch time, cure time to 90% or 95% of maximum stiffness, and a cure-rate index.

The most widely used cure meter is the oscillating-disk rheometer (ASTM D2084), in which the specimen is contained in a sealed test cavity under a positive pressure and maintained at an elevated temperature. A biconical disk, embedded in the specimen, is oscillated through a small arc. The autographically recorded force is proportional to the shear modulus of the rubber. An envelope of a typical cure curve is given in Fig. 5.4. The minimum torque (M_L), maximum torque (M_{HF}), scorch time (t_{s2}) and time to 90% cure ($t_c(90)$) are indicated. Also obtainable[25] are initial viscosity at zero time, minimum viscosity, and reversion (time after M_{HF} to reach 98% of M_{HF}).

Many illustrations of rheometer use have been described[26,27] as well as of the Viscurometer,[28] which works on a similar principle. The biconical rotor used in each instrument produces a shear rate that is uniform over the entire rotor surface. The Vulkameter[4] subjects the specimen to linear shear rather than the torsional shear applied by the rheometer. In instruments such as the Mooney viscometer (ASTM D1646) that use a flat disk, the shearing rate is near zero at the central shaft and a maximum at the disk periphery. If any cure meter is used as the sole instrument of cure, there remains the danger that other important properties will be overlooked.

Other Methods

Any property of a rubber which changes in a regular manner with time or temperature of cure may be used to evaluate the state of cure. Methods based on different properties, however, will not in general give results that agree with

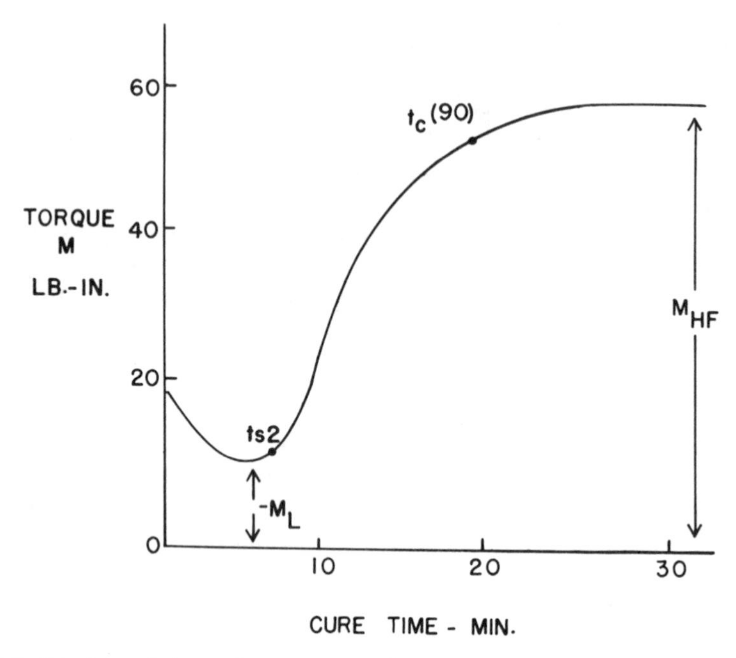

Fig. 5.4. Envelope of rheometer cure curve that attains an equilibrium torque (ASTM D2084). Torque may be obtained in nN/m as well as lb/in.

each other. For example, a sample that is overcured for best tear resistance might be grossly undercured for optimum hysteretic properties. Indices that have been used[22] include free and combined sulfur, tensile product (tensile strength multiplied by the ultimate elongation), permanent set, equilibrium modulus, swelling in solvents, crosslink density, and hardness.

HARDNESS

Hardness, as applied to rubber, may be defined as the resistance to indentation under conditions that do not puncture the rubber. Different instruments designed to measure hardness do not usually agree well with each other for any of several reasons: definition of scale end points, shape and size of indentor point, total load applied, rate and time of load application. Hardness is a property of rubber that is usually expressed in terms of instrument parameters rather than basic units. A high-modulus rubber is, of course, also hard, but the relationship is not often required.

The spring-loaded pocket durometer is the most common instrument for measuring hardness of elastomers. The durometer (ASTM D2240) in particular is known and used world-wide. In this instrument, the scale runs from zero-hardness for a liquid to 100 for a hard plane surface such as glass. The Type A durometer is used for soft stocks, up to a reading of 90. Above 90, the Type D durometer, having a different indentor shape and different stiffness spring, is used. One reason for the popularity of this instrument is its portability for field use and adaptability to fairly irregular surfaces. Difficulties arise, however, in reproducibility of results by different operators; differences of 5 units and practical tolerances of 10 to 12 points are not rare.

Better reproducibility is obtained by dead-weight loading, as in ASTM D531, than in the spring-loaded, hand-held durometer. These particular instruments, however, are not as widely used as is D1415, in which results are expressed as International Rubber Hardness Degrees. These units are directly related to Young's modulus while still being approximately equivalent to durometer readings. The test consists of a measurement of the difference between the depths of penetration of a ball point into the rubber specimen under a small initial load and a large final load. This is essentially the same as the ISO R48 method.

The initial reading of a durometer depends on the rate of loading and the final reading on the duration of loading, both because of creep in the rubber. Some portable instruments, such as the Rex or some Shore models, register the maximum reading. Most bench models, however, are designed to be read after some loading period such as 30 seconds. Since different materials have different creep characteristics, it is very important to precision that the rate of loading be standardized and the duration of loading be specified.

Since so many hardness meters exist, the choice of which to use is difficult. The Shore is undoubtedly the most widely used and understood but is not as precise as bench models. For referee purposes, the ASTM committee on physical testing of rubber recommends D1415 for ordinary soft-rubber compounds and D530 for hard-rubber products.

In ordinary SBR rubber compounds, hardness increases with increased cure. In natural-rubber or butyl-rubber compounds, hardness increases to a maximum and then decreases because of reversion as the cure time is increased.

TRANSITIONS IN ELASTOMERS

Types of Transitions

Two types of reversible, temperature-dependent transitions are important concepts in rubber technology: first-order and second-order. In a first-order transition, there is an abrupt change in level of a property, such as volume, specific heat, or modulus, within a small temperature range. In a second-order transition, there is no abrupt change of level, but rather a change in slope of the line representing that property plotted against temperature. For example, there is no change in volume at a second-order transition temperature, but a definite change in coefficient of thermal expansion. First-order changes are usually attributed to crystallization and second-order changes to vitrification, i.e., to becoming glass-like. More than one second-order transition temperature may exist for a given polymer, however, only one of which is truly a glass transition (T_g). The main difference between a rubber and a rigid plastic is that rubber has a glass transition below room temperature while a plastic has a glass transition above room temperature.

As is detailed in ASTM D832, crystallization, vitrification, and simple temperature effects have many distinguishing characteristics, each of which could lead to a definitive test. Some of these tests are described in the following sections. First, however, let us consider some of the principles involved.

Effect of Rate of Deformation

According to the kinetic-molecular theory of rubber elasticity, the modulus of a rubber increases as its temperature is increased. This presumes, however, an equilibrium deformation, which is probably unattainable; some creep inevitably persists. Modulus tests made at a finite deformation usually, but not always, show a decrease in modulus with an increase in temperature. This occurs because stress relaxation is a time-dependent process and more time is required for a given amount of stress relaxation as the temperature is lowered toward the glass-transition temperature. An effect of this principle is that if a deformation is forced at a greater rate than can be accommodated by the elasticity of the specimen, it breaks. A low-temperature brittle point is, therefore, very sensitive to rate of deformation. Another consequence is the difficulty of getting the same results from different types of low-temperature testing equipment, which usually have different time rates of sample deformation. An illustration of this difficulty was demonstrated in a comprehensive study of stiffness testing at low temperatures,[29] which included the following reasons for variation in results from different instruments and different laboratories in an interlaboratory program:

1. Rate of testing.
2. Geometry of the apparatus.
3. Nonlinear characteristics of the elastomers.
4. Variance in the way the materials reacted to temperature.
5. Variations in testing conditions.
6. Variations in methods.

Differential Thermal Analysis (DTA)

Differential thermal analysis is a technique for studying the thermal behavior of materials as they undergo physical and chemical changes during heating and cooling. The name is derived from the differential thermocouple arrangement, consisting of two thermocouples wired in opposition as shown in Fig. 5.5.[30]

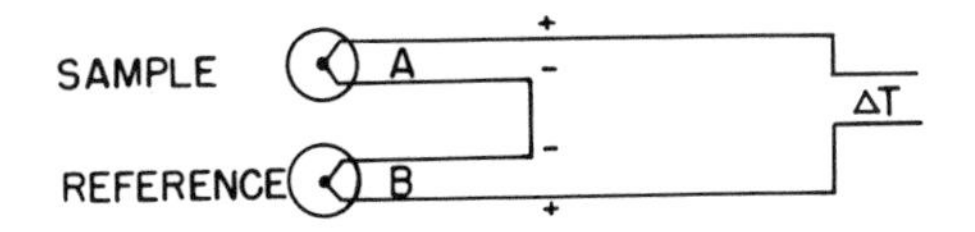

Fig. 5.5 Differential thermocouple.

Thermocouple A is placed in a sample of the material to be analyzed. Thermocouple B is placed in an inert reference material, which has been selected so that it will undergo no thermal transformations over the temperature range being studied. When the temperature of the sample equals the temperature of the reference material, the two thermocouples produce identical voltages and the net voltage output is zero. When sample and reference temperatures differ, the resultant net voltage differential reflects the difference in temperature between sample and reference at any point in time.

Physical properties of elastomers that may be studied by DTA include first-order transitions (crystallization) and second-order transitions (e.g., glass transition). In either case, the sample and a material of comparable heat capacity and thermal conductivity, each containing a thermocouple, are cooled rapidly in the DTA chamber to below the suspected transition temperature. The chamber temperature is then raised at a given rate and a plot obtained of ΔT against sample temperature. At transition points, the sample will interchange heat with the chamber without a change in its own temperature until the transition is complete, giving a plot of the nature shown in Fig. 5.6 to illustrate crystal melting. In crystal formation, the peak would be above the baseline while at a second-order transition point, there would normally be a change in baseline to a lower level at the higher temperature.

In addition to its use in locating transition temperatures, DTA has many applications to rubber testing which might be considered chemical rather than

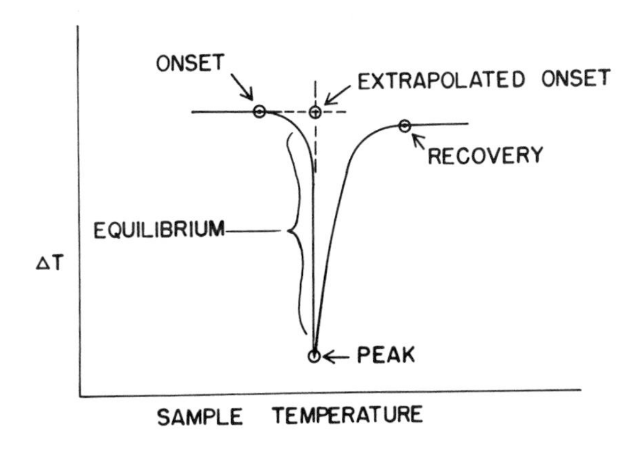

Fig. 5.6. Thermogram for melting of crystals in an elastomer.

physical. These include identification, composition, solvent retention, thermal stability, oxidative stability, polymerization, curing, and thermochemical constants.

Tests for Crystallization

The DTA test discussed in the previous section is usually satisfactory for measuring the temperature at which a polymer crystallizes most rapidly. Sometimes, however, crystallization rates are too slow to fit well into the normal time scale of the test, for example, the rate for unstretched natural rubber. In such cases, the specimen may be held for long times in the temperature range of interest with periodic measurements of some property such as modulus. ASTM D797 is an example of such a test, one in which simple beam stiffness in bending is the measured property. The Young's modulus increases, of course, with an increase in degree of crystallinity.

Crystallization rate of an elastomer is increased by stretching, a fact which is used in the T–R (temperature–retraction) test for crystallization (ASTM D1329). In this test, the specimen is first elongated to 250–350%, and then frozen to a practically nonelastic state and released. As the temperature is raised at a uniform rate, the recovery attained is measured at regular temperature intervals. The difference between the temperature of 10% retraction (TR10) and that of 70% retraction (TR70) increases as the tendency to crystallize increases.

The DTA method for evaluating crystallinity is based on a thermal property: heat of crystallization. The ASTM methods are based on a mechanical property: modulus. An optical property, the polarizing effect of crystals, is used in the photomicroscope method. A density change is the basis for dilatometer methods and X-ray reflection properties for X-ray methods; diffuse crystals give rings in the diffraction patterns and aligned crystallites give spots. This roster of tests is a good illustration of the principle that a measurement of any material property that changes in a regular manner with change in test conditions may serve as the basis for a test.

Low-temperatures Stiffness

Rubber stiffness, as measured by nonequilibrium deformation, increases with a decrease in temperature, but the relationship between modulus and temperature is by no means linear. It is rather characterized by plateaus such as shown in Fig. 5.7. Outside the "rubbery" plateau, a rubber cannot serve its intended purpose. Various tests have been devised to measure the temperature at which an elastomer becomes inserviceable. For some uses, a "leathery" response is still satisfactory, but for other uses, a lower limiting modulus must be specified. One proposed value[31] is a Young's modulus of 69 MPa (10,000 psi) as measured with a 10-second loading time. Since a value in this region occurs at about

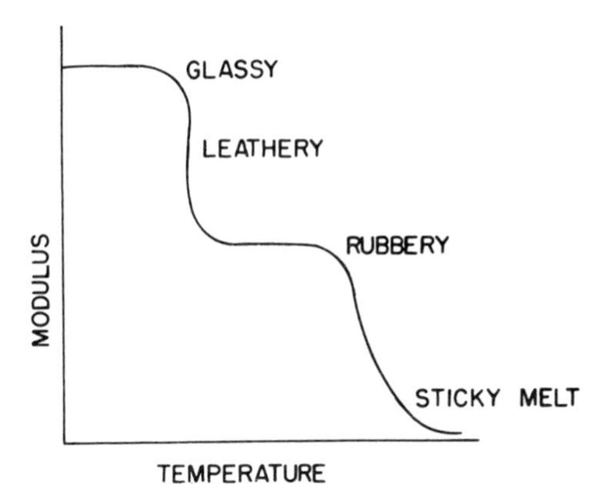

Fig. 5.7. Characteristic temperature-modulus curve for polymers. (Ref. 13)

30°C above the glass transition temperature, T_g is a possible serviceability index. More direct measurements are usually chosen, however.

Perhaps the most widely used low-temperature test for elastomers is that based on the Gehman torsional wire apparatus, ASTM D1053. A schematic diagram of the working parts of the instrument is shown in Fig. 5.8. Multiple specimens F on the rack D can be rotated to clamp individually to the base of the calibrated torsion wire B. A 180° rotation of the torsion head A rotates the top of the specimen by an amount measured on the protractor. Since the torque is known, the apparent modulus of rigidity is calculable. If tests are made at a series of temperatures, plots of relative modulus as functions of temperature are readily obtained. Commonly used indices are temperatures at which the modulus reaches 2, 5, 10, and 100 times its value at room temperature. The apparatus is also usable for tests of continued stiffening at a given temperature, e.g., crystallization or plasticizer incompatibility.

Another ASTM test for low temperature stiffening is D797, Young's modulus in flexure. The specimen is supported as a simple beam and a known load applied centrally. Young's modulus is calculable from the specimen and support geometry, the added load, and the resultant deflection. Modulus is normally plotted as a function of temperature. Indices such as the temperature at which the modulus reaches 69 MPa (10,000 psi) can be read from the graph. This apparatus is also suitable for studies of crystallization or low-temperature creep.

Other ASTM tests for low-temperature stiffening include the Temperature–Retraction (TR) test, D1329; and a bend test for rubber-coated fabrics, D2136. Many other tests have also been used commercially, ten of which were included in the interlaboratory study of reference 29. No "best" test resulted from this study. Poor agreement among the results from different methods forced the conclusion that the term "modulus" as applied to rubberlike materials is rather vague unless the method of obtaining that modulus is also given.

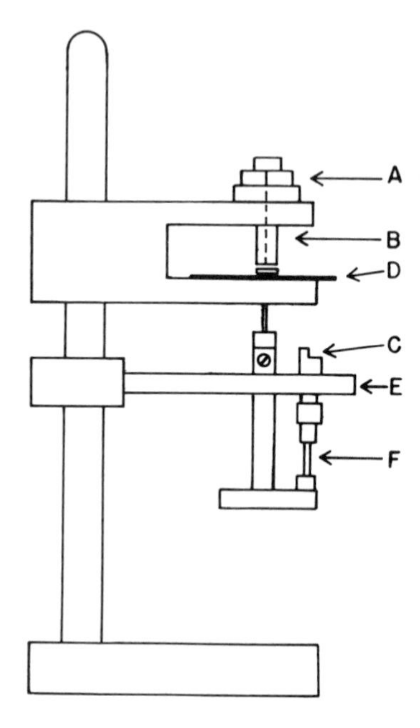

Fig. 5.8. Gehman low-temperature testing apparatus (ASTM D1053). A-torsion head, B-torsion wire, C-clamp stud, D-movable protractor, E-specimen rack, F-specimen.

Low-temperature Brittleness

When rubber is forced to deform faster than its relaxation time permits, it breaks. Extremely high speed would be required to produce a brittle fracture at room temperature, but as the temperature is lowered, a point is reached at which a specimen breaks under a given deformation at a given speed. In ASTM D746 (Vol. 9.02), specimens 1.9 mm (0.075 in.) thick are bent sharply by impact with a striker arm moving at 6 to 7 feet per second. When multiple specimens are used, care must be taken that the breaking energy is not high enough to slow the striker arm below 6 feet per second *after* the impact. Otherwise, too low a brittle temperature would be indicated.

As with any low-temperature test, either a gaseous or liquid heat-transfer medium may be used if it has been shown to give equivalent results on specimens of a material having similar composition to that of the test material. This means that a liquid medium should not corrode or swell the rubber and that the rubber should not crystallize in the range of test temperatures. The longer times required for thermal equilibrium in the gaseous medium would promote more crystallization than would the liquid medium.

Low-temperature stiffening and low-temperature embrittlement do not correlate well enough with each other to permit either index to be inferred from the other. The choice of test to be run must be made on the basis of anticipated service conditions. Brittle fracture is visually distinguishable from tensile failure in that the failure surfaces have a glassy rather than a ragged appearance.

Other Low-temperature Tests

The principal "rubbery" characteristics that an elastomer must maintain to remain serviceable at low temperatures are a low modulus and freedom from embrittlement. Another property that sometimes becomes important, however, is that of recovering original dimensions after removal of a deflecting force at low temperatures. This requires a test such as ASTM D1229. In this test, the specimen, in the form of a cylinder, is compressed 25% at room temperature, cooled to the test temperature, and kept there for 22 or 94 hr. It is then released and the thickness measured after 10-second and 30-minute recoveries at the test temperature.

Compression-set measurement is often used to evaluate seal retention at low temperatures, although a more direct test would be a stress-relaxation measurement. This is analogous to the set test except that compression is maintained and decay of restoring force with time is measured. Crystallization is especially dangerous for seal retention, but low-temperature stiffening and second-order transition are also important.

Tension-set measurement at low temperature is not as common as compression-set because rubber is seldom used in tension, but the TR test (ASTM D 1329) could be used for this purpose. The room-temperature tension-set test of ASTM D412 could also be adapted to low-temperature use.

Most static rubber properties measurable at elevated or room temperatures could also be made at low temperatures with proper regard for the time scale of deformation. In a hardness test, for example, the loading time would be much more critical than it would be at higher temperatures. Dynamic properties are hard to measure at low temperatures because of the complications introduced by heat generation. A necessary decision is whether to measure the property during the first few oscillations before appreciable heating has occurred, or after thermal equilibrium has been reached.

DYNAMIC MECHANICAL TESTS

None of the tests that have been developed for measuring dynamic properties of rubber can be regarded as universally used. The examples cited here are some for which operating procedures have been standardized, but this does not necessarily imply widespread usage. Extensive reviews of this subject have been written which include much theoretical basis for the types of tests that have been developed.[9,12,24,25]

Terminology for dynamic-properties testing has been fairly well standardized. The definitions given in ASTM D2231, for example, are in essential agreement with those adopted by ISO TC/45. Four more general terms have also been defined:[12]

Resilience: In a rubber-like body subjected to and relieved of stress, the ratio of energy given up on recovery from deformation to the energy required to produce the deformation. Resilience for these materials is usually expressed as a percent.
Hysteresis: The percent energy lost per cycle, or 100 minus the resilience percentage.
Dynamic Modulus: The ratio of stress to strain under vibratory conditions. It is usually expressed in pascals or pounds per square inch for unit strain.
Damping: Refers to the progressive reduction of vibrational amplitude in a free vibration system. Damping is a result of hysteresis, and the two terms are frequently used interchangeably.

A set of consistent testing procedures has also been formulated. Measured values of dynamic properties of rubber may be influenced by the particular apparatus and method used, so it is not sufficient to specify a certain dynamic modulus, for example, unless the details of its measurement are also specified. The types of tests used can be classified generally as impact rebound, forced vibration in both resonance and nonresonance conditions, and free vibration.

Rebound Tests

The simplest test for resilience is the falling ball rebound.[25] If the drop height of the steel ball is divided into 100 equal parts, the rebound height is equal to the resilience. The same principle is used in the Bashore Resiliometer (ASTM D2632), a compact instrument in which the plunger follows a guide rod in its fall and rebound.

In a pendulum rebound test, the specimen, held at the rest position of the pendulum (zero degrees), is impacted by the center of percussion of the arm. The angle of rebound is followed on a scale and resilience calculated by the formula:

$$R = \frac{1 - \cos\text{ (angle of rebound)}}{1 - \cos\text{ (angle of fall)}} \times 100$$

Many designs for the pendulum rebound apparatus have been used. In the Goodyear-Healy design (ASTM D1054), provision is also made for measuring the depth of penetration of the pendulum head. European laboratories prefer the

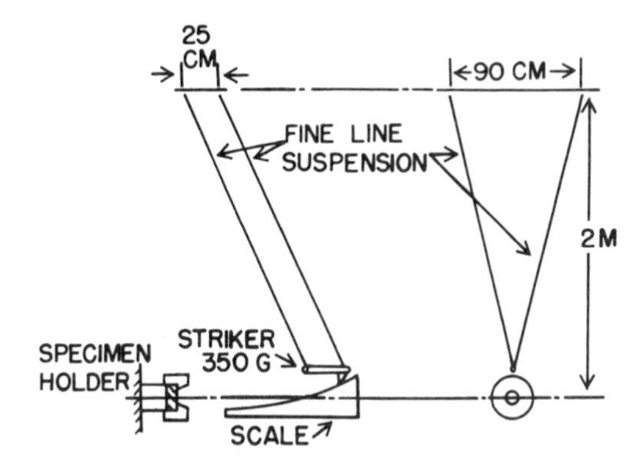

Fig. 5.9. Lüpke pendulum.

Lüpke pendulum, adopted by ISO TC/45, or the Schob pendulum. The Lüpke design is interesting in that the impacting weight is held by four flexible suspension lines (Fig. 5.9). This minimizes absorption of energy in the instrument. As with any rebound test, however, care must be taken that the specimen support is very rigid.

Interlaboratory tests have shown that high correlation should not be expected between results from two different types of rebound testers. Results are influenced by plunger weight, design of the impacting head, drop height, penetration, and energy absorbed in the apparatus. In any rebound test, the specimen should be preconditioned with about six impacts before a reading is taken. After this, the rebound height is usually stabilized.

Rebound tests are very sensitive to bulk temperature of the specimen but, fortunately, not to surface temperature. This allows tests to be made in the open laboratory of specimens conditioned at either low or elevated temperatures, with very little error. As shown in Fig. 5.10, as the specimen temperature is lowered, the resilience decreases to a minimum; then increases again. This minimum occurs when the rubber is in the "leathery" state of Fig. 5.7.

Rebound is increased by an increase in modulus or by a decrease in hysteretic loss. In fact, both the storage modulus and loss modulus (see section on forced vibration) may be calculable from rebound data coupled with contact time of a pendulum with the vulcanizate specimen.[34,35] Dynamic properties are usually measured, however, by the more direct methods of free or forced vibration.

Free-vibration Tests

In free-vibration tests, the rubber specimen forms the spring in a mechanical system with inertia chosen so that a damped oscillation of the desired frequency results from release of a deformation. Deformation in compression, shear, torsion, tension, or torsion plus extension have been used. Perhaps the most com-

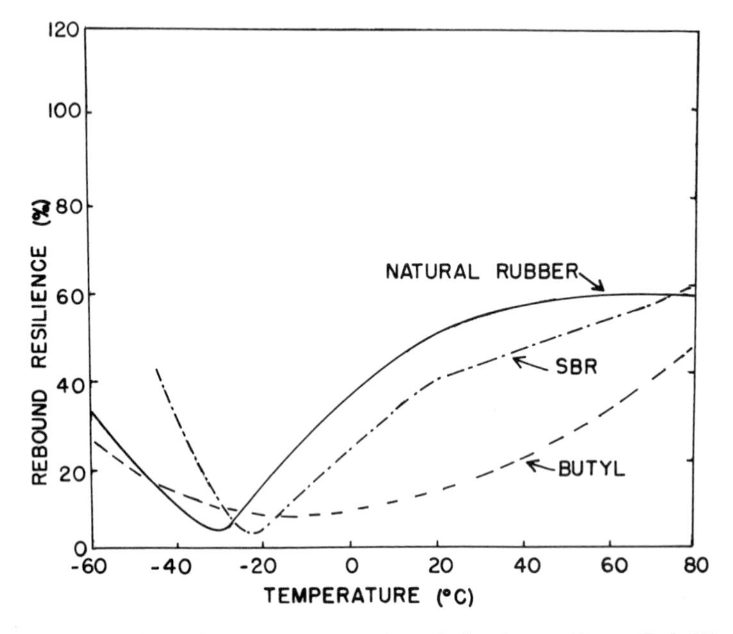

Fig. 5.10. Effect of temperature on rebound of various rubbers. (Ref. 33)

mon instrument of this type is the Yerzley Oscillograph, ASTM D945, of which a sketch is shown in Fig. 5.11. A balanced beam is supported at its center and designed so that its motion is controlled by a rubber specimen strained in either compression or shear. A pen mounted at one end of the beam draws a trace on a recorder drum rotating at a constant speed. The natural frequency of free vibration is measured from this trace and used to calculate the effective dynamic modulus K of the rubber when deformed in either compression or shear.

A number of other properties can be obtained from the trace. For example, the average ratio of heights of successive oscillations gives an arbitrary measure

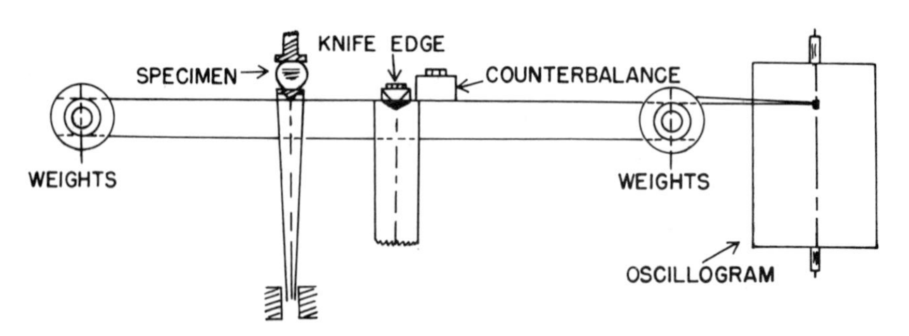

Fig. 5.11. Yerzley oscillograph (ASTM D945).

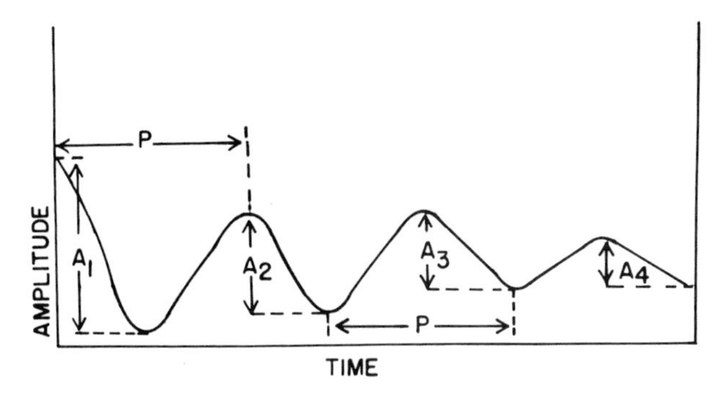

Fig. 5.12. A typical damped oscillation curve.

of resilience of the rubber which, expressed as a percentage, is termed the *Yerzley resilience*. In a trace such as that of Fig. 5.12, the Yerzley resilience equals

$$\left(\frac{A_3}{A_2} + \frac{A_4}{A_3}\right) 50$$

Another type of free-vibration instrument is the torsion pendulum,[13,33] a classic for studying dynamic properties of rubber. Many designs of apparatus have been used; that in Fig. 5.13 will illustrate the principles. The bottom of the ribbon-shaped specimen is rigidly attached while the top is fastened, either solidly or through a torsion wire, to a structure having an adjustable moment of

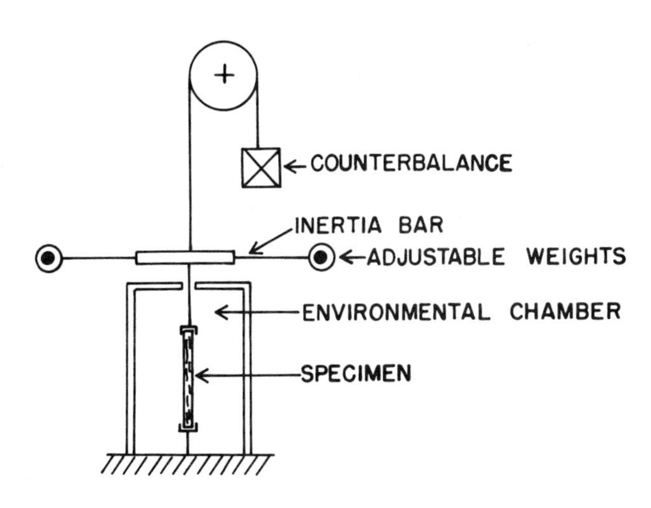

Fig. 5.13. A torsion pendulum device.

inertia. The weight of this device is counterbalanced through a cord having negligible torsional stiffness.

When the test is performed, the inertial system is displaced through a small angle (about 1.5°) and released. The shear modulus may be calculated from the period;[13] the shorter the period the greater the modulus. A measure of damping is given by a quantity called the *logarithmic decrement* (see Fig. 5.12):

$$\Delta = \log_e \frac{A_3}{A_2} = \log_e \frac{A_2}{A_3} = \ldots$$

An objection to any free-vibration method is that amplitude is not maintained so that a calculated modulus, which depends on amount of deformation in a polymer, is not consistent from one material to another. This can be overcome in the torsional pendulum by exerting a mechanical moment (by magnetic means) that is equal in magnitude but reversed in sign to that caused by internal friction in the specimen. The required force is a measure of damping. In effect, however, this then becomes a nonresonance-forced vibration test at low frequency.

Forced-vibration Tests

A rubber is placed in forced vibration when the disturbing force is periodic and continuing. The system is in a *steady-state* condition when the hysteresis energy exactly balances the impressed energy and in a *resonance* condition when the exciting frequency equals the natural frequency of the spring and mass system. Many forced-vibration methods have been developed with perhaps the greater number being resonant.[13,32,33,36] These include forced vibrators in both shear and compression, reed vibrators where several modes may be excited during a single test, and resonant-beam tests for low-frequency studies. Calculations are simpler in resonance than in nonresonance conditions.

The theory of forced-vibration testing is too extensive to enter into here; the reader is referred to ASTM D2231 for pertinent definitions, concepts, and factors affecting dynamic measurements. Each type of dynamic mechanical test yields a different damping term. In order to compare data obtained by different methods, it is necessary to convert all to a common basis. This is done by using complex moduli, G^* and E^*:

$$G^* = G' + iG'' \text{ (shear)}$$

$$E^* = E' + iE'' \text{ (Young's)}$$

G' and E' are the "real" parts of the shear modulus and Young's modulus respectively, i.e., the elastic parts. These are sometimes called storage moduli. The quantity i is equal to $\sqrt{-1}$, so G'' and E'' are the "imaginary parts" of the moduli, i.e., those due to damping forces, often called the *loss moduli*. A very useful damping term, called the *dissipation factor* or *loss tangent*, is defined as G''/G' or E''/E'. The dissipation factor is proportional to the ratio of energy dissipated per cycle to the maximum potential energy stored during a cycle.

Dynamic-properties testing provides a good example of the modern systems approach to physical testing in which a single apparatus performs the tests and computes the results, or different modules can be used in a single basic unit to extend its capability over a number of different tests. Commercial units are available, for example, which both test and compute spring rate and damping coefficient for selected values of mean load, cyclic amplitude, frequency, and temperature. This test process is based on the fundamental differential equation for a spring-mass system:

$$F = Kx + C\frac{dx}{dt} + M\frac{d^2x}{dt^2}$$

where x is the instantaneous displacement of the mass from its rest position, F is the displacing force at that time, K is the spring rate, C is the damping coefficient, M is the vibrating mass, and t is the time at which the displacement is measured. When a load cell is used, only its platen, the specimen, and its clamp constitute the mass, so M can be ignored in the above equation. The fundamental material properties K and C are then easily calculable in the nonresonant electrohydraulic closed loop system.

Other systems are available that can perform an amazing variety of tests and basically require only changing of clamps. The commercial MTS electro-hydraulic system, for example, includes the following tests on a single basic apparatus:

1. Conventional stress-strain relationships; tension or compression tests, and so forth.
2. High-strain-rate testing (tension or compression).
3. Short-term- or long-term-creep testing.
4. Fatigue testing—constant amplitude; programmed amplitude, mean level and/or frequency; and random amplitude/frequency.
5. Service simulation tests (sample trace testing from service records).
6. Component testing.
7. Low cycle fatigue using constant strain, constant deflection, constant load.
8. Testing in other loading modes (e.g., bending, torsion).
9. Environmental tests.
10. Biaxial and triaxial testing.

In tests for dynamic properties either the amplitude or frequency or both are kept low so that heat build-up in the specimen is minimized. In many European methods the test results are obtained during the first few cycles before the specimen heats up much; the argument being that the specimen is then at a known uniform temperature. American methods, however, usually specify attainment of thermal equilibrium before taking readings, believing that a slight temperature gradient is not as important as being sure that stress softening (Mullins effect) has reached an equilibrium.

Dynamic tests can distinguish differences between compounds that "static" or low-speed tests cannot, because the frequency of the dynamic test is such that there is less time for creep or stress relaxation. Also, time dependence can be evaluated by running at different frequencies. Furthermore, most rubber articles are used in dynamic service.

Constant amplitude tests give relatively higher heat build-up for stiff compounds than do tests at constant maximum force. This is because, in constant amplitude tests, the heat generated is the product of hysteresis and dynamic modulus; in constant maximum force tests, it is a direct function of the ratio of hysteresis to dynamic modulus. The dynamic stiffness governs the amplitude in this instance, and thereby the heat generated.[20]

Heat Build-up Tests

Damping properties as measured in rebound or vibration tests provide a measure of heat-generating potential under the usually mild conditions of those tests. More severe tests are used to measure actual head build-up and resistance of the sample to deterioration at the attained temperature.

Perhaps the best-known of these tests are those performed with the Goodrich flexometer and the Firestone flexometer (ASTM D623). In the Goodrich apparatus, a definite compressive load is applied to the cylindrical specimen while a high-frequency cyclic compression of definite amplitude is superposed. The increase in temperature at the base of the specimen is measured with a thermocouple. Specimens may be tested under a constant applied load or a constant initial compression. Change in specimen height can be measured continuously during flexure. Comparing this change in height with the observed set after test enables an estimate of the degree of stiffening (or softening).

In the Firestone flexometer, an eccentric rotary motion is applied to one end of the specimen, which is shaped like the frustrum of a rectangular pyramid, while it is held under a constant compressive load. The time required for a definite change in height of the specimen is determined. At the end of the test, the temperature may be measured by inserting a needle thermocouple into the center of the specimen. This is perhaps the more common use for the apparatus. It can also be used for a *blow-out* test, however, with the use of a heavy load and large throw and running to failure. Either flexometer can be used at elevated temperatures as well as at room temperature.

Flexometers are not recommended for use in establishing purchase specifications because both correlation with service life and reproducibility between different laboratories are uncertain. As with so many other physical tests, comparisons between compounds are more valid when they have similar composition than when they are widely different as, for example, compounds based on different polymers.

Flex Resistance

In compression fatigue tests by the flexometer methods, discussed in the preceding section, rubber failure is mostly due to heat. In fact, the "blow-out" condition is accompanied by pyrolitic decomposition with gaseous products. Tests for fatigue failure caused principally by mechanical action are usually made in the bending mode. The term *flexure* generally means "deformation in bending" although *flexometers* compress or shear a specimen and *flexible foams* are those that are easily compressed. The mechanical action in flexure is, of course, extension on the outside of the bend and compression on the inside. The *neutral plane*, near the center of an isotropic specimen, is neither extended nor compressed.

Flexural fatigue failure in a rubber-fabric composite is usually manifested by delamination. The ASTM test for this type of durability is the Scott Flexing Machine (D430 Method A), in which strip specimens are stretched over a rotatable hub and clamped at a bend angle of 135°. After a prescribed load is applied on one of the clamps, thus placing the specimen in tension, the test strip is pulled back and forth over the hub at about 160 cycles per minute. Separation between fabric and rubber is indicated by melting of a wax coating caused by the excess heat generated at the failure point. This type of test is applicable to either belts or tire bodies, and samples may be specially molded for the test or cut from finished products.

The Du Pont flexing machine, also described in ASTM D430, provides an action similar to that of the Scott machine except that a reverse curvature is also included in the flex cycle. A total of 21 test specimens form a test belt that is run over a series of 4 pulleys arranged in a V-formation. Failure is indicated by visual inspection. Normally, the test can be continued until all 21 specimens have failed.

Flexural fatigue failure in rubber or in the rubber part of composites is usually divided into two categories: crack initiation and crack growth. The Du Pont apparatus can be used for crack initiation tests, in which case failure is indicated by nicks or pinholes in the corrugations that are molded on the rubber face of the specimen.

Crack Initiation Tests

Perhaps the most widely used test for both crack initiation and crack growth, however, is the De Mattia (National Flexing Machine), described in ASTM

D430 for initiation and in D813 for growth tests. Either tensile or flexure deformation may be used for crack initiation tests. In the tensile tests, normal dumbbell specimens are alternately stretched and completely relaxed so that a slight bend is induced. This is an important point because flex life in any type of deformation is sometimes much shorter if zero strain is included in the cycle than if all the flexing is entirely on one side of zero, especially for crystallizable materials.

The De Mattia specimen most commonly used is 6 inches long, 1 inch wide, and 0.25 inch thick, with a groove molded across its width. The flex cycle is such that the specimen is bent almost double at the groove, then almost straightened out; no gross extension occurs. This type of bending, without a mandrel, can allow progressively greater strain concentration as the bending line becomes hotter and weaker. The energy absorbed in the critical area is thus maintained or even increased as the test proceeds. This should lead to more reproducible end points than occur in mandrel tests such as the Du Pont or Scott, where bending energy may actually decrease at a hot or weak point.

Results of crack initiation tests may be expressed in any of several ways: (1) a severity comparison of the various samples at a definite number of flexing cycles; (2) ascertaining the number of cycles required to attain a definite severity rating; (3) comparison of the number of cycles required to attain progressive degrees of severity ratings; (4) after a suitable number of cycles, depending on the compound, examining and rating the specimens according to the degree of cracking by comparison with a set of standard specimens graded 0 (no cracking) to 10 (completely cracked through).[25] The latter two methods give combinations of crack-initiation and crack-growth ratings.

Crack-growth Tests

The ranking of a group of compounds for crack initiation is often quite different from the ranking for crack growth, so separate evaluations are needed. Compounds of SBR, for example, normally resist crack initiation much better than do compounds of natural rubber, but once a crack is started in each, it proceeds much faster in the SBR sample. Some types of apparatus may be used for either test, however. The De Mattia method for testing crack growth is described in ASTM D813. The grooved specimen is used and a crack is started by piercing with a specified tool at the center of the groove. Crack length is measured at regular intervals with the end point being the number of cycles required to extend the crack to 0.5 or 0.75 inches. Alternatively, the average rate of crack growth over the entire period or a certain portion of the period may be reported.

Another standardized test for cut growth is the Ross flexing machine, ASTM D1052, in which the pierced section of the specimen is bent over a mandrel through a 90° angle. The number of cycles for the cut length to increase 100%, 200%, 300%, 400%, and 500% of its original length is reported.

Effect of Test Conditions

Both crack initiation and crack growth are speeded in oxygen, especially in ozone, and slowed in a nitrogen atmosphere. Crack growth of SBR is speeded by increase in temperature, but that of natural rubber may be either speeded or slowed. All cracking tests are constant amplitude, so sample modulus and thickness variations affect actual stresses and hence the rate of cracking.

Flexing tests do not necessarily correlate well with service performance, a situation that occurs with many accelerated tests. For this reason, flexing tests should not be specified unless close correlation can be shown with service.

Advanced Tests for Fatigue

Major rubber laboratories in both industry and academia are developing apparatus and methods for fatigue measurements that are much more sophisticated than those described here. Testing principles are shifting from purely empirical to those based on results of theoretical studies on the nature of fatigue. These are not yet ready for standardization, but some results of the efforts are appearing as commercial machines.

AGING TESTS

Aging tests are performed to evaluate, within a relatively short period of time, the susceptibility of a rubber to deterioration of physical properties because of environmental effects. As with any test in which service conditions are exaggerated for economy of testing time, care is required to ensure that the degree of acceleration is uniform. For example, the temperature of an aging test should not be increased beyond the point where a sudden change in tensile strength occurs unless it can be shown that the same change occurs over a longer time at lower temperatures.

All ASTM aging methods carry the warning that no correlation with natural aging should be implied. The user must determine whether or not useful correlation exists.

Heat Aging

Perhaps the most common aging test is a comparison of tensile and hardness properties before and after heating in air. Standard methods include the Geer oven (ASTM D573), in which all the specimens being tested at a given time are held in a common air space. This is satisfactory if it can be shown that air contamination does not influence the results. Careful cleaning between tests is also needed to avoid contamination.

Cross-contamination among specimens having different compositions may be prevented by using the test-tube method (ASTM D865). Test tubes, holding no more than three specimens each, are heated in an oil bath or an aluminum block. Convective air circulation is permitted through glass tubes inserted in the tube stoppers. Self-contamination is effectively prevented by use of a tubular oven (ASTM D1870) in which heated fresh air enters one end of the tube and is exhausted from the system without being recirculated. Royo discusses effects of air-change rate.[37]

A more severe test is the air-pressure heat test (ASTM D454), in which the specimens are held at 127°C (260°F) and 552 kPa (80 psi) air pressure. This was originally designed for use on inner-tube compounds and is not now widely used. The community aging is an objectionable feature of this test, also.

Aging tests for specific products include ASTM D622 for automotive air-brake and vacuum hose, D1055 for latex foam rubbers, D1056 for expanded rubbers, and D296 for fire hose.

Oxygen Aging

In a method analogous to the air-pressure heat test, oxygen may be used instead of air (ASTM D572). The pressure is 2.07 MPa (300 psi) and temperature 70°C (158°F). Tensile properties are again used to judge the aging effects.

Many instruments have been devised to measure directly the rate of combination of oxygen with rubber: gravimetric (measures increase in weight of rubber), manometric (measures drop in oxygen pressure), and volumetric (measures drop in oxygen volume). These are used more in research and development than in control or specification testing. Oxygen absorption, although usually associated with degradative processes, may, especially in the early stages of cure, contribute to the development of desirable properties. In fact, oxygen may compete with sulfur as a vulcanizing agent. Thus the interpretation of oxygen absorption curves requires study in terms of the experimental conditions.

Ozone is a form of oxygen which has a particularly severe effect on rubber, especially if it is stretched. In ASTM D1149, bent, stretched, straight, or tapered specimens are exposed to ozone concentration of 50 parts per 100 million in air. Effects are judged by visual examination and comparisons with standards. Other tests such as ASTM D3395 provide for flexing or stretching of the specimen during at least part of its exposure time. Other ASTM tests exist for evaluation of specific products: D3041 for rubber-coated fabric; D1373 for insulating tape; and D1352, D574, D2526, and D470 for wire and cable insulation. Increasing ozone pollution of the atmosphere is showing the need for practical testing of its effects.[38]

Weather Resistance

Laboratory tests for ozone resistance do not necessarily correlate well with outdoor weathering tests. The outdoor tests introduce such additional variables as light-catalyzed oxidation, water leaching, changing temperatures, and ozone concentration. In ASTM D750, the specimens are exposed, either continuously or intermittently, to carbon-arc lights and water spray. Ozone may also be present. Exposure effect is judged by tensile tests and visual examination.

In ASTM D518, detailed instructions are given for mounting specimens for either laboratory or outdoor weather exposure. Effects are judged by visual examination. In the D1171 test for weathering of automotive compounds, triangular specimens are stretched around a mandrel. Specimen cracks are rated against a set of standard photographs. Cracks appear perpendicular to the direction of stretch.

Discoloration due to sunlight and heat is the subject of ASTM D1148. An important feature of this type of test is the emphasis on control of light intensity. Wavelengths between 2000 and 2500 angstroms are especially critical. Sunlight produces crazing in either stretched or unstretched rubbers, whereas ozone produces cracks.

An aging index called *fractional strain energy at break* has been used effectively for weathering tests.[39] This expression describes an aged vulcanizate in terms of its integrity (retained tensile strength) and its extensibility (retained elongation) as follows:

$$\text{Fractional strain energy} = (\text{Aged tensile strength} \times \text{aged elongation}) \div (\text{Original tensile strength} \times \text{original elongation})$$

Control of specimen thickness and elongation is critical for achieving repeatable results on effect of weathering at different locations.[40]

Exposure to Liquids

Another deteriorating factor for rubber products is exposure to various liquids such as fuels, oils, chemicals, and even water. Any rubber may be swollen by certain chemicals. Although an "oil-resisting" rubber may resist swelling by certain oils, there are other oils or solvents that will cause it to swell, thus causing a deterioration in physical properties. Detailed procedures for evaluating change in properties of elastomeric vulcanizates resulting from immersion in liquids is given in ASTM D471.

CREEP, STRESS RELAXATION, AND SET TESTS

The term *creep* in testing parlance refers to an increase in deformation with time of a specimen held under a constant deforming force; *stress relaxation* is the decrease in recovery force with time of a specimen held under a constant deformation; and *set* is the deformation remaining after removal of an applied force or deformation. Test purposes are usually better served by expressing the test conditions and indices as *stress and strain* rather than *force and deformation*, but this distinction has led to confusion in the definition for creep index.[2] Creep in rubber has been traditionally defined as the increase in deformation of a specimen during a specified time interval after loading, expressed as a percentage of its deformation at the start of that interval. In other industries, however, creep is usually defined as the increase in deformation expressed as a percentage of the unstressed dimension. The latter usage is consistent with the ASTM D1566 definition for creep as ''the time-dependent part of a strain resulting from stress.''

Although creep and stress relaxation can be mathematically related, they are usually evaluated separately. A creep test would be indicated for a material that is to be used in a product requiring dimensional stability, such as a motor mounting. A stress-relaxation test is indicated for a usage that requires a stable restoring force, such as a seal or a rubber spring.

Any type of deformation might be used in creep, relaxation, or set tests. Compression is used in ASTM D1390 for evaluating stress relaxation. There is no ASTM test designated especially for creep, but provision is made in D945 for use of the Yerzley Oscillograph for testing creep in compression or shear. A method is described in reference 22 for use of the apparatus in ASTM D797 for measurig creep in bending, especially at low temperatures. The British Rubber and Plastics Research Association (RAPRA) has a fairly sophisticated creep apparatus that can operate in either compression or shear.[41]

Stress relaxation in tension is often used to evaluate age resistance, but there is no general standard for this application. Two types of relaxation in tension are used—continuous, in which the specimen is held stretched throughout the test, and intermittent, in which it is stretched only periodically for short times so that measurements can be made. The continuous method provides a measurement of degradative reactions in the polymer-chain network, while the intermittent measures the net effect of both degradative and crosslinking reaction.[2,42,43]

Methods for evaluating set are given in ASTM D412 for tension and D395 for compression. The usual measurement of set is intended to measure delayed elastic recovery, but if chemical changes have occurred, the residual deformation may be permanent. Low-temperature set tests are discussed in the section ''Transition in Elastomers.''

Some tests for set, relaxation, or creep that are made in compression may specify lubrication between the rubber specimen and the confining plates, and some may specify bonding. Results from the two conditions are, of course, not comparable.

TEAR TESTS

Tear tests results are strongly dependent on the type of specimen used, the rate of tearing, and the temperature. As described in the scope of ASTM D624,

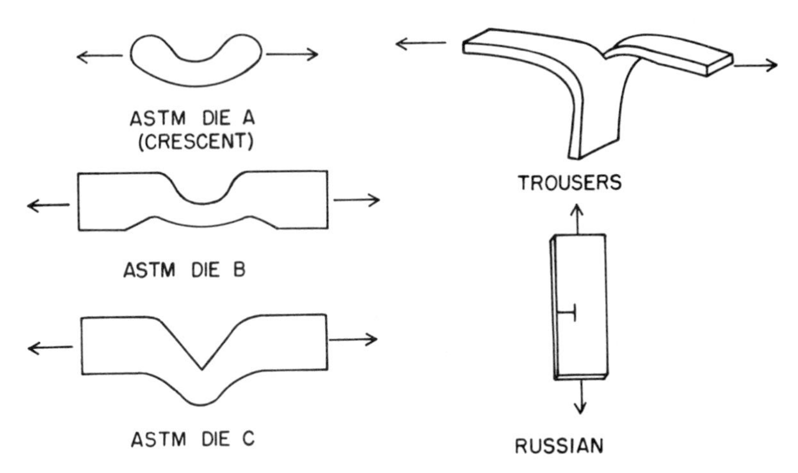

Fig. 5.14. Types of tear specimens.

"The method is useful, therefore, only for laboratory comparisons and is not applicable for service evaluations, except when supplemented by additional tests, nor for use in purchase specifications." Nevertheless, many such tests are run, perhaps because they seem to be logical extensions of the hand-tear evaluation so useful to the old-time compounder.

Three types of tear specimens are classified by Buist: indirect tearing as in the trousers specimen of Fig. 5.14, tearing perpendicular to the direction of stretching as in the ASTM methods, and tearing in the direction of stretching as in the Russian test piece.[44] Except in the ASTM Die C specimen, nicks of prescribed lengths are cut into the region of desired stress concentration.

Rate of stretching in the ASTM method is 20 inches per minute. An increase in rate normally decreases the tearing energy for SBR rubbers, but gives a more complicated effect in natural rubber. On the theory that at least a portion of tire wear is a result of high speed tearing, such a test has been considered for evaluation of abrasion resistance.

Tear test results are usually expressed as the force required to tear a specimen of given thickness, but the required pull on specimens of the ASTM type is dependent on width as well as thickness. It has been suggested that results on such specimens should be called *tear strength*, whereas those on the trousers specimen should be called *tear resistance*.[44]

Tear tests sometimes yield smooth plots of force against strain but other times, notably with natural rubber, the curve is oscillatory in nature. In such cases, called *knotty tear*, the high force points on the oscillatory curve are usually recorded as the required pull.

Of the three types of specimens, the lowest force values are given by the trousers specimen, next by ASTM Die C, next by Die A or B, and highest by the Russian specimen. Low correlation of results from different specimens should be expected.

Tear resistance of black-loaded general-purpose elastomers is quite sensitive to state of cure, often showing a sharp ''optimum'' curing time, which is usually lower than the optimum as indicated by modulus and tensile properties. Tear in black-loaded stocks often progresses from the smooth type at low states of cure to a jagged fibrous type at tighter cures. Low vulcanization temperatures and short cures are desirable for high tearing resistance.

ADHESION TESTS

ASTM D413 presents two standard tests for rubber adhesion to fabric: (1) the dead-weight method measures the rate of separation per unit width under constant load, and (2) the machine method measures the required force for constant rate of separation. Since the resulting plots are often jagged, average results are given. As in tear strength, resistance to separation is decreased sharply by an increase in test temperature. Constant-temperature rooms are suggested.

ASTM D429, for testing adhesion of rubber to metal, also provides two methods: (1) rubber part assembled between two parallel metal plates and separated by straight pull, and (2) rubber part assembled to one metal plate and stripped at 90°. In the first case, results are expressed in force per unit area of adhered surfaces and in the second case as force per unit of specimen width.

In addition to these general tests, special methods for products such as friction tape, hose, insulating tape, and belting have been standardized by ASTM. A common problem in adhesion testing is failure in the rubber part rather than at the bond. In such cases, the bond can be presumed to be satisfactory if product stress concentration at the bond is no greater than that of the test specimen. Otherwise, a specimen must be devised with sufficient concentration of stress to provide bond failures.

ABRASION TESTS

Abrasion resistance is a property of rubber that can be only incompletely evaluated by laboratory tests. Three principal reasons may be cited for the well-known lack of validity of such tests, especially as applied to tire-tread wear: (1) any acceleration of the normal wear process introduces new wear mechanisms, so that the original mechanism is not being truly evaluated; (2) no satisfactory method has been found to maintain a constant sharpness of a laboratory abrasive surface and simultaneously remove abraded rubber from the working surface; and (3) relative wear rating depends upon severity of test conditions. Road service includes a greater range of conditions than could reasonably be duplicated in a laboratory.

Wear involves removal of surface material that, for tires at least, includes such diverse mechanisms as removal of chunks by cutting, high-speed tear, degradation of surface to low-molecular-weight material that can transfer to an opposing surface, and even degradation to gaseous products. Temperature is, of course, critical to each of these mechanisms and the temperature cycle at the tread surface during a tire's rotation on pavement is very inconsistent. In spite of these problems, abrasion resistance tests can be useful if the range of validity is recognized. For example, correlation with service rating is often high if tests are restricted to compounds based on a single polymer, which is probably a modulus effect. This restricted testing is often adequate, e.g., in evaluation of carbon black.

The Pico Abrader (ASTM D2228) shows good agreement both within and among laboratories, i.e., its precision and reproducibility are high, but no claim is made for validity. Its correlation with road-wear tests is best when the latter are made under high-severity conditions.[20] The method involves rubbing a pair of tungsten carbide knives of specified geometry and controlled sharpness over the surface of the specimen in a rotary fashion under controlled conditions of load, speed, and time. A dust is used on the specimen and at the interface between the knives and the specimen to engulf the removed particles and to maintain the cutting knives free from contamination. Weight loss of the specimen is determined.

FRICTION TESTS

Friction requirements for rubber range from the very low coefficients in antifriction bearings to the very high coefficients desired in tire treads or rubber heels. Several extensive reviews of this subject have been made, in one of which it was pointed out that there are two components to rubber friction—adhesive and hysteretic.[45] Except in very-well-lubricated sliding, the adhesive component dominates; but in tire traction, for example, wet pavement often presents a critical situation in which the hysteretic properties of rubber become important.

Perhaps the most common rubber friction apparatus is the British Portable Skid Tester (ASTM E-303, Vol. 04.03), which was actually developed as a pavement-evaluation instrument. This is a portable device in which the specimen is attached to the base of a pendulum arm and contacts the opposing surface during a swing of the pendulum. The weighted pendulum head is free to move vertically on the pendulum arm so that the swing amplitude is determined by the friction of the rubber against the pavement surface. This instrument is very well adapted to testing on a wet surface.

NONDESTRUCTIVE TESTING

Nondestructive testing of rubber products is a growing field that is now mostly limited to specific items. With their proven capabilities, however, more general tests are likely to appear. A summary of their range and potentialities has been given by Halsey.[46] These methods include holography, ultrasonics, infrared radiation, microwaves, X-rays, and gauges based on nuclear, capacitive, optical, or magnetic principles.

Ultrasonics have been used for thickness measurements, including that of a coating of one material on another, and for locating voids. Infrared radiation permits noncontacting surface temperature measurements to be made on stationary or moving objects. Microwaves can be used for noncontacting thickness measurements without producing radiation hazard. Both the loss tangent and the dielectric constant of the material can be measured independently. X-rays, including both xeroradiography and fluoroscopy, are sensitive to flaws of various kinds—especially to inclusion of foreign material that is more opaque to X-rays than is the test material.

COMPUTER USAGE

Any test instrument which produces an electrical signal that can be calibrated in terms of the measured property level is a potential candidate for the tie-in with a computer. This may be an on-line instrument, in which the coded results are fed directly to the computer for immediate processing, or the results may be stored and introduced to the computer at a later time. Depending at least partly on the speed at which data must be logged, storage can be by punched card, paper tape, or magnetic tape.

An *on-line* (real-time) computerized system of four tensile-testers as used at Du Pont has been described by Stanton.[47] This system included a *dedicated* computer (for that purpose only) which was programmed to accommodate nine tests performed on the tensile-testing machine: stress-strain, Finch tear, ASTM Die B tear (Winkleman), ASTM Die C tear (Grains), flexural modulus of elasticity, compression-deflection, paper tensile, peel adhesion, and trouser tear. Four of these programs had alternate routines, making thirteen tests in all. The computer not only processed the data and printed out the calculated results but also controlled a segment of the electronic circuit in each unit. This application illustrates the principal reasons for computerizing physical tests: saving employees' time, accurate logging and calculating of large amounts of data, consistent test control, and rapid operation.

TEST DESIGN

Apart from logging of data, the use of a computer can make certain types of testing programs practical that would otherwise require excessive computation time. An example is the so-called designed experiment. By this, we mean a system of tests in which the factor levels of each variable are distributed according to a design table that permits calculation of main effects and interaction effects of each variable. A great deal more information is available from this type of program than from the one-variable-at-a-time traditional test, especially in research or development testing. An example has been given by Derringer[48] for evaluating effect of type and amount of acceleration in a silica-filled natural rubber. Control plots were used to analyze trends in 300% modulus, 500% modulus, tensile, elongation at break, Mooney scorch, and 95% Rheometer cure time.

Another example of the combined use of designed experiment and computer, in this case a special purpose analog computer, has been given by Claxton et al.[49-51] Means are described for maximizing a chosen response (e.g., economy) while maintaining other responses (e.g., durometer hardness, cure rate constant, cure induction time) at acceptable levels in a compounding study involving various levels of sulfur, Bismate, Santocure NS, Vultrol, and Paraflux. The method and computer are adaptable to any compounding study involving up to five factors and eight reponses.

TRENDS IN PHYSICAL TESTING

With the development of new polymers and more versatile tests, the rubber compounder is more nearly able than formerly to tailor-make an elastomer for each application. Consequently, less reliance is placed on meeting arbitrary specifications and more on achieving properties that demonstrably relate to the applications. A new polymer usually requires a new interpretation of test results since its compounds will have a new combination of properties that may meet a specific requirement as well as would previously recognized combinations.

The compounder's responsibility for product performance is increasing, requiring, again, extra care in selecting tests and evaluating the results. Since composition and design of a rubber product are often interrelated in achieving optimum performance and reliability at minimum cost, more tests are being made on actual products under operating or accelerated conditions. There is, consequently, a decline in testing for fundamental properties, such as tensile strength, ultimate elongation, or hysteresis, unless the need for certain levels of such properties can be demonstrated. Both static and dynamic modulus tests, however, are assuming even greater roles.

The new tensile testers now available commercially reflect the increased demand for versatility in physical testing for research and development purposes. Such items as a large range of crosshead speeds, environmental chambers, high-speed recording, computer interfacing, or even minicomputer components are more important for such applications than is high-volume output. The versatility of basic instruments is further increased by use of modules for particular applications, by provisions for cycling, and by choice of constant rate of deflection or constant rate of load application. Often, either stress-relaxation or creep tests are also possible on the same apparatus. Principles proven on such multi-use instruments may then be applied to development of specialized high-volume testers for control applications.

As Murtland states, a big push is on to expand quality-control testing onto the production line.[52] This will require computerized control to amass and analyze the flood of acquired data. Instruments will continue the trend toward programmed cycles, since shortage of skilled operators will be further compounded by the great increase in the number of instruments installed. End-product testing will see a great increase of nondestructive testing applications developed for use on the production line.

Each new consumer-protection law requires additional testing for quality control, which sometimes requires duplicate or replicate testing because each processor of a material must assume responsibility for each previous processor. The amount of physical testing per unit of production is thus continually increasing, and it is more important than ever that tests be used as effectively as possible.

REFERENCES

1. American Society for Testing and Materials (ASTM), 1916 Race Street, Philadelphia, PA 19103.
2. R. P. Brown, *Physical Testing of Rubbers*, Applied Science Publishers, London, 1979.
3. R. P. Brown and B. E. Read, eds., *Measurement Technique for Polymeric Solids*, Elsevier, New York, (1984).
4. P. Kainradl, "Critical Reflections on Physical and Technological Test Methods in the Rubber Industry," *Rubber Chem. Technol.* **52,** 232 (1979).
5. P. K. Freakley and A. R. Payne, *Theory and Practice of Engineering with Rubber*, Applied Science Publishers, London, 1978.
6. C. Hepburn and R. J. W. Reynolds, *Elastomers: Criteria for Engineering Design*, Applied Science Publishers, London, 1979.
7. "Rubber: Sampling and Testing," Federal Test Method Standard No. 601. General Services Administration, Business Service Center, Region 3, Seventh and D Streets, S.W., Washington, DC 20024.
8. *Index of Federal Specifications and Standards*, Superintendent of Documents, U.S. Government Printing Office, Washington, DC, 20402.
9. ISO publications are available from American National Standards Institute, 1430 Broadway, New York, NY 10018.
10. Collaborative Testing Services, Inc. 8343-A, Greensboro Drive, McLean, VA 22102.
11. A. E. Juve, "On Testing of Rubber," (Goodyear Medal Awards Lecture), *Rubber Chem. Technol.* **37,** xxiv (April–June 1964).
12. L. A. Wood, "Tables of Physical Constants of Rubber," *Polymers Handbook*, J. Brandrup and E. H. Immergut, eds., Interscience (John Wiley & Sons), New York, 1965; *Rubber Chem. Technol.* **39,** 132–142 (1966).
13. L. E. Nielson, *Mechanical Properties of Polymers*, Reinhold, New York, 1962, Ch. 7.
14. Ibid, Ch. 1.
15. A. Z. Wiriat, "How to Determine the Tensile Strength of Vulcanizates After Immersion in Liquids," *Rubber Age*, p. 65, Feb. 1972.
16. F. N. B. Bennett, "Wallace Optical Extensiometer," H. W. Wallace & Co., Croydon, Surrey, U.K.
17. F. S. Myers and J. D. Wenrick, "A Comparison of Tensile Stress-Strain Data from Dumbbell, Ring, and Oval Specimens," *Rubber Chem. Technol.* **47,** 1213 (1974).

18. Takeru Higuchi, H. M. Leeper, and D. S. Davis, ''Determination of Tensile Strength of Natural Rubber and GR-S. Effect of Specimen Size,'' *Anal. Chem.* **20,** 1029 (1948); *Rubber Chem. Technol.* **22,** 1125 (1949).
19. *Handbook of Extruded Rubber*, The Goodyear Tire and Rubber Co., Akron, OH 44316, 1969.
20. J. R. Beatty and M. L. Studebaker, ''Physical Properties Required of a Rubber Compound,'' *Elastomerics*, p. 33, Aug. 1977.
21. C. L. M. Bell, D. Stinson, and A. G. Thomas, ''Measurement of Tensile Strength of Natural Rubber Vulcanizates at Elevated Temperatures,'' *Rubber Chem. Technol.* **55,** 66 (1981).
22. F. S. Conant, ''The Effect of State of Cure on Vulcanizate Properties,'' in ''*Vulcanization of Elastomers*'', G. Alliger and I. J. Sjothun, eds., Reinhold, New York, 1964.
23. ''Lectures from a Course in Basic Rubber Technology,'' given by Philadelphia Rubber Group with Villanove University, 1955, Lect. IV.
24. ''Continuous Measurement of the Cure Rate of Rubber,'' ASTM Special Technical Publication No. 383, 1965.
25. ''Test Methods,'' in *The Vanderbilt Rubber Handbook*, George G. Winspear, ed., R. T. Vanderbilt Co., New York, 1968.
26. Joseph R. Weber and Hector R. Espinol, ''Cure State Analysis,'' *Rubber Age* **100** (3), 55 (1968).
27. G. E. Decker, R. W. Wise, and D. Guerry, Jr., ''Oscillating Disc Rheometer,'' *Rubber World* **147** (3), 68 (1962); *Rubber Chem. Technol.* **36,** 451 (1963).
28. A. E. Juve, P. W. Karper, L. O. Schroyer, and A. G. Veith, ''The Viscurometer—An Instrument to Assess Processing Characteristics,'' *Rubber World* **149** (3), 43 (1963).
29. F. S. Conant, ''A Study of Stiffness Testing of Elastomers at Low Temperatures,'' ASTM Bulletin, July 1954, pp. 67–73 (TR 145).
30. ''Du Pont 900 Differential Thermal Analyzer,'' Instruction Manual, E. I. du Pont de Nemours & Co., Instrument Products Div., Quillan Building, Concord Plaza, Wilmington, DE 19898.
31. F. S. Conant and J. W. Liska, ''Some Low Temperature Properties of Elastomers,'' *J. Appl. Phys.* **15,** 767 (1944).
32. A. C. Edwards and G. N. S. Farrand, ''Elasticity and Dynamic Properties of Rubber,'' in *The Applied Science of Rubber*, W. J. S. Nauton, ed., Edward Arnold Ltd., London, 1961.
33. S. D. Gehman, ''Dynamic Properties of Elastomers,'' *Rubber Chem. Technol.* **30,** 1202 (1957).
34. A. C. Bassi, ''Dynamic Modulus of Rubber by Impact and Rebound Measurements,'' *Polymer Eng. and Sci.* **18,** 750 (1978).
35. R. D. Stiehler, G. E. Decker, and G. W. Bullman, ''Determination of Hardness and Modulus of Rubber with Spherical Indentors,'' *Rubber Chem. Technol.* **52,** 255 (1979).
36. J. H. Dillon, I. B. Prettyman, and G. L. Hall, ''Hysteretic and Elastic Properties of Rubberlike Materials Under Dynamic Shear Stresses,'' *J. Appl. Phys.* **15,** 309 (1944).
37. J. Royo, ''Effect of Air Change Rate on Rubber Accelerated Ageing Tests in Hot Air,'' Pts. 1 and 2, *J. Polymer Testing* **3,** 2 (1982).
38. J. Dlab, ''Results of Investigation on Vulcanized Rubber Aging by the Effect of Atmospheric Ozone and Service Life Determination,'' *Proc. Int. Rubber Conf.*, 1979, pp. 980–989.
39. E. W. Bergstrom, ''Environmental Aging of Elastomers,'' *Elastomerics* **109,** 21 (Mar. 1977).
40. A. R. Dedrick and D. A. Paterson, ''Accelerated Aging Tests on Elastomeric Sheeting: An Inter-location Correlation,'' *J. Elastomers Plast.* **7,** 3, 315–28 (1975).
41. M. M. Hall and D. C. Wright, *RAPRA Members J.*, No. 5 (Sept./Oct. 1976).
42. R. Clamroth and L. Rueta, ''Stress Relaxation Measurement as a High Speed Method for Predicting the Long Term Behavior of Vulcanized Rubber,'' *Rubber Chem. Technol.* **56,** 31 (1983).
43. R. P. Brown, ''Stress Relaxation of Rubber in Tension—the Simple Approach,'' *J. Polymer Testing* **1,** 1 (1980).
44. J. M. Buist, ''Physical Testing of Rubber,'' in *The Applied Science of Rubber*, W. J. S. Nauton, ed., Edward Arnold Ltd., London, 1961.

45. F. S. Conant and J. W. Liska, "Friction Studies on Rubberlike Materials," *Rubber Chem. Technol.* **33,** 1218 (1960).
46. G. H. Halsey, "Non-Destructive Testing," *Rubber Age* **100** (2), 62 (1968).
47. J. L. Stanton, "Computerized Physical Testing of Elastomers," Paper No. 20, ACS Div. of Rubber Chemistry, Los Angeles, Apr. 30, 1969.
48. G. C. Derringer, "Predicting Rubber Properties with Accuracy," Paper No. 8, ACS Div. of Rubber Chemistry, Los Angeles, Apr. 30, 1969.
49. W. E. Claxton, "Basics for Blend Optimization," *Rubber World* **159** (3), 42 (1968).
50. W. E. Claxton, H. C. Holden, and J. W. Liska, "A Blend Optimizer for the Rubber Industry," *Rubber World* **159** (3), 47 (1968).
51. W. E. Claxton, "Use of Computers in Formulating Rubber Compounds," International Rubber Conference, Paris, Jun. 1, 1970.
52. W. Murtland, "Rubber Testing Sparks Instrument Boom," *Rubber World* **162** (1), 53 (1970).

6

NATURAL RUBBER

A. SUBRAMANIAM
Rubber Research Institute of Malaysia
Kuala Lumpur, Malaysia

INTRODUCTION

Nearly 2000 species of trees, shrubs, or vines of the tropical and temperate regions produce latex from which natural rubber or a closely related substance can be obtained. However, the latex from the trees of *Hevea brasiliensis* is the only important commercial source of natural rubber. The tree is indigenous to the Amazon valley.

HISTORY OF RUBBER[1,2]

Natural rubber has been known to the inhabitants of South America for centuries. Christopher Columbus is considered to be the first European to discover it, during his second voyage in 1493–6. He found the natives in Haiti playing with balls made from the exudate of a tree called ''cau-uchu'' or ''weeping wood.'' The term ''rubber'' was coined by John Priestly in 1770, when he found that the material could erase pencil marks.

Rubber was introduced to the Western world by Charles de la Condamine, who sent samples to France from Peru in 1736 and published the results of his observations in 1745. By the end of the eighteenth century, Europe and America were using a few tons of rubber per year. However, users found it difficult to work with solid rubber. Moreover, articles made from natural rubber turned sticky in hot weather and stiffened in the cold.

Two important developments in the nineteenth century enabled these problems to be solved and laid the foundation for the multibillion dollar modern rubber industry. In 1820, Thomas Hancock invented a machine called the ''masticator'' that allowed solid rubber to be softened, mixed, and shaped. In 1839, Charles Goodyear discovered the process of vulcanization. He found that heating a mixture of rubber and sulfur yielded products that had much better properties than the raw rubber.

Soon a variety of articles from rubber started to come into the market and the demand for rubber grew rapidly. Exports of raw rubber from Brazil increased from a few hundred tons in 1846 to almost 10,000 tons by 1880.

It was soon apparent that Brazil would not be able to meet future demand. The British considered the possibility of cultivating rubber in Asia. In 1876, Henry Wickham collected 70,000 seeds from Brazil and sent them to Kew Gardens for germination. Of the more than 2000 seeds that germinated, most were sent to Ceylon (Sri Lanka) and some to Singapore and Malaya (Malaysia). Later shipments were made to the East Indies (Indonesia). By 1880, *Hevea* seedlings were widely distributed in Asia.

Plantation Rubber

Ten years elapsed before *Hevea* plantations began to make their appearance; commercial exploitation began only at the end of the century. The growth of the industry was mainly due to the efforts of one man: Henry N. Ridley. In the 1890's, working in Singapore, he devised an effective and economic method of obtaining latex from *Hevea* trees by an excision of the bark. This method, called "tapping," caused less damage to the tree and produced more latex than the South American practice of cutting into the bark. The use of acid to coagulate latex and produce sheets which were then dried in smoke was the discovery of John Perkins in 1899. The invention of the pneumatic tire in 1888 by John Dunlop led to the booming demand for rubber with the beginning of the motoring era early in the twentieth century.

Plantation rubber production in Asia grew rapidly and outstripped the production of wild rubber from Brazil by about 1913. Southeast Asia has remained the predominant natural rubber-producing region since then.

MODERN RUBBER INDUSTRY

The land under rubber cultivation and the production of natural rubber have grown steadily except during World War II. In 1983, more than 7.5 million hectares of land in the world were under rubber cultivation and about 4 million tonnes (metric tons) of rubber were produced. The Southeast Asian region accounted for about 80% of the total production. Malaysia was the biggest producer (39%), followed by Indonesia and Thailand (Table 6.1)[3].

The cultivation of rubber is carried out either in estates (plantations more than 40 hectares in size) or in smallholdings (farms less than 40 hectares in size). The average size of estates in Malaysia is 700 hectares while the average smallholding is only 2 hectares. About 76% of the total planted acreage in Malaysia, 80% in Indonesia and 95% in Thailand are smallholdings. In Malaysia, the productivity of the estate sector is 1500 kg of rubber per hectare per year compared to about 1000 kg in the smallholdings.

Table 6.1. World Production of Natural Rubber (1983)[a]

Country	*Tonnes ('000)*	*TSR[b] as % of Total NR*
Malaysia	1562	46
Indonesia	997	71
Thailand	587	13
India	168	1
China	159	–
Sri Lanka	140	4
Philippines	75	–
Liberia	65	–
Others	257	–
Total	4010	

[a]Source: Reference 3 [b]TSR = Technically Specified Rubber

BOTANY AND AGRICULTURE[4,5]

The *Hevea* Tree

The *Hevea* tree grows best in tropical regions with an average temperature of 25–30°C and a rainfall of at least 2 m per year distributed throughout the year. However, China has in recent years succeeded in growing rubber in a colder climate. The tree grows satisfactorily in most soils at an altitude below 300 m. It can grow to a height of more than 40 m if left untapped. The economic life of the tree is about 25–30 years, by which age its height is under 20 m. The leaves are trifoliate and are shed once a year in the process called "wintering." The fruit is a pod of three seeds, each slightly larger than acorns.

Latex is found in the latex vessels contained in the cortex, especially in the layer 2–3 mm thick nearest to the cambium. They have internal diameters of about 20 μm. They turn spirally up the tree at an angle of about 4° to the vertical, thus forming a right-handed spiral. The trees produce latex all the year round, but the yield usually drops during the wintering season.

Tapping of Latex

Latex is obtained from the latex vessels without damaging the trees by the process called tapping. A slice of bark is shaved off with a special knife to a depth just short of the cambium layer. The cut is made at an angle of 25–30° to the horizontal to sever the maximum number of latex vessels (Fig. 6.1). Tapping is done before sunrise, when the turgor pressure in the tree is maximum and the yield of latex highest. The latex flows from the cut along a metal spout and into a cup made of glass or glazed earthenware. Tapping is continued at regular intervals by reopening the cut down the tree.

Fig. 6.1. Tapping of latex from *Hevea* using (top) a conventional knife; (bottom) a mechanized knife.

Many variations have been used in the size and shape of the cuts and in the frequency of tapping.[6] A very common method is to tap each tree halfway round the circumference every other day (denoted by $\frac{1}{2}$S. d/2, or half-spiral, alternate daily). After about 5 years, the tapped panel is rested and the bark allowed to regenerate. A new panel is opened on a different area of the tree.

Immediately after tapping, the latex flows rapidly, then declines to a steady rate, then slows down and finally stops. The stoppage of the flow is due to "plugging" of the latex vessels by the coagulum formed at the opening of the cuts.

Breeding of *Hevea* and Cultivation

The breeding of *Hevea* is carried out by vegetative reproduction from proven high-yielding trees and by the use of selected seeds from such trees derived by cross-fertilization. Vegetative reproduction is usually carried out by bud grafting. In this technique, a bud from a high-yielding tree is removed and inserted under the bark of the lower stem of a young seedling—the rootstock. After the bud has taken, the stem of the rootstock is removed above the growth point. The bud begins to grow and eventually becomes the trunk of the mature tree. All trees that are derived by vegetative reproduction from a single mother tree are known as *clones*. Each clone has its own characteristics of growth, yield, resistance to disease, and so forth. Some of the properties of latex and rubber are also clonal characteristics.

The young plants are allowed to grow in a nursery in the first year and are then transplanted to the field. A mature planting may have a density of 250 to 400 trees per hectare. Soil fertility is maintained by the use of fertilizers and by growing cover crops such as legumes. Measures are also taken to prevent damage to the trees by pests and disease.

The common diseases of *Hevea* involve the roots and leaves. All these can be controlled except for the "South American Leaf Blight," a fungal disease for which there is no known cure. This disease destroyed the *Hevea* plantations in Brazil early this century. Fortunately, Asia and Africa are free from this disease.

The trees are ready for tapping in 5–7 years, when the girth reaches about 50 cm when measured about 60 cm from the ground.

The original plantings of *Hevea* yielded only about 500 kg of rubber per hectare per year. Through breeding and selection programs carried out at research institutes in Indonesia and especially in Malaysia, productivity of the tree has been increased tremendously over the years. Modern *Hevea* clones are capable of yielding over 2500 kg per hectare per year.[7] The vast majority of plantation acreage in Malaysia has been planted with high-yielding clones. Such replanting programs are now being implemented in the smallholder sector as well as in the other producing countries.

Further horticultural and physiological developments are continuously being made. For example, by bud-grafting another clone at a later stage, a "crown-budded" or three-part tree with good yield and secondary characteristics is produced.[8] Certain properties of rubber, such as Mooney viscosity, from these composite trees are intermediate in value between those of the component clones. Thus by a suitable choice of trunk and crown clones, it is possible to produce a tree capable of yielding rubber with specified viscosity values.

A further increase in the yield of latex is achieved by the use of chemicals called *yield stimulants*.[9] These stimulants when applied to the bark prolong the flow of latex by delaying the plugging mechanism. The most effective yield stimulant is 2-chloroethane phsosphonic acid ("Ethrel" or "Ethephon"), the active principle of which is ethylene. In some clones, yields can be doubled over short periods using this stimulant. By the use of yield stimulants together with shorter tapping cuts (1–2 mm long or even pin-pricks) or reduced frequency of tapping, some clones can yield as much latex as the unstimulated trees with conventional tapping.[10] Such modern methods of tapping and yield stimulation can give good yields with decreased labor requirements and bark consumption.

A mechanized tapping knife which allows more trees to be tapped per hour is also currently under investigation (Fig. 6.1).

PROCESSING OF LATEX

About 3 to 4 hours after tapping, the latex is collected from the tree, treated to prevent premature coagulation and brought to a factory or a smallholder processing center. Ammonia (about 0.05%) is the most common stabilizer added to the latex though others such as sodium sulfite and formaldehyde are also still used. Mixed stabilizers such as boric acid-ammonia and hydroxylamine-ammonia are used to make certain special grades of rubber for export or local processing. Where latex itself has to travel a long distance, higher concentrations of ammonia are used. In the initial collection (3 to 4 hours), about 80–85% of the latex produced by the tree is collected as latex (field latex).

The latex continues to exude very slowly for several hours after the initial collection. This latex is left to coagulate spontaneously in the cup to form "field coagulum," or "cuplump;" a small amount of latex also coagulates as a thin sheet on the tapping cut to form "tree lace." These are collected on the next tapping day and constitute about 15–20% of total yield.

On arrival at the factory, the latex is sieved and blended. Field latex is either concentrated by removing part of the water to give "latex concentrate," or it is deliberately coagulated and processed into solid dry rubber. All cuplumps are processed into dry rubber.

About 7–10% of the world's natural rubber is converted into latex concentrate; more than 75% of this comes from Malaysia. Concentration is achieved by centrifugation (most common), by creaming, or by evaporation. The centrifuged latex is shipped as latex concentrate containing 60% dry-rubber content.

The remainder of the latex and field coagulum are processed into conventional types of rubber such as ribbed smoke sheets (RSS), pale crepes, and brown crepes, or into the newer forms of technically specified "block" rubbers (TSR).

DRY-RUBBER PRODUCTION

Conventional Grades

Ribbed Smoke Sheets (RSS). The bulk of the field latex is converted into RSS. The blended latex is diluted with water to about 15% dry-rubber content and coagulated with formic acid. Before coagulation sets in, aluminium partitions are inserted vertically in slots in the coagulating tank. After storage for 1 to 18 hours, the soft thick gelatinous slabs are compressed by passage through four to six rollers to remove water and produce sheets of about 5 mm thickness. The last pair of rollers are grooved and thus produce the characteristic criss-cross rib markings on the sheet. This increases the surface area and facilitates drying. The rubber sheets are dried for 4–7 days in sheds called ''smoke houses,'' which are heated to about 60°C by the smoke-fire of burning wood. The dried sheets are packed into bales of 113 kg (250 lb); the bales are coated with talc to prevent bale-to-bale adhesion.

Air-dried Sheets. These are produced in a similar manner but are dried in a current of hot air without the use of smoke. ADS are light amber in color.

Michelin Sheets. These are prepared and dried like RSS but the wet sheets are allowed to dry at ambient temperature for 1–2 days before they are taken into the drying sheds. This process of ''maturation'' is believed to cause changes in the nonrubber substances in natural rubber and lead to beneficial effects during vulcanization.

Pale Crepe. The natural color of crepe rubber is pale yellow due to the presence of β-carotenes. Pale crepe is a light-colored premium grade of rubber from which the yellow pigments have been removed. It is used in products for which lightness of color is important.

Pale crepes are made from selected latices producing high-viscosity rubber and with a low-pigment content and low tendency to darken by enzymatic reactions. Such latices are first diluted to about 20% dry rubber content and the yellow pigments are bleached with 0.05% of tolyl mercaptan. Alternatively, the pigments are removed by fractional coagulation. A small amount of acid is added at first; after about 3 hours, about 10% of the rubber coagulates and this contains the bulk of the β-carotenes. In practice, a combination of both methods is used. The treated latex is then coagulated and set into slabs that are then passed eight or nine times through grooved differential rollers with liberal washing. The resultant thin crepes, about 1–2 mm thick, are dried in hot air at about 40°C for 2 weeks. Pale crepe is packed in talc-coated bales of 102 kg (224 lb) or wrapped in polyethylene films and enclosed in paper bags.

Sole Crepes. These are used by the footwear industry and are made by laminating plies of pale crepe to the required thickness, and consolidated by hand-rolling and finally through even-speed rollers.

Brown and Blanket Crepes. Brown crepes are made from cuplumps and tree lace. These two materials are soaked in water to remove surface dirt and are then thoroughly cleaned and blended by milling through a battery of rollers driven at friction speed. Proper blending is carried out on the rollers to obtain a uniform color. The crepes are dried and packed as in pale-crepe production.

Blanket crepes are made from a mixture of RSS cuttings, wet slabs of coagulum formed during the handling of latex, unsmoked sheets, cuplumps, and so forth.

The raw materials used for brown-crepe production are also used to make TSR; these grades are now becoming scarcer.

Grading of Conventional Rubber

The conventional forms of natural rubber are graded according to the ''Green Book,'' which sets out the International Standards of Quality and Packing for Natural Grades.[11] Grading is by visual examination and is based on the presence or absence of extraneous foreign matter (dirt), bubbles, uniformity and intensity of color, mold and rust spots, and so on. In this way, the conventional forms of natural rubber are classified into 8 types made up of a total of 35 different grades.

The dirt content of the premium grades is usually low, and this is generally reflected in their better technological properties. However, most of the other visual criteria have no technological basis.

Technically Specified Rubbers (TSR)

Technically specified rubbers were first introduced into the market by Malaysia in 1965 as the Standard Malaysian Rubber (SMR). This adherence to standards has been followed by other producing countries such as Indonesia with Standard Indonesian Rubber (SIR), and so on.

TSR is graded not visually, but according to the source of the rubber (latex or field coagulum) and on its properties. The bale size is reduced to a more convenient weight of $33\frac{1}{3}$ kg (75 lb). The bales are packed in polyethylene sheets to prevent contamination. The rubber is shipped in one-tonne pallets made of wood or shrink-wrapped plastic.

The introduction of TSR has necessitated revolutionary changes in methods of processing.[6,12] The rubber is prepared in granulated or crumb form rather than as sheets or crepes; this allows for easier cleaning, especially of the lower grades. The new process is more automated and the increasing use of machinery requires factories with large output, which imparts more uniformity to the final product. Drying is carried out at a temperature of 100°C or slightly higher; this reduces the processing time from a week to 2 days. The method also allows better control of the rubber properties by suitable chemical treatment.

There are two main processes employed in the size reduction of the coagulated rubber to crumbs:

The *Heveacrumb* process is based on a mechano-chemical method using an incompatible oil such as castor oil as the crumbling agent.[13] About 0.5% or less of castor oil is added to the latex before coagulation or it is sprayed on the wet coagulum as an emulsion during creping. On passing through creping rollers, the rubber breaks up into fine crumbs.

The *Comminution* process uses no additives. Size reduction to crumbs is by mechanical means such as dicing machines, rotary cutters, granulators, pre-breakers, hammer-mills, extruders, shredders, or similar equipment.

SMR L. This is a light-colored rubber. Field latex is preserved with ammonia or a mixture of ammonia and boric acid. About 0.05% of sodium metabisulfite is added soon after bulking to prevent enzymic darkening. The latex is coagulated without dilution at a pH value of 5 using formic acid. The coagulum is left to mature for 6–12 hours and then converted to granular form. Drying of the crumbs is carried out in deep-bed through circulation driers at 100°C for 4–5 hours. The dry "biscuits" are compressed into standard-size ($33\frac{1}{3}$ kg) blocks measuring 66 × 33 × 18 cm, wrapped in polyethylene and placed in one-tonne crates ready for shipment.

SMR CV. This is viscosity-stabilized rubber. It is made by adding 0.15% by weight of hydroxylamine neutral sulphate to ammonia-preserved field latex before coagulation with acid. The rubber is processed and dried as for SMR L. The hydroxylamine inhibits the storage hardening reaction in natural rubber (see below) that leads to the increase in its Mooney viscosity. Three grades are available in the Mooney range of 44–55, 55–65 and 65–75 units. A lower-viscosity SMR LV grade containing 4% of a nonstaining process oil is also available with the Mooney range of 45–55 units.

SMR WF. This is similar to SMR L but is of a darker color. Usually, SMR L that fails to pass the Lovibond color test* is graded as SMR WF.

SMR GP.[14] This is a general-purpose, viscosity-stabilized grade of rubber suitable for use in tires without blending with other grades. It is made from a mixture of 60% latex-grade rubber and 40% field coagulum. The 60% latex-grade content can be either latex or unsmoked sheets, or a mixture of the two. Hydroxylamine is usually added as a spray during creping.

Except for the L grade, the foregoing rubbers are available only under the SMR label. The following are available under the general TSR scheme.

TSR 5. This is made from latex as in the SMR L process (without treatment with sodium metabisulfite). For *SMR 5*, however, the source of the material is

* The color of a pressed disc of raw rubber of standard thickness is compared and matched as closely as possible with that of standard colored glasses. The latter are calibrated in color index units according to the intensity of their color using the "Lovibond color scale."

restricted to sheets obtained from the conventional process; for example, RSS prepared by conventional methods can also be classified as SMR 5 if it passes the specifications.

TSR 10, TSR 20, and TSR 50. These are made from field coagulum material but may also include RSS cuttings. In a typical process, the raw materials are soaked in water and then initial-size reduction is carried out. They are soaked again to clean and blend the materials. Final-size reduction is by means of crepers, hammer-mills and extruders. The crumbs are dried at 100°C or at a slightly higher temperature. If the rubbers are known to contain excessive amounts of pro-oxidants (e.g., copper ions), the crumbs are dipped in phosphoric acid solution before drying. Final grading of the dried rubbers is based on the results of specification tests.

Specifications for TSR

The basic specifications for SMR at present (1984) are given in Table 6.2.[15] The properties specified are dirt, ash, nitrogen content, volatile matter, Plasticity Retention Index (PRI), and Wallace ''rapid'' plasticity. In addition, color is specified for SMR L and Mooney viscosity for the SMR CV, LV, and GP grades.

The quality of the rubber is assured by a test certificate issued with each consignment.

Table 6.2. Standard Malaysian Rubber Specifications (from 1979)

Parameter	*SMR L*[a] *SMR WF*[a] *SMR CV*[a,d] *SMR LV*[a,d]	*SMR 5*[b]	*SMR 10* *SMR GP*[c,d]	*SMR 20*	*SMR 50*
Dirt retained on 40 μm mesh (max. % wt.)	0.03	0.05	0.10	0.20	0.50
Ash content (max. % wt.)	0.50	0.60	0.75	1.00	1.50
Nitrogen content (max. % wt.)	0.60	0.60	0.60	0.60	0.60
Volatile matter (max. % wt.)	0.80	0.80	0.80	0.80	0.80
Wallace rapid plasticity, P_0 (min.)	30[e]	30	30	30	30
PRI (min. %)	60	60	50	40	30
Color (Lovibond max.)	6.0[f]	-	-	-	-
Mooney viscosity (ML 1 + 4, 100°C)	-[g]	-	-[h]	-	-
Cure	R[i]	-	R[i]	-	-

[a]From latex. [b]Sheet material. [c]Blend.
[d]Viscosity stabilized; SMR LV contains 4% process oil.
[e]Only for SMR L and SMR WF. [f]Only for SMR L.
[g]Three grades SMR CV 50, CV 60, CV 70 with viscosity limits 45–55, 55–65, 65–75 units respectively; SMR LV with producer viscosity limits at 45–55 units.
[h]SMR GP producer viscosity limits at 58–72 units.
[i]Cure information provided for SMR CV, L, WF and GP in the form of rheograph (R).

One of the important parameters specified is the Plasticity Retention Index. This is measured by the ratio P_{30}/P_0, where P_0 is the initial Wallace plasticity and P_{30} is the Wallace plasticity after aging for one-half hour at 140°C. This ratio is usually expressed as a percentage. It is a measure of the resistance of raw rubber to oxidation on heating.

In high-temperature mixing, the breakdown behavior of rubber is influenced both by initial viscosity and by PRI. The latex grades usually have PRI values above 60%. For PRI values below 60%, PRI is correlated with the heat build-up and ageing of vulcanizates[16] and with reversion during cure.[17]

Advantages of TSR over Conventional Grades

Manufacturers have found advantages in overall mastication/mixing times by the use of TSR since these generally have lower viscosities than conventional grades.[18] By the use of viscosity-stabilized grades such as SMR CV and SMR GP, premastication can be dispensed with in many cases, thus giving savings in mixing costs. TSR's are also more uniform in their properties. It is sometimes claimed that some vulcanizate properties, such as tensile strength at elevated temperatures, of the conventional grades of RSS, are better than those of the equivalent TSR grades. The differences are usually marginal and are not significant. The production of TSR continues to increase yearly. In 1983, 46% of Malaysian rubber and 71% of Indonesian rubber were sold as TSR.

OTHER FORMS OF NATURAL RUBBER

Many other forms and types of natural rubber have been produced to meet the specific requirements of consumers. These have usually involved chemical modification of the polymer chains or simple changes in the production process. Some of the commercially more important ones are described below.

Technically Classified Rubbers (TCR)[19]

These are natural-rubber samples having variable cure rates, depending on the technique used in converting latex to dry rubber. Three types by label are available: red-, yellow- and blue-circle rubbers, corresponding to slow-, medium- and fast-curing rubbers, respectively.

Oil-extended Natural Rubber (OENR)

OENR is a rubber containing 20% to 30% of an aromatic or naphthenic process oil.

Tire Rubber[20]

This is made from a mixture of 30 parts rubber (as latex), 30 parts sheet rubber, 40 parts cuplumps and 10 parts of an aromatic process oil. It is viscosity stabilized and requires no premastication: it is similar to SMR GP except for the oil.

Deproteinized Natural Rubber (DPNR)[21]

DPNR is prepared by diluting latex and reacting it with an enzyme that removes most of the proteins from natural rubber. DPNR's properties of low creep and stress relaxation and highly reproducible modulus make it suitable for use in engineering components.

Peptized Rubber

This is a low-viscosity rubber prepared by adding a peptizer to latex before coagulation or to the dry rubber itself.

Powdered Rubber

Spray-dried natural rubber containing about 8–10 parts of suitable partitioning agents is available as a free-flowing powder. Its present end use is in solution adhesives.

Skim Rubber

Skim latex is the by-product of latex centrifugation. It is coagulated (usually with sulfuric acid) and made into sheet, thick-crepe or granulated rubber. It contains a higher proportion of nonrubbers, especially of proteins, than the average grades and is fast-curing.

Superior Processing Rubbers (SP)

These are made by blending normal latex and prevulcanized latex before coagulation, and drying by conventional methods. Prevulcanized latex is made by adding a curative dispersion containing sulfur, zinc oxide and accelerator to natural-rubber latex and heating to 82–85°C for a few hours. PA 80 contains 80% prevulcanized rubber and 20% unvulcanized rubber. PA 57 is similar to PA 80, but also contains 30% of a nonstaining processing oil. SP rubbers confer superior extrusion properties such as better surface smoothness and lower die swell when blended with natural and synthetic rubbers.

Heveaplus MG Rubbers (MG)

These are made by polymerizing methyl methacrylate monomer in the presence of natural-rubber latex. The polymethylmethacrylate (PMMA) chains are thereby grafted to the rubber molecules. The resultant latex is coagulated and made into crepe. Two products are available: MG 30 and MG 49 containing 30% and 49%, respectively, of PMMA. They are used in adhesives and rigid moldings.

Epoxidized Natural Rubber (ENR)[22,23]

This is perhaps the most important modification of natural rubber ever made. It is produced by epoxidizing natural rubber in the latex stage by reaction with formic acid and hydrogen peroxide. ENR 10, ENR 25 and ENR 50 contain 10 mole %, 25 mole % and 50 mole %, respectively, of epoxide groups. The introduction of epoxide groups progressively increases the glass-transition temperature of natural rubber and changes some of its properties markedly. For example, ENR 50 has low air permeability, comparable to butyl rubber and good oil resistance at room temperature, comparable to medium nitrile rubber. Large-scale commercial production on ENR has not yet started.

Thermoplastic Natural Rubbers (TPNR)[24]

These are physical blends of natural rubber and polypropylene, mixed in different proportions to give rubbers with different stiffness properties. Mixing is done at 180°C or higher. TPNR may also contain a third polymer, viz. polyethylene. They are not available commercially but can be made easily at the consumer's factory. They are suitable for injection molding into products for automotive applications such as flexible sight shields and bumper components.

Many other chemically modified forms of natural rubber were available in the past. These included cyclized rubber, chlorinated rubber, hydrochlorinated rubber, isomerized or anticrystallizing rubber, and depolymerized rubber. These are now not commercially significant since synthetic equivalents are available at cheaper prices.

PROPERTIES OF RAW NATURAL RUBBER

Natural-rubber latex consists of particles of rubber hydrocarbon and nonrubbers suspended in an aqueous serum phase. The average dry-rubber content of latex may range between 30% and 45%. A typical composition of fresh latex is shown in Table 6.3.

Table 6.3. Typical Composition of Fresh Latex and Dry Rubber

	Latex, %	*Dry Rubber, %*
Rubber hydrocarbon	36	93.7
Protein	1.4	2.2
Carbohydrates	1.6	0.4
Neutral lipids	1.0	2.4
Glycolipids + phospholipids	0.6	1.0
Inorganic constituents	0.5	0.2
Others	0.4	0.1
Water	58.5	-

Nonrubber Substances

During the preparation of dry rubber, much of the water-soluble nonrubber substances are lost, but most of the lipids are retained together with more than half the proteins and small quantities or inorganic salts and other substances. Each of these classes of nonrubbers consists of many individual substances. The lipids, for example, consist of mono-, di- and triglycerides, sterol and sterol esters, tocotrienols and their esters, free fatty acids, glycolipids and phospholipids.[25] Natural rubber easily contains more than 100 individual chemical compounds. Some of these nonrubber substances greatly influence the properties of natural rubber, in both the raw and vulcanized states.

The properties affected by the nonrubber substances are summarized in Table 6.4.

Physical Properties

The rubber hydrocarbon in freshly tapped latex is almost completely soluble (more than 95%) in common solvents such as toluene and tetrahydrofuran. It has a chemical structure of almost 100% *cis*-1,4-polyisoprene units.

Storage Hardening. The Mooney viscosity (V_R) of freshly prepared rubber ranges from 50 to 90 units (ML 1 + 4 at 100°C) depending on the mixture of clonal latices used in the preparation. On storage, the rubber hardens or stiffens spontaneously and the viscosity increases. The storage hardening reaction is increased by conditions of higher temperature and lower humidity. This reaction is believed to be due to abnormal groups, probably aldehydic, present in the rubber molecules.[31] These groups react with the amino groups of free amino acids and proteins to give crosslinks. The concentration of these groups is only about 5–10 $\times$ 10^{-6} moles per g rubber, equivalent to only 1 or 2 groups per molecule.[32] These groups can be deactivated and the storage hardening action

Table 6.4. Properties of Natural Rubber Influenced by Nonrubber Substances

Property	*Influence of Nonrubbers*
Latex stability	Carbohydrates act as substrates for bacterial growth—lead to increased volatile acid formation and lower stability.
Color	Yellow—caused by β-carotenes. Dark—enzymic reaction of polyphenol oxidase.
Cure	Phospholipids and some proteins are natural accelerators; fatty acids are activators.
Oxidation	Tocotrienols are natural antioxidants.[25] Copper, managanese and iron ions are pro-oxidants.
Storage hardening	Proteins and free amino acids react with abnormal groups in rubber.
Crystallization	Unstrained crystallization rate increased by stearic acid,[26] some water-soluble substances retard rate.[27]
Creep and stress relaxation	High contents of proteins and ash lead to moisture absorption, which results in high creep and stress relaxation in vulcanizates.[28]
Modulus	Increased by proteins.[28]
Filler effect	Proteins act as fillers. One part of protein is equivalent to 3 parts of HAF black.[29]
Heat build-up	Heat build-up in the Goodrich flexometer test is decreased by fatty acids and increased by proteins.
Tear strength	Increased by proteins.[30]
Dynamic crack growth	Resistance increased by proteins.[30]

[25-30]See references for chapter.

inhibited by reacting rubber with 0.15% of hydroxylamine salt.[31] Such viscosity stabilized CV grades show an average increase in V_R of only 4–8 units after 4–5 years of storage at ambient temperature.

Gel Content. Freshly prepared natural rubber has a low gel content, of about 5–10%. On storage, the gel content increases and may reach 50%, or even higher on long storage. The increase in gel content mainly involves storage hardening but may also partly be due to free radical reactions. The gel is not a true network and the gel content depends on the solvent used, being lower in good solvents. It is a loosely crosslinked structure with a very high M_c value (Table 6.5).

Gel in rubber easily breaks down on mastication. Masticated rubber is completely soluble in common aliphatic and aromatic solvents, chlorinated hydrocarbons, tetrahydrofuran and carbon disulfide. Lower ketones, alcohols, and lower esters are nonsolvents.

Molecular Weight.[33] The molecular weight distribution of the rubber hydrocarbon in freshly prepared rubber is a bimodal or a skewed unimodal distri-

Table 6.5. Swelling Ratios and M_c Values of Gel in Natural Rubber

Grade	*Gel, %*	*Swelling Ratio in Toluene*	$M_c \times 10^{-6}$
RSS 1	11	84	2.6
SMR CV	3	111	4.1
SMR 10	28	84	2.6

bution, depending on the clone. A random blend of the common clonal rubbers would have a weight-average molecular weight of 1–1.5 × 10^6 and a number-average of 3–5 × 10^5. The molecular weight of commercial grades of rubber is a little uncertain because of the presence of gel. The average molecular weights and typical distributions of the soluble portion of some commercial grades of rubber are shown in Table 6.6 and Fig. 6.2, respectively. Old samples of rubber except the CV grades usually have a skewed unimodal distribution.

Crystallization. Due to its high stereoregularity, natural rubber crystallizes spontaneously when stored at low temperatures or when it is stretched.

Unstrained rubber has a maximum rate of crystallization at about −26°C. But even at 0°C, natural rubber can crystallize in a few weeks. The maximum degree of crystallinity reached is only about 25–30%.

Crystallized unstretched rubber melts within a range of a few degrees. The melting point depends on the crystallization temperature. An equilibrium melting temperature of over 40°C is estimated for unstretched rubber. Crystallization leads to a stiffening of the rubber. This is different from storage hardening and is reversible by heating the rubber. Natural-rubber samples stored in warehouses at ambient temperatures in temperature climates crystallize and must be

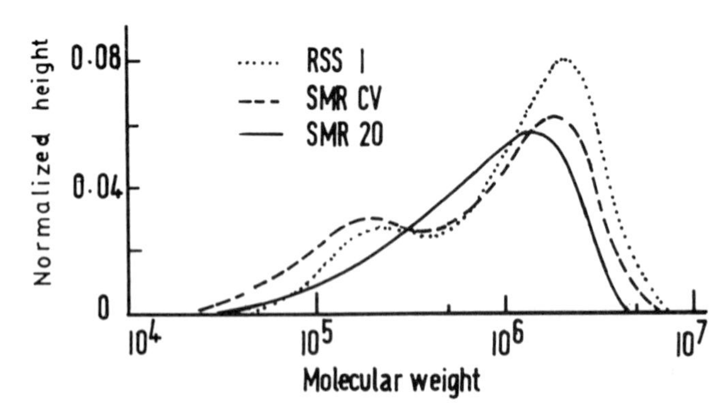

Fig. 6.2. Molecular-weight distribution curves for commercial grades of natural rubber.

Table 6.6. Molecular Weights of Soluble Natural Rubber

Grade	$\bar{M}_w \times 10^{-6}$ *from GPC*	$\bar{M}_n \times 10^{-5}$ *from GPC*
RSS 1	1.0–2.3	2.3–4.5
SMR 5	.95–2.7	2.2–5.6
SMR CV	.80–2.0	1.8–3.7
SMR 20	.92–2.2	2.1–4.0

melted in a hot room before they can be mixed and compounded. Crystallization can be inhibited by isomerizing a few of the *cis*-groups to *trans*-groups by treating the rubber with sulfur dioxide.[34]

Rapid crystallization on stretching gives natural rubber its unique high tensile strength and tear resistance in pure gum or in nonreinforced vulcanizates. Crystallization occurring on extrusion through a die, even at temperatures above 100°C, may sometimes lead to processing problems.

VULCANIZATION OF NATURAL RUBBER

Sulfur Vulcanization

Sulfur vulcanization is still the most widely used method of crosslinking natural rubber. In natural-rubber compounds, except ebonite, the amount of sulfur used can be varied from 3.5 phr down to only 0.4 phr. The accelerated sulfur vulcanization systems can be classified into three types:

1. conventional systems containing high sulfur/accelerator ratios;
2. efficient (EV) systems containing high accelerator/sulfur ratios; and
3. semi-EV systems that are intermediate between (a) and (b).

An arbitrary division of these is given in Table 6.7.[35]

The structural features of the network obtained in the accelerated sulfur vulcanization of natural rubber are shown in Fig. 6.3. Typical vulcanizate structures at optimum cure times and some properties are shown in Table 6.8.[35]

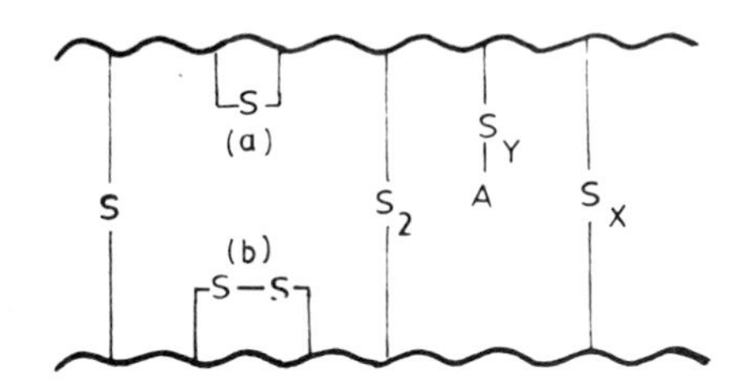

Fig. 6.3. Network structure of natural rubber vulcanized with sulfur.

Table 6.7. Sulfur (S) and Accelerator (A) of Different Vulcanization Systems for Natural Rubber[a]

	S phr	*A phr*	*A/S*
Conventional	2.0–3.5	1.2–0.4	0.1–0.6
Semi-EV	1.0–1.7	2.5–1.2	0.7–2.5
EV	0.4–0.8	5.0–2.0	2.5–1.2

[a]Source: Reference 35.

Conventional systems give vulcanizates which possess excellent initial properties like strength, resilience and resistance to fatigue and abrasion, and are satisfactory for many applications. However, their heat-aging resistance, creep, and stress-relaxation properties are less satisfactory. For good heat-aging resistance and low compression set, an EV system is essential; or a semi-EV system may be chosen as a compromise between cost/performance. EV systems can also be vulcanized at higher temperatures (180–200°C) and are less antagonistic to antioxidants than conventional systems.[36]

Soluble EV and semi-EV systems use zinc 2-ethylhexanoate rather than stearic acid, the latter leading to insoluble zinc stearate formation in the vulcanizates and affects properties. Also, sulfur is limited to 0.8 phr. Soluble EV systems overcome some of the problems of EV systems such as high physical creep and low resilience. They are ideal for use in engineering components that require low compression set, low creep, low stress relaxation, and high reproducibility in modulus and strength.

Urethane Vulcanization[35,37]

Vulcanization of natural rubber can be carried out by a new class of reagents marketed under the trade name Novor. They are basically diurethanes that are

Table 6.8. Vulcanizate Structures and Properties[a]

	Conventional	*Semi-EV*	*EV*
Poly- and disulfidic crosslinks, %	95	50	20
Monosulfidic crosslinks, %	5	50	80
Cyclic sulfide concentration	High	Medium	Low
Low-temperature crystallization resistance	High	Medium	Low
Heat-ageing resistance	Low	Medium	High
Reversion resistance	Low	Medium	High
Compression set, 22 hr at 70°C, %	30	20	10

[a]Source: Reference 35

stable at processing temperatures, but which dissociate into their component species of nitrosophenols and diisocyanates at the vulcanization temperatures. The free nitrosophenols react with rubber molecules to give pendant aminophenol groups that are then crosslinked by the diisocyanate.

Urethane-crosslinked, natural-rubber networks show a marked increase in fatigue resistance on ageing compared to most sulfur systems. Vulcanization can be carried out at 200°C without loss of modulus and tensile properties. However, these systems are expensive and may also have scorch problems. In practice, they are used as covulcanization agents with sulfur systems.

PROCESSING OF NATURAL RUBBER

Natural rubber is usually considered to have good processing properties. Although it is tough and "nervy" at temperatures well below 100°C, it breaks down easily to a useable plasticity. Generally it can be adapted to any fabrication technique of the rubber factory. The viscosity-stabilized grades of natural rubber do not generally require premastication before the incorporation of fillers and other compounding ingredients. For the nonstabilized grades, a short mastication time before compounding is a common practice.

The efficiency of mastication is lowest at around 100°C. Mastication is best carried out below 80°C (well-cooled open mill) or above 120°C (internal mixer).[38] Chemical peptizers allow mastication to be carried out at lower temperatures and are thus useful for increasing mastication throughput. The viscosity of the masticated rubber is strongly dependent on that of the bale rubber used. Rubbers with high initial viscosities tend to break down faster. The breakdown behavior also depends on the PRI values of the rubbers. Thus it is generally found that latex-grade rubbers such as RSS 1 and SMR L (with high PRI values) break down more slowly than cuplump grades such as SMR 20 (with lower PRI values) and therefore require more mastication.

Although the processability of natural rubber cannot be predicted by any single parameter, viscosity still remains the most widely used measure of processing quality. During mixing, good control of compound viscosity within fairly narrow limits is essential to insure smooth operation during further processing, such as extrusion and injection molding. For natural rubber, a relatively good correlation exists between mixed-batch viscosity and raw-rubber viscosity if the viscosities are not too low or too high (Fig. 6.4).[38]

In extrusion of a fully mixed batch, the batch viscosity is the main factor controlling the die swell and the stress developed. The grade of rubber has a lesser influence. Thus viscosity is a useful guide to processing behavior for masticated rubbers at intermediate viscosity levels. However, this is not so for the viscosity-stabilized grades. A masticated rubber has better extrusion properties than a nonmasticated rubber of the same viscosity.

Unvulcanized compounds of natural rubber have superior green strength and building tack compared to other elastomers.[39]

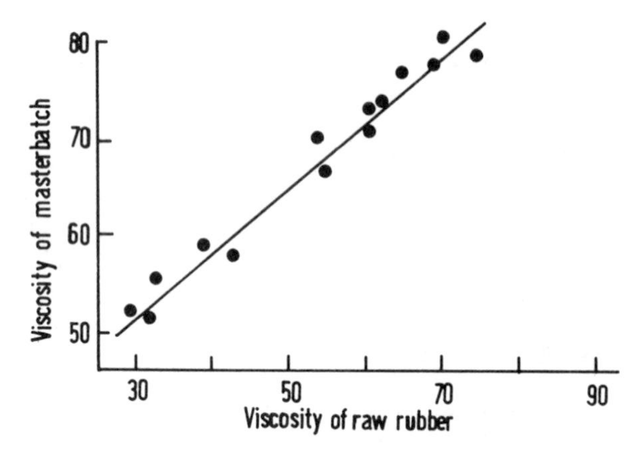

Fig. 6.4. Viscosity (ML 1 + 4 at 100°C) of tread masterbatch as a function of raw-rubber viscosity. (Ref. 38)

VULCANIZATE PROPERTIES

The physical properties of natural rubber vulcanizates are dependent, like other elastomers, on several variables such as compound viscosity, type, and amount of fillers, degree of filler dispersion, degree and type of crosslinking, and so forth.

Strength

Natural rubber is well-known for the strength properties of its vulcanizates. The tensile strengths of gum vulcanizates range from 17 to 24 MPa while those of black filled vulcanizates range from 24 to 32 MPa. Strength can also be characterized as tear resistance (i.e., the force needed to induce tear) or as cut-growth resistance, in both of which natural rubber is excellent. This high strength of natural rubber is certainly due to its ability to undergo strain-induced crystallization. The strength drops rapidly with increase in temperature but is still better than in other elastomers.

Abrasion and Wear

Natural rubber has excellent abrasion resistance, especially under mild abrasive conditions. The abrasion resistance of natural-rubber compounds is improved by blending with a small proportion of polybutadiene. The wear resistance of tire treads depends on the tire-surface temperature. Below about 35°C, natural rubber shows better wear than SBR, while above 35°C, SBR is better.[40]

Skid Resistance

The skid resistance of tread compounds of the same hardness depends on their viscoelastic properties. Thus the highly resilient natural rubber compounds show lower skid resistance than SBR at temperatures above 0°C. The wet-skid resistance of natural rubber above 0°C can be improved by extending with oil during mixing or by use of OENR. Treads from OENR are as good as OESBR on wet roads and are superior on icy roads.[41,42] Thus OENR blended with polybutadiene is suitable for treads of winter tires.

Dynamic Properties

Natural rubber has high resilience, with values exceeding 90% in well-cured gum vulcanizates. However, except for DPNR, the values are 1–2% lower than those of synthetic *cis*-1,4-polyisoprene.[13] In black-filled vulcanizates, the differences are smaller.

At large strains, the fatigue life of natural rubber is superior to that of SBR; the reverse is true for small strains. Good resistance to flexing and fatigue together with high resilience makes natural rubber useful in applications where cyclic stressing is involved.

Compression Set

Compression set and related processes, such as creep, are poorer in natural rubber than in synthetic polyisoprene. This advantage for the latter is due to the effect of the nonrubber substances, and DPNR shows much improved properties. Compression set is reduced by a good cure.

Aging

Natural-rubber vulcanizates can be given adequate heat-aging resistance by a suitable choice of vulcanization systems and by use of amine or phenolic antioxidants. Similarly, the poor ozone resistance under both static and dynamic conditions can be improved with waxes and antiozonants of the *p*-phenylenediamine type.

PRODUCT APPLICATIONS

With its wide range of useful properties, natural rubber can be used in a large variety of applications. Despite this, the share of natural rubber in the elastomer market has decreased progressively since World War II (Table 6.9).[2,3] This is partly due to the higher prices of natural rubber relative to SBR and partly due to inadequate supplies. The increase in the share of natural rubber in the last

Table 6.9. World Consumption of Natural Rubber[a]

Year	*NR ('000 Tonnes)*	*NR Share of Total Elastomers (%)*
1940	1130	100
1950	1750	75
1960	2100	52
1970	2990	35
1978	3730	29.8
1980	3760	31.2
1983	3970	32.5

[a]Sources: References 2 and 3.

Table 6.10. Use of Natural Rubber in Products

Product	*Percent[a]*
Tire and tire products	70–72
Mechanical goods	9–10
Latex products	7–8
Footwear	4–5
Engineering products	3–4
Adhesives	1–2
Others	2–3

[a]Estimated from different sources.

few years is due to the large switch to radial tires in the United States and elsewhere. The approximate usage of natural rubber by products is shown in Table 6.10.

Tires

Almost no natural rubber is used in the treads of passenger car tires in the USA, Western Europe, and Japan due to the better wet-skid resistance and wear characteristics of OESBR/BR blends. There is a small use of OENR in winter-tire treads. In passenger-car bias-ply tires, natural rubber is used only in the carcass, where the hot tear resistance, adhesion, and tack of natural rubber are used to advantage. In passenger-car radial-ply tires, natural rubber is used in the carcass as well as the sidewall, the latter due to the superior fatigue resistance and low heat build-up of natural rubber.

In commercial vehicles, the amount of natural rubber used increases with the size of the tire. In large earthmover tires, for example, almost 100% natural rubber is used due to the requirements of low heat generation and high cutting resistance.

Natural rubber is used in blends with halobutyl rubbers in the inner liner of tubeless tires.

Mechanical Goods (Industrial Products)

These include a large variety of products such as hose, conveyor belts, rubber linings, gaskets, seals, rubber rolls, rubberized fabrics, etc. In these products, the choice of elastomer is made on the best compromise between price and performance. Natural rubber is used in some products only because it has certain properties that cannot be matched by any other rubber.

Engineering Products

Engineering products deserve special mention as a special class of industrial products. Rubber is a unique engineering material because, unlike other engineering solids, it has high elastic deformability and an almost theoretical value for Poisson's ratio (0.5). The stiffness of a natural-rubber component in different directions may be varied independently by the judicious use of shape effects.[44] In dynamic applications such as springs, antivibration mountings, bushings, and so forth, high fatigue resistance, good strength, and durability are additional points in favor of natural rubber. However, natural rubber is not suitable if it has to come in contact with oil. In other applications such as bridge bearings, factors such as weathering, ozone resistance, and low-temperature flexibility are also important. Natural rubber is now accepted as suitable for use in bridge bearings, in place of, for example, neoprene.

Latex Goods

Natural-rubber latex has now been largely replaced by polyurethane in foam for upholstery and bedding. The main uses of latex are in dipped goods, foam, carpet backing, thread and adhesives.

Footwear

Natural rubber is extremely suitable for rubber footwear manufacturing. Its use is limited only by cost.

PRODUCT COMPOUNDING FORMULATIONS

When natural rubber is received in the factory, its cure properties are usually tested in a rheometer, using one of the standard test recipes shown in Table

Table 6.11. Standard Recipes for Cure Study of Natural Rubber

Ingredients	*ACS 1*	*TBBS*	*ISO Black*
Rubber	100	100	100
ZnO	6	5	5
Stearic acid	0.5	2	2
Sulfur	3.5	2.25	2.25
MBT	0.5	–	–
TBBS	–	0.70	0.70
IRB No. 4	–	–	35

Table 6.12. Formulations for Tires

		Passenger Car	
	Truck Tread A	*Carcass B*	*Sidewall & Shoulder C*
Natural rubber	100	100	40
OESBR	–	–	55
Polybutadiene	–	–	27.5
Aromatic oil	5	–	5
Coumarine resin	–	–	2
ISAF black (N-220)	45	–	–
GPF black (N-660)	–	–	75
FEF black (N-550)	–	45	–
Zinc oxide	5	8	3
Silica	–	10	–
Bonding agent	–	3.5	–
Stearic acid	3	0.8	2
Antioxidant	1[a]	1.0[b]	2.5[a]
Antiozonant	1	–	–
Wax	2	–	1
Hexamethylenetetramine	–	1.5	–
CBS	–	–	1.25
OBS	0.6	0.7	–
TMTD	–	–	0.1
Sulfur	2.5	2.8	2.4
Retarder	–	0.1	–
Total	165.1	173.4	216.75
Cure, minutes/°C	35/140	15/150	15/150
Vulcanizate Properties			
Hardness, IRHD	67	69	58
Tensile strength, MPa	31.6	22.0	16.2
Elongation at break, %	601	350	450
Modulus 300%, MPa	11.1	17.5	10.6

[a]N-Isopropyl-N′-phenyl-*p*-phenylenediamine.
[b]N-(1,3-Dimethybutyl)-N-*p*-phenylenediamine.

6.11. The so-called ACS-1 recipe has been the most widely used formulation and still remains popular. However, the use of MBT is not typical of factory formulations. The TBBS-based system is a good alternative but is relatively less established as a test recipe. In gum formulations, the ACS-1 recipe shows greater sensitivity in discriminating between grades of natural rubber having small differences in cure behavior.

The compounding formulations used in specific products vary slightly from factory to factory. Some typical formulations for use with natural rubber in a variety of products are given in Tables 6.12–6.17. The products include tires, conveyor belts, hose, general mechanicals, engineering components, and footwear. These are taken mainly from the Natural Rubber Technical Information Sheets published by the Malaysian Rubber Producers' Research Association.[45–49]

Table 6.13. Formulations for Conveyor Belts[45]

	*Cover**	*Friction**	
	A	*B*	*C*
Natural rubber	100[a]	100[b]	100[b]
Process oil[c]	4	–	–
Coal tar oil	–	9	–
Coumarone resin	–	15	–
Pine tar	–	–	7.5
HAF-LS black (N-326)	45	–	–
SRF-LM-NS black (N-762)	–	18	20
Coated calcium carbonate	–	12	10
Zinc oxide	5	10	15
Stearic acid	2	1	2
Antioxidant[d]	2	1	1
CBS	1.4	1	0.5
TMTD	0.4	–	0.5
Sulfur	0.35	2.5	0.5
Total	160.15	169.5	157.0
Cure, minutes/°C	35/140	25/141	25/141
Vulcanizate Properties			
Hardness, IRHD	64	50	40
Tensile strength, MPa	28.8	20.9	18.0
Elongation at break, %	605	595	680
Modulus 300%, MPa	7.8	5.3	2.6

*Formulation A is a heat-resistant cover suitable for sharp, abrasive materials, e.g. for conveying hot sinter and hot foundry sand. Formulations B and C are suitable as general purpose and heat-resistant friction mixes, respectively. These can be used with natural or synthetic fabrics, but nylon, rayon and polyester textiles must first be treated to promote adhesion.

[a]Latex grade SMR WF, SMR CV. [b]SMR 10, SMR 20. [c]Low-volatility oil.

[d]Polymerized 2,2,4-trimethyl-1,2-dihydroquinoline.

Table 6.14. Rubber Hose[46]

	*Cover**	*Lining**	
	A	*B*	*C*
Natural rubber (SMR 10, SMR 20)	100	100	50
SBR 1500	–	–	50
Process oil[a]	5[a]	–	5[b]
Factice[c]	–	25	–
FEF black (N-550)	60	–	50
Zinc oxide	5	5	10
Stearic acid	2	2	2
Antioxidant	2[d]	2[e]	4[f]
Wax	2	–	–
TBBS	1	–	2.1
CBS	–	0.6	–
TMTD	–	–	1
DTDM	1	–	–
Sulfur	1.2	3	0.25
Total	179.2	137.6	174.35
Cure, minutes at 140°C	45	30	60
Vulcanizate Properties			
Hardness, IRHD	70	42	70
Tensile strength (TS), MPa	22	23	18
Elongation at break (EB), %	410	740	450
Modulus 300%, MPa	17	2.4	12
Change in TS after 3 days at 100°C, %	−20	–	+10
Change in EB after 3 days at 100°C, %	−25	–	−5
Change in TS after 7 days at 70°C, %	–	−10	–
Change in EB after 7 days at 70°C, %	–	−15	–

*Formulation A is designed for a high-strength, general-purpose black-hose cover and may be suitable for intermittent exposure at elevated temperature. Formulation B is suitable for the lining of a grit- and sand-blasting hose. Formulation C is designed for a heat-resistant hose lining and may be suitable for use with low-pressure steam.

[a]Any process oil is suitable. [b]High-viscosity aromatic type.

[c]e.g., Whitbro 844 (Anchor).

[d]N-(1,3-Dimethylbutyl)-N-phenyl-*p*-phenylenediamine.

[e]Polymerized 2,2,4-trimethyl-1,2-dihydroquinoline.

[f]2 parts of 2-mercaptobenzimidazole and 2 parts of (e).

Table 6.15. Formulations for General Mechanicals[47]

	*A**	*B**	*C**
Natural rubber (SMR 10)	100	100	100
Process oil[a]	10	10	10
GPF black (N-660)	5	35	60
Zinc oxide	10	10	10
Stearic acid	2	2	2
Antidegradant[b]	1.5	1.5	1.5
Wax	1.5	1.5	1.5
CBS	0.6	0.6	0.6
Sulfur	2.8	2.8	2.8
Total	133.4	163.4	188.4
Cure, minutes at 153°C	20	20	20
Vulcanizate Properties			
Hardness, IRHD	39	50	61
Tensile strength, MPa	21.0	20.9	20.1
Elongation at break, %	690	605	475
Modulus 300%, MPa	1.9	5.6	11.4
Compression set, 1 day at 70°C, %	21	24	23

*Formulations A–C span the hardness range 40–60 and are suitable for use in a wide range of molded goods in which resistance to heat-ageing is not critical.
[a]Any process oil.
[b]e.g. N-Isopropyl-N′-phenyl-*p*-phenylenediamine.

Abbreviations

The following standard abbreviations have been used for the common accelerators:

CBS	N-Cyclohexylbenzothiazole-2-sulfenamide
DTDM	N,N′-Dithiobismorpholine
MBT	2-Mercaptobenzothiazole
OBS	N-Oxydiethylene benzothiazole-2-sulfenamide
TBBS	N-*t*-Butylbenzothiazole-2-sulfenamide
TBTD	Tetrabutylthiuram disulfide
TMTD	Tetramethylthiuram disulfide

REFERENCES

1. Fordyce Jones, in *History of the Rubber Industry*, P. Schidrowitz and T. R. Dawson, eds., Heffer, Cambridge, 1952, Ch. 1.
2. P. W. Allen, *Natural Rubber and the Synthetics*, Crosby Lockwood, London, 1972.
3. IRSG, *Rubber Statistical Bulletin* **38** (12), 2 (1984).

Table 6.16. Formulations for Engineering Components[48]

	A	*B*
Natural rubber (SMR CV)	100	100
Process oil[a]	9	-
SRF-LM-NS black (N-762)	10	25 ↔ 80
Zinc oxide	5	5
Zinc 2-ethylhexanoate	2	2
Antioxidant[b]	2	2
OBS	1.44	1.44
TBTD	0.6	0.6
Sulfur	0.6	0.6
Total	130.64	136.64 ↔ 191.64
Cure, minutes at 150°C	35	35
Vulcanizate Properties		
Hardness, IRHD	36	45 ↔ 65
Tensile strength (TS), MPa	23	29 ↔ 20
Elongation at break, (EB), %	710	625 ↔ 385
Compression set, 1 day at 70°C, %	11	11
Resilience, Dunlop, 50°C, %	92	88 ↔ 71
Change in TS after 28 days at 70°C, %	+4	−2 ↔ −7
Change in EB after 28 days at 70°C, %	−10	−9 ↔ −24
Change in TS after 3 days at 100°C, %	−3	−14 ↔ −9
Change in EB after 3 days at 100°C, %	−13	−16 ↔ −24

[a]Paraffinic or naphthenic oil.
[b]Polymerized 2,2,4-trimethyl-1,2-dihydroquinoline. Antiozonants can be added if desired.
The above formulations are the soluble EV system and span the hardness range 35–65 IRHD. They are suitable, with minor modifications, for a wide range of engineering applications including springs, couplings, mountings, bearings, seals, and bushes.

4. M. J. Dijkman, *Hevea: Thirty Years of Research in the Far East*, University of Miami Press, Coral Gables, FL 1951.
5. D. C. Blackley, *High Polymer Latices*, Vol. I, Maclaren, London, 1966, Ch. IV.
6. P. De Jonge, *J. Rubb. Res. Inst. Malaysia* **21,** 283 (1969).
7. B. C. Sekhar, *Malaysian Rubber Review* **2** (2), 5 (1979).
8. W. Leong and P. K. Yoon, *Proceedings of the Rubber Research Institute of Malaysia Planters' Conference*, Kuala Lumpur, 1976, p. 87.
9. P. D. Abraham et al., *J. Rubb. Res. Inst. Malaysia* **23,** 90 (1971).
10. Ismail Hashim, T. C. P'ng, O. K. Chew and J. L. Anthony, *Proceedings of the Rubber Research Institute of Malaysia Planters' Conference*, Kuala Lumpur, 1978, p. 128.
11. *International Standards of Quality and Packing for Natural Rubber Grades* ("The Green Book"), The Rubber Manufacturers' Association, New York, 1969.

Table 6.17. Formulations for Shoe Soles[49]
(for Injection-compression Molding)

	Heavy Duty	Sports or Casual	
	A*	B*	C*
Natural rubber	100[a]	90[b]	80[b]
High styrene resin[c]	20	–	20
Process oil	–	10	10
Mineral rubber[d]	5	–	–
HAF-LS black (N-326)	75	–	–
Precipitated calcium carbonate	–	69	69
Aluminum silicate	–	30	30
Zinc oxide	5	5	5
Titanium dioxide	–	10	10
Stearic acid	3	1	1
Antioxidant	1[e]	1[f]	1[f]
Paraffin wax	1	1	1
OBS	0.8	–	–
TMTD	0.3	–	0.5
CBS	–	2.0	2.0
Sulfur	2.5	1.2	1.2
Total	212.6	210.2	230.7
Cure, minutes at 160°C	5	5	5
Vulcanizate Properties			
Hardness	85	62	66
Tensile strength, MPa	20.2	16.1	14.3
Elongation at break, MPa	345	570	530
Modulus 200%, MPa	11.5	–	–
Modulus 300%, MPa	–	5.1	5.8
Compression set, 1 day at 70°C, %	20	43	40

*Formulation A is suitable for molded-on soling for heavy-duty applications such as army or working boots. Formulations B and C are suitable for good-quality, white sports-shoe soles, or, with modifications, for colored soles.
[a]SMR CV [b]SMR L [c]*Polysar SS 250* (Polysar)
[d]*MRX* (Anchor) [e]Polymerized 2,2,4-trimethyl-1,2-dihydroquinoline
[f]Non-staining phenolic antioxidant

12. D. J. Graham, *J. Rubb. Res. Inst. Malaysia* **22,** 14 (1969).
13. J. E. Morris, *op. cit.* **22,** 39 (1969).
14. "SMR Bulletin No. 10, Rubber Research Institute of Malaysia," Kuala Lumpur, 1981.
15. "SMR Bulletin No. 9, Rubber Research Institute of Malaysia," Kuala Lumpur, 1979.
16. L. Bateman and B. C. Sekhar, *Rubber Chem. Technol.* **39,** 1608 (1966).
17. S. Semegen and Cheong Sai Fah, in *The Vanderbilt Handbook*, Robert O. Babbit, ed., Vanderbilt, Norwalk, CO, 1978, Ch. 5, p. 35.
18. H. W. Greensmith, *NR Technol.* **4,** 13 (1973).
19. *Plrs' Bull. Rubb. Res. Inst. Malaysia* **35,** 28 (1958).

20. S. W. Sin and K. K. Leong, in *Proceedings of the Tyre Rubber Seminar 1972*, J. C. Rajarao and Cheong Sai Fah, eds., Rubber Research Institute of Malaysia, Kuala Lumpur, 1974, p. 65.
21. C. M. Lau and W. P. Chang, *Technology Series Report 3*, Rubber Research Institute of Malaysia, Kuala Lumpur, 1976.
22. "Epoxidised Natural Rubber", Rubber Research Institute of Malaysia, Kuala Lumpur, 1983.
23. C. S. L. Baker, I. R. Gelling, and R. Newell, presented at the 125th Meeting of the Rubber Division, American Chemical Society, Indianapolis, May 1984, Paper No. 11.
24. D. J. Elliott, *NR Technol.* **12,** 59 (1981).
25. H. Hasma, "Lipids in the Latex and Rubber of *Hevea brasiliensis* Muell. Arg. and their Effects on Some Properties of Natural Rubber," Thesis, Rijksuniversiteit Ghent, 1984, p. 128.
26. E. H. Andrews and A. N. Gent in *The Chemistry and Physics of Rubberlike Substances*, L. Bateman, ed., Mclaren, London, 1963, Ch. 9. p. 235.
27. Ong Chong Oon, Private Communication, Rubber Research Institute of Malaysia, 1972.
28. G. T. Knight and A. S. Tan, *Proceedings of the International Rubber Conference 1975*, Kuala Lumpur, Vol. 4, p. 115.
29. H. S. Lim, Private Communication, Rubber Research Institute of Malaysia, 1974.
30. E. J. Gregg and J. H. Macey, *Rubber Chem. Technol.* **46,** 47 (1973).
31. B. C. Sekhar, *Proceedings of the Natural Rubber Research Conference 1960*, Kuala Lumpur, p. 512.
32. A. Subramaniam, *Proceedings of the International Rubber Conference 1975*, Kuala Lumpur, Vol. 4, p. 3.
33. A. Subramaniam, "Rubber Research Institute of Malaysia Technology Bulletin No. 4," Kuala Lumpur, 1980.
34. J. I. Cunneen and G. M. C. Higgins in reference 26, Ch. 2, p. 19.
35. D. J. Elliott in *Developments in Rubber Technology*, Vol. I, A. Whelan and K. S. Lee, eds., Applied Science Publishers, London, 1979.
36. J. I. Cunneen, *Rubber Chem. Technol.* **41,** 182 (1968).
37. M. Porter, *NR Technol.* **4,** 76 (1973).
38. G. M. Bristow, *NR Technol.* **10,** 53 (1979).
39. James D. D'Ianni, *J. Rubb. Res. Inst. Malaysia* **22,** 214 (1969).
40. K. A. Grosch, *J. Inst. Rubber Ind.* **1,** 40 (1967).
41. K. A. Grosch, *Rubber Age* **99,** (10), 63 (1967).
42. A. Schallamach, *NR Technol.* **1,** No. 3, (1970).
43. G. M. Bristow, J. I. Cunneen and L. Mullins, *NR Technol.* **4,** 16 (1973).
44. P. B. Lindley in *Rubber in Engineering 1973*, P. B. Lindley and H. G. Rodway, eds., NRPRA, London, 1973, Paper A.
45. MRPRA, *NR Technical Information Sheet* D 58, D 80 (1979).
46. MRPRA, *NR Technical Information Sheet* D 8, D 9 (1976).
47. MRPRA, *NR Technical Information Sheet* D 105, D 117, D 118 (1982).
48. MRPRA, *NR Technical Information Sheet* D 72 (1979).
49. MRPRA, *NR Technical Information Sheet* D 43 (1978).

7

STYRENE-BUTADIENE RUBBERS

JAMES NEIL HENDERSON
Tire Materials Research
The Goodyear Tire & Rubber Company
Akron, Ohio

More than half of the world's synthetic rubber is styrene-butadiene rubber (SBR). World usage of SBR is about the same as that of natural rubber: between eight and nine billion pounds of each (see Fig. 7.1 and Tables 7.1 and 7.2). The factors accounting for this dominance are economic and technological: (1) the availability of styrene and butadiene precursors in fossil hydrocarbons makes these two monomers preeminent among the possible sources of synthetic rubber, and (2) they can be combined in rubber compounds that can be processed conveniently in tire molds, where thermal treatment converts them into tires. Much of the modern world's transportation relies on these durable, tough elastic composites. In fact, the tire industry has come to depend on SBR more than on natural rubber, partly because the latter has at times been harder to get, and partly because SBR often is better for the purpose required. Similarly, other synthetic rubbers have competed on the basis of both availability and technological utility; in some cases, they have even found niches, but lesser ones. It is instructive to examine history, searching for the reasons SBR has won its dominance, especially at a time when its share of the market is declining.

HISTORICAL BACKGROUND

Shortages of natural rubber have repeatedly been the major spur for the development of synthetic rubbers. At such times, production of synthetics has been able to get established even if their properties are significantly inferior to those of the natural product. During World War I, Germany used methyl rubber, made from dimethylbutadiene, because the Allied blockade almost completely cut off the supply of natural rubber from the tropical regions where it was produced. Methyl rubber was so inadequate, given the level of compounding and tire technology of that time, that its production was dropped at war's end. In

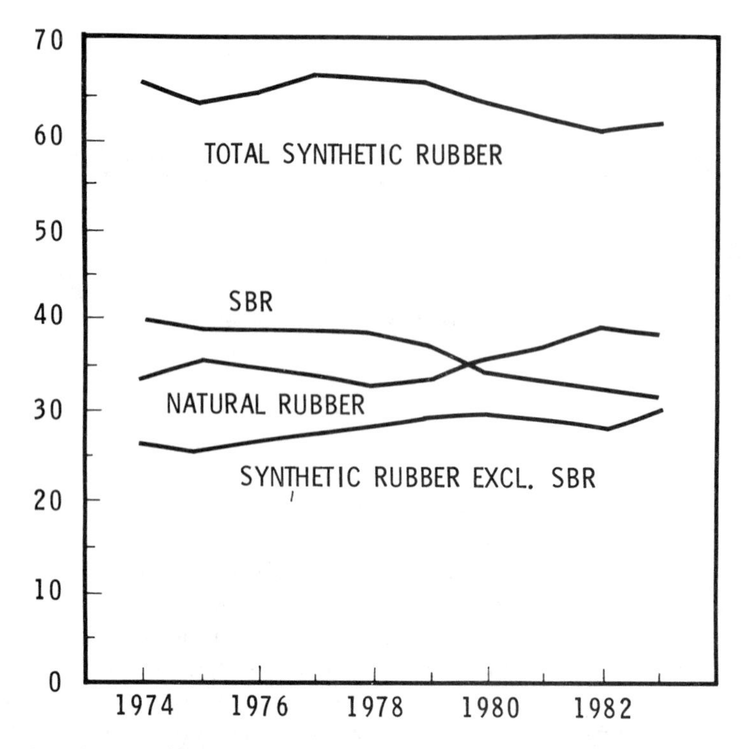

Figure 7.1. Percent of free-world rubber supply. (Adapted from *Worldwide Rubber Statistics, January 1985*, International Institute of Synthetic Rubber Producers, Inc., Houston TX. Table IV, p. 13.)

the 1920's, the ability of emulsion systems to achieve both high production rates and high-molecular-weight products using free radical catalysts was recognized. German chemists experimented with various monomers during that decade and the next and, by the early 1930's, had developed butadiene-styrene copolymer (Buna S) and butadiene-acrylonitrile copolymer (Buna N). This was all part of a national effort to make Germany independent of foreign sources of rubber.

Similarly, the massive American effort of World War II was driven by the shortage of natural rubber and was accelerated when the Japanese took control of Asian plantation areas. The U.S. program chose the butadiene-styrene copolymer as the major synthetic rubber to replace natural. In America it was then called GR-S (Government Rubber-Styrene). The name SBR superceded GR-S after the sale of the plants to private industry in 1955. No significance remains in the order in which the monomers styrene and butadiene are named in these copolymers. SBR is a generic term covering a wide variety of rubbers differing not only in the ratio of styrene/butadiene and their content of other substances

Table 7.1. Free-World Synthetic Rubber Production.[a]

Thousands of metric tons	*1973*	*1974*	*1975*	*1976*	*1977*	*1978*	*1979*	*1980*	*1981*	*1982*
Styrene-butadiene										
Solid	3269	2877	2577	2750	2950	2936	2915	2583	2467	2254
Latex	247	246	237	386	379	405	425	310	288	253
Polybutadiene	783	757	699	784	895	919	981	910	898	832
Polyisoprene	290	220	196	210	208	223	236	221	167	128
Ethylene-propylene	180	198	179	239	290	310	348	316	341	279
Polychloroprene	329	313	289	311	319	328	354	314	317	322
Butyl	349	336	316	353	386	415	441	415	406	374
Nitrile	201	195	138	181	190	208	236	207	199	182
Acrylates	—	—	28	29	29	30	32	32	31	25
Other	74	72	36	48	60	73	80	70	73	59
Total	5722	5214	4695	5291	5706	5847	6048	5378	5187	4708

[a]Adapted from *Chemical & Engineering News*, Apr. 30, 1984, p. 38.
Note: Production of communist countries not included. Source: International Institute of Synthetic Rubber Producers.

Table 7.2. U.S. Synthetic Rubber Production.[a]

Thousands of metric tons	*1974*	*1975*	*1976*	*1977*	*1978*	*1979*	*1980*	*1981*	*1982*	*1983*
Styrene-butadiene	1466	1179	1333	1395	1395	1378	1074	1032	879	905
Polybutadiene	310	290	353	361	378	397	311	342	288	333
Ethylene-propylene	126	84	130	158	174	176	145	178	123	175
Polychloroprene	163	144	165	165	161	183	151	143	119	115
Nitrile	88	55	73	70	73	75	63	66	45	55
Other[b]	406	229	305	334	379	319	327	373	381	396
Total	2559	1981	2359	2483	2560	2528	2071	2134	1835	1979

[a]Adapted from *Chemical & Engineering News*, April 30, 1984, p. 48.
[b]Includes such elastomers as polyisoprene, butyl, silicone, and urethane rubbers.
Source: Department of Commerce, industry sources.

such as soap and extender oil, but also in the type of polymerization by which they are made.

In the 1950's, the powerful synthetic utility of Ziegler-Natta and organolithium catalysis was discovered. Within a few years, it was possible to make not only synthetic replicas of natural rubber but also Solution SBR and many other controlled-structure elastomers that were entirely new. This revolution seemed likely to threaten the dominance of both NR and Emulsion SBR. In fact, however, the production of both rubbers continued to increase. One reason is that the new "stereorubbers" had to be made in solvents, which are more expensive and less expendable than water, although this factor may on occasion have been more than compensated for by other process differences and by property benefits. More significantly, in the U.S. and other industrially developed countries, existing plants were designed for water emulsions and were already paid for. Similarly, existing recipes and processes for mixing and curing had been optimized for NR and emulsion SBR. Finally—and in the long run it has to be seen as the really important factor—NR and emulsion SBR do have property advantages in many applications.

Glass-transition Temperature

One property that had been well-appreciated was the transition from rubber to "glass," which occurs when the temperature drops below a characteristic value for the polymer, known as the *glass-transition temperature*, T_g. Germany's methyl rubber had a T_g high enough so that tires made from it became glassy overnight in winter weather, a failing that was burned into the memory of a succeeding generation of rubber chemists. They called it poor "freeze resistance." The German chemists who developed Buna S and the American chemists in the Rubber Reserve program of World War II realized that they were controlling the freeze resistance, or T_g, when they manipulated the proportions of styrene and butadiene. It was not accidental that the tire grades of SBR had a range of freeze resistance bracketing that of NR. What was not so well-understood was that too low a T_g also had its disadvantages. When solution polybutadiene became available, with a T_g below $-100°C$, it was quickly found to have excellent durability in tire treads because of its resistance to abrasion. However, tires with all-polybutadiene treads were in use only briefly before it was discovered that improved treadwear was accompanied by decreased traction on wet roads at moderate temperatures.

The trade-off between traction and treadwear became recognized during the 1960's as a general rule for tread rubbers. A consequence is that any rubber with a T_g *outside* the range of about $-70°C$ to $-50°C$ has to be blended with at least one other rubber in general-purpose tread compounds. On the other hand, either SBR or NR can be used as the sole elastomer in a tire compound.

Processing

Processing is another advantage that emulsion SBR has over many synthetics. It mixes readily with carbon black and other compounding ingredients, readily forms a continuous band in milling, and extrudes smoothly. The reasons for its good processing are not fully understood. Molecular weight and molecular weight distribution are involved, and also the relatively large number of long branches in SBR molecules. The nature of the styrene monomer unit is perhaps also advantageous, compared to the monomer units found in other polymers, even after their diverse contributions to T_g have been taken into account.

Raw-material Sources

The hydrocarbon shortages of the 1970's wrote new pages into the story of SBR. To begin with, there was a year-long crisis in the supply of styrene, which temporarily upset the economics of SBR versus competitive rubbers such as NR and polybutadiene, particularly "medium vinyl polybutadiene." This rubber was designed to cover the same T_g range as SBR. Although none of it became commercially available during the styrene shortage, its development was given a boost. Then, the oil shortage and the rise of OPEC shook the economic foundation of the petrochemical industry, causing reevaluations of all the synthetic rubber processes and stimulating the development of nonpetrochemical routes. Mexico's guayule rubber, an alternative to natural rubber that does not require a tropical climate, and South Africa's coal-based polyisoprene, both arose from this stimulus, although in both cases the political motivation of assuring a domestic source of cis-polyisoprene has made them feasible regardless of economics. Also in both cases, these developments might affect SBR some day, although hardly at all so far.

The most important consequence of the world-wide realization that fossil hydrocarbon supplies are finite is the world-wide drive towards more efficient use. In the rubber industry, this has meant smaller growth, because automobiles began to be made smaller, with smaller wheels and lower loads. But most particularly, there has been an effort to design tires that waste less energy. This single factor, the loss of energy as heat from tires rolling on the road, has been estimated to account for 6 percent of the fuel consumed by medium-sized automobiles. The term most widely used for it in the rubber industry is *rolling resistance*. Even a small improvement would be highly significant because the amount of fuel consumed in transportation is such a large quantity. Efforts to reduce rolling resistance are having a big effect on the consumption of emulsion SBR, largely because of the perception that it is a poor rubber with respect to rolling resistance. Solution SBR, on the other hand, has the great advantage that its properties may be controlled by process changes during manufacture to give much improved rolling resistance. Consequently, there has been in recent years a trend toward replacement of emulsion SBR by solution SBR.

Radial Tires

Another technological factor has been the conversion to radial tires, a gradual process that began in Europe about 1970 and is still continuing. The building of radial tires entails a greater stress, tending to separate adjacent plies in the "green" (uncured) tire. Higher "green strength" in the rubber compound is thus required. Consequently, radialization has been accompanied by a shift towards more NR and less SBR in tire manufacture. Concurrently, economic trends have reinforced the return to NR. The price of all petroleum-based polymers has risen with the rise of OPEC. Both the quality and the supply of NR have improved with advances in the cultivation and marketing of Hevea, particularly in Malaysia.

Fig. 7.1 shows the actual trends in the world's supply of new rubber in the years from 1974 through 1983.

SOURCES AND MANUFACTURE OF EMULSION SBR

The methods developed in Germany in the 1930's and in the U.S. in the 1940's are in use in many parts of the world today. Raw material sources have undergone some changes.

Raw Materials

The monomers butadiene and styrene are the principal raw materials. Other raw materials in the manufacture of SBRs are extending oils, soaps, antioxidants, antiozonants, short-stopping agents, coagulating agents, molecular weight modifiers, catalysts, catalyst modifiers, water, and hydrocarbon solvents. Any of these materials or their residues may be present in the finished rubber; extender oil ranges from none to as much as 30%, and soap may amount to 5% or more in an emulsion SBR. The total nonrubber constituents are rarely more than 10%, except in oil-extended rubbers, and are usually less than 1 or 2% in a solution SBR. With the exception of water and solvents, the raw materials tend to end up in the finished rubber in amounts approaching their actual usage.

Butadiene. This monomer originates now mainly as a by-product in the operation of naphtha crackers for the production of ethylene. Three billion pounds were used in the U.S. in 1983, half of this amount for SBR and styrene-butadiene latexes. Nearly 60% of the total was domestic by-product butadiene, 25%

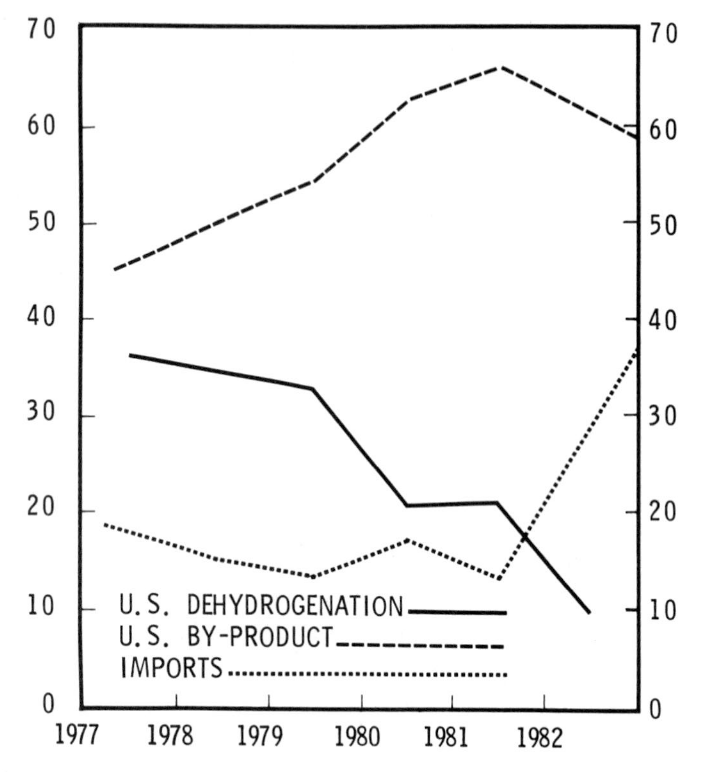

Figure 7.2. Butadiene supply by source, percent of total. (Adapted from *Data Resources Inc., Chemical Service*, Data Resources Inc., Lexington, MA, 1984. Chart II-3, p. 18.)

imported by-product, and the rest was made by the old method, dehydrogenation of other C_4 hydrocarbons (see Fig. 7.2). Dehydrogenation is competitive now only when there is strong demand for butadiene. The amount available from naphtha crackers is dependent not only on the demand for ethylene but also on the feedstock going into the crackers. Light feedstocks (rich in ethane) give as little as 1.5% butadiene and heavy naphtha feedstocks as much as 15%. This situation suggests a continuing instability in the supply and price of co-product butadiene, with dehydrogenation providing a moderating buffer. Past periods of by-product abundance, however, have driven dehydrogenators out of the business.

Styrene. This monomer is a larger commodity than butadiene. Whereas the rubber industry consumes most of the butadiene available (more than 75% in the U.S. in 1984, of which styrene-butadiene polymers accounted for 46%—see Table 7.3), the rubber industry's use of styrene has become a relatively minor part of the styrene market. In 1984, the rubber industry used roughly

Table 7.3. Outlook for SBR and Other Butadiene Derivatives.[a]

Butadiene Consumption by Derivative, millions of pounds	*1983*	*1984*	*1985*	*1986*	*1987*
Styrene butadiene rubber	1166	1266	1265	1259	1288
Polybutadiene rubber	680	749	738	761	804
Hexamethylenediamine	431	438	441	486	544
Styrene butadiene latexes	295	313	317	330	336
Acrylonitrile butadiene styrene	212	227	222	237	253
Neoprene rubber	196	228	231	230	235
Nitrile rubber	79	91	88	89	91
Other	138	149	152	156	164
Total	3197	3461	3454	3548	3715
		Percent of Total Consumption			
Styrene butadiene rubber	36.5%	37	37	36	35
Polybutadiene rubber	21.2	22	21	21	22
Hexamethylenediamine	13.5	13	13	14	15
Styrene butadiene latexes	9.2	9	9	9	9
Acrylonitrile butadiene styrene	6.6	6	6	7	7
Neoprene rubber	6.1	7	7	6	6
Nitrile rubber	2.5	3	2	2	2
Other	4.3	4	4	4	4

[a]Adapted from "Chemical Service, Butadiene and Derivatives," Data Resources, Inc., (March 1985), Table II-1, p. 28.

13% (SBR alone, 7%) of the 6 billion pounds styrene consumed in the United States (see Table 7.4). Nonrubber uses for styrene continue to grow, but rubber uses are nearly stable. Among the reasons for this stability are the trends toward smaller tires and toward radial tires, which give better mileage.

The availability of pure styrene by dehydrogenation of ethylbenzene, using either the Dow Chemical process or the Badische Anilin-und-Soda Fabrik process, was a factor leading to the selection of SBR as the major synthetic rubber of World War II. In the 1940's and 1950's, styrene was manufactured mainly for SBR. Soon, however, polystyrene plastics overshadowed styrene-based elastomers. If styrene growth projections of about 6% per year through 1990 are fulfilled, it will be because of the growth of plastics rather than of rubbers. Styrene operating capacity in the U.S. is about 9.5 billion pounds, but current consumption is 6 billion pounds.

MANUFACTURE OF EMULSION SBR

Free-radical-emulsion processes for the polymerization of styrene/butadiene developed before 1950 are still in widespread use. In a typical process (see Table 7.5 and Fig. 7.3), a soap-stabilized water emulsion of the two monomers is converted in a train of, for example, ten continuous reactors of 4000 gallons

Table 7.4. Outlook for SBR and Other Styrene Derivatives.[a]

Styrene Consumption by Derivative, millions of pounds	*1983*	*1984*	*1985*	*1986*	*1987*
Polystyrene (crystal)	1866	2090	2069	2175	2321
Polystyrene (impact)	1668	1720	1859	1968	2099
Acrylonitrile butadiene styrene	555	643	647	664	698
Styrene butadiene latexes	407	435	446	462	468
Unsaturated polyester resins	407	465	481	495	522
Styrene butadiene rubber	363	393	398	398	405
Miscellaneous styrene resins	196	233	238	254	271
Styrene acrylonitrile	69	73	76	80	88
Other	130	141	150	154	160
Total	5661	6193	6364	6650	7032
		Percent of Total Consumption			
Polystyrene (crystal)	33.0%	34	32	33	33
Polystyrene (impact)	29.4	28	29	30	30
Acrylonitrile butadiene styrene	9.8	10	10	10	10
Styrene butadiene latexes	7.2	7	7	7	7
Unsaturated polyester resins	7.2	8	8	7	7
Styrene butadiene rubber	6.4	6	6	6	6
Miscellaneous styrene resins	3.5	4	4	4	4
Styrene acrylonitrile	1.2	1	1	1	1
Other	2.3	2	2	2	2

[a]Adapted from "Chemical Service, Benzene and Derivatives," Data Resources, Inc. (March 1985), Table III-2, p. 32.

capacity each. Water, butadiene, styrene, soaps, initiators, buffers, and modifier are fed continuously; the temperature is maintained at 5°C to 10°C; and conversion is allowed to proceed until it is about 60% in the last reactor. Downstream, shortstop is mixed into the emulsion to stop the process at that conversion; unreacted butadiene is flashed off with steam and recycled; unreacted styrene is stripped off in a subsequent step and also recycled; and the rubber is recovered from the latex in a series of operations. These may include introduction of antioxidants, blending with oil, blending for molecular weight control, dilution with brine in preparation for the coagulation step, coagulation, dewatering, drying, and packaging the rubber.

Polymerization at 5°–10°C is called the *cold process*, and the product is called *cold SBR*, in distinction from *hot SBR*, which is made at about 50°C (see Table 7.5). The older hot process uses initiators such as potassium peroxydisulfate (or potassium persulfate), whereas cold initiators are redox combinations such as *p*-methane hydroperoxide/ferrous sulfate, which are much more active and provide higher molecular weight at lower temperature. In the hot process, conversion is allowed to proceed to about 70%; one result is a higher degree of branching and incipient gelation. Cold SBR gives better abrasion resistance and consequently better treadwear, as well as better dynamic properties. The higher

Table 7.5. Typical SBR Recipes.[e]

	Cold SBR 1500	*Cold SBR 1502*	*Hot SBR 100*
Composition, phm[a]			
Butadiene	71	71	71
Styrene	29	29	29
p-Menthane hydroperoxide	0.12	0.12	-
Potassium peroxydisulfate	-	-	0.3
t-Dodecylmercaptan	0.2	0.18	-
n-Dodecylmercaptan	-	-	0.5
Emulsifier makeup, phm[a]			
Water (final monomer-water ratio adjusted to 1:2)	190	190	190
Sodium stearate (soap flakes)	-	2.5	5
Disproportionated rosin acid soap	4.5-5	2.55	-
Trisodium phosphate dodecahydrate (buffer electrolyte)	0.50	-	-
Tripotassium phosphate (buffer electrolyte)	-	0.40	-
Tamol N[b] (secondary emulsifier)	0.02–0.1	0.02–0.1	-
Versene Fe-3[c] (iron complexing agent)	0.01	0.01	-
Sodium dithionite (oxygen scavenger)	0.025	0.025	-
Sulfoxylate activator makeup, ppm[a]			
Ferrous sulfate heptahydrate			
Versene Fe-3[c]	0.04	0.04	-
Sodium formaldehydesulfoxylate	0.10	0.10	-
Water	10	10	-
Shortstop Makeup, phm[a]			
Sodium dimethyldithiocarbamate	0.10	0.15	-
Sodium nitrite	0.02	0.02	-
Sodium polysulfide	0.05	-	-
Hydroquinone	-	-	0.1
Water	8.0	8.0	-
Polymerization conditions			
Temperature, degrees C	5	5	50
Final conversion, %	60–65	60–65	72
Coagulation	salt-acid	Salt-acid	Salt-acid
Antioxidant (about 1 phr[d])	non-staining	Staining	Staining
Properties of coagulated dried rubber			
Organic acid content, wt %	5–7	5–7	4–6
Styrene content, wt %	24	24	24
Mooney viscosity at 100 °C, ML-4, min	46–58	48	46–58

[a]Parts per hundred parts of monomer.
[b]Sodium salt of naphthalenesulfonic acid-formaldehyde condensate (W. R. Grace).
[c]Disodium salt of ethylenediaminetetracetic acid (EDTA) (The Dow Chemical Company).
[d]Parts per hundred parts of rubber.
[e]From R. G. Bauer, *Kirk-Othmer Enc. of Chem. Tech.*, 3d Ed., vol. 8, John Wiley & Sons, 1979, p. 611.

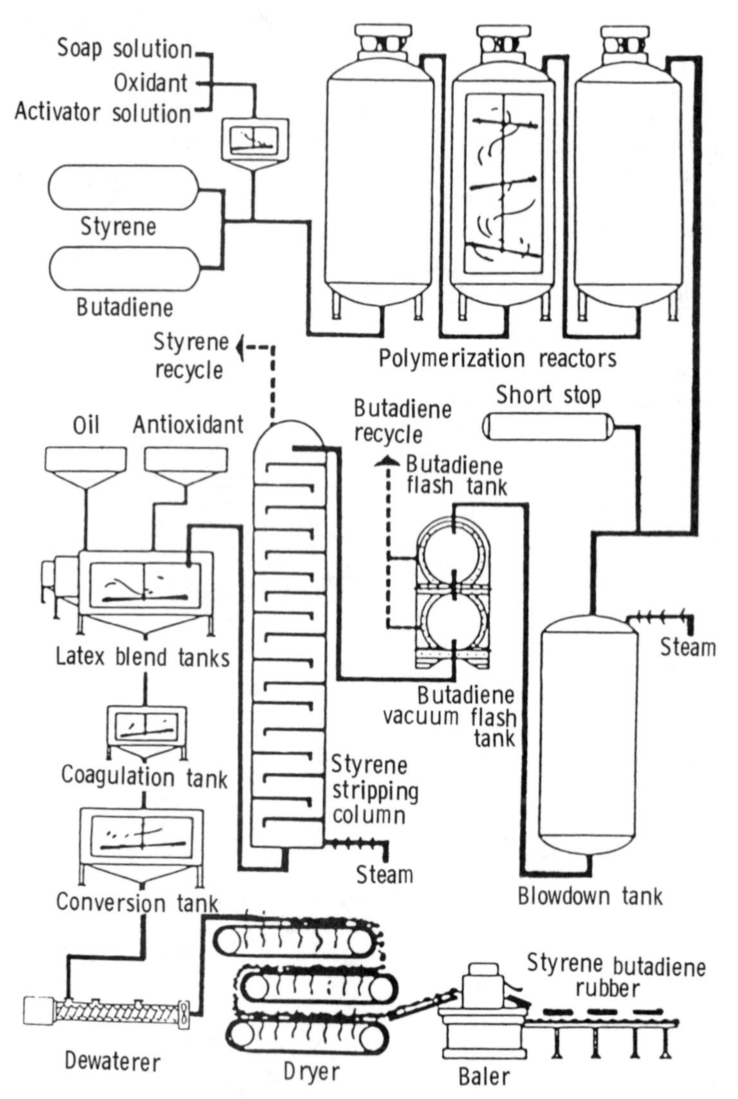

Figure 7.3. Manufacture of emulsion SBR. (Adapted from R. G. Bauer, "Styrene-Butadiene Rubbers," in *Vanderbilt Rubber Handbook*, R. O. Babbit, ed., R. T. Vanderbilt Company, Inc., Norwalk, CT, 1978. Fig. 1, p. 59.)

molecular weights obtainable also permit higher levels of oil extension. For these reasons, hot SBRs have been replaced in tire applications.

Black masterbatch SBR is also supplied by synthetic rubber producers, primarily for fabrication of rubber goods in locations where mixing of carbon black is not convenient. Carbon black has many advantages as a component in SBR vulcanizates:

1. It is a filler, with a cost usually lower than that of the elastomer itself.
2. It is a reinforcing agent, the black-loaded vulcanizate being many times stronger than the gum (unfilled) vulcanizate.
3. It stiffens and hardens the vulcanizate.
4. It usually improves aging, especially in sunlight.
5. It can provide electrical and thermal conductivity.

Effective mixing, however, requires large, expensive equipment, and the process is environmentally messy. Fabricators often prefer elastomer with the black premixed—black masterbatch. It can be further mixed with other materials in extruders or simpler equipment.

Carbon black for black masterbatch is usually added as a water slurry before the coagulation step. Pelletized black is pulverized by steam in a micronizer and quenched with a water spray, which produces a slurry to be fed to the stripped latex. The latex is then coagulated with acid and the rubber dried and packaged. Oil-extended black masterbatches are also supplied.

SOLUTION SBR

Solution SBR has comprised a minor proportion of SBR since the early 1960's. It has always enjoyed certain property advantages over emulsion SBR, including the following:

1. Easily controlled molecular weight and a narrow distribution of molecular weight.
2. Little or no branching.
3. A higher concentration of pure elastomer since there is no soap.
4. Considerably more control over the structural details of its molecules.

These differences were not always perceived as advantages. For example, it was necessary to broaden the molecular weight distribution and introduce some branching in order to improve the processing and correct the tendency for the raw rubber to flow during storage. Rubber scientists were not at first able to provide much guidance as to how the structural details of SBR should be changed in order to improve properties. In recent years, however, better results have been achieved with solution SBR than with emulsion in making tire rubber blends that can simultaneously meet the competing demands of the different parts of vehicle tires. For example, a better balance of treadwear, traction, and rolling resistance has been reported with various versions of solution SBR in tread compounds. Consequently, there has been some acceleration in the replacement of emulsion SBR with solution SBR, and a more rapid trend is projected.

Vinyl Content and Blockiness

The term ''solution SBR'' has come to mean, almost exclusively, a copolymer of styrene/butadiene prepared in hydrocarbon solution with a lithium initiator. Polar modifiers such as diethyl ether or tetrahydrofuran may be added. They modify not only the butadiene microstructure but also the rates at which the monomers enter the polymer. This ''modified lithium'' process enables the manufacturer to tailor a product far beyond the options available in the emulsion processes. For example, the vinyl content can be controlled from about 8 to 80%. In emulsion, about 15 to 20% of the butadiene unit will be vinyl and not much can be done to change this proportion. In addition, the manufacturer of a solution SBR can distribute styrene and butadiene units more or less randomly through the polymer chain (much as they are distributed in emulsion SBR), or construct polymer chains that are quite ''blocky'' in their monomer distribution. In recent years, many patents have been issued in Japan, Europe, and America describing improvements made in tire compounds by including solution SBR in the blend of elastomers. On a variety of grounds, raising the vinyl to some value higher than 15% or introducing a certain degree of blockiness is often claimed to be essential for improvements in tire performance.

Molecular Weight and Branching

Molecular weight, molecular weight distribution, and branching are three more structural features (in addition to vinyl content and blockiness) that the lithium processes control but emulsion processes do not. The reason is that a lithium polymerization proceeds by the linear addition of monomer units without branching or termination; the molecules grow until they run out of monomer or until some optional reagent is added. If more monomer is supplied, per lithium species, they grow to a higher molecular weight. If all the polymer molecules begin to grow at the same time, as they do in batch polymerization with a very rapid initiator, they all grow to the same size, and the molecular weight distribution is very narrow. If a ''branching agent'' is present, such as a reagent with more than one function of similar reactivity, the molecules will branch; otherwise, they will be linear. In this context, butadiene is effectively monofunctional because the residual double bond is far less reactive towards the propagating lithium species. A branching agent such as silicon tetrachloride is capable of converting a solution of ''live'' linear lithium polymer into a star-branched (four-armed) polymer with molecular weight four times the original lithium polymer. Under practical conditions, a fraction of linear polymer of the original molecular weight will also remain.

The profoundly different nature of free-radical emulsion polymerization does not permit these degrees of control. Unlike lithium species, propagating radical species terminate by coupling and in other ways. The essence of the emulsion process is that it tends to isolate the growing polymeric radicals, each within its own particle, where they continue to grow by the accretion of the monomer diffusing in until termination overtakes them. As the polymerization proceeds, the polymer/monomer ratio in the particles increases. Radical reactions other than monomer addition become more likely, such as (1) abstraction of hydrogen from within the polymer chain and creation of branched radicals, (2) crosslinking, and (3) termination by disproportionation. Since the emulsion process permits the growth of molecules beyond the desired range of molecular weight, modifiers are included in the recipe; these mercaptans limit molecular weight by terminating polymer molecules and transferring the radical to monomer. As a result of these processes, most of the polymer molecules at any time during the polymerization are dead. Successful polymerization requires continuous reinitiation, provided mostly by the slow decomposition of peroxide initiator. In contrast to lithium polymerization, molecular weight is controlled mostly by an added modifier, and not by the monomer/initiator ratio. Molecular weight distribution is quite broad because of transfer and termination. Branching is considerable, complex, and not amenable to control. Because of the faster rate of the radical process versus the anionic at the molecular level, the emulsion-polymer microstructure is more mixed and not very sensitive to reaction conditions. About 20% of the butadiene units are vinyl. Reactivity ratios, which express the relative tendencies of butadiene and styrene to enter the polymer, are much closer to unity in emulsion than in anionic, so that distribution of monomer units is closer to a random distribution and not controllable.

MANUFACTURE OF SOLUTION SBR

Solution SBR is typically made in hydrocarbon solution with n-butyllithium initiator. The reactors are often used interchangeably for the preparation of lithium polybutadiene. They may be jacketed and agitated, tubular, or adiabatic. The process may be continuous or batch. In principle, every polymer molecule remains live until a deactivator or some other agent capable of reacting with the anion intervenes. In practice, depending on the temperature at which polymerization is carried out or concluded, there is more or less extensive thermal termination. Trace impurities, or intentionally added poisons, may also terminate chains before the polymerization is complete. In any case, the manufacturer is still able to control molecular weight by controlling the ratio of monomer to initiator. Branching agents such as silicon tetrachloride are sometimes added in a subsequent reactor or mixing chamber. Similarly, terminating agents such as methanol may be added to terminate surviving chains. Antioxidants are also added while solvent is still present to protect the polymer during drying. Finally, any extending oil is added at this stage. The first step in the removal of solvent is usually steam stripping, followed by dewatering and expulsion through orifices at relatively high temperature.

Several kinds of solution SBR are offered to tire manufacturers. (See *IISRP Synthetic Rubber Manual*, 9th Ed.)

Property Control

Manufacturers have taken advantage of the design flexibility of solution SBR to control SBR properties in both drastic and subtle ways. With the object of maximizing the rolling resistance without sacrificing treadwear and traction in tires, they have manipulated the following:

1. Styrene/butadiene composition
2. Proportions of vinyl/cis/trans in the butadiene units
3. Sequence distribution of the monomer units (blockiness)
4. Molecular weight
5. Molecular weight distribution
6. Both the kind and amount of branching

Emulsion SBR manufacturers also control styrene/butadiene composition and thereby the T_g, which is the single most important determinant of the trade-off between traction and wear over the range that is practical for tire elastomers. Rolling resistance likewise has considerable dependence on T_g. Since hysteresis, the unrecoverable energy loss on deformation, increases with T_g, rolling resistance losses also increase with T_g. Thus, emulsion SBR does have some design flexibility in the important area of T_g.

Similarly, molecular weight (but not molecular weight distribution) can be controlled in emulsion just about as conveniently as it can in solution. Molecular weight affects strength, processing, and rolling resistance. There are ways, however, to control the molecular weight distribution of solution SBR not available in the manufacture of emulsion SBR. This possibility should be an advantage because the low molecular fraction always present in an emulsion system (because of chain transfer to modifier, for example) can be avoided in solution. Hysteresis and rolling resistance should thereby be improved (fewer chain ends) over that obtainable with emulsion, and this improvement is without any change in T_g and therefore without penalty in either wear or traction. There is a catch, however. The very narrow distributions possible in solution SBR do not process sufficiently well in Banbury mixing unless the average molecular weight is quite low. In that case, strength properties and wear *do* suffer. Similarly, the high-molecular-weight fraction found in emulsion SBR is believed advantageous. When all these considerations are weighed, little advantage accrues to the solution polymer through management of the molecular weight distribution per se.

Branching

Branching is manipulated in solution SBR. Since branches have a more drastic effect on processing than molecular weight distribution does, it is possible to optimize the processibility at a higher average molecular weight than with linear polymers. And this can be accomplished with a very discrete introduction of branches, avoiding the large number of chain ends usual in an emulsion polymer. The result is better rolling resistance without sacrificing wear or traction, as compared with the more randomly branched emulsion polymers of the same styrene and vinyl composition. One of the ways of achieving this result is to introduce to the unterminated "living" lithium polymer a multifunctional agent, such as silicon tetrachloride, in an amount somewhat less than the number of lithium atoms on the ends of polymer chains. This gives a large fraction of four-armed star molecules, which greatly improve the processing over what it would be at the same molecular weight of linear polymer. At the same time, the number of chain ends per molecular weight is minimized in star-branching compared with more random branching.

In practice, branching a solution polymer also broadens the molecular weight distribution. This broadening, which would tend to work against the goal of minimizing hysteresis, turns out to be the minor effect. Judicious branching improves the balance of tread properties in spite of it.

Thus, solution SBR can be made with better hysteretic properties (better rolling resistance) than the corresponding emulsion polymer without any corresponding disadvantage in treadwear, traction, or processing.

Blending

In at least two other ways, the ability to control the structure and architecture of SBR molecules may provide advantages. One is the intramolecular effect, from the choice of vinyl in place of styrene, and the other is an intermolecular effect, from the compatibility of different kinds of polymer molecules and consequently their tendencies to segregate in different phases when blended in the mill or mixer. There have been a number of patents aiming to exploit such effects in Solution SBR's.

The intramolecular effect is even more subtle. Mixtures of polymers may have very wide ranges of interdispersion. The average particle size may range from visible (mixtures by any criterion) to molecular (solutions by any criterion). Particle size may depend on the temperature, mixing history, and the frequency of stresses to which the particles may be subject. The degree of dispersion matters because it determines whether the mixture exhibits a single T_g or more than one T_g. All the important properties of the rolling tire have some dependence on the T_g or plurality of T_g's. Unlike emulsion SBR, solution SBR can be designed so that the sequences of monomer units are very different in different parts of the same molecule. One end can be primarily polybutadiene while the other is all polystyrene. The midsection can be more or less graded, or "tapered." Such polymers are said to be "blocky." They usually have more than one T_g, depending on the molecular weight, temperature, and frequency involved. Blockiness, with its attendant effects on T_g, is usually disadvantageous and avoided (though not always) in the manufacture of solution SBR. Even without blockiness, the great variety of structures that can be produced results in a greater range of compatibilities when solution SBR is blended with other rubbers. Many patents claim to exploit these effects for advantages in various parts of tires.

PROPERTIES

The single most important property of any elastomer is the glass-transition temperature, T_g. As has been discussed above, T_g is controlled in all SBR's by choosing the monomer ratio, and, in solution SBR's, by choosing the microstructure of the butadiene units as well. For SBR polymers prepared by emulsion polymerization at 50°C, T_g can be calculated from the styrene content (S = weight fraction of styrene) using the following empirical equation:

$$T_g = (-85 + 135S)/(1 - 0.5S)$$

For a similar copolymer prepared at 5°C, T_g is given by

$$T_g = (-78 + 128S)/(1 - 0.5S)$$

For solution copolymers, account must also be taken of the variable vinyl microstructure in the butadiene units. A general equation is based on the assumption that the reciprocal of the copolymer T_g is the sum of the reciprocals of the component homopolymer T_g's in proportion to their weight fractions in the copolymer. When these component T_g's are taken to be 100°C for polystyrene, −100°C for 1,4-polybutadiene, and 0°C for all-vinyl polybutadiene, the equation simplifies to the following:

$$T_g = [1/(0.00578 - 0.0031S - 0.00212V + 0.00212VS)] + 273$$

where S is the weight-fraction of styrene in the polymer and V is the weight-fraction of vinyl units in the butadiene portion of the polymer. If the polymer analysis is reported as % vinyl on the total polymer, V should be calculated from this analysis by simple proportionality as follows:

$$V = (\%\text{ vinyl in total polymer} \times 100)/(\%\text{ butadiene in polymer})$$

The general equation may also be used for emulsion polymers but is usually less exact than the empirical equations that are fitted to real data in a particular range. Solution SBR's that have significant blockiness, as a result of runs of styrene units adjacent to each other, do not obey any of the equations. The most readily observed glass transition in these blocky polymers is lower than one would calculate because it reflects all of the butadiene units but not all of the styrene units.

T_g Measurement

The importance of the glass-transition temperature for tire applications is hard to overstate. It defines the minimum temperature above which rubbery properties may be expected; below its T_g, SBR, like any other noncrystalline polymer, is glassy and brittle. T_g is the most important determinant of the abrasion process and consequently of tire wear. Similarly, it controls friction, and consequently tire traction. It also has a strong influence on hysteresis and consequently on the rolling resistance or fuel consumption caused by tires.

The glass-transition temperature may be readily measured by differential scanning calorimetry (DSC) or differential thermal analysis (DTA). The Gehman low-temperature torsion flex test (ASTM D 1053) is less convenient but is still useful in some cases. Various dynamic mechanical methods (see subsequent discussion) also provide values for T_g with much more precise control of frequency. As a cautionary note, it should be emphasized that any of these practical methods of measuring T_g produces values that are dependent on rates of change occurring during the measurement. Thus, T_g's by different methods may not agree.

Molecular-weight Measurement

Molecular weight and molecular-weight distribution (MWD) are important determinants of SBR properties, as has been discussed. Gel permeation chromatography (GPC) is widely used for the measurement of both molecular weight and MWD. Methods based on viscosity effects are also still used for molecular weight, via dilute solution viscosity (as inherent, or intrinsic, viscosity) or via bulk viscosity (as Mooney viscosity). None of these methods gives molecular weight directly. Each requires calibration against known standards and/or the determination of constants that are characteristic of the polymer. In particular, Mooney is widely used as an index of processing ease without any attempt to relate it to molecular weight.

Dynamic Mechanical Measurements

Perhaps the most important improvement in the science of characterizing SBR and other rubbers is in the area of dynamic mechanical measurements. These methods attempt to get at the problem of accounting for rubber behavior in actual service, where deformations occur cyclically as the tire rolls on the road. The rubber responds to some extent inelastically—absorbing or wasting (as heat or permanent distortion) part of the energy of deformation—and to some extent elastically—giving back work done on it during the deforming part of the cycle. The proportions depend on (1) the temperature and the T_g, (2) the frequency of the deformation; and (3) structural features such as molecular weight, distribution, and branching. The ''loss factor,'' tan δ [tan δ = (loss modulus)/(storage modulus)] is a related property that can be measured on dynamic mechanical testers such as the Rheovibron. Tan δ at 60°C is widely taken as an indicator of rolling resistance attributable to the rubber compound; tan δ at 0°C has some correlation with T_g and with skid resistance.

COMPOUNDING AND PROCESSING

The compounding of styrene-butadiene rubbers is similar to that of natural rubber and other unsaturated hydrocarbon rubbers. The most convenient and effective compounds in large-scale usage, such as tires, are all based on fillers, such as carbon black; extending oils; zinc oxide; sulfur; accelerators, such as mercaptobenzothiazole; and protective agents, such as antioxidants, antiozonants, and waxes. Processing these complex mixtures into smooth compounds that can be quickly pressed, sheeted, calendered, or extruded is a most important step for manufacture. Emulsion SBR is the prototype of a ''general-purpose'' rubber because of its ability to be blended with any other such rubber into compounds that process and cure homogeneously. Thus, it is said to have ''cure-compatibility'' and excellent ''processing'' or ''processibility.'' Besides T_g, and possible chemical attributes that may be important but are poorly understood, the good processing of SBR results from its favorable combination of molecular weight and molecular weight distribution and the considerable proportion of long branches in its molecules. The main advantage of solution SBR is that it can be constructed so as to have just enough branching and molecular weight distribution for adequate processing, while at the same time providing maximized molecular weight for rolling resistance and wear.

Compounding recipes with low sulfur or with only organically bound sulfur, as in thiazoles, lead to vulcanizates with better aging but slower curing. Zinc stearate, or zinc oxide plus stearic acid, is the most common activator for SBR. There are many accelerators that speed up slow-curing stocks and retarders that slow down ''scorchy'' ones. Recipes may also contain plasticizers, softeners, tackifiers, and other ingredients that have given evidence of solving some compounding problem or other.

Recipes that might be suitable for tires, and the expected properties, are given in Tables 7.6 and 7.7. Recipes for some nontire uses are given in Tables 7.8 and 7.9.

Table 7.6. Representative Recipe for Passenger Tire Treads.[a]

	Radial	*Bias*
SBR 1712	82.5	82.5
BR 1252	55	55
ISAF-HS black N-234	70	-
HAF-HS black N-339	-	70
REOGEN	1	1
Stearic acid	2	2
AGERITE RESIN D	2	2
ANTOZITE 67	1	1
Sunolite 240	3	3
Zinc oxide	3	3
Sulfur	1.75	1.5
DURAX	1	1
Total	222.25	222
Monsanto Rheometer at 160°C, 1° arc, 3 cpm		
t_{S1} minutes	4.5	4.8
$t_C(90)$ minutes	13.3	14.2
Press Cures at 160°C		
Minutes	13	13
Physical Properties		
Stress at 300%, MPa/psi	5.4/780	4.5/700
Tensile strength, MPa/psi	16.6/2410	17.4/2520
Elongation at break %	580	690
Shore durometer A	60	56
Goodyear—Healy Rebound, % 22°C	48.3	50.5
Dynamic Properties 22°C, 12 Hz, 0.44 MPa load, 3.5% ampl[b]		
E_E storage modulus, MPa	7.36	6.73
E_L loss modulus, MPa	2.04	1.73
Tan δ	0.279	0.263
Relative heat generation	0.349	0.0361
Ozone Cracking[b]		
Static	0	0
Dynamic cracks, mm	0.5 to 1.5	0.5 to 1.5
Flex Fatigue Life—(Monsanto)[b]		
kc at 2.19 extension ratio	205	125
kc at 2.36 extension ratio	239	175

[a]From *Vanderbilt Rubber Handbook*, R. T. Vanderbilt Company, Inc., Norwalk, CT, 1978, p. 650.
[b]Cured 20 min at 160°C.

Table 7.7. Representative Recipe for Truck Tire Treads and Sidewalls.[a]

	Radial	*Bias*	*Bias*	*SW Radial*
SMR 5	–	50	–	65
Tire rubber[b]	–	50	40	35
BR 1220	–	–	82.5	
SBR 1712	100	–	–	
ISAF-HM black N-220	55	55	70	
FEF black N-550	–	–	–	50
AGERITE SUPERFLEX	1	1	1	1
ANTOZITE 67	2	2	2	2.5
Sunolite 240	1.5	2	2.5	2
REOGEN	–	2	2	
Highly aromatic oil	6	10	8	
Stearic acid	2	2	2	2
Zinc oxide	4	3	3	3
Sulfur	2.2	1.8	1.5	1.8
AMAX	0.65	0.7	–	0.7
MORFAX	–	–	1	
Total	174.35	179.5	215.5	163
Monsanto Rheometer at 150°C, 1° arc, 3 cpm				
t_{S1} minutes	3.0	5.6	5.5	4.4
$t'_C(90)$ minutes	13	19	25	13
Press Cures at 150°C				
Minutes	22	22	22	13
Physical Properties				
Stress at 300%, MPa/psi	9.7/1410	5.1/740	4.6/670	7.5/1090
Tensile strength, MPa/psi	24.8/3600	19.4/2810	19.8/2810	18.6/2700
Elongation at break, %	580	700	760	530
Shore durometer A	58	54	54	–
Goodyear-Healy[c] rebound % 22°C	62.5	61.5	55.5	
Dynamic Properties 22°C, 12 Hz, 0.36 MPa load, 3.5% ampl[c]				
E_E storage modulus, MPa	6.23	5.39	6.01	–
E_L loss modulus, MPa	1.26	1.14	1.48	–
Tan δ	0.199	0.210	0.246	–
Relative heat generation	0.0303	0.0371	0.0386	–
Ozone Cracking				
Static	0	0	0	0
Dynamic, mm	0.5–1.5	0.5–2.5	0.5–2.5	0.25–1.0
Flex Fatigue Life (Monsanto)[d]				
kc at 2.19 extension ratio	98	301	81	71
kc at 2.36 extension ratio	57	63	45	53

[a]From *Vanderbilt Rubber Handbook*, R. T. Vanderbilt Company, Inc., Norwalk, CT, 1978. p. 652.
[b]90% rubber hydrocarbon—10% oil; calculated as 100% vulcanizable hydrocarbon.
[c]Cured 32 min at 150°C.
[d]Cured 30 min at 150°C.

Table 7.8. Representative Compounds for Conveyor Belts.[a]

Friction and Skim Coat	*1st Grade*		*Heat Resistant*
SBR 1500	75	SBR 1500	100
Smoked sheet	25	Smoked sheet	-
REOGEN	2	REOGEN	2
Stearic acid	1	Stearic acid	1
Zinc oxide	5	Zinc oxide	5
AGERITE RESIN D	1	AGERITE RESIN D	1.5
RIO RESIN N	-	RIO RESIN N	10
Paraflux	7	Paraflux	7
Sundex 790	7	Sundex 790	3.5
FEF black (N-550)	20	FEF black (N-550)	20
SRF black (N-770)	25	SRF black (N-770)	25
AMAX	1.25	AMAX	-
Sulfur	2.25	Sulfur	0.15
METHYL TUADS	0.2	METHYL TUADS	1.5
ETHYL TUADS	-	ETHYL TUADS	1.5
ALTAX	-	ALTAX	1
Totals	171.7		179.15
Press Cures: 15 min @ 153°C (307°F)			
Physical Properties			
Stress @ 300%, MPa/psi	5.58/810	Stress @ 300%, MPa/psi	2.07/300
Tensile strength, MPa/psi	15.65/2270	Tensile strength, MPa/psi	11.03/1600
% elongation	570	% elongation	880
Shore A hardness	52	Shore A hardness	46
Mooney Scorch and Plasticity @ 121°C (250°F)			
Min. to 5-point rise	46	Min. to 5-point rise	28
Plasticity	26	Plasticity	28

[a]From *Vanderbilt Rubber Handbook*, R. T. Vanderbilt Company, Inc., Norwalk, CT, 1978, p. 705.

Preparation of SBR compounds is similar to that of other rubbers. The ingredients are mixed in internal mixers or on mills and may then be extruded, calendered, molded, and cured in conventional equipment. Mixing procedures vary with the compound, In general, the rubber, zinc oxide, antioxidants, and stearic acid are mixed; then the carbon black is added in portions with the oil. This "nonproductive" mix may be considered a black masterbatch. It may be desirable at this point to dump, sheet out, and cool the batch. The next phase involves mixing in all the other ingredients, with the accelerator and sulfur being added last. The mix is now considered "productive," since it will "cure"—i.e., crosslink irreversibly—when held for a carefully regulated time at a selected temperature, such as 150°C.

Table 7.9. Representative Compounds for Washing Machine Hose[a]

SBR 1503	60
SBR 1018	40
REOGEN	2
Stearic acid	2
Zinc oxide	5
AGERITE GEL	2
Coumarone-indene resin	20
Sunproofing wax	7
Whiting	175
DIXIE CLAY	25
Titanium dioxide	5
Lithopone	10
Sulfur	3
ALTAX	1.5
METHYL ZIMATE	0.125
Total	357.625
(Press Cure: 10 min @ 153°C/307°F)	
Physical Properties	
Stress @ 300%, MPa/psi	1.24/180
Tensile strength, MPa/psi	2.21/320
% elongation	550
Shore A hardness	52
Mooney Scorch @ 121°C (250°F)	
Min to 5-point rise	0.60

[a]From *Vanderbilt Rubber Handbook*, R. T. Vanderbilt Company, Inc., Norwalk, CT, 1978, p. 721.

Mixing procedures for rubber stocks vary with different companies. Some plasticize the rubber in a separate operation before blacks and other ingredients are added. Some make a masterbatch first, without a preliminary plasticizing step. If additional plasticizing is required, the masterbatch is remilled, and self-plasticization of the rubber is avoided. Excessive remilling of SBR compounds may lead to gel formation and poorer extrusions.

Not only does modern styrene-butadiene rubber have extrusion properties superior to those of natural rubber, but its stocks have less tendency to scorch in processing. Although cold SBR is often preferable to hot for optimum physical properties, hot SBR can be better for both processing and product properties. Hot SBR breaks down more rapidly to a desirable molecular weight on the mill, develops less heat, and accepts more filler in processing. All types of SBR require less sulfur than natural rubber does for curing. The usual range is about 1.5 to 2.0 parts per hundred rubber. On the other hand, SBR requires more accelerator because of the lower unsaturation to achieve the same rate of cure.

Tire tread-wear and aging properties are superior to those of natural rubber; resistance to abrasion and resistance to crack initiation are better. Building tack is still poor and dynamic properties are such that heavy-duty tires become too hot in use. Without reinforcing fillers such as carbon black or silica, the physical properties of SBR are much inferior to those of natural rubber. Similarly, its green strength properties—for example, the tensile strength of the fully mixed compound before cure—are distinctly inferior to those of natural rubber. This is the principal reason it was necessary to go back to a higher proportion of natural rubber in radial tires, which would otherwise deform in transit from the tire-building machines to the curing presses.

APPLICATIONS

Of all the SBR made, about 75% goes into tires. The rest goes into shoes and other footwear, mechanical goods, sponge and foamed products, waterproofed materials, hose, belting, adhesives, and other miscellaneous uses. Styrene-butadiene latexes constitute a more or less independent 10% of the market, mostly nontire. SB latexes typically have much more styrene than SBR and are supplied in latex form to customers, who use them principally in carpet-back coatings. The construction industry is the next largest user.

BIBLIOGRAPHY

1. R. O. Babbit, ed., *Vanderbilt Rubber Handbook*, R. T. Vanderbilt Company, Norwalk, CT, 1978.
2. *Synthetic Rubber Manual*, 9th Ed., International Institute of Synthetic Rubber Producers, Houston, TX, 1983.
3. *Worldwide Rubber Statistics, January 1985*, International Institute of Synthetic Rubber Producers, Houston, TX, 1985.
4. Chemical Service, Data Resources Inc., Lexington, MA, 1984.
5. W. M. Saltman, ''Styrene-Butadiene Rubbers,'' in *Rubber Technology*, M. Morton, ed., 2d Ed., Van Nostrand Reinhold, New York, 1973.
6. R. G. Bauer, *Kirk-Othmer Encyclopedia of Chemical Technology*, 3rd Ed., Vol. 8, John Wiley & Sons, New York, 1979, pp. 608–625.
7. J. D. D'Ianni, ''Styrene-Butadiene Rubbers,'' in *Introduction to Rubber Technology*, M. Morton, ed., Reinhold, New York, 1959.
8. K. H. Nordsiek, Paper No. 48, ACS Rubber Division Meeting, Indianapolis, May 1984.
9. R. Bond, G. F. Morton, and L. H. Krol, *Polymer* **25,** 132–141 (1984).
10. K. M. Schur, *Elastomerics*, Jan. 1984, pp. 18–24.
11. C. R. Wilder, J. R. Haws, and T. C. Middlebrook, Paper No. 81, ACS Rubber Division Meeting, Houston, Oct. 1983.
12. M. Ogawa and M. Ikegami (Bridgestone Tire), U.S. Pat. 4,387,756 (Jun. 14, 1983).
13. T. Fujimaki, T. Yamada, and S. Tomita (Bridgestone Tire), U.S. Pat. 4,396,743 (Aug. 2, 1983).
14. L. Y. Chang and J. S. Shackleton, *Elastomerics*, Mar. 1983, pp. 18–26.
15. T. A. Chavchich, L. B. Nikitina, V. V. Senik, and Kh. N. Borodushkina, *Int. Polymer Sci. & Tech.* **10** (9), T/29–T/31 (1983).

16. J. Diamant, D. Soong, and M. C. Williams, *Polymer Eng. & Sci.* **22** (11), 673 (1982).
17. Y. Takeuchi, Y. Yoshimura, N. Ohshima, M. Sakakibara, and A. Tsuji (Japan Synthetic Rubber), UK Appl. 2,085,896 A (May 6, 1982).
18. R. E. Bingham, R. R. Durst, H. J. Fabris, I. G. Hargis, R. A. Livigni, and S. L. Aggarwal (General Tire & Rubber), U.S. Pat. 4,355,156 (Oct. 19, 1982).
19. C. Freppel (Michelin), U.S. Pat. 4,361,682 (Nov. 30, 1982).
20. R. J. Blythe, R. Bond, and G. Engelbertus (Dunlop), UK Appl. 2,071,117 (Sept. 16, 1981).
21. A. Y. C. Lou, *Tire Sci. & Tech.* **6** (3), 176–188 (1979).
22. T. C. Bouton and S. Futamura, *Rubber Age*, Mar. 1974, pp. 33–39.
23. H. E. Railsback, W. S. Howard, and N. A. Stumpe, *Rubber Age*, Apr. 1974, pp. 46–55.
24. G. Kraus, C. W. Childers, and J. T. Gruver, *J. Appl. Polymer Sci.* **11,** 1561–1591 (1967).
25. J. N. Short (Phillips Petroleum), U.S. Patent 3,094,512 (June 18, 1963).

8

POLYBUTADIENE AND POLYISOPRENE RUBBERS

L. J. KUZMA
Nordson Corporation
Westlake, Ohio

HISTORY AND DEVELOPMENT

The history and development of both polyisoprene and polybutadiene rubbers closely parallel each other. Each of these rubbers rose from meager beginnings to become keystones in the rubber and tire industries. The highlights of the progress made in these synthetic elastomers since their discovery have been reviewed in several articles.[1-7]

In 1879, Bouchardat undertook what was perhaps the first preparation of a synthetic rubberlike material by treating isoprene with hydrochloric acid.[8] Report of this work quickly triggered the search for other potential treatments capable of converting isoprene to a useful rubber. Progress in this area was slow and not fully successful. The first efficient catalysts discovered for polymerizing isoprene were the alkali metals. Polybutadiene was first prepared and studied by Lebedev, a Russian, in 1910 using alkali metals as the initiator in diene polymerizations.[9,10] His work eventually led to the first industrial production facility for synthetic rubber.

During World War I, Germany actively pursued the idea of replacing natural rubber supplies, cut off from the Far East by the British naval blockade, with a synthetic polybutadiene produced with an alkali metal catalyst. In the USSR, the rod polymerization of butadiene by metallic sodium was developed on a commercial scale and remained a mainstay of Soviet synthetic rubber production until the mid-1940's.[11]

In the 1920's, the use of emulsion polymerization to prepare high-molecular-weight elastomers at high production rates was demonstrated. In the 1930's, Germany concentrated heavily on these emulsion rubbers (Buna-series), an interest that continued into World War II. During that conflict, both the United States and Germany turned to emulsion rubbers, produced by free-radical catalysis, to meet their wartime rubber needs. Emulsion styrene-butadiene copolymers (GR-S) and acrylonitrile-butadiene copolymers (GR-N), produced in Government owned and operated synthetic rubber plants (The Rubber Reserve Company), temporarily filled the rubber void. Yet the production of polyisoprene and polybutadiene by these technologies gave products that were inferior, respectively, to natural rubber and to emulsion styrene-butadiene copolymers.

One major problem with emulsion polybutadienes is that they exhibit poor processing characteristics. In order to aid processibility, the emulsion rubber was prepared at a low Mooney viscosity, resulting in poor cold-flow properties. To compensate for this deficiency, the emulsion rubber was masterbatched with carbon black, the end product showing improved treadwear and resistance to cracking in tire applications. Despite arduous efforts, emulsion polybutadiene was not accepted as a large-scale commercial prospect during the 1940's.

It was not until the mid 1950's that interest in synthetic polyisoprene and polybutadiene was again renewed, the result of the discovery by Ziegler and coworkers[12] that ethylene could be polymerized to a high-molecular-weight polymer in the presence of a hydrocarbon solvent and a triethyl aluminum/titanium tetrachloride (Al–Ti) catalyst using relatively mild temperature and pressure conditions. As a result of this discovery, a myriad of synthetic elastomers soon followed.

Horne and coworkers[13, 14] polymerized isoprene to an essentially all *cis*-1,4-polyisoprene using the Al–Ti catalyst. This work was quickly followed by the synthesis of highly stereo-regular *cis*-1,4-polybutadiene by Phillips Petroleum Company.[15] The commercial development of these rubbers required only that a source of monomers be made available at a relatively low cost.

While this search for economic routes to monomer production proceeded, other technical aspects related to the commercial production of these stereo-regular-solution polymers were being tackled in several laboratories around the world. This work included studies on monomer purity requirements, the optimization of polymerization conditions, catalyst optimization, and the complex finishing processes required in production. The intensity and dedication of this research and development effort soon became evident, for by 1960, just a few years after the initial discovery of stereospecific catalysts, both polyisoprene and polybutadiene were being produced on a commercial scale.

POLYMER STRUCTURE

Polyisoprene

The polymerization of isoprene and butadiene are examples of addition polymerization in which the repeating structural unit within the polymer backbone has the same molecular weight as the entering monomer unit.

With isoprene, the building of this polymer backbone can occur in several

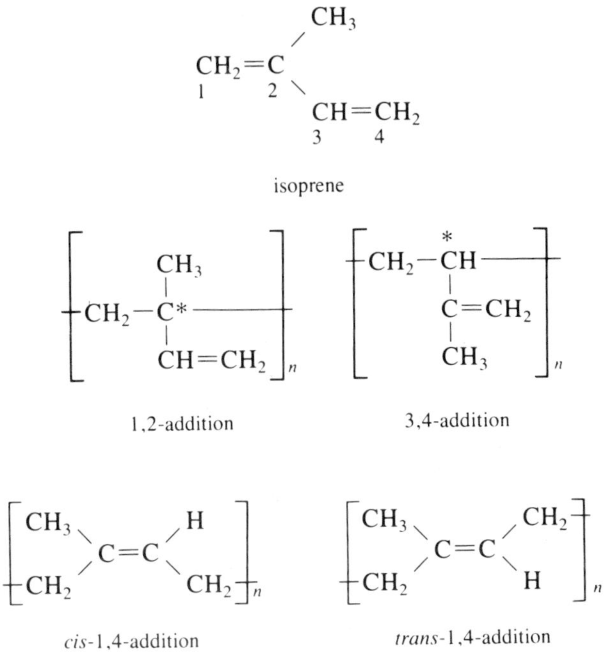

$$\underset{1}{CH_2}{=}\underset{2}{C}(CH_3){-}\underset{3}{CH}{=}\underset{4}{CH_2}$$

isoprene

$$\left[-CH_2-\overset{CH_3}{\underset{CH=CH_2}{C^*}}-\right]_n \qquad \left[-CH_2-\overset{*}{CH}(-C(CH_3){=}CH_2)-\right]_n$$

1,2-addition   3,4-addition

$$\left[\begin{matrix} CH_3 & & H \\ & C{=}C & \\ -CH_2 & & CH_2- \end{matrix}\right]_n \qquad \left[\begin{matrix} CH_3 & & CH_2- \\ & C{=}C & \\ -CH_2 & & H \end{matrix}\right]_n$$

cis-1,4-addition   *trans*-1,4-addition

Fig. 8.1. Isoprene and polyisoprene structures.

ways, depending upon where the addition occurs. The polymerization addition process can be a reaction involving the 1,2-, 3,4- or 1,4- positions of isoprene to give the structures shown in Fig. 8.1.

In the case of 1,2- and 3,4- addition, an asymmetric carbon is formed (see asterisks) that can have either an R or an S ("d" or "l") configuration. In general, equal numbers of R and S configurations are produced during an addition polymerization, which results in no net optical activity for the polymer. The disposition of the R and S configurations along the polymer backbone, however, results in diastereomeric isomerism. Although many combinations of sequences are possible, only three arrangements are commonly considered in polymers. These diastereomeric isomers are referred to as isotactic, syndiotactic, and atactic (see Fig. 8.2).

In the case of the isotactic polymer, all of the monomer units add in the same configuration (R or S). In the syndiotactic case, the polymer chain is composed of alternating configurations, whereas in the atactic polymer, the disposition of configurations is random.

The structures of isotactic and syndiotactic 3,4- and 1,2-polyisoprene are given in Fig. 8.3.

Of the eight possible arrangements and configurations discussed for poly-

—R—R—R—R—R—R—R—R—

isotactic

—R—S—R—S—R—S—R—S—

syndiotactic

—R—R—S—S—S—R—R—S—

atactic

Fig. 8.2. Schematic representations of tacticity.

isoprene, only three have been isolated: *cis*-1,4-, *trans*-1,4-, and atactic-3,4-polyisoprene.

A further structural complication arises in that the isoprene monomer units are capable of linking together in head-to-tail, head-to-head, or tail-to-tail arrangements (see Fig. 8.4).

The use of analytical procedures, such as pyrolysis and ozonalysis, to determine these sequences has been worked out only for linkages involving 1,4-isoprene units.[16] The higher stereo-regular polyisoprenes have been determined to contain mainly head-to-tail linkages.[17]

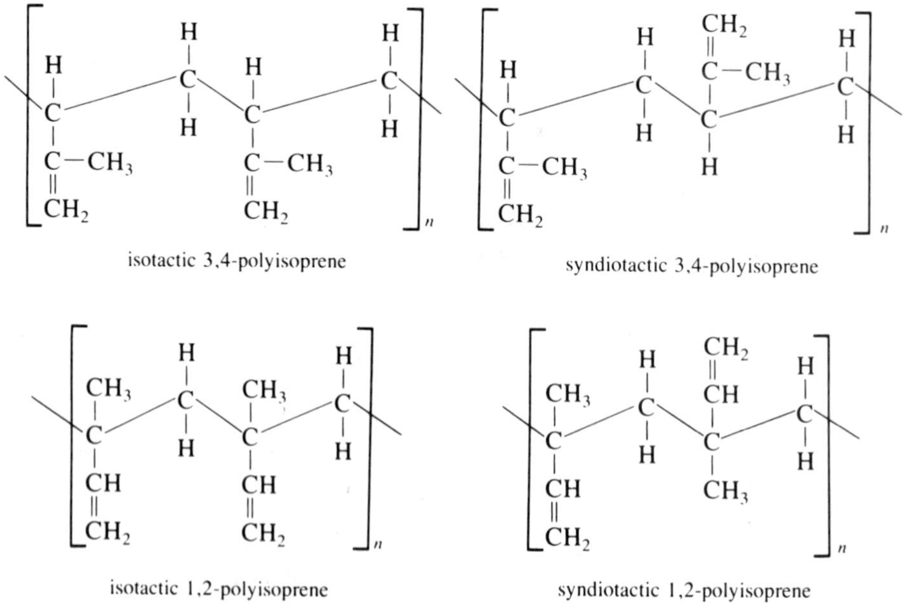

isotactic 3,4-polyisoprene

syndiotactic 3,4-polyisoprene

isotactic 1,2-polyisoprene

syndiotactic 1,2-polyisoprene

Fig. 8.3. Structures of isotactic and syndiotactic polyisoprene.

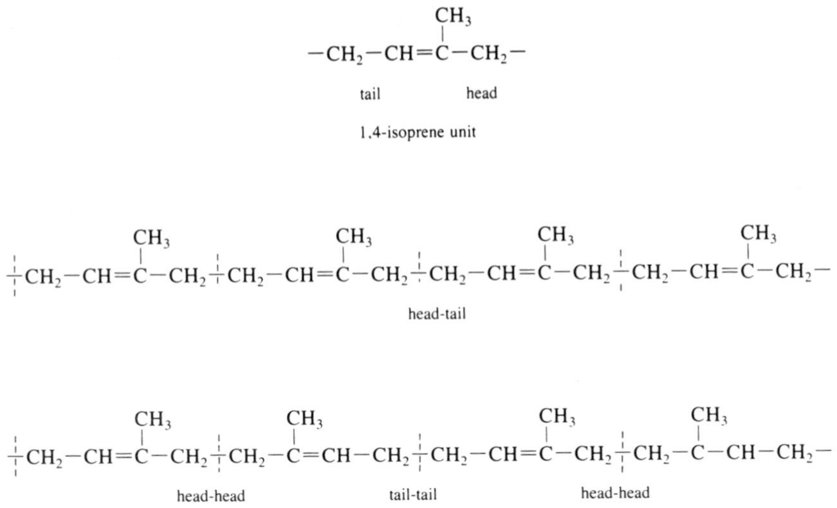

$$-CH_2-CH{=}\overset{\overset{\displaystyle CH_3}{|}}{C}-CH_2-$$

tail head

1,4-isoprene unit

$$\vdots CH_2-CH{=}\overset{\overset{\displaystyle CH_3}{|}}{C}-CH_2 \vdots CH_2-CH{=}\overset{\overset{\displaystyle CH_3}{|}}{C}-CH_2 \vdots CH_2-CH{=}\overset{\overset{\displaystyle CH_3}{|}}{C}-CH_2 \vdots CH_2-CH{=}\overset{\overset{\displaystyle CH_3}{|}}{C}-CH_2-$$

head-tail

$$\vdots CH_2-CH{=}\overset{\overset{\displaystyle CH_3}{|}}{C}-CH_2 \vdots CH_2-\overset{\overset{\displaystyle CH_3}{|}}{C}{=}CH-CH_2 \vdots CH_2-CH{=}\overset{\overset{\displaystyle CH_3}{|}}{C}-CH_2 \vdots CH_2-\overset{\overset{\displaystyle CH_3}{|}}{C}-CH-CH_2-$$

head-head tail-tail head-head

Fig. 8.4. Head-tail and head-head, tail-tail 1,4 linkages in 1,4-polyisoprene.

Cyclopolyisoprene

Another possible polyisoprene structure can result from the cyclopolymerization of isoprene. When isoprene is polymerized by certain Ziegler-Natta or cationic catalysts, a cyclic, saturated structure forms that is similar to that obtained by cyclization of 1,4- and 3,4-polyisoprenes.[18, 19] Figure 8.5 shows the commonly accepted structure for the cyclic portion of the cyclopolyisoprene chain.[20] These polymers are insoluble, powdery polymers because of inter and intramolecular crosslinking and also the rigidity resulting from the cyclic segments.

Crystallinity

The more highly stereoregular polyisoprenes demonstrate varying degrees of crystallinity. A high degree of crystallinity requires that the polymer possess long sequences in which the structure is completely stereoregular, that is, it

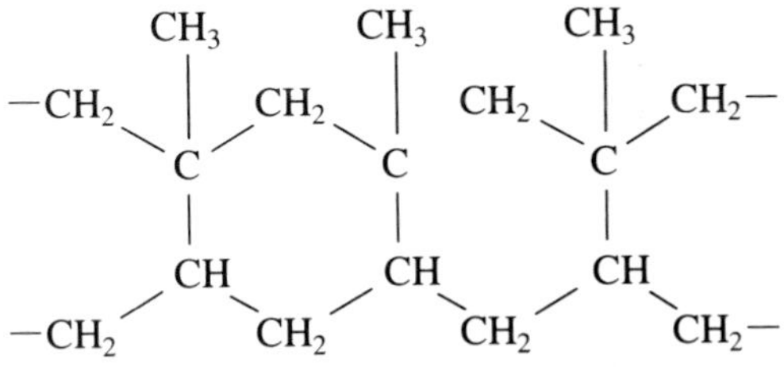

CH₃ CH₃ CH₃
$-CH_2$ CH_2 CH_2 CH_2-
C C C
CH CH CH
$-CH_2$ CH_2 CH_2 CH_2-

Fig. 8.5. Structure of cyclic portion of cyclopolyisoprene.

must contain linear segments composed exclusively of 1,4-, 1,2-, or 3,4-polyisoprene units. In addition, all the monomer units must be linked in the head-tail configuration.[17]

Natural rubber (hevea) crystallizes on standing at room temperature or below, with maximum crystallization rates occurring in the range, −11°C to −35°C.[21] At 0°C, complete crystallization occurs in 10 days, whereas at −25°C only a couple of hours are required. No crystallization occurs at temperatures below −50°C.[22] Crystallization rates are markedly affected by any stretching of the rubber. At high elongations of the rubber, the crystallization process is complete in a matter of seconds.[22]

The synthetic *cis*-1,4-polyisoprenes, prepared with Ziegler-type Al–Ti catalysts, are similar to natural rubber in their crystallization behavior, having identical crystal structures.[14,23]

Polyisoprene prepared with lithium catalysts show significant crystallization only upon stretching.[24–26] This is not surprising since recent data[26a] on the microstructure has shown that the *cis*-1,4 content may be as low as 70 percent, depending on the reaction conditions, i.e., type of solvent, lithium concentration, etc.

Gutta percha, a natural *trans*-1,4-polyisoprene, has two crystalline forms: the stable, high-melting α form and the less stable, lower-melting β form. The α form melts at 58 to 67°C and the β form at 50 to 57°C.[27] The synthetic *trans*-1,4-polyisoprene prepared with a Ziegler-type catalyst gives essentially identical crystallization behavior to its natural counterpart.[27]

Gel and Branching. The structure and behavior of polyisoprene is further complicated by the possibility of branching and gel. The degree of gel and branching is dependent upon the source of the *cis*-1,4-polyisoprene. For Al–Ti catalyzed *cis*-1,4-polyisoprene, the gel content is 5 to 25%, most of which is termed "loose" gel since it breaks down readily upon mastication.[28] The gel content in natural rubber is usually greater than that in the Al–Ti catalyzed polymer. This gel is also a "loose" gel. Lithium polymerized *cis*-1,4-polyisoprene usually contains no gel. Both natural rubber and Al–Ti-catalyzed *cis*-1,4-polyisoprene also contain large quantities of microgel.[29]

The determination of the nature and degree of branching in natural rubber and Al–Ti catalyzed *cis*-1,4-polyisoprene are complicated by the presence of the gel and microgel. Removal of the gel/microgel fraction from the soluble portion can have a marked effect on the resulting physical properties such as viscosity, molecular weight, and branching.

Polybutadiene

The configurations of polybutadiene are *cis*-, *trans*-, and vinyl. Since either or both of the double bonds in butadiene can be involved in the polymerization mechanism, the resulting polymer may have a variety of configurations. These result from the fact that the spatial arrangement of the methylene groups in the polybutadiene backbone allow for geometric isomerism to occur along the polymer chain. The different polymer structures of polybutadiene are given in Fig. 8.6.

Participation of both double bonds in the polymerization process gives rise to a 1,4-addition, which can be either *cis*-1,4- or *trans*-1,4-, depending upon the disposition of groups about the polymer double bond. Participation of only one double bond results in a vinyl, or 1,2-addition, that can have three possible structures, just as 1,2- and 3,4-polyisoprene have: isotactic, syndiotactic, and atactic.

The four structurally different configurations of polybutadiene give rise to notably different behavior.[30] Since the pure stereoisomers are highly regular, they are likewise highly crystalline. The *cis*-1,4-isomer crystallizes upon stretching over 200%. The *trans*-1,4 and vinyl isomers are crystalline without elongation.

The high *cis*-1,4-polybutadiene is a soft, easily solubilized elastomer that exhibits excellent dynamic properties, low hysteresis, and good abrasion resistance. A glass-transition temperature of $-102°C$ is reported for this rubber.[31] This low glass temperature is believed to account for its excellent low-temperature performance, abrasion resistance, and high resilience.

The *trans*-1,4-polybutadiene, in contrast, is a tough elastomer, having a reported glass-transition temperature of $-107°C$[32] and $-83°C$ (94% *trans*-).[33] In addition to its high hardness and thermoplasticity, it is sparingly soluble in most solvents.

The 1,2-isotactic and 1,2-syndiotactic polybutadienes are rigid, crystalline materials with poor solubility characteristics. A glass-transition temperature value of $-15°C$ is reported for these rubbers.[32] The atactic polymers are soft elastomers possessing poor recovery characteristics.

Another important factor influencing polymer properties is the interaction or relationship of a polymer chain unit of given microstructure with its neighboring units. An example of this is 1,2-polybutadiene, where both isotactic and syndiotactic configurations are possible.

The monomer unit may also be capable of adding unsymmetrically. In 1,2-addition, the polymer chain is irregular, depending as it does upon which end of the incoming monomer unit adds to the growing chain. In 1,4-addition, the modes of addition are all equivalent.

Other polymer-chain variations, such as crosslinking, may also lead to differences in polymer physical properties.

POLYMERIZATION PROCESS

Most commercial processes for the preparation of polyisoprene and polybutadiene employ solution polymerization. In general, these systems are based upon

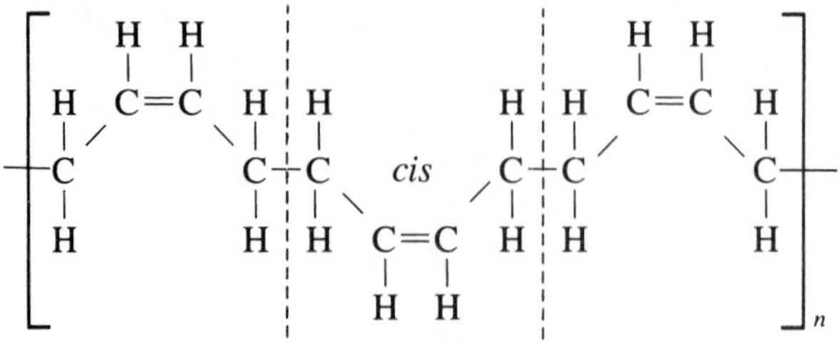

cis-1,4-polybutadiene

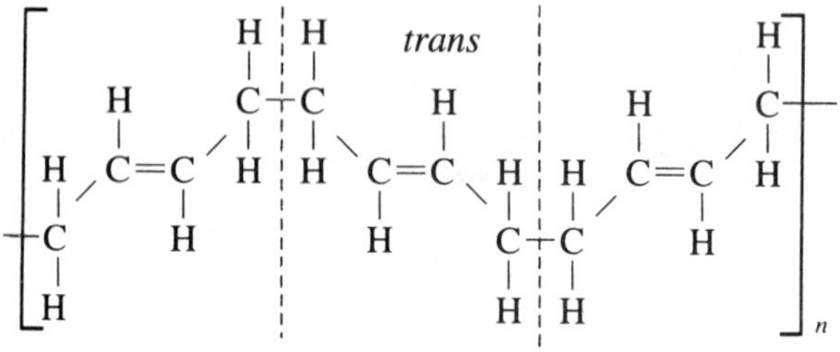

trans-1,4-polybutadiene

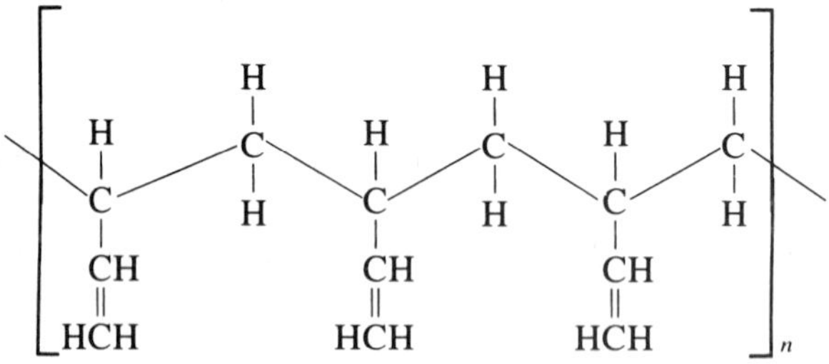

isotactic 1,2-polybutadiene

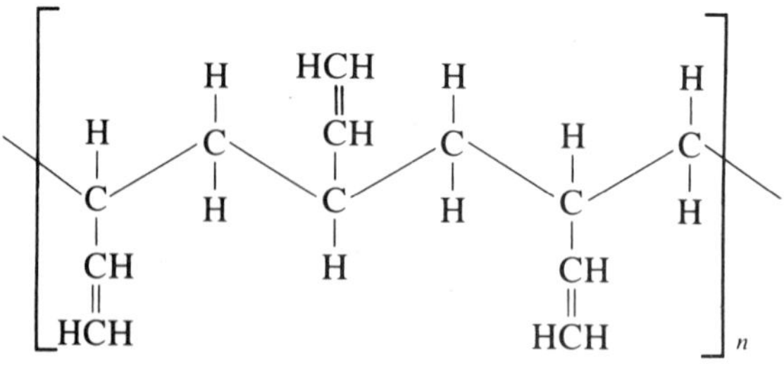

syndiotactic 1,2-polybutadiene

Fig. 8.6. Polybutadiene structures.

organic lithium compounds or coordination catalysts containing metals in reduced valence states. On the laboratory scale, these polymerizations are relatively straightforward. Special attention and care must be given to the elimination of even the smallest traces of air, moisture, and deleterious impurities. Examples of common impurities include acetylene, hydrocarbons, cyclopentadiene, sulfur, and nitrogen-containing compounds. The presence of even low levels may be not only detrimental to polymer conversion but can also strongly influence the physical properties of the polymer.

Thus, polymerizations are carried out using essentially pure dry monomer and solvent such as aromatic, aliphatic, and alicyclic hydrocarbons. Catalysts are charged under some inert atmosphere, e.g., nitrogen or argon. The polymerization is then carried out at a preselected temperature and time with adequate agitation of the reaction media. When the desired conversion of polymer is achieved, the catalyst is deactivated and the polymer stabilized against oxidation. The polymer is then recovered from the unreacted monomer and solvent, washed to remove the catalyst residues, and dried.

Fig. 8.7 shows a flow sheet for the commercial production of polybutadiene. The polymerization is carried out continuously in a series of large reaction vessels, equipped with efficient agitation and cooled by either reflux or jacket cooling. The catalyst system is introduced into the monomer-solvent premix either as separate catalyst components or as a preformed catalyst complex.

Polymerization residence time can be adjusted through changes in the charge rate of the premix. Monomer concentration in the premix is determined by the heat-exchanging capacity of the system and the ability of the polymerization process to transport the viscous cements. Monomer concentrations in aromatic solvents are usually 10 to 12 wt. %; in aliphatic solvents, they may be greater than 16 wt. %.

After polymerization is complete, the viscous cement is simultaneously deactivated and stabilized with antioxidants and then washed with water to remove catalyst residues. Solvent and unreacted monomer are steam-stripped. More than 95% of the solvent and unreacted monomer are recovered after azeotropic drying and distillation and subsequently recycled.

By adding oil with adequate mixing to the polymer cement, the rubbers can be oil-extended with only minor process modification. The uniform dispersion of the oil in the polymer is of utmost importance to the final product. It can then be recovered and washed and dried in a manner similar to the oil-free polymer. Another modification is the addition of carbon black to the oil-extended solution of the rubber, thereby forming a black masterbatch.

CATALYSIS

Isoprene

Of the many catalysts that are capable of polymerizing isoprene, only four have attained commercial importance, as follows:

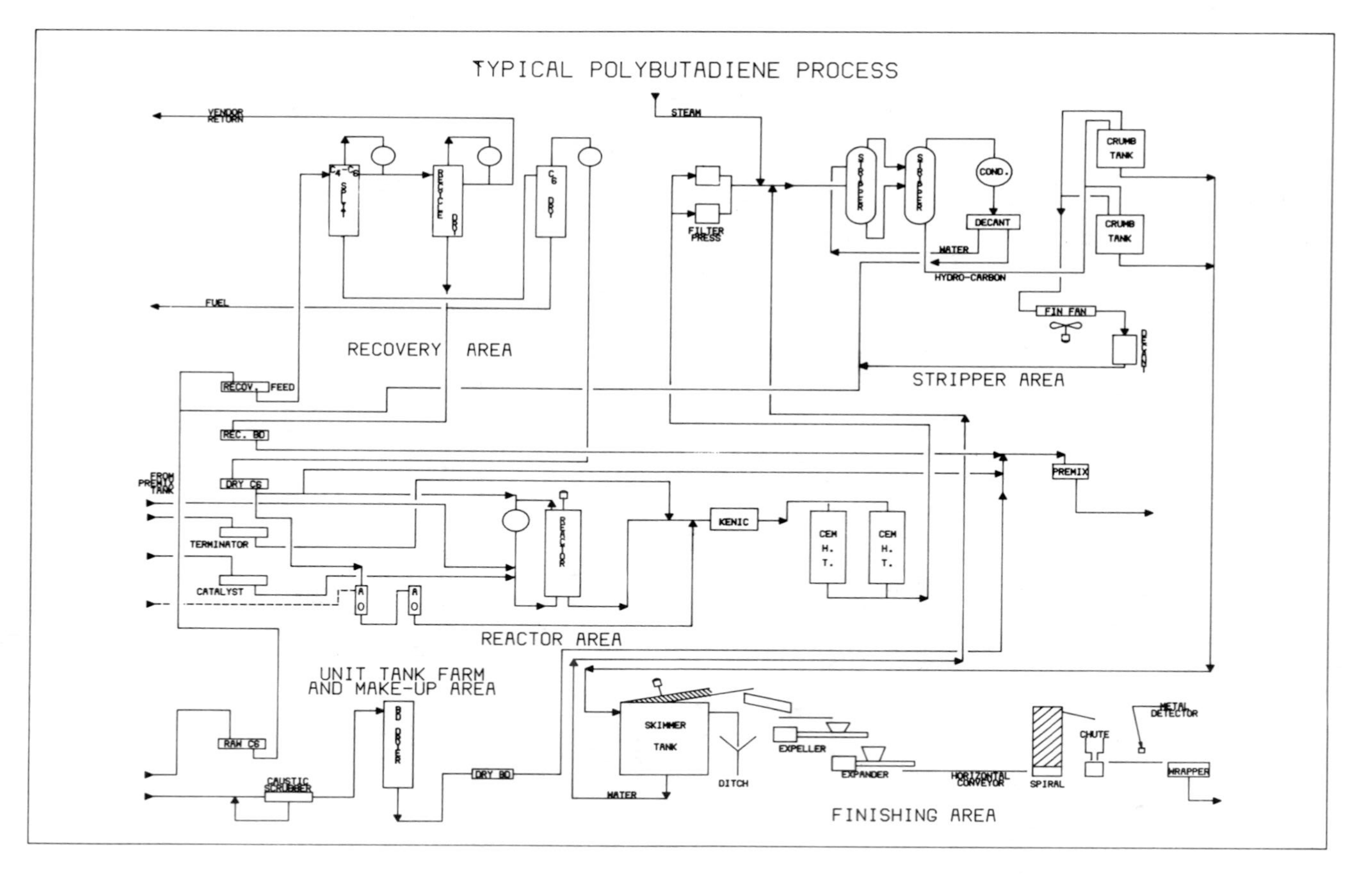

Fig. 8.7. Flow sheet for the production of high-*cis*-1,4-polybutadiene in solution. (Courtesy of the Goodyear Tire & Rubber Company)

1. A coordination catalyst containing trialkyl aluminum and titanium tetrachloride (Al–Ti)
2. A coordination catalyst of alane (aluminum-hydride derivative) and titanium tetrachloride
3. Lithium alkyl
4. A coordination catalyst of aluminum alkyl and a vanadium salt for making *trans*-1,4-polyisoprene.

The first three all yield *cis*-1,4-polyisoprene.

The reaction of equimolar amounts of an aluminum-alkyl compound with titanium tetrachloride in heptane at room temperature results in a two-phase catalyst system, consisting of a finely divided brown solid precipitate and a yellowish liquid. The brown solid is the crystalline form of β-$TiCl_3$, with aluminum present in very small amounts.[34] Other crystalline forms of $TiCl_3$, like the α and γ modifications, are not suitable for preparing *cis*-1,4-polyisoprene. The liquid portion of the catalyst consists essentially of diethylaluminum chloride.

Active $TiCl_3$-containing catalysts can be made with other combinations (e.g., by the reduction of $Ti(OR)_4$ with $RAlCl_2$).[35] But however made, the catalysts active for polymerizing isoprene contain a heterogeneous Ti-Cl_3-rich solid phase. The concentration of active centers is estimated from polymer yields to be only a few percent of the total $TiCl_3$ present.[36]

The most important single variable in the Al–Ti catalyst is the mole ratio in which the two components, aluminum alkyl and titanium tetrachloride, are combined.

Other variables also play a role, however, in determining the structure and physical properties of the polyisoprene, including: (1) catalyst preparation temperature, (2) choice of aluminum alkyl (nature of R group), (3) catalyst aging, (4) catalyst modification with other reagents (e.g., electron donors), (5) polymerization temperature, (6) monomer concentration, and (7) solvent.

The polymerization of isoprene has also been carried out in the presence of alkali metals or organoalkali compounds (other than organolithium).[37–39] These are heterogeneous reactions both in bulk and hydrocarbon solvents. A comprehensive overview of the influence of solvent and positive counterion on polymer structure has been given.[40] Only the lithium-based initiators in hydrocarbon give rise to high *cis*-1,4-structures. A trend of decreasing *cis* structure with increased solvent basicity has been observed. The other alkali metals usually yield polymers of mixed structure with little or no *cis*-1,4-content.

The use of alkali metals was quickly phased out in favor of the use of organolithium compounds. These initiators can be handled in easy-to-use homogeneous systems with excellent reproducibility. For this reason organolithium initiators find use in the commercial production of *cis*-1,4-polyisoprene and *cis*-1,4-polybutadiene.

In the absence of inhibiting impurities, these polymerizations proceed to essentially 100% conversion. There is essentially no chain termination nor transfer, provided compounds having an active hydrogen are not present. The molecular weight (number average) is calculated by the weight of monomer consumed divided by the moles of initiator. This polymerization results in the preparation of linear polymers having controllable molecular weights and possessing a very narrow molecular weight distribution.

Other polymerization catalyst systems that have been screened for isoprene polymerization include (1) soluble nickel complexes involving a monometallic π-complex with a monomer,[41,42] (2) Zr salts or Mg alkyls,[43] (3) cerium-based catalysts,[44,45] (4) uranium and thorium-based catalysts,[46] and (5) several lanthanide and actinide rare-earth-series catalysts.

As previously mentioned, isoprene catalysts based upon alanes and titanium tetrachloride have shown good performance. In these catalysts, the aluminum alkyl is replaced by derivatives of alane (AlH_3). These systems are unique in that no organometallic compound is involved in producing the active species from titanium tetrachloride. The substituted alanes are usually complexed with Lewis acid-type donors, the resulting complexes being soluble in aromatic solvents.[47]

Most coordination catalysts that produce high *cis* structure for polybutadiene result in mixed microstructures with polyisoprene. Only in the case of vanadium co-catalysts, which produce high *trans*-1,4-polyisoprene, have coordination catalysts reached commercial status with isoprene.

Progress in the emulsion polymerization of isoprene has been at a stalemate for over 20 years. Therefore, there are no commercial radical-catalyzed polyisoprenes.

In aqueous systems, the rhodium catalysts, which give very high *trans*-1,4-polybutadiene, yield polyisoprenes with mixed structure.[48,49]

Butadiene

In the case of polybutadiene, polymerization catalysts similar to those discussed for polyisoprene have been evaluated. The oldest ionic butadiene polymerization catalyst is *metallic sodium*. It was used as early as 1910 by the Russians and by the Germans during World War I for the polymerization of dimethylbutadiene. The early German Buna rubbers were polybutadienes prepared in the presence of sodium or potassium catalysts.

Many reagents have been found that serve as activators for sodium polymerizations, including CO_2, hydroxyl compounds, certain metal oxides, and others. The sodium polymerization is highly exothermic and occurs in two stages, the first stage of the polymerization being very slow and a second stage rapidly going to completion.[50] Microstructure varies greatly, depending upon the alkali metal used (see Table 8.1).[51–53]

Table 8.1. Effect of Alkali Metal on Polybutadiene Microstructure.

		Composition, %			
Metal	*Ionization Potential, Volts*	cis-*1,4*	trans-*1,4*	*1,2*	*Reference*
Li	5.4	35	55	10	51
Na	5.1	23(10)[a]	45(25)[a]	32(65)[a]	52
K	4.3	13(15)[a]	49(40)[a]	38(45)[a]	52
Rb	4.2	7	31	62	53
Cs	3.9	6	35	59	53

[a]Values obtained in Ref. 51.

Similar results have also been observed for organoalkali metal catalysts. The ability of the alkali metals to form 1,4-linkages decreases in order of the decreasing ionization potential of the alkali metal.

Lithium and its organic derivatives form 1,4-linkages, and, in particular, *cis*-1,4-linkages (35%), with the greatest ease. All the other alkali metal catalysts give a predominantly vinyl structure (45 to 65% vinyl). Mechanistically, lithium metal first adds to a monomer unit, resulting in a dilithium species that then proceeds to add other monomer units in a manner similar to organolithium compounds.[54–56] The electron deficiency of lithium facilitates the formation of coordination complexes with electron-rich molecules, like butadiene. This deficiency also results in highly associated organic compounds of lithium in solution, the carbon-lithium bond being essentially covalent.

Butyllithium is the most commonly used organoalkali metal catalyst. Polymers prepared with butyllithium in heptane possess 40 to 50% *trans*-, 40 to 50% *cis*- and 5 to 10% vinyl structure,[57] similar to the figures obtained with lithium metal.

The polymer structure is independent of conversion and reaction temperature but can be influenced by initiator and monomer concentration[57] and is dramatically affected by the introduction of electron donors—such as ethers, amines, and sulfides—that solvate the alkali metal[58, 59] and thus increase the vinyl structure.

Medium- and high-vinyl 1,2-polybutadiene resins (vinyl content between 20 and 80%) are produced commercially by solution polymerization of butadiene in the presence of a butyl lithium or sodium catalyst and an anionic complexing substance such as diglyme, diethyl ether, tetramethylethylenediamine, or tetrahydrofuran/heptane.

Polymerization of butadiene by alkyl lithium proceeds via a living polymer in a way similar to the polymerization of polyisoprene discussed in an earlier section. No termination mechanisms are present, and the molecular weight of the polymers is proportional to the amount of butadiene monomer in the system, one polymer chain growing per molecule of monomeric butyllithium. Since all the growing chains have an equal probability of chain growth, the polymers have a narrow molecular weight.

Alfin catalysts,[60-62] which are related to the alkali metal catalysts, have been used to prepare both polyisoprene and polybutadiene. These catalysts, based upon sodium, consist of a complex of NaCl, sodium alkoxide, an alkenyl sodium such as sodium isopropoxide, and an alkyl sodium. Alfin polymers, however, have molecular weights up to several million and are unprocessable by conventional techniques. Nevertheless, the molecular weight of alfin polymers can be modified to approximately 300,000 by the addition of modifiers, such as 1,4-dihydrobenzene and 1,4-dihydronaphthalene.[63] Control occurs through hydrogen transfer. The polymer microstructure for polybutadiene is 60% *trans*-1,4, 17% *cis*-1,4, and 16% vinyl.

Free-radical polymerizations involving polybutadiene and copolymers containing polybutadiene have been used for years. Free-radical polymerizations, whether performed in emulsion or solution, are initiated by an active radical, $R^{\cdot}$, formed by the decomposition of a peroxide, peroxysulfate, or similar free-radical forming reactions. The free-radical, $R^{\cdot}$, takes part in the initial step of polymer growth only, eventually becoming far removed from the continuing site of polymerization.

Emulsion systems, in general, contain water, monomer(s), an initiator, and an emulsifier. Polymerization is initiated when the monomer and free radical come into contact within a soap micelle in aqueous phase.

The free-radical polymerization of butadiene is a function of the reaction temperature. As the polymerization temperatures decrease, the *cis*-1,4 content also decreases, with no *cis*-1,4 content being observed at temperatures below −15°C. This same observation is made for polybutadiene segments of copolymers with styrene, acrylonitrile, and others.

Although the literature reports a wide variety of possible catalysts for polymerizing butadiene, only catalysts containing titanium, cobalt, and nickel have been successfully developed on a commercial scale for the production of stereoregular polybutadienes.

Titanium-based catalysts for polymerization of butadiene are divided into four groups: (1) AlR_3 with $TiCl_4$, (2) AlR_3 with TiI_4, (3) AlR_3 (or aluminum alkyl hydrides) with $TiCl_4$ and an iodine-containing compound, and (4) other organometallic compounds with TiI_4.

The stereospecificity of the titanium-based catalysts depends largely upon the nature of the halogen atom attached to titanium. In the series $TiCl_4$, $TiBr_4$, and TiI_4, the *cis*-1,4 content increases from 60 to 90% proving that iodine is essential for the formation of high *cis*-1,4-linkages.

In preparation of $TiCl_4$ catalysts, minor changes in experimental conditions, e.g., the Al/Ti molar ratio, can affect not only the catalyst activity but also the nature of the final product.[65]

The AlR_3/TiI_4 system has been the most extensively studied, since it produces high (>90%) *cis*-1,4-, linear, gel-free polybutadiene.[64,66–69] This system has been successfully scaled up to a commercial production process. The catalyst becomes active at an Al/Ti molar ratio of 1 or slightly higher. As the Al/Ti ratio increases, a maximum is reached between 3 and 5 and then decreases slowly with increasing AlR_3. The titanium tetraiodide, being insoluble in hydrocarbon media, requires special treatment before catalyst makeup. With increased catalyst concentration in the TiI_4 system, the polymerization rate and conversion increase while the molecular weight decreases. The molecular weight is also a function of the polymerization temperature, monomer concentration, and ratio of the catalyst components.

TiI_4 and AlR_3 form a heterogeneous catalyst. The solid and liquid portions of this catalyst have been separated, but neither portion shows catalytic activity by itself.[69,70]

Catalyst systems using titanium compounds with mixed halogens have also been reported. The use of TiI_2Cl_2 and $TiICl_3$[69] both produce a catalyst comparable in activity to TiI_4 but lack control in optimizing the I/Cl ratio.

Other organometallic compounds, besides aluminum alkyls, can be used in butadiene polymerizations as reducing agents for titanium, including organometallic compounds of lithium,[71] sodium,[72] and zinc and lead.[73]

A method of preparing high-*cis*-1,4-polybutadiene using a catalyst consisting of $CoCl_2$ and $(i\text{-}C_4H_9)_3Al$ was patented in 1956.[74] Much of the preliminary polymerization with cobalt involved heterogeneous catalysts containing alkyl aluminum halides and cobalt compounds, such as CoX_2 (X = halide), CoS, $CoSO_4$, $CoCO_3$, and $Co_3(PO_4)_2$.[75,76] Homogeneous catalysts are made from soluble Co(II) and Co(III) salts (octanoates, naphthenates, and acetylacetonates) and certain donor complexes, such as $CoCl_2 \cdot$ pyridine. Only the soluble cobalt catalysts are of any practical importance.

The polymerization of butadiene in benzene with R_2AlCl–Co catalyst gives high conversions of >99% *cis*-1,4-polybutadiene. The molecular weight of this polymer varies between 10,000 and 1,000,000, depending upon experimental parameters. The high stereospecificity associated with this catalyst is limited to butadiene. With isoprene, polymers of mixed microstructures are prepared that have only a 65% *cis*-1,4 structure.

Much of the very early published work on cobalt catalysts is in disagreement, and many reported experiments were irreproducible. These discrepancies were later found to be the result of the presence of catalytic quantities of certain electron donors in these systems. In fact, one peculiarity of the cobalt system is that polymerization can occur at a rapid rate only in the presence of small quantities of water or certain other compounds. Other examples of successful activators are $AlCl_3$, metallic Al, alcohols, reactive organic halides, halogens, hydrogen halides, and organic hydroperoxides.[77,78]

The number of polymer chains formed per cobalt atom in the polymerization system is very high. Since the molecular weight of the polymers changes very little with increasing conversion,[79] chain transfer must be considered an important process in cobalt systems. The polymer molecular weight can be regulated by the addition of certain chain-transfer reagents, including H_2,[80] ethylene or propylene, allene, methylallene, and certain unsaturated compounds (e.g., cyclooctadiene).

The study of nickel catalysis of butadiene polymerization has in recent years spurred not only considerable industrial interest as a commercial process but also academic interest in the mechanistic pathways of these reactions. A large number of nickel compounds and complexes, both homogeneous and heterogeneous, give excellent yields at relatively low reaction temperatures of high (98%) *cis*-1,4-polybutadiene, either by themselves or in conjunction with other organometallic components.

Catalytic activity has been observed in one-, two-, and three-component systems containing nickel. These systems can be divided into those based upon nickel salts and those based upon π-allyl nickel complexes; however, polymerization mechanisms proposed for both groups of catalysts include π-allylic complex formation as an intermediate.

In general, nickel-salt catalyst systems consist of (1) an organic complex compound of nickel, (2) a Lewis-type acid (e.g., boron trifluoride etherate),[81–84] and (3) an organometallic compound. The most notable example is the ternary system R_3Al–Ni octanoate-$BF_3 \cdot$ etherate.[81] The polybutadiene produced with this catalyst has a better than 98% *cis*-1,4-content and possesses a high molecular weight, a broad molecular-weight distribution, and no substantial gel content. The catalyst efficiency for nickel is high, yielding 10^3 to 10^4 g of polybutadiene per gram of nickel. As in cobalt systems, chain transfer is the main mechanism of polymer-chain termination. Polymerizations can be carried out in aromatic or aliphatic solvents.

Fluoride is essential in these systems for preparing high-molecular-weight, high-*cis*-polybutadiene, and many fluorine-containing organic and inorganic compounds, including HF,[85] have been successfully used as the Lewis acid Likewise, Li alkyls have been used to replace aluminum alkyls as the organometallic component.[86]

All three components of these ternary systems are indispensable for catalytic activity. The order of addition of the components, preparation temperature, aging history, and so forth, greatly influence the activity of the resulting catalyst systems.

The second main group of nickel catalysts involve a unique class of hydrocarbon derivatives of nickel. Classified as π-allyl complexes, they involve the polycentric bonding of the delocalized π electrons of three carbon atoms with suitable bonding orbitals in the nickel atom.[87–92]

The π-allynickel halides have shown catalytic activity with butadiene. Polymer configuration depends upon the halide within the complex. A *cis*- polymer is obtained with the chloride, whereas the bromide and iodide give *trans*-configurations.[90] The trend to form the *trans*- polymer apparently increases with decreasing electronegativity of the halogen used. Speculation as to the reason for this effect range from steric considerations to polar effects, although conclusive experimental evidence is lacking.[93]

Many examples of polymerization using nickel in a reduced state are reported for binary catalyst systems. Among them, π-complexes of cyclooctadienyl and bis(cyclootadienyl) nickel (0) with metallic halides, acidic metal salts, and protonic acids have yielded highly stereospecific 1,4-polybutadienes.[93–97] The activity of these catalyst systems depends strongly on the nature of the substance used as cocatalyst.

Bis(cyclopentadienyl) nickel in association with various Lewis acids polymerizes butadiene to stereospecific products that depend upon the nature of the acid.[95–98] A bimetallic complex is proposed in which the halogens act as bridging groups. Addition of butadiene to this complex is believed to result in a π-allyl system.

The reaction of π-crotylnickel halides with Lewis acid (B, Al, Ti, V, Mo, W halides and so on), yields products insoluble in hydrocarbons. These complexes initiate polymerization of butadiene to a high-*cis*-1,4 structure regardless of the nature of the halogen used.[98–102] The catalyst activity of these reaction products is higher than for the corresponding π-crotylnickel halides alone.

The π-allylic type bonds with nickel are not only instrumental in catalyst formation but are believed to be the basis of the propagation mechanism in butadiene polymerizations,[91] with butadiene π-bonded to nickel as an intermediate in the propagation step. The effect of solvent and counter-ions on this mechanism are by no means clear. It is known, however, that counter-ions are very important in determining the polymer microstructure as proved by the necessity of having fluoride or fluoride-containing ligands to prepare high-*cis*-polybutadiene. The electron affinity of the counter-ion can directly influence the *d*-electron density of the nickel; the more electron withdrawing the counter-ion, the higher the *cis*- content of the polymer.[103] Both cerium and uranium catalyst systems have been shown recently to give high-*cis*-1,4-polybutadiene (up to 99% *cis*-).

A typical cerium catalyst system[104] consists of a cerium compound (e.g., the octanoate salt), together with a mixture of an aluminum alkyl and an alkyl aluminum halide. The microstructure of the resulting polymers appears to be independent of the catalyst components' concentrations and the nature of the halogen attached to the aluminum.

Tris- (allyl) uranium halides[105] and uranium alkoxides[106] have both been successfully used to prepare high-molecular-weight *cis*-1,4-polybutadiene (up to 99% *cis*-). The microstructure of the polymer is again independent on which halogen is used. Activity of these catalysts can be enhanced by the addition of Lewis acids (e.g., $TiCl_4$).[107]

Crystalline high-*trans*-1,4-polybutadienes (mp = 145°C) have been prepared with catalysts containing titanium, vanadium, chromium, rhodium, iridium, cobalt, and nickel. High-*trans*-1,4-polybutadiene was first prepared in 1955[108] with a heterogeneous catalyst containing $(C_2H_5)_3Al$ and α-$TiCl_3$. Other heterogeneous catalysts displaying similar activity were prepared with $TiCl_4$, TiI_4, VCl_4, VCl_3, and $VOCl_3$. The highest content *trans*-1,4-polymers, however, were prepared with homogeneous vanadium catalysts.[109]

Among the heterogeneous catalysts, VCl_3 shows the highest stereospecificity, giving 98 to 99% *trans*-1,4 content.[110] The microstructure in this system is unaffected by the Al–V molar ratio; VCl_4 and $VOCl_3$[111] give a 94 to 98% *trans*-1,4-polymer containing a 10% amorphous fraction.

The preparation of crystalline syndiotactic 1,2-polybutadiene (mp = 156°C) was first reported in a patent in 1955.[112] The method used homogeneous catalysts consisting of an aluminum alkyl, together with an acetylacetonate or alcoholate complex of titanium, vanadium, or chromium. Other halogen-free complexes of these metals likewise produced the syndiotactic polymer.[114]

With halogen-free vanadium complexes—e.g., vanadium tri(acetylacetonate)—polymers having 78 to 86% vinyl content can be prepared. This vinyl content can be increased to greater than 90 percent by employing higher Al–V molar ratios (3/1–30/1) and/or by aging the catalyst.[113]

Isotactic 1,2-polybutadiene (mp = 126°C) has been prepared with homogeneous catalysts obtained by the reaction of aluminum alkyl with several soluble compounds of chromium, such as the acetylacetonate, $Cr(CNR)_6$, chromium hexacarbonyl, and others.[115–117] Polymer conversion for these systems is low.

Atactic 1,2-polybutadiene can be prepared with alkali metal catalysts in the presence of electron donors such as ethers or amines.

Processing and Curing of Polybutadiene and Polyisoprene

The processing characteristics of polybutadiene are influenced by the polymer microstructure, molecular weight, molecular-weight distribution, and degree of branching.[3,4,6,118–124] The polymer undergoes mastication, mixing, molding, and curing. A Banbury mixer and a roll mill are employed for mastication and mixing; a calender and extruder, for molding.

Most polybutadienes are highly resistant to breakdown and have poor mill-banding characteristics and rough extrusion appearance compared to SBR elastomers. The solution polybutadienes process satisfactorily when blended with other elastomers such as SBR. Emulsion polybutadiene processes better than solution polymer but not as well as SBR; it is commonly blended with other elastomers for enhanced processing.

Certain chemical peptizers slightly increase breakdown and improve processing.[3] A lower Mooney viscosity also improves processing but may lead to cold-flow[4] problems. In addition, a broad molecular-weight distribution and branching[120] both improve milling and extrusion behavior as compared to a linear polymer. Excessively high-molecular-weight polybutadiene tends to crumb on the roll mill.

Very-high-*cis*-polybutadiene prepared with a uranium catalyst is reported to have greatly improved millability, calenderability, tack, green strength, and adhesion to fabric.[106] Such processability would eliminate the need for blending with other elastomers. The mill processabilities of polybutadienes comprised of variable microstructures and catalyst are presented in Table 8.2. The processability improves with increasing *cis*- content.

Curing and compounding recipes depend upon the choice of the individual manufacturer. Polybutadienes are cured employing conventional sulfur recipes or peroxide systems. Normally, the polymer is blended with another elastomer, then mixed with filler (HAF, ISAF blacks, for example); typically an aromatic processing oil, wax, antioxidant, antiozonant; and curing ingredients at a later stage. For curing, thiazoles and sulfenamides are generally employed alone or in combination with secondary accelerators. Cure rates are close to SBR using similar loads of sulfur.

An example of a polybutadiene test recipe, this one employed at Goodyear,[125] consists of the following:

Budene 1207	90.00 phr
No. 1 RSS	10.00
ISAF black	50.00
Circosol	10.00
Paraffin	2.00
Wingstay 100	1.00
Stearic acid	3.00
Zinc oxide	3.00
Altax	0.45
DPG	0.75
Sulfur	1.25

Typical vulcanizate properties are given in Table 8.3.

Railsback and Stumpe[121] have published the comparative vulcanizate properties of medium vinyl polybutadiene/SBR blend, SBR, and an SBR/*cis*-polybutadiene blend (see Table 8.4). A test recipe for tire treads was employed since the primary use of these blends is in tread stocks.

The trend in Table 8.4 shows the normal pattern based on polymer composition. SBR blended with *cis*- polybutadiene has significantly improved abrasion resistance and tear resistance, but the skid resistance is slightly decreased. The SBR/medium vinyl polybutadiene blend has enhanced blow-out resistance, improved abrasion and decreased skid resistance. Elastomers with high mobility and low glass-transition temperatures have high elasticity, very good abrasion, and poor skid, and vice versa.

Medium vinyl polybutadiene is a potential partial or full replacement for SBR

Table 8.2. Processability of Poly(butadiene)s.[a]

		25°C		40°C		60°C		75°C	
Catalyst	*Composition,[b] percent*	*Raw Polymer*	*Black Stock*	*Raw Polymer*	*Black Stock*	*Raw Polymer*	*Black Stock*	*Raw Polymer*	*Black Stock*
Uranium	98.5/1.0/0.5	Good	Good	Good	Good	Good	Good	Good	Fair
Nickel	97.0/2.0/1.0	Good	Good	Good	Good	Good	Fair	Fair	Fair
Cobalt	97.0/1.5/1.5	Good	Good	Good	Good	Fair	Fair	Bad	Bad
Titanium	91.0/4.5/4.5	Good	Good	Fair	Fair	Bad	Bad	Bad	Bad
Lithium	32.6/55.4/12.0	Bad	Bad	Bad	Bad	Bad	Bad	Bad	Bad

[a]Ref. 121.
[b]1,4-*cis*/1,4-*trans*/1,2.

Table 8.3. Poly(butadiene) Vulcanizate Properties.

300% modulus, kg/cm^2	70
Tensile strength, kg/cm^2	165
Elongation, %	540
Shore A hardness	59
Tear strength, die C, kg/cm^2	50.3

and has only recently become commercially important to the rubber industry. Use of medium vinyl polybutadiene would compensate for styrene, should a shortage of this monomer occur in the petrochemical industry. This would inevitably result in a price increase for butadiene monomer, since it would be in demand for both the existing conventional, as well as the new general-purpose medium vinyl polybutadiene rubbers.

Synthetic polyisoprenes containing gel are generally characterized by high initial Mooney viscosities, but they break down rapidly upon mastication. There is no premastication requirement for Al-Ti polyisoprenes, and the compounding ingredients can be readily incorporated with proper mixing. Li polyisoprenes often require premastication in the presence of peptizers for better incorporation

Table 8.4. Comparison of PBD Blends.

Compound	*A*	*B*	*C*
45-percent vinyl PBD	68	—	—
cis-4 1203	—	—	35
SBR 1712	32	100	65
Styrene, %	7.5	23.5	15.3
	Vulcanizate Properties		
300% modulus, kg/cm^2	91	91	98
Tensile strength, kg/cm^2	202	207	197
Elongation, %	560	550	510
Shore A hardness	59	58	59
ΔT, °C at 15 min	42	42	39
Yerzley resilience	58	55	59
Blow-out time, min	>60	17	15
Crescent tear 100°C, kg/cm^2	37	37	44
	Abrasion Index		
12875 km	106	100	137
	Wet-Skid Index		
40 km/hr	102	108	104

of compounding ingredients. Green strength (cohesive strength) is a prime requirement in fabrication applications involving stress (e.g., tire building). Natural rubber possesses the highest green strength, Al-Ti synthetic polyisoprenes have low green strength, and Li polyisoprenes have little to no green strength after processing. Chemical additives, however, can be used to improve the green strength of Al-Ti: polyisoprenes, thus allowing it to be used as a replacement for natural rubber in tires.

The cured properties of *cis*-1,4-polyisoprene vulcanizates are in general similar to those for natural rubber. Both show high gum tensile strength, good mechanical strength in filled compounds, high resilience and low heat build-up in related applications.

REFERENCES

1. G. S. Whitby, ed., *Synthetic Rubber*, John Wiley & Sons, New York, 1954.
2. E. Tornqvist, Ch. 2 in *Polymer Chemistry of Synthetic Elastomers*, J. P. Kennedy and E. Tornqvist, eds., John Wiley & Sons, New York, 1968.
3. R. S. Hammer, and M. E. Railsback, in *Rubber Technology*, M. Morton, ed., Van Nostrand Reinhold, New York, 1973.
4. J. F. Svetlik, in *Vanderbilt Rubber Handbook*, G. G. Winspear, ed., R. T. Vanderbilt Company, New York, 1968.
5. W. Cooper, Chapt. 2 in *Stereo Rubbers*, W. M. Saltman, ed., John Wiley & Sons, New York, 1977.
6. E. Ceausescu, ed., *Stereospecific Polymerization of Isoprene*, Pergamon Press, Oxford, 1983.
7. W. M. Saltman, E. Schonberg, H. A. Marsh, and S. J. Walters, *Rubber Chem. Technol.* **52,** 526 (1979).
8. G. Bouchardat, *Comp. Rend.* **89,** 1117 (1879).
9. S. V. Lebedev, *Zh. Russ. Fiz. Khim. Ova.* **42,** 949 (1910).
10. Ibid., **45,** 1249 (1913).
11. A. Talalay and M. Magat, *Synthetic Rubber from Alcohol*, Interscience, New York, 1945.
12. K. Ziegler, E. Holzkamp, H. Breil, and H. Martin, *Angew. Chem.* **67,** 541 (1955).
13. S. E Horne, assigned to Goodrich-Gulf Company. U.S. Pat. 3,144,743 (Dec. 17, 1963).
14. S. E. Horne et al., *Ind. Eng. Chem.* **48,** 784 (1956).
15. *Wall Street Journal*, East. Ed., Apr. 26, 1956.
16. M. J. Hackathorn and M. J. Brock, *Rubber Chem. Technol.* **45,** 1295 (1972).
17. G. Natta, *Mod. Plast.* **34,** 169 (1956).
18. J. K. Clark, J. Lal, and J. N. Henderson, *J. Polym. Sci.*, **B9,** 49 (1971).
19. M. A. Golub, Ch. 10B in *Polymer Chemistry of Synthetic Elastomers*, Part II, J. P. Kennedy and E. Tornqvist, eds., Interscience Publishers, New York, 1969.
20. I. Kossler and J. Vodehal, *Anal. Chem.* **40,** 825 (1968).
21. G. S. Whitby in *The Chemistry and Technology of Rubber*, C. C. Davis, ed., J. T. Blake, assoc. ed., Reinhold, New York, 1937.
22. L. R. G. Treloar, Chs. 1 and 11 in *The Physics of Rubber Elasticity*, Oxford University Press, New York, 1958.
23. G. Natta and F. Danusso, eds., *Stereoregular Polymers and Stereospecific Polymerizations*, Pergamon Press, New York, 1967, p. 140.
24. K. W. Scott, G. S. Trick, R. H. Mayor, W. M. Saltman, and R. M. Pierson, *Rubber Plast. Age* **42,** 175 (1961).

25. A. DeChirico, P. C. Lanzani, M. Piro, and M. Bruzzone, *Chim. Ind.* **54,** 35 (1972).
26. M. J. Brock and M. J. Hackathorn, *Rubber Chem. Technol.* **45,** 1303 (1972).
26a. M. Morton, *Anionic Polymerization*, Academic Press, New York, 1983; M. Morton, and J. P. Rupert, ACS Symposium Series No. 212, American Chemical Society, Washington, 1983.
27. W. Cooper and G. Vaughan, *Polymer* **4,** 329 (1963).
28. G. R. Himes, Ch. 11 in *Rubber Technology*, 2d Ed., M. Morton, ed., Van Nostrand Reinhold, New York, 1973.
29. J. R. Purdon, Jr., and R. D. Mate, *J. Polym. Sci.*, **A-1** (8), 1306 (1970).
30. G. H. Stempel, *Polymer Handbook*, 2d Ed., vol. 1-5, John Wiley & Sons, New York, 1975.
31. G. S. Trick, *J. Appl. Polym. Sci.* **3,** 253 (1960).
32. W. S. Bahary, D. I. Sapper, and J. H. Lane, *Rubber Chem. Technol.* **40,** 1529 (1967).
33. T. S. Daintan, D. M. Evans, F. E. Hoare, and T. P. Melia, *Polymer* **3,** 297 (1962).
34. G. Natta, P. Corrandine, I. W. Bassi, and L. Porri, *Rend. Anad. Noz. Lincei* **24** (8), 121 (1958).
35. S. Cucinella, A. Mazzei, W. Marconi, and C. Busetto, *J. Macromol. Sci.-Chem.*, **A-4** (7), 1549 (1970).
36. Y. B. Monokov, N. K. Minchenkova, and S. R. Rafikav, *Dokl. Akad. Nauk. SSSR* **236** (5), (1977); *Chem. Abstr.* **88,** 07511 (1978).
37. F. E. Matthews, and E. H. Strange, Brit. Pat. 24,790 (Oct. 25, 1911).
38. C. H. Harries, *Ann.* **383,** 184 (1911).
39. I. H. Labhardt, Ger. Pat. 255,786 (Jan. 27, 1912) and 255,787 (Apr. 9, 1912), assigned to BASF Badische Anilin.
40. A. V. Tobolsky and C. E. Rogers, *J. Polym. Sci.* **40,** 73 (1959).
41. P. Teyssie and F. Dawans, Ch. 3 in *The Stereo Rubbers*, W. M. Saltman, ed., Wiley-Interscience, New York, 1977.
42. B. A. Dolgoplosk, *Kinet. Katal* **18,** 1146 (1977).
43. W. Cooper, in *The Stereo Rubbers*, W. M. Saltman, ed., Wiley-Interscience, New York, 1977, pp. 50–53.
44. W. C. VanDohlen, T. P. Wilson, and E. G. Caflisch, U.S. Pat. 3,297,667 (Jan. 10, 1967), assigned to Union Carbide.
45. M. C. Throckmorton and W. M. Saltman, U.S. Pat. 3,541,063 (Nov. 17, 1970), assigned to Goodyear Tire & Rubber Co.
46. M. C. Throckmorton, and W. M. Saltman, U.S. Pat. 3,676,411 (Jul. 11, 1972), assigned to Goodyear Tire & Rubber Co.
47. W. Marconi, A. Mazzei, S. Cesca, and M. DeMalde, *Chim. i Ind.* (Milan) 51, 1084 (1969); W. Marconi, A. Mazzei, S. Cucinelli, and M. DeMalde, *Makromol. Chem.* **71,** 118, 134 (1964).
48. E. Matsui and T. Tsuruta, *Polym. J.* **10** (2), 133 (1978); E. Matsui et al. *J. Macromol. Sci. Chem.* **A 11,** 999 (1977).
49. A. A. Entezeami, A. Deluzarche, B. Kaempf, and B. Schue, *Eur. Polym. J.* **13** (3), 203 (1977).
50. C. S. Marvel, W. J. Bailey, and G. E. Inskeep, *J. Polym. Sci.* **1,** 275 (1946).
51. H. E. Adams, R. L. Bebb, L. E. Forman, and L. B. Wakefield, *Rubber Chem. Technol.* **45,** 1252 (1972).
52. R. V. Basova, A. A. Arest-Yakubovich et. al. *Dokl. Akad. Nauk SSSR* **149** (5), 1067 (1963).
53. C. E. Bawn, *Rubber Plast. Age* **42,** 267 (1961).
54. M. Morton, *Rubber Plast. Age* **42,** 397 (1961).
55. S. Bywater, *Pure Appl. Chem.* **4,** 319 (1961).
56. C. E. Bawn and A. Ledwith, *Quart. Rev.* (London) **16,** 361 (1962).
57. M. Morton, *Anionic Polymerization*, Academic Press, New York, 1983; M. Morton and J. P. Rupert, ACS Symposium Series No. 212, American Chemical Society, Washington, 1983.

58. I. Kuntz and A. Gerber, *J. Polym. Sci.* **42,** 299 (1960).
59. V. A. Krapochev et al., *Dokl. Akad. Nauk SSSR* **115,** 516 (1957); *Rubber Chem. Technol.* **33,** 636 (1960).
60. A. A. Morton, *Ind. Eng. Chem.* **42,** 1488 (1950).
61. A. A. Morton, J. Nechilow, and E. Schoenberg, *Rubber Chem. Technol.* **30,** 326 (1957).
62. A. A. Morton and E. J. Lanpher, *J. Polym. Sci.* **44,** 233 (1960).
63. *Chem. Eng. News* **47,** 46 (1969).
64. W. Franke, *Kautsch. Gummi Kunstst.* **11,** 254 (1958).
65. G. Natta, *Chim. Ind.* (Milan) **42,** 1207 (1960).
66. I. Ja. Poddubnyi, V. A. Grechanovsky, and E. G. Ehrenburg, *Makromol. Chem.* **94,** 268 (1966).
67. W. Cooper, G. E. Vaughan, D. E. Eaves, and R. W. Madden, *J. Polym. Sci.* **50,** 159 (1961).
68. R. P. Zelinski and D. R. Smith, Belg. Pat. 551,851 (Oct. 17, 1956), assigned to Phillips Petroleum Co.
69. W. M. Saltman and T. H. Link, *Ind. Eng. Chem. Prod. Res. Dev.* **3** (3), 199 (1964).
70. P. H. Moyer and M. H. Lehr, *J. Polym. Sci.* **A3,** 217 (1965).
71. M. H. Lehr and P. H. Moyer, *J. Polym. Sci.* **A3,** 231 (1965).
72. Brit. Pat. 920,244 (Oct. 3, 1960), assigned to Phillips Petroleum Co.
73. Brit. Pat. 931,440 (Apr. 7, 1961), assigned to Phillips Petroleum Co.
74. C. E. Brockway, and A. F. Ekar, U.S. Pat. 2,977,349 (Nov. 7, 1956), assigned to Goodrich Gulf Chem. Inc.
75. C. Longiave, R. Castelli, and G. F. Croce, *Chim. Ind.* **43,** 625 (1961).
76. C. Longiave, G. T. Grace, and R. Castelli, Ital. Pats. 592,477 (Dec. 6, 1957) and 588,825 (Dec. 24, 1957), assigned to Montecatini S.p.A.
77. M. Gippin, *Ind. Eng. Chem., Prod. Res. Dev.* **1,** 32 (1962).
78. V. N. Zgonnik et al., *Vysokomol. Soedin.* **4,** 1000 (1962).
79. B. A. Dolgoplosk et al, *Dokl. Akad Nauk SSSR* **135,** 847 (1960).
80. C. Longiave, R. Costelli, and M. Terraris, *Chem. Ind.* **44,** 725 (1962).
81. K. Ueda U.S. Pats. 3,170,904, 3,170,905, 3,170,906, and 3,170,907 (Feb. 23, 1965), assigned to Bridgestone Tire and Rubber Co.; K. Ueda et al. 3,178,403 (Apr. 13, 1965), assigned to Bridgestone Tire and Rubber Co.
82. S. Kitagawa and Z. Harada, *Jpn. Chem. Q.* **IV-1,** 41 (1968).
83. Brit. Pats. 905,099 and 906,334 (Dec. 31, 1959), assigned to Bridgestone Tire and Rubber Co.
84. T. Matsumoto and A. Onishi, *Kogyo Kagaku Zasshi* **71,** 2059 (1968).
85. M. C. Throckmorton and F. S. Farson, *Rubber Chem. Technol.* **45,** 268 (1972).
86. C. Dixon, et al., *Eur. Polym. J.* **6,** 1359 (1970).
87. G. Wilke et al., *Angew. Chem. Int. Ed.* **5,** 151 (1966).
88. G. Wilke, *Angew. Chem. Int. Ed.* **2,** 105 (1963).
89. L. Porri, G. Natta, and M. C. Gallazzi, *J. Polymer. Sci.* **C-16,** 2525 (1967).
90. L. Porri, M. C. Gallazzi, and G. Vitulli, *Pol. Lett.* **5,** 629 (1967).
91. F. Dawans and P. Teyssie, *Ind. Eng. Chem., Prod. Res. Dev.* **10,** 261 (1971).
92. E. O. Fischer and H. Werner, *Metal π-Complexes*, Elsevier, New York, 1966.
93. F. Dawans and P. Teyssie, *C. R. Acad. Sci.* (Paris) **263 C,** 1512 (1966).
94. J. P. Durand, F. Dawans, and P. Teyssie, *J. Polym. Sci.* **B 5,** 785 (1967).
95. F. Dawans and P. Teyssie. *J. Polym. Sci.* **B 3,** 1045 (1965).
96. J. P. Durand, F. Dawans, and P. Teyssie, *J. Polym. Sci.* **B 6,** 757 (1968).
97. F. Dawans and P. Teyssie, *C. R. Acad. Sci.* (Paris) **261 C,** 4097 (1967).
98. E. I. Tinyakova et al. *J. Polym. Sci.* C **16,** 2625 (1967).
99. B. D. Babitskii, et al. *Dokl. Akad. Nauk SSSR* **161,** 282 (1965).
100. Ibid., 583 (1965).

101. B. D. Babitskii et al. *Vysokomol. Soedin* **6,** 2202 (1964).
102. B. D. Babitskii et al. *Izv. Akad. Nauk. SSR*, 1507 (1965).
103. T. Matsumoto and J. Furukawa, *J. Macromol. Sci. Chem.* **A 6,** 281 (1972).
104. M. C. Throckmorton, *Kautsch. Gummi Kunstst.* **22,** 293 (1969).
105. A. deChirico, P. C. Lanzani, E. Raggi, and M. Bruzzone, *Makromol. Chem.* **175,** 2029 (1974).
106. M. Bruzzone, A. Mazzei, and G. Giuliani, *Rubbers Chem. Technol.* **47,** 1175 (1974).
107. G. Lugli, A. Mazzei, and S. Poggio, *Makromol. Chem.* **175,** 2021 (1974).
108. G. Natta, L. Porri, and M. Mozzanti, Ital. Pat. 545,952 (Mar. 12, 1955), assigned to Montecatini S.p.A.
109. G. Natta, L. Porri, and A. Carbonaro, *Rend. Accad. Nax. Lincei* **31** (8), 189 (1961).
110. G. Natta, L. Porri, P. Corradini, and D. Morero, *Chem. Ind.* **40,** 362 (1958).
111. G. Natta, L. Porri, and A. Mazzei, *Chem. Ind.* **41,** 116 (1959).
112. B. Natta and L. Porri, Belg. Pat. 549,544 (July 15, 1955), assigned to Montecatini S.p.A.
113. B. Natta, L. Porri, G. Zanini, and L. Fiore, *Chem. Ind.* **41,** 526 (1959).
114. G. Natta, L. Porri, and A. Carbonaro, *Makromol. Chem.* **77,** 126 (1964).
115. G. Natta, L. Porri, and A. Palvarini, Ital. Pat. 563,507 (Apr. 19, 1956), assigned to Montecatini S.p.a.
116. G. Natta, L. Porri, G. Zanini, and A. Palvarini, *Chem. Ind.* **41,** 1163 (1959).
117. B. Natta, *Mat. Plast.* **1,** 3 (1956).
118. K. Ninomiya and G. Yasuda, *Rubber Chem. Technol.* **42,** 714 (1969).
119. G. Alliger and F. C. Weissert, in *High Polymers*, Vol. 33, J. P. Kennedy and E. Tornqvist, eds., Wiley-Interscience, New York 1968.
120. N. Tokita and I. Pliskin, *Rubber Chem. Technol.* **46,** 1166 (1973).
121. H. E. Railsback and N. A. Stumpe, Jr., *Rubber Age* **107** (12), 27 (1975).
122. E. W. Duck, *Eur. Polym. J.* **155,** 38 (1973).
123. J. R. Haws, L. L. Nash, and M. S. Wilt, *Rubber Ind.* **9,** 107 (1975).
124. K. H. Nordsiek, *Polymer Age* **4,** 332 (1973).
125. *Elastomeric Materials,* The International Plastics Selector, Inc., San Diego, 1977.

9

ETHYLENE-PROPYLENE RUBBER

E. K. EASTERBROOK AND R. D. ALLEN
Uniroyal Chemical Co.
Naugatuck, Connecticut

INTRODUCTION

Ethylene-propylene rubber—EPM and/or EPDM—was first introduced in the United States in limited commercial quantities in 1962. Though commercial production began only in 1963, EPM/EPDM rubber is now the fastest growing elastomer (ca. 6% per year), and there are presently five manufacturers in the United States, four in Europe, and three in Japan. The reasons for its rapid growth are its excellent properties, especially in an ozone environment, and its ability to be highly extended.

Nomenclature

There are two types of ethylene-propylene rubber, EPM and EPDM. These designations follow a nomenclature convention endorsed by the American Society for Testing and Materials, the International Institute of Synthetic Rubber Producers, Inc., and the International Standards Organization. The designation EPM applies to the simple copolymer of ethylene and propylene [''E'' for ethylene, ''P'' for propylene, and ''M'' for the polymethylene ($-(CH_2)_x-$) type backbone]. In the case of EPDM, the ''D'' designates a third comonomer, a diene, which introduces unsaturation into the molecule.

History

In 1951, Karl Ziegler discovered a new class of polymerization catalysts, a discovery that opened up a whole new frontier in polymer chemistry involving many types of monomers. These catalyst systems are of a coordinated anionic structure and consist of a transition metal halide and an organometallic reducing agent, usually an aluminum alkyl. Ziegler obtained patent protection on his invention[1] and primarily employed it in the polymerization of ethylene. Previous to this discovery, only branched, low molecular weight, low density polyethylene was available, from a high pressure, free radical type of reaction. Using these new catalysts, it was possible to prepare high density, high molecular weight, linear polymers at low temperatures and pressure.

This invention was extended significantly in Italy by Guilio Natta. Previously, it had been impossible to prepare a high molecular weight linear polymer of propylene by any means. Not only did it now become possible to do so with the Ziegler type catalyst but it was also possible to vary the stereospecific structure of the resultant chains (isotactic, syndiotactic, and atactic). Of more significance to the elastomeric field, however, was Natta's discovery of catalyst systems that could make amorphous, and thus elastomeric, copolymers of ethylene and propylene.[2]

The catalyst systems for the preparation of these polymers consist primarily of soluble (homogeneous) types rather than insoluble (heterogeneous) types employed for the synthesis of stereospecific polypropylenes. They consist of a vanadium halide along with a halogenated aluminum alkyl and, in some instances, an activator. The polymerization is usually conducted in a hydrocarbon solvent, although slurry-type copolymerizations are sometimes employed (by Polysar and Montecatini).

In 1963, Karl Ziegler and Giulio Natta were jointly awarded the Nobel prize for chemistry in recognition of their discoveries, which have led to several commerically important new plastics and elastomers, including ethylene-propylene rubbers.

POLYMER STRUCTURE

The structure of the regular, alternating amorphous copolymer of ethylene

$$H_2C{=}CH_2$$

and propylene

$$CH_3{-}\overset{\overset{\large H}{|}}{C}{=}CH_2$$

can be written

$$-(CH_2{-}CH_2{-}\overset{\overset{\large CH_3}{|}}{C}H{-}CH_2){-}_n$$

As one can see, this structure is the same as one would get from the hydrogenation of natural rubber (*cis*-1,4 polyisoprene).

Because of the saturated nature of this polymer molecule (EPM), the normal type of sulfur curatives cannot be employed for vulcanization. Peroxide cures in the presence or absence of a coagent are used instead. Since these cures are expensive, malodorous, and difficult to handle on standard rubber equipment, EPDM polymers were developed and are now more widely used.

EPDM's are EPM's that possess unsaturation. The unsaturation is introduced by copolymerizing the ethylene and propylene with a third comonomer, which is a nonconjugated diene. The dienes are so structured that only one of the double bonds will polymerize and the unreacted double bond acts as a site for sulfur crosslinking. This latter unsaturation is also so designed that it does not become part of the polymer backbone but a side group. As a consequence, the terpolymer retains the excellent ozone resistance that the copolymer possesses.

The three comonomers employed in industry to introduce unsaturation are the following:

1. Dicyclopentadiene (DCPD):[3]

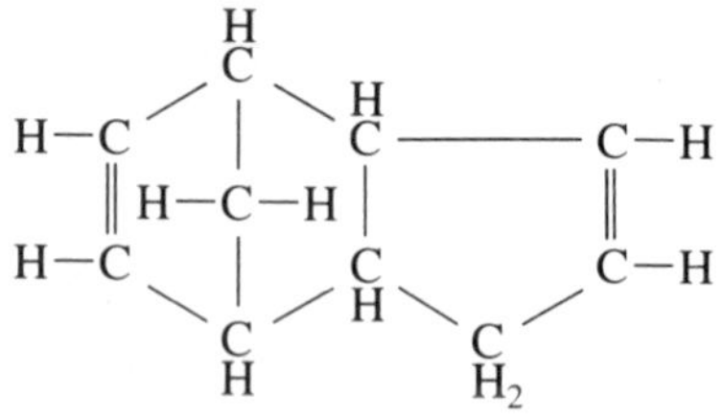

```
            H
            C
         /     \   H
   H-C      |     C ———————— C-H
     ‖  H-C-H  |              ‖
   H-C      |     C           C-H
         \     /  H  \      /
            C           C
            H           H2
```

2. Ethylidene norbornene (ENB):

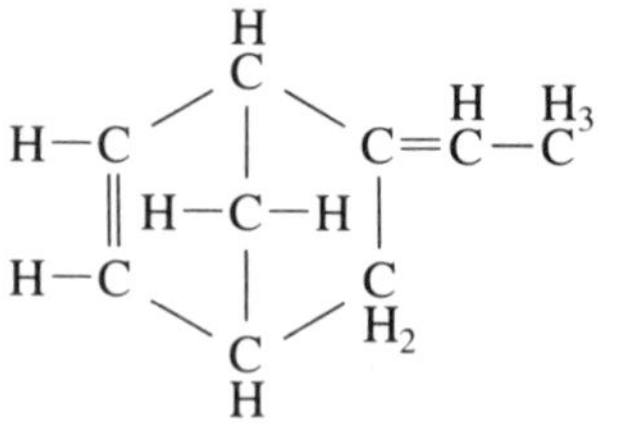

```
            H
            C
         /     \      H  H3
   H-C      |      C=C-C
     ‖  H-C-H  |
   H-C      |      C
         \     /   H2
            C
            H
```

3. 1, 4 hexadiene (1, 4 HD):[4]

$$CH_2{=}CH{-}CH_2{-}CH{=}CH{-}CH_3$$

In all three cases, the polymerizable double bond is the one illustrated on the left side of the molecule. The most common termonomer employed is ethylidene norborne because of its ease of incorporation and its greater reactivity toward sulfur vulcanization.

The structure of the ethylene–propylene terpolymer (EPDM) can be illustrated as follows:

$$\left[\left[-\overset{H}{\underset{H}{C}}-\overset{H}{\underset{H}{C}}- \right]_x \left[-\overset{H}{\underset{H}{C}}-\overset{H}{\underset{CH_3}{C}}- \right]_y \left[-\overset{H}{C}\underset{H}{-}\overset{H}{C}- \;\; \begin{matrix} HC - \overset{H}{\underset{H}{C}} - CH \\ H_2C \overset{H}{-} C \\ \qquad \| \\ \qquad CH \\ \qquad | \\ \qquad CH_3 \end{matrix} \right]_z \right]_n$$

POLYMER VARIABLES AND PROPERTIES

In addition to variations resulting from the introduction of the different termonomers, several other variations are possible in EPDM's. All of them affect the resultant properties of the polymers.

(E-P) Ethylene-propylene Composition

The contents of E-P elastomers are generally reported as weight-percent ethylene and vary from 75 to 45%. The monomers are randomly distributed, resulting in amorphous type copolymers. The higher ethylene-containing polymers, which contain some crystallinity, are beneficial in that they possess higher green strength (shape retention), can be more highly loaded with fillers/oil, result in higher compounded tensile strengths, can be more readily pelletized, and possess better extrusion properties. The disadvantages of the higher ethylene contents, however, are their poorer mill processing behavior at lower temperatures, inferior low-temperature properties, and their difficulty in being mixed. Since crystallinity tends to increase with decrease in temperature, it is appropriate to store high ethylene terpolymers under sufficient warmth to minimize dispersion problems.

Molecular Weight

The molecular weight of an elastomer is commonly reported as the Mooney viscosity (ML). In the case of EPDM's, these values are obtained at elevated temperatures, usually 125°C. The primary reason for this is to melt out any effect that high ethylene content could produce (crystallization), thus masking the true molecular weight of the polymer. Mooney viscosities can vary from a low of 20 to a high of 100. Polymers of even higher molecular weight exist but are extended with oil (25 to 100 phr). Such polymers should be stored in the absence of light (UV) since the oils will tend to accelerate oxidative degradation (gelation).

Advantages of higher molecular weights are similar to those of higher ethylene contents in that tensile and green strength improve. In addition, these polymers can be more highly loaded with fillers and oil. Disadvantages are their poorer processing and dispersibility.

Molecular-weight Distribution (MWD)

Molecular-weight distribution is not normally reported as a polymer variable but in most applications it is a very important property. A measure of the MWD is now commonly obtained through the use of Gel Permeation Chromatography at elevated temperatures (135°C). The reported value is the ratio of the weight-average molecular weight to the number-average molecular weight ($\overline{M}_w/\overline{M}_n$). This value can usually vary from 2 to 5. For broad MWD polymers, there is usually a variation in E-P composition as well. The end with the higher molecular weight possesses a higher ethylene content than the end with the low molecular weight.[5]

Polymers with broad molecular weight possess excellent mill processing and calendering capabilities and higher green strength. These polymers are extensively used for applications that don't allow large loadings of filler and oil. Their disadvantages are their slower cure rate and poorer state of cure. Narrow molecular weight distributions are more common for EPDM's since they result in faster cure rates, better state of cure, and smoother extrusions.

Diene Type

As has been mentioned, three termonomers are employed in the manufacture of EPDM's: ethylidene norbornene (ENB), 1,4 hexadiene (1,4 HD), and dicyclopentadiene (DCPD). Even though all three introduce unsaturation into the EPDM backbone as vulcanization sites, they impart varying characteristics to the elastomers because of their different structure.

Ethylidene Norbornene (ENB). ENB is the most widely used termonomer employed even though it is the most expensive, the reasons being that it is the most readily incorporated during copolymerization and the double bond introduced has the greatest activity for sulfur vulcanization. This activity is also of such a nature that EPDM's containing ENB have the greatest tendency to be cocured with diene elastomers.[6] Another unique characteristic of this termonomer is that it makes it possible to prepare linear as well as branched polymers by varying the conditions under which the polymers are synthesized.[5] Branching has an important role in establishing the rheological properties of a polymer.[7] Under proper control, it can introduce properties to the EPDM that are beneficial in certain applications.

1,4 Hexadiene (1,4 HD). Polymers containing 1,4 HD exhibit a slower cure rate than ENB but possess certain properties that are superior. One such property is its excellent heat characteristic, which is closest to EPM. Such polymers exhibit a good balance of chain scission and crosslinking reactions.[8] Polymers prepared with 1,4 HD are normally linear in structure and possess excellent processing characteristics.

Dicyclopentadiene (DCPD). The main advantages of DCPD are its low cost and relative ease of incorporation, in which it is similar to ENB. Of the three termonomers, it has the slowest cure rate. All polymers prepared with it are branched as a result of the slight polymerizability of its second double bond. As mentioned previously, this branching can be beneficial, for example, by imparting ozone resistance to diene rubber blends.

Diene Level

In addition to the changes introduced by the type of termonomer, the level of each can cause variations. Usually three different levels are employed: low (Iodine Number = 2 to 5), medium (Iodine Number = 6 to 10), and high (Iodine Number = 16 to 20). The expected effect is an increase in cure rate with increase in iodine number, which can be illustrated by Monsanto Rheometer curves. High-iodine-number (≥ 18), ENB terpolymers possess cure rates similar to those of diene rubbers and thus cocurability. Lower iodine numbers are usually employed with the norbornene (ENB, DCPD) type polymers in order to improve their ageing characteristics.

Miscellaneous Properties

Table 9.1 illustrates some, physical constants and/or properties of EPM/EPDM. Since these values were obtained on specific samples, they may vary in accordance with the composition of the polymer. For instance, the density and refractive index for a terpolymer containing DCPD or ENB will be greater than they are for the copolymer or a terpolymer containing 1,4 HD. In fact, it is possible to establish quantitatively the content of an ENB-type polymer by measuring its refractive index at 90°C.[10] The higher temperature is necessary in order to melt out any crystallinity effects that might result with copolymers of high ethylene content.

The glass transition temperature and thus the brittle point are other properties

Table 9.1. Physical Constants of Ethylene-Propylene Rubbers.[9]

Constant	*Units*	*Value*
Density	g/cc	0.854
Expansivity $(1/V)(dV/dT)p$	$(°C)^{-1}$	7.5 and 8.8×10^{-4}
$(1/L)(dL/dT)p$	$(°C)^{-1}$	2.5×10^{-4}
Specific heat, Cp	cal/g/°C	0.53
Thermal conductivity	cal/s/cm/°C	4.2×10^{4}
Refractive index		$1.4740^{23°C}$
		$1.4524^{90°C}$
		$1.4423^{120°C}$
Dielectric constant	10^3 Hz	2.5
Dissipation factor	10^3 Hz	0.2
Diffusion constant, $D_{(0)}$	cm^2/s	
benzene 23°C		6.1×10^{-8}
n-hexane 23°C		4.1×10^{-8}
CH_2Cl_2 23°C		1.3×10^{-7}
$CHCl_3$ 30°C		1.2×10^{-7}
Permeability coefficient	cm^2/s/atm°C	
CH_2Cl_2 30°C		2.8×10^{-7}
H_2O 37.5°C		4.5×10^{-8}
Glass-transition temperature	°C	−60

that will vary slightly with the ethylene and termonomer content of the polymer under study. Polymers possessing a higher ethylene content will exhibit higher brittle points, again as a result of crystallinity. In the case of the termonomers, the 1,4 HD is advantageous in that it imparts slightly better low-temperature properties than does ENB or DCPD. In general, however, all of them possess excellent low-temperature properties.

Table 9.1 also reveals one property that a practical compounder can consider for reducing unit costs in the fabrication of finished ethylene/propylene rubber products in addition to their ability to be highly extended. This property is their low density as compared to other natural and synthetic elastomers. Elastomers are bought by weight but used by volume. As a result, for a given weight of polymer, more finished articles can be produced with an EPM or EPDM.

MANUFACTURERS OF EPDM

Unlike the development of SBR, EPDM was developed independently by different manufacturers. Not only the process of manufacture (solution vs. slurry) but also the termonomer and catalyst system employed can vary considerably. These factors, as has already been mentioned, can affect molecular-weight distribution, branching, and cure rate. As a consequence, EPDM's from various suppliers possessing similar composition (E/P) and Mooney viscosity may possess considerably different rheological and curing properties.

EPDM's polymers are employed in a wide range of applications and processes. In order to maintain competitiveness, each of the suppliers has developed polymers employing their own technical expertise. As a consequence, the number of varieties of EPM and EPDM has grown consierably since their initial production. At the present time, the 1983 edition of the *Elastomer Manual* published annually by the International Institute of Synthetic Rubber Producers, lists some 125 different types. Before employing a particular EPDM it would be advantageous to become familiar with the literature from the various suppliers.

Of the twelve manufacturers, five are in the United States, four in Europe, and three in Japan. These companies are listed in Table 9.2, along with their trade name, plant location, and estimated plant capacity. According to the International Institute of Synthetic Rubber Producers, the plant capacity of EPM/EPDM has grown from 359,000 metric tons in 1975 to 502,000 metric tons in 1984.

APPLICATIONS

Over the last ten years, the production of EPM/EPDM has grown at an average rate of 6% per year (1973 to 1982). The last two years' growth has been even higher. This has been primarily due to the use of EPDM in rubber membranes for roofing, agriculture, and water distribution. EPDM compounds have been developed for a great variety of applications, among which automotive and appliance uses have been particularly significant. Major applications are shown in Table 9.3.

PROCESSING

Conventional rubber equipment is used for the processing of EPDM compounds. Selection of the polymer or blend of polymers is an essential consideration in order to obtain optimum processibility.

Mixing

Mixing usually is accomplished with an internal mixer, such as a Banbury™ mixer. In some instances, it may be desirable or necessary to use the mill mixing process. For mill mixing, the low-viscosity, higher-propylene EPDM types are preferred. The higher-viscosity and higher-ethylene polymers are considered too difficult for practical mill mixing. Furthermore, compounds based on these polymers typically contain such high amounts of filler and oil that mill mixing is usually impractical from this standpoint alone.

Although all types of EPDM are suitable for the internal mixer, no single mixing procedure is satisfactory for all compounds. The variety of polymers,

Table 9.2. World EPDM Manufacturers.

Company Name	*Plant Location*	*Plant Start-up*	*Plant Capacity, Thousands of Metric Tons*	*Tradename*
E. I. du Pont de Nemours & Co.	Beaumont, TX (U.S.)	1963	85	Nordel
Exxon Chemical Co.	Baton Rouge, LA (U.S.)	1963	65	Vistalon
Montecatini Edison S.pA.	Ferarra, Italy	1963	55	Dutral
Mitsui Petrochemical Industries, Ltd.	Iwakuni-Ohtake, Japan	1963	33	Mitsui EPT
Uniroyal, Inc.	Geismar, LA (U.S.)	1964	55	ROYALENE®
N. I. Nederlandse Staatsmijen (DSM)	Geleen, Netherlands	1966	55	Keltan
Copolymer Rubber & Chemical Corp.	Addis, LA (U.S.)	1968	45	Epsyn
Enoxy Chemical, Ltd.[a]	Grangemouth, UK	1969	16	Intolan
Sumitomo Chemical Co., Ltd.	Chiba, Japan	1969	36	Esprene
Japan EPR Co.	Yokkaichi, Japan	1970	33	JSR EP
Polysar, Ltd.	Orange, TX (U.S.)	1971	28	Polysar
Chemische Werke Huls	Maarl, W. Germany	1971	28	Buna AP

[a]Presently idle.

Table 9.3. Applications for EPDM.

Tire sidewalls
Inner tubes
Automotive
Weatherstripping (sponge and dense) for doors, windows, trunk lids
Radiator and heater hose
Air emission hose
Tubing
Brake components
Isolators and mounts
Grommets, body rubber
Appliances
Inlet and drain hose
Boots
Seals
Mounts
Building and construction
Glass sealers,
Curtain wall gaskets and tapes
Rubber sheeting for roofing
Pond and ditch liners
Reservoir liners
Agricultural equipment
Hoses
Seed tubes
Cushioning
Sheeting for grain storage
Liquid fertilizers
Silos
Tank linings
Wire and cable
Mechanical goods
Dock fenders
Belting
Gasketing
Seals
O-rings
Faucet washers
Roll covers

the range of compound qualities (which affect choice of filler type), and the condition of the particular mixer are some of the factors that influence the mixing technique.

The most popular procedure is to load the ingredients in "one shot," and often the batch is loaded "upside-down" (rubber put in last, after fillers, oils, and so forth). Depending upon the particular machine (its mechanical condition, speed, and so forth) mixing cycles can range from 3 to 7 minutes. Using this kind of procedure, the batch is usually dumped when the temperature reaches 115°C to 130°C. This procedure may be limited to compounds that are based on medium- and low-ethylene polymers loaded with mineral fillers and/or semi-reinforcing types of carbon black (SRF, GPF, FEF).

For compounds utilizing high-ethylene or very-high-molecular-weight EPDM polymers and for compounds containing reinforcing carbon blacks (HAF, ISAF), the ''one-shot'' or ''upside-down'' loading procedure often results in poor dispersion of black or rubber or both. These polymers and blacks will usually not disperse well if processing oil is added early in the mixing cycle. Better dispersion is attained by an incremental procedure in which rubber and part of the black are loaded into the mixer with little or no oil. The withheld black and oil may then be added in one, two, or more increments.

If allowable from the standpoint of providing required properties, an easy-to-disperse softer black or higher-structure black, should be used in combination with the more difficult-to-disperse, highly reinforcing black. Then the easy-to-disperse black can be added with the plasticizer as the second increment in the mixing procedure. As was mentioned previously, high ethylene EPDM's should be stored under warm warehouse conditions to minimize dispersion problems.

Other variations in mixing procedures are successful. For example, a modified ''upside-down'' technique in which part of the EPDM polymer is added on top of fillers and plasticizers provides rapid mixing and good dispersion. This procedure may also be further modified by withholding part of the oil, which is then added incrementally. In addition, two-stage mixing is frequently employed when better dispersion or batch-to-batch consistency is needed.

Extrusion

The selection of polymer is a prime consideration if the EPDM compound is to be used for extrusion processing. The high-ethylene types are most useful for this process because they provide good green strength for shape retention. In addition, some green strength seems necessary for good feeding into the extruder; in fact, lack of green strength in the uncured compound is usually associated with poor feeding characteristics. If compounds do not feed easily, and if feed is erratic, surging of the extrudate results in its size varying. Hence a well-designed extrusion compound should contain enough high-ethylene polymer to provide the desired processing behavior.

Since excessive green strength (tough compound) is a possibility, one of the lower ethylene types of EPDM should be part of the polymer blend to avoid it.

A wide variety of extrusion equipment is to be found. Most of the older extruders still in use are the short-barrel, hot-feed type, and for these the stock must usually be ''warmed'' on a mill and then fed to the extruder. This process, therefore, requires a compound that is reasonably easy to band on the warm-up mill. Excessive green strength and/or high compound viscosity will cause difficulty in banding and may also cause bagging. These results may be avoided by using a lower-ethylene and/or lower-viscosity polymer blended with the high-ethylene type in the compound.

In general, the newer cold feed extruder is preferred. Not only does it obviate the need for a warm-up mill, it also provides better control of processing conditions so that more uniform extrudates are obtained. For most compounds, the extruder screw should be operated at a higher temperature than the barrel. Conditions should be adjusted to provide a smooth surface and optimum rate of extrusion. In general, extrusion temperature conditions are somewhat higher for EPDM compounds than for other types of rubber compounds.

Molding

Molded rubber parts are made by either a compression, transfer, or injection process, for all of which polymers having medium to high propylene are usually preferred. In applications involving highly extended compounds, however, it may be desirable to use a high-ethylene polymer, either exclusively or in part, to take advantage of its greater extendability. The prime consideration is good flow characteristics, and polymer selection should be made accordingly.

EPDM compounds have proved to be particularly suitable for injection molding. When properly compounded, EPDM provides a fast cure rate and good flow and is not prone to reversion at the high curing temperatures common to this process. Compounding considerations for injection molding are not appreciably different from those for compression or transfer molding. The curing system is usually the point of most concern and the plasticizer should be selected for low volatility at the processing temperature.

Calendering

The EPDM polymer selected for calendering should be in the medium- to high-propylene range. If the compound is to be lightly loaded, the lowest-viscosity, highest-propylene type is most suitable. Also, a polymer with broad molecular-weight distribution and medium- to high-propylene content is a better choice. For medium to highly loaded compounds, the higher-viscosity types are required. The desirable processing is also often attained by a proper selection of two types of EPDM polymers.

ELEMENTS OF COMPOUNDING EPDM

Curing System for EPDM

If one substitutes one EPDM for another in an existing formulation without changing the cure system, the rate of cure and the properties will almost certainly not be the same, particularly if the two polymers utilize different third monomers. The properties also can be affected by differences in molecular weight and other variations in polymer composition. In general, polymers with DCPD or 1,4 HD will require somewhat more active accelerators and/or higher levels to provide satisfactory cure rates.

As already mentioned, EPDM has pendant unsaturation that allows sulfur vulcanization. Like most other synthetic rubbers, common accelerators may be used. Since the particular combination selected depends upon many considerations—such as processing methods, the properties desired, cost, and compatibility—few generalizations are possible. Usually, however, the cure system will contain a thiazole accelerator (MBT, MBTS, or the like) in combination with a thiuram and/or a dithiocarbamate. Sulfur donor-type accelerators may replace elemental sulfur, if heat resistance and/or compression set requirements are severe. For some molded-goods applications, accelerator bloom may be unacceptable. To assure a nonblooming compound, it is necessary to maintain the levels of the various chemicals below their solubility limits.

Approximate limits for most commonly used accelerators are as follows:

MBT, MBTS, CBS, ZMBT, ZDBDP*	3.0 phr
ZDBDC, DTDM	2.0 phr
ZDEDC, ZDMDC, TDEDC, TMTD	0.8 phr
DPTT, TMTM, TETD, FDMDC	0.8 phr

It has long been known that low sulfur or sulfur donor cure systems give good heat resistance and improved compression set. By using a cure system with 3 to 4 phr of a thiazole (MBT, MBTS, or CBS) in combination with a thiuram and a dithiocarbamate and a level of sulfur below one part, it is possible to obtain outstanding heat resistance. For exposure temperatures above 150°C, a selected antioxidant should be added to enhance the heat resistance of this system. Another sulfur-donor, low-sulfur system that provides good heat resistance but better compression set than that obtained with a "high thiazole" system entails using 2- to 3-phr levels of a thiuram, of two different dithiocarbamates, of a dithiomorpholine, and a low level of sulfur.

In black, steam-cured extrusions, bloom is usually not a problem. Some cure systems that bloom when press cured may not bloom when cured in steam. Since steam curing cycles are typically longer and take place at lower temperatures than molding cycles, the cure system may be quite simple, for example 1 phr MBTS (or MBT), 1.5 phr TMTD, and 1.5 phr sulfur. A dithiocarbamate may be added if an increased cure rate is desired.

---

*See Appendix on p. 281 for symbol designation.

Reinforcement

As is true of other noncrystallizing polymers, EPDM requires reinforcement to be of practical value since the mechanical properties of the unfilled rubber are quite poor. Carbon black is the most useful material for this reinforcement, but silica, clay, talc and some other mineral fillers may also be used. To attain their full effectiveness as reinforcing agents in EPDM rubbers, carbon black and other fillers must be well dispersed. High tensile strength, good tear resistance, and improved abrasion resistance are usually associated with good reinforcement. Well-mixed batches also provide better and more uniform processing for extrusion, calendering, molding, and so forth.

Plasticizers and Processing Aids

Naphthenic oils have been the most widely used plasticizers for EPDM compounds because they provide the best compatibility at reasonable cost. For applications at higher temperatures or in colored compounds, paraffinic oils are usually chosen because of their lower volatility and improved UV stability. Some paraffinic oils tend to bleed from cured, high-ethylene EPDM compounds. If such oils must be used, it is usually advisable to replace part (20–25 phr) of the high-ethylene rubber with one having somewhat lower ethylene content.

Aromatic oils have an adverse effect on some compound properties and, of course, must not be used in conjunction with peroxide curing systems.

Stearic acid, zinc stearate, or other internal lubricants are often included in a compound to aid processing.

EPDM compounds are inherently not tacky. Should there be a need for building tack, it would be necessary to add a tackifier to the compound. Some care must be taken in making the selection because of compatibility problems and the effects on cure rate of tackifiers with excessive unsaturates. Some tackifiers have been developed specifically for EPDM. Suggestions in EPDM and tackifier manufacturers' literature should be used as a guide for specific problems involving tack.

COMPOUNDING EPDM FOR VARIOUS APPLICATIONS

The inherent properties of EPDM polymers have led to their usage in a considerable variety of applications. We will discuss some of the more significant uses and show some formulations to illustrate concepts in compound development for specific applications.

Sheeting

Rubber sheeting has been used for some years in roofing, but in rather limited quantities because of its cost compared to that of cheap asphalt. The situation has changed dramatically, and the economics for EPDM sheeting are now much more favorable. The formulation suggested for this application is based on a polymer that offers good physical properties at fairly high extension, good green strength, and good calendering. The filler loading is designed to provide good processibility and reinforcement to meet the required tensile and tear strength properties. Because of long-term aging conditions, a low-volatility, paraffinic oil and a sulfur donor, low-sulfur cure systems are employed. Since the application involves making lap-seam splices in the field, the cured compound, to facilitate adhesion, is nonblooming. Test data show the characteristics of green strength, cured physical properties, and heat, water, and ozone resistance that are required for good performance in roof sheeting applications. See Table 9.4 for a typical sheeting recipe.

Automotive

When EPDM was introduced as a commercial polymer, the automotive industry was quick to recognize areas where it might improve part performance. Major

Table 9.4. Typical Recipe for Sheeting.

EPDM A	100.00
N-347 black	120.00
Talc	30.00
PARAFFINIC oil type 104B	95.00
Zinc oxide	5.00
Stearic acid	1.00
MBTS®	2.20
TMTD	0.65
TETD	0.65
Sulfur	0.75
	355.25
Mooney viscosity, ML-4 @ 100°C	53
Mooney scorch MSR—135°C, min	16.0
Rheometer 150°C, 1° arc, 1.7 Hz, micro-die	
Torque, min.	0.6
(N.m) max.	2.6
time, ts_1	5.0
(min) tc_{50}	10.3
tc_{90}	27.0

Table 9.4. (*Continued*): Physical Properties

	Calendered Sheet	*Specification ASTM D-8 (tentative)*
	Uncured	
Tensile strength, max., KPa	965	-
100% modulus, kPa	690	-
Elongation, %	650	-
	Cured	
Dusted/wrap cure/3h/60 psi g. steam		
	Unaged	
Tensile strength, MPa	17.4	9.0 min.
Elongation, %	450	300 min.
200% modulus, MPa	8.1	
Hardness, durometer A	65	60 ± 5 min.
Tear strength, die C, N/mm	36.8	22 min.
Tensile set—50% extension, %	3.0	5 max.
	Aged 7 days/125°C Air Oven	
Tensile strength, MPa	17.6	8.3 min.
Elongation, %	290	225 min.
Hardness, durometer A	71	—
Linear shrinkage, %	0.5	2.0 max.
	Aged 7 days/70°C Water	
Weight increase, %	1.6	2.0 max.
	Ozone Resistance	
100 pphm, 50% extension, 7 days, no cracks	Pass	Pass

Table 9.5. Radiator Hose Compound.

EPDM B	105.0
EPDM A	25.0
Zinc oxide	3.0
N-650 black	130.0
N-762 black	95.0
Ground whiting	40.0
High-viscosity, paraffinic oil type 104B	130.0
Stearic acid	1.0
TMTD	3.0
DTDM	2.0
ZDBDC	2.0
ZDMDC	2.0
Sulfur	.5
	538.5

uses have included sponge and dense weatherstripping for doors, windows, and trunk lids. Radiator and heater hoses, air emission hoses, and various tubing specifications have profited from EPDM. Brake components—including boots, diaphragms, many grommets, isolators, mounts, and other suspension parts—have also taken advantage of it.

A typical radiator hose compound (see Table 9.5) is based on a blend of polymers, the major portion of which is a high-ethylene type that offers high green strength for desired shape retention. The use of a small amount of a lower-ethylene type assures oil compatibility. Carbon blacks provide the desired smooth extrudates and modest physical properties required. Ground whiting is incorporated to lower the cost. A nonvolatile oil is necessary to meet the required heat resistance. The cure system is chosen to provide an acceptable rate of cure and meet the specification requirements for heat resistance, compression set, and coolant resistance. By these means, satisfactory performance at a reasonable cost is provided for the radiator hose application (see Table 9.6)

Table 9.6. Radiator Hose.

Unaged Physical Properties, 30′ @ 160°C Cure	
Hardness, durometer A	72
Modulus at 100%, MPa	3.9
Tensile strength, MPa	9.9
Elongation, %	300
Heat-aged, ASTM D573, 70 hr @ 121°C	
Hardness change, pts.	+4
Tensile strength change, %	+13
Elongation change, %	−29
Compression set, ASTM D395 Method B, 70 hr @ 121°C	
Plied specimens	58
Immersion in Ethylene Glycol–Water (55:45), 70 hr @ Boiling Temp.	
Tensile strength change, %	+8
Elongation change, %	−17
Volume change, %	+2.0
Immersion in 2% Soluble Oil in Water, 70 hr @ Boiling Temp.	
Tensile strength change, %	+13
Elongation change, %	−17
Volume change, %	+10.7

Table 9.7. Master Cylinder Reservoir Seal.

EPDM C	100.0
Zinc oxide	5.0
N774 black	55.0
Type 104B Oil	5.0
Substituted amine A.O.	0.8
Process aid	1.5
Trimethylol propane trimethacrylate	2.0
Dicumyl peroxide, 40% active	7.0
	176.3

A master cylinder reservoir seal (see Table 9.7) typifies an automotive brake component that requires outstanding performance. The applicable compound employs a low-molecular-weight polymer with an E/P ratio near 50/50 to provide good processibility. Since seal shrinkage as a result of plasticizer being extracted by brake fluid is undesirable, the level of oil must be kept to a minimum, which in turn limits the level of carbon black. The combination of dicumyl peroxide and the coagent, trimethylol propane trimethacrylate imparts excellent heat resistance and very low compression set. A relatively expensive EPDM compound is needed to assure adequate performance in this part.

To illustrate the extreme extent to which some EPDM polymers can be extended and still be practical for certain uses, a formulation for a low-cost extrusion compound is given in Table 9.8. A blend of a high molecular weight and a high-ethylene, oil-extended polymer is chosen to provide the desired processibility. Oil-extended types facilitate mixing. This compound can be used for low-cost hoses, tubing, or weatherstripping. Its green strength is adequate for shape retention unless the profile in question is extremely intricate or delicate. The total oil content, including the oil present with the polymers plus that added during compounding, is 220 phr. The oil combined with the high loading of carbon black and ground whiting is responsible for the low cost. Even at this level of extension, the compound has excellent resistance to ozone—the essential performance requirement for the applications intended.

Automotive weatherstripping for doors and trunk lids ordinarily uses closed-cell extruded EPDM sponge. A typical compound utilizes an EPDM polymer with high ENB to provide a fast cure rate. (See Table 9.9 for a recipe.) Its high molecular weight imparts good green strength for cross-sectional shape retention. This compound is fairly highly extended in the interest of low cost and rapid extrusion at relatively low temperature settings on the extruder. The combination of blowing agent-activator and cure system provides a desirable cure-blow rate, allowing satisfactory continuous extrusion/curing operation. The cured sponge has the excellent physical properties and low water absorption required for satisfactory performance in this application.

Table 9.8. Highly Extended, Low-cost Compound (Grade 8BA705C12).

EPDM D	50.0
EPDM B	105.0
Zinc oxide	3.0
N-650 black	250.0
N-774 black	100.0
Ground whiting	200.0
Type 103 Oil	165.0
Zinc stearate	1.5
MBTS®	3.0
TMTD	.8
ZDBDC	1.5
Sulfur	2.0
	881.8

	30′ @ 160°C Cure	*8′ @ 182°C Cure*
Unaged Physical Properties		
Hardness, durometer A	75	75
Tensile strength, MPa	5.9	5.3
Elongation, %	150	180
Heat-aged, D573, 70 hr @ 100°C		
Hardness change, pts.	+4	+5
Tensile strength change	+5	+23
Elongation change, %	−27	−33
Compression Set, D395 Method B, 22 hr @ 70°C		
% Set	30[a]	38[b]
Suffix C12—Ozone Resistance, D1171, 72 hr @ 50 pphm @ 40°C		
Quality retention rating	100	100

[a]Solid specimen.
[b]Plied specimen.

Wire and Cable

EPDM has found wide acceptance in wire and cable applications because of its inherently excellent electrical properties, combined with resistance to ozone, heat, cold, and moisture. Aside from these advantages, the polymer chosen in Table 9.10 provides good extrusion processing with a relatively low level of oil and clay filler. Red lead is used to impart stable, long term, wet electrical properties. A peroxide-coagent cure system is needed for best aging.

Table 9.9. Recipe for Continuous Extruded Closed-cell Automotive Sponge.

EPDM E	100.0
Catalpo clay	50.0
N-762 black	75.0
N-550 black	15.0
York whiting	30.0
Zinc oxide	3.0
Stearic acid	1.0
Circosol 4240	65.0
BIK®	1.0
CELOGEN® AZ-130	8.0
MBT	2.0
TDEDC	1.0
ZDBDC	2.0
DPTT	2.0
Sulfur	2.0
	367.0

Recommended Continuous Curing Conditions

Precure—3 min @ 310°F	
3 min @ 390°F	
Expanded density: lb/cu ft	19.6
kg/m^3	313.9
Water absorption, % change in weight	0.3
25% compression deflection: psi	10.7
KPa	73

MISCELLANEOUS EPM/EPDM APPLICATIONS

EPM/EPDM was originally considered as a specialty rubber. However, it has found extensive usefulness in a wide variety of applications. The only major rubber application area it has not fully penetrated is automotive tires. The wide usage of this elastomer is connected to its multifunctional nature.

Some areas outside typical rubber applications in which EPM/EPDM has found usefulness in significant quantities are viscosity index improvers for lubricating oils, impact modifiers for polyolefins, and thermoplastic elastomers. The latter materials are primarily blends with polyolefins where the elastomer is the major constituent.

As with most polymers, EPM/EPDM polymers also lend themselves to chemical modification. Significant work has been done with halogenation; however, no commercialization has yet resulted. A significant new application now developing involves the production of various grafts on the polymer, opening the way for the modification of many other plastics such as polystyrene,

Table 9.10. Typical Formulation for an EPDM Insulation Compound (90°C).

This EPDM insulation compound has excellent physical, ageing, and electrical properties and meets the requirements of UL 44–RHW and RHH, IPCEA 3.13 and 3.16 and ASTM D 1523 and D 1679 specifications. Data is from a No. 14 solid-tinned copper conductor with 0.047-in. insulation cured 60 sec at 250 psi steam in a CV unit.

EPDM F	100	Burgess KE	60
Zinc oxide	5	Sunpar 2280	15
NAUGARD® Q	1	Sartomer SR 350	2
Red lead	5	Di Cup 40%	7

Unaged

Tensile strength, psi	1430
Elongation, %	295
M200%, psi	940
UL set, %	5

14 days Air Oven, 121°C—% Retained

Tensile strength	106%
Elongation	95

21 days Air Oven, 121°C—% Retained

Tensile strength	96%
Elongation	91

Electrical Properties

	75°C Water			*90°C Water*		
		Power Factor			*Power Factor*	
	Sic	*40V/M*	*80V/M*	*Sic*	*40V/M*	*80V/M*
1 day	2.68	1.28	1.36	2.67	1.58	1.64
7 days	2.69	.97	1.08	2.69	.94	1.04
14 days	2.71	.83	.93	2.71	.82	.89
1 month	2.72	.68	.77	2.74	.82	.91
2 months	2.76	.67	.74	2.80	1.05	1.12
3 months	2.79	.73	.81	2.87	1.15	1.23
4 months	2.81	.67	.80	2.92	1.15	1.22
5 months	2.84	.61	.78	2.96	1.14	1.23
6 months	2.87	.72	.81	2.98	1.10	1.19
7 months	2.89	.76	.78	3.00	1.06	1.13
8 months	2.90	.79	.85	2.98	.96	1.03

(*cont.*)

Table 9.10. (*Continued*)

	Insulation Resistance, megohms/1000 ft	
Weeks		
1	9000	5500
5	11500	5500
9	8200	9500
13	13500	7500
17	8200	6000
21	8500	5500
25	8000	6000
29	10500	6000
33	8200	6000
36	7500	6000

poly(acrylonitrile-costyrene), nylon, and so forth, where improved aging and weathering are desired.

REFERENCES

1. K. Zeigler, H. Martin, and E. Holzkamp, U.S. Pat. 3,113,115 (Dec. 3, 1963), assigned to Karl Zeigler.
2. G. Natta and G. Boschi, U.S. Pat. 3,300,359 (Jan. 24, 1967), assigned to Montecatini Edison S.p.A.
3. S. Adamek, E. A. Dudley, and R. T. Woodhams (Dunlop Rubber Co.), U.S. Pat. 3,211,709 (Oct. 12, 1965), assigned to Hercules Powder Co.
4. W. F. Gresham and M. Hunt, U.S. Pat. 2,933,480 (Apr. 19, 1960), assigned to E. I. duPont de Nemours and Co.
5. E. K. Easterbrook, et al., ''A Discussion of Some Polymerization Parameters in the Synthesis of EPDM Elastomers,'' XXII IUPAC, Macro Molecular Preprint, Vol. II, 712 (1971).
6. K. H. Wirth, U.S. Pat. 3,492,370, and R. E. Barret, U.S. Pat. 3,492,371 (both Jan. 27, 1970), assigned to Copolymer Rubber & Chemical Corp.
7. K. P. Beardsley and C. C. Ho, ''Rheological Properties as Related to Structure from EPDM Polymers,'' *Jour. of Elastomers and Plastics* **16,** 20 (1984).
8. V. R. Landi and E. K. Easterbrook, ''Scission and Crosslinking During Oxidation of Peroxide Cured EPDM,'' *Poly. Eng. and Sci.* **18,** (15), 1135 (1978).
9. F. P. Baldwin and G. VerStrate, ''Polyolefin Elastomers Based on Ethylene and Propylene,'' *Rubber Chem. Technol.* **45,** 768 (1972).
10. I. J. Gardner and G. VerStrate, ''Determination of Ethylidene Norbornene in EPDM Terpolymers,'' *Rubber Chem. Technol.* **46,** 1019 (1973).

APPENDIX

Accelerator Codes

CBS	N-cyclohexyl-2-benzothiazole sulfenamide
MBT	2-mercaptobenzothiazole

MBTS	2,2′ dibenzothiazyl disulfide
TMTD	Tetramethylthiuram disulfide
TETD	Tetraethyl thiuram disulfide
ZDMDC	Zinc dimethyldithiocarbamate
ZDBDC	Zinc dibutyldithiocarbamate
DTDM	4′,4 dithiodimorpholine
TDEDC	Tellurium diethyldithiocarbamate
DPTT	Dipentamethylene thiuram hexasulfide
TMTM	Tetramethylthiuram monosulfide
FDMDC	Ferric dimethyldithiocarbamate
ZMBT	Zinc salt of MBT
ZDBDP	Zinc-0,0-dibutylphosphorodithioate
ZDEDC	Zinc diethyldithiocarbamate

Polymer Reference

Designation	*Trade Name*	*Description*
EPDM A	ROYALENE 512	68/32 E/P, ML-4 @ 125°C = 60 medium ENB
EPDM B	ROYALENE 622	75/25 E/P, ML-4 @ 100°C = 55 40 phr Type 103B Oil, medium ENB
EPDM C	ROYALENE 521	51/49 E/P, ML-4 @ 100°C = 45 medium ENB
EPDM D	ROYALENE 400	68/32 E/P, ML-4 @ 100°C = 38 100 phr Type 103B oil, medium DCPD
EPDM E	ROYALENE 525	57/43 E/P, ML-4 @ 125°C = 66 high ENB
EPDM F	ROYALENE 301T	68/32 E/P, ML-4 @ 125°C = 42 medium DCPD

Glossary of Products

Trade Name	*Description*	*Supplier*
BIK	Surface coated Urea	Uniroyal, Inc.
CELOGEN® AZ	Azodicarbonamide	Uniroyal, Inc.
Circosol 4240	Naphthenic oil	Sun Petroleum Pdts. Co.
Catalpo Clay	Aluminum silicate	Freeport Kaolin Co.
Chemlink 30	Trimethylol propane trimethacrylate	Synthetic Pdts. Co.
Sartomer SR-350	Trimethylol propane trimethacrylate	Sartomer Co.
DELAC® S	N-cyclohexyl-2-benzothiazole sulfenamide	Uniroyal, Inc.
Di Cup 40 KE	Dicumyl peroxide, 40% Active	Hercules, Inc.

Glossary of Products (*Cont.*)

Trade Name	*Description*	*Supplier*
AMINOX®	Diphenylamine acetone reaction product	Uniroyal, Inc.
NAUGARD® Q	Polymerized 1,2 dihydro-2,2,4 trimethylquinoline	Uniroyal, Inc.
Sunpar 2280	Paraffinic oil	Sun Petroleum Pdts. Co.
TUEX®	Tetramethylthiuram disulfide	Uniroyal, Inc.
TE-80	Proprietary composition	Technical Processing, Inc.
Vul Cup 40 KE	*a*,*a*′-bis(*t*-butylperoxy) diisopropyllbenzene	Hercules, Inc.
York White	Natural calcium carbonate	R. E. Carroll, Inc.

10

BUTYL AND HALOBUTYL RUBBERS

J. V. Fusco and P. Hous
Exxon Chemical Company
Linden, New Jersey

INTRODUCTION

Butyl rubber, introduced in 1942, is commercially produced by cationically copolymerizing isobutylene with small amounts of isoprene. The halogen derivatives, chloro- and bromo-, were introduced in the early 1960's and have been commercially available since then. The halogenated derivatives of butyl rubber provide greater vulcanization flexibility and enhanced cure compatibility with other, more unsaturated general-purpose elastomers. Butyl polymers are among the most widely used synthetic elastomers in the world, ranking third in total synthetic elastomers consumed. As speciality elastomers, the butyl-based polymers have found their most important application in the tire industry although a host of other applications have evolved and continue to utilize their unique properties. There are two producers in the free world: Exxon Chemical Company, an affiliate of Exxon Corporation; and Polysar, Ltd., of Canada. The polymer is also produced in the USSR for internal consumption.

BUTYL RUBBER

Commercial butyl-rubber grades poly(methylpropene-co-2-methyl-1,3 butadiene) or poly(isobutylene-co-isoprene), are prepared by copolymerizing small amounts of isoprene, 1 to 3% of the monomer feed, with isobutylene, catalyzed by $AlCl_3$ dissolved in methyl chloride. The extremely rapid reaction is unique, proceeding via cationic polymerization at $-100°C$ to completion in less than a second. Monomer purity is important to achieve the desired polymer molecular weights. The methyl chloride diluent and monomer feed must be carefully dried.

Historical

Butyl rubber had its origin in the work of the researchers, Gorianov and Butlerov (1870) and Otto (1927), who found that oily homopolymers of isobutylene could be produced in the presence of boron trifluoride. It wasn't until the 1930's that the I. G. Farben Company of Germany produced high-molecular-weight polyisobutylenes that possessed rubber-like properties but could not be vulcanized by normal methods because of their saturated hydrocarbon structure. These homopolymers are available today in a variety of molecular weights from Badischer of Germany and Exxon Chemical Company under the trade names OPPANOL® and VISTANEX®, respectively. Their unique age-resistant properties have made them particularly suitable for uncured rubber sheeting, caulks and sealants, adhesives and chewing gum.

Subsequent research in the 1930's led W. J. Sparks and R. M. Thomas[1-5] of Exxon Research and Engineering Company (then known as the Standard Oil Development Company) to advance the state of the art of isobutylene polymerization, and in 1937 produced the first vulcanizable isobutylene-based elastomer by incorporating small amounts of a diolefin, and particularly isoprene, into the polymer molecule. This introduced the first concept of limited olefinic functionality for vulcanization in an otherwise saturated copolymer. Corresponding vulcanizates of these new copolymers (now known as "butyl rubber") were found to possess a set of unique and desirable properties, e.g., low gas-permeability, high hysteresis, outstanding resistance to heat, ozone, chemical attack and tearing. Subsequent development to commercialization was spurred during World War II by the U.S. Government as part of its rubber-procurement program. Exxon built and operated the first butyl commercial facility in cooperation with the Government's Rubber Reserve Board. In 1955, the commercial butyl plants were purchased by Exxon Chemical from the U.S. Government.

Butyl-rubber Manufacture

A schematic diagram of a typical butyl plant is shown in Fig. 10.1. The feed, which is a 25% solution of isobutylene (97–98%) and isoprene (2–3%) in methyl chloride, which is the diluent, is cooled to −100°C in a feed tank. At the same time, aluminum chloride is also being dissolved in methyl chloride. Both of these streams are then continuously injected into the reactor. Because the reaction is exothermic and is practically instantaneous, cooling is very important. To remove the heat of reaction, liquid ethylene is boiled continuously through the reactor cooling coils, keeping the reaction at −100°C. As the polymerization proceeds, a slurry of very small particles is formed in the reactor. This slurry overflows into a flash drum that contains copious quantities of hot water. Here the mixture is vigorously agitated, during which time the diluent and unreacted hydrocarbons are flashed off overhead.

At this point, an antioxidant and zinc stearate are introduced into the polymer. The antioxidant is added to prevent breakdown of the polymer in the subsequent finishing section. Zinc stearate is added to prevent the agglomeration,

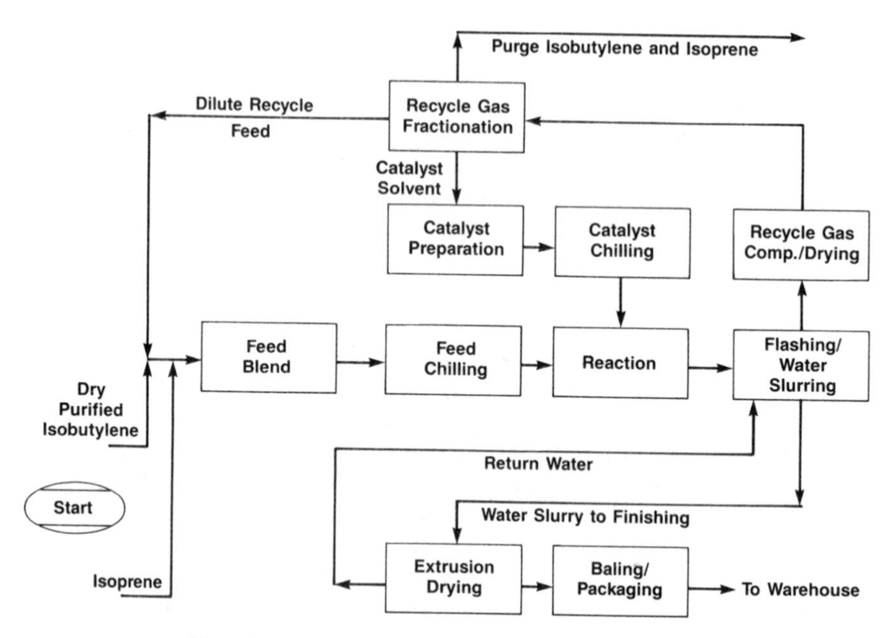

Fig. 10.1. Basic components of butyl-rubber process.

or sticking together, of the wet crumb. The slurry is then vacuum-stripped of residual hydrocarbons.

In the finishing operation, the butyl-rubber slurry is dewatered in a series of extruders to bring the water content to 5–10% in the rubber. Final drying is accomplished in a third extruder by allowing the compressed polymer melt to expand through a die to form an exploded crumb. The crumb is air conveyed to an enclosed fluidized bed conveyor, where water vapor is removed and the crumb is cooled and baled.

Butyl Physical Properties

The most widely used commercially available butyl rubbers are copolymers of isobutylene and isoprene and are described in Table 10.1. Grades are distinguished by molecular weight (Mooney viscosity) and mole % unsaturation. The mole % unsaturation is the number of moles of isoprene per 100 moles of isobutylene. This is illustrated in Fig. 10.2, in which n represents the moles of isobutylene that are combined with m moles of isoprene. Thus in our case, n would equal 98, and m would equal 2. In comparison, natural rubber would have n equal to 0 and m equal to 100, since this molecule is 100% polyisoprene. It could be called 100-mole-percent unsaturated.

Fig. 10.3 is an attempt to place the unsaturation of butyl and natural rubber

Table 10.1. Commercial Butyl-rubber Grades.

Typical Range for Mooney Viscosity		*Unsaturation (Ave. Mole %)*	*Supplier and Typical Grades*	
M_L @ 100°C (1 + 8)	*M_L @ 125°C (1 + 8)*		*Exxon Chem. Co.*	*Polysar Limited (Canada)*
41–49	-	0.7 to 1.0[a]	Exxon Butyl 065	Polysar Butyl 100
41–49	-	2.0 to 2.2[a]	Exxon Butyl 365	Polysar Butyl 402
41–49	-	1.1 to 1.4[a]	Exxon Butyl 165	-
-	50–60	1.5 to 1.8[a]	Exxon Butyl 268	Polysar Butyl 301
-	46–60	0.7 to 1.0[a]	Exxon Butyl 077	Polysar Butyl 101
-	42–58	0.7 to 1.0[b]	-	Polysar Butyl 101−3

[a]All grades stabilized with non-staining antioxidant.
[b]This grade contains *no* antioxidant.

on a spacial basis. For example, a butyl rubber with 1 mole percent of unsaturation would have a molecular weight between points of unsaturation of well over 5000. Natural rubber, on the other hand, has a molecular weight between points of unsaturation of only 68. This is the molecular weight of an isoprene molecule.

Unsaturation is determined by an iodine absorption technique[6] or by ozone degradation of the polymer followed by subsequent viscosity measurements of the degraded fragments,[7] or, more recently, by nuclear magnetic resonance (NMR). According to Rehner,[8] the isoprene unit enters the chain randomly in a *trans*-1,4 configuration, since chemical analysis shows little evidence for 1,2 and 3,4 modes of entry at the unsaturation levels present in commercial grades

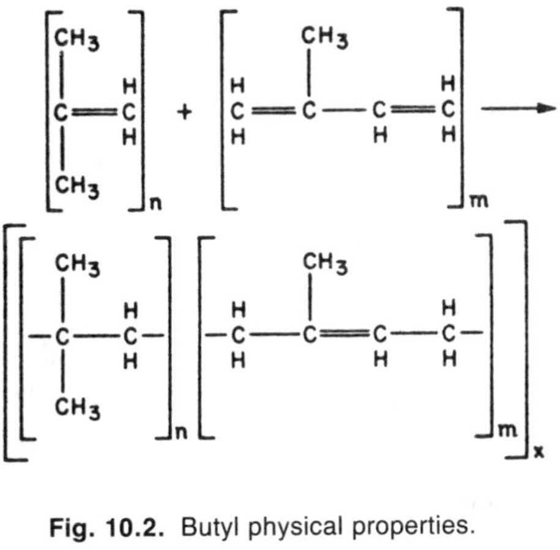

Fig. 10.2. Butyl physical properties.

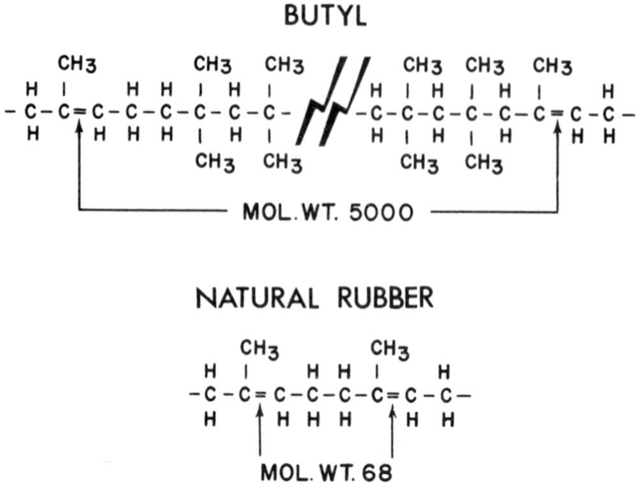

BUTYL

```
    CH3          CH3   CH3   //   CH3   CH3   CH3
 H   |      H  H  |  H  |   //  H  |  H  |  H  |        H
-C - C = C - C - C - C - C - C -//- C - C - C - C - C - C = C - C -
 H   ↑   H  H  H  |  H  |  //   H  |  H  |  H     ↑   H  H
     |           CH3   CH3      CH3   CH3         |
     |_______________ MOL.WT. 5000 _______________|
```

NATURAL RUBBER

```
       CH3         CH3
 H      |     H  H  |      H
-C  -  C = C - C - C - C = C  -  C -
 H      ↑   H  H  H     ↑  H   H
        |_______________|
           MOL. WT. 68
```

Fig. 10.3. Unsaturation of butyl and natural rubber on special basis.

of butyl. The above three isomer forms for the isoprene unit in the polymer are depicted below:

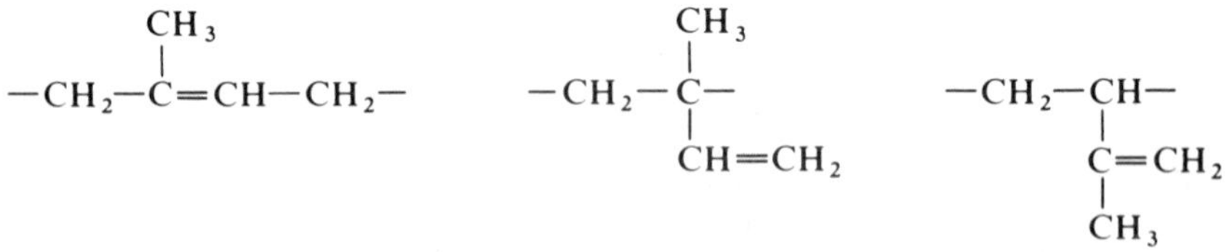

$$-CH_2-\overset{\overset{\displaystyle CH_3}{|}}{C}=CH-CH_2- \qquad -CH_2-\underset{\underset{\displaystyle CH=CH_2}{|}}{\overset{\overset{\displaystyle CH_3}{|}}{C}}- \qquad -CH_2-\underset{\underset{\displaystyle \underset{\underset{\displaystyle CH_3}{|}}{C}=CH_2}{|}}{CH}-$$

trans-1, 4   1, 2   3, 4

The discovery and development of butyl rubber not only furnished rubber technology with a new elastomer, but, as mentioned earlier, it also provided a new principle. This is the concept of *low functionality.* A low-functionality elastomer has sufficient chemical unsaturation so that flexible low-modulus vulcanized networks can be produced, but the larger inert portions of the chain contribute towards oxidation and ozone resistance. Today the concept of low functionality is well-established. For example, vulcanizable ethylene-propylene terpolymers (EPDM) now extends the family of low-functionality elastomers.

A low-molecular-weight semiliquid variety of butyl rubber (40,000 viscosity-average molecular weight) was developed by Exxon[9] and is available from the Hardman Company in Belleville, New Jersey. This polymer is used primarily in coatings and sealants applications. An aqueous dispersion of butyl (55% to 65% solids) utilizing an anionic emulsifier also developed by Exxon[9] is available from the Burke and Palmason Chemical Company of Pompano Beach,

Florida. The latex is also used in adhesives and coatings. A partially crosslinked variety of butyl utilizing divinylbenzene is available from Polysar Ltd. of Canada[10] and is used in sealant/tape applications.

Butyl-rubber Properties, Vulcanization, and Applications

The molecular characteristics of low levels of unsaturation between long segments of polyisobutylene produce unique elastomeric qualities that find application in a wide variety of finished rubber articles. These special properties can be listed as (1) low rates of gas permeability, (2) thermal stability, (3) ozone and weathering resistance, (4) vibration damping and higher coefficients of friction, and (5) chemical and moisture resistance.

The more chemically inert nature of butyl rubber is reflected in the lack of significant molecular-weight breakdown during processing. This allows one to perform operations such as heat treatment or high-temperature mixing to alter the vulcanizate characteristics of a compound. With carbon black-containing compounds, hot mixing techniques promote pigment-polymer interaction,[11] which alters the stress-strain behavior of a vulcanizate. The shape of the stress-strain curve of the vulcanizate from a heat-treated mixture is a reflection of a more elastic network, and heat treatment has resulted in more flexible vulcanized compounds for a given level of a carbon black type. More flexible butyl-rubber compositions have also been prepared with certain types of mineral fillers such as reinforcing clays, talcs, and silicas that contain appropriately placed OH groups in the lattice. This enhancement of pigment-polymer association has usually been accomplished in the presence of chemical promoters.[12,13]

The heat treatment of filler-butyl rubber masterbatches can be accomplished at 300–350°F for 5 minutes in a Banbury mixer with the aid of promoters (for example *p*-quinone dioxime). In general, the processing of butyl rubber follows accepted factory operations. Banbury mixing in conventional formulations requires no longer times than with other natural and synthetic elastomers. Of course, there is no premasticating operation to produce a degree of molecular-weight breakdown, as in the case of natural rubber, but prewarming of the butyl rubber prior to mixing reduces mixing times.

Gas Permeability

The permeability of elastomeric films to the passage of gas is a function of the diffusion of gas molecules through the membrane and the solubility of the gas in the elastomer.[14] The polyisobutylene portion of the butyl molecule provides a low degree of permeability to gases and is a familiar property, leading to an almost exclusive use in inner tubes. For example, the air permeability, at 65°C, of SBR is about 80% that of natural rubber, while butyl shows only 10% perme-

Table 10.2. Air Loss of Inner Tubes during Driving Tests.

Inner Tube	*Original Pressure (psi)*	*Air Pressure Loss (psi)*		
		1 Week	*2 Weeks*	*1 Month*
Natural rubber	28	4.0	8.0	16.5
Butyl	28	0.5	1.0	2.0

ability on the same scale. The difference in air retention between a natural-rubber and a butyl inner tube can be demonstrated by data from controlled road tests on cars driven 60 mph for 100 miles per day. Under these conditions, it was shown (Table 10.2) that butyl is at least 8 times better than natural rubber in air retention. Other gases such as helium, hydrogen, nitrogen, and carbon dioxide are also well retained by a butyl bladder membrane. While the significance of these properties in inner tubes is waning, it is of importance in air barriers for tubeless tires, air cushions, pneumatic springs, accumulator bags, air bellows, and the like. A typical formulation for a butyl-rubber passenger-tire inner tube is given in Table 10.3.

Thermal Stability

Butyl-rubber sulfur vulcanizates tend to soften during prolonged exposure to elevated temperatures of 300–400°F. This deficiency is largely the result of the sulfur crosslink, coupled with low polymeric unsaturation which allows no compensating oxidative (crosslinking) hardening. However, certain crosslinking systems, and specifically the resin cure[15] of butyl, provide vulcanized networks of outstanding heat resistance. This has found widespread use in the expandable bladders of automatic tire-curing presses.

Two tire-curing bladder formulations are given in Table 10.4. In compound

Table 10.3. Composition, Butyl Inner Tube.

Exxon Butyl 268	100
GPF Carbon Black	70
Paraffinic process oil	25
Zinc oxide	5
Sulfur	2
TMTDS[a]	1
MBT[b]	0.5
Cure range 5′ at 350°F − 8′ at 330°F	

[a]TMTDS = tetramethyl thiuram disulfide.
[b]MBT = mercaptobenzothiazole.

Table 10.4. Compounds, Tire-curing Bladders.

	1	2
Exxon Butyl 268	100	100
Neoprene GN	5	–
HAF Carbon Black	50	50
Process oil	5	5
Zinc oxide	5	5
Reactive phenol Formaldehyde resin[a]	10	–
Brominated phenol Formaldehyde resin[b]	–	10

[a]Amberol ST 137—Rohm and Haas.
[b]SP-1055—Schenectady Chemical Co.

1, the neoprene serves as a halogen-containing activator, while compound 2 uses a partially brominated resin[16] that does not require an external source of halogen. Butyl-rubber tire-curing bladders have a life of 300–700 curing cycles at stream temperatures of 350°F or higher for approximately 20 minutes per cycle. Other applications would be conveyor belting for hot-materials handling, and high-temperature service hoses.

Ozone and Weathering Resistance

The low level of chemical unsaturation in the polymer chain produces an elastomer with greatly improved resistance to ozone when compared to polydiene rubbers. Butyl with the lowest level of unsaturation (Exxon Butyl 065) produces high levels of ozone resistances, which are also influenced by the type and concentration of vulcanizate crosslinks.[17] For maximum ozone resistance, as in electrical insulation, and for weather resistance, as in rubber sheeting for roofs and water management application, the least unsaturated butyl is advantageously used. A typical butyl-rubber sheeting compound is given in Table 10.5.

The ozone resistance of butyl rubber coupled with the moisture resistance of its essentially saturated hydrocarbon structure finds utility as a high-quality electrical insulation. A cable-insulation formulation for use up to 50 KV employs the lowest unsaturated butyl and the *p*-quinone dioxome (GMF) cure system, as shown in Table 10.6.

Vibration Damping

The viscoelastic properties of butyl rubber are a reflection of the molecular structure of the polyisobutylene chain. This molecular chain with two methyl side groups on every other chain carbon atom possesses greater delayed elastic response to deformation. The damping and absorption of shock have found wide application in automotive suspension bumpers.[18] An elastomer with higher

Table 10.5. Butyl Sheeting Compound.

Exxon Butyl 065	100
HAF Carbon Black	48
SRF Carbon Black	24
Zinc oxide	5
Petrolatum	3
Wax	4
Sulfur	1
TDEDC[a]	0.5
MBT[c]	0.5
ZDMDC[b]	1.5

[a]TDEDC = Tellurium diethyldithiocarbamate.
[b]ZDMDC = Zinc dimethyldithiocarbamate.
[c]MBT = Mercaptobenzothiazole.

damping characteristics also restricts vibrational force transmission in the region of resonant frequencies. *Transmissibility* is the ratio of output force to input force under impressed oscillatory motion.

In the region of resonance where the impressed frequency of vibration is equal to the natural frequency of the system, the more highly damped butyl compositions more effectively control vibrational forces.[19] These frequencies are in the region of vehicular-body vibration, and as a result butyl compositions are employed in the fabrication of automotive body mounts. In theory, more highly damped systems will less effectively isolate vibration at very high frequencies. However, the damping coefficients of viscoelastic polymers decrease with increasing frequencies in actual practice. This effect partially alleviates higher-frequency deficiencies of butyl systems. In addition at higher frequencies, dynamic stiffness becomes a controlling factor governing transmissibility, and this dynamic behavior can be greatly influenced by compounding variations as well as by the size and the shape of the molded part. For this reason, it is

Table 10.6. Cable-insulation Formulation.

Exxon Butyl 065	100	Masterbatch-mixed at 300°F
Zinc oxide	5	
Calcined clay	100	
Pb_3O_4	5	
130°F m.p. wax	5	
Low-density polyethylene	5	
SRF carbon black	10	
MBTS[a]	4	Added as a separate cooler mix
P-quinone dioxime	1.5	

[a]MBTS = Mercaptobenzothiazyl disulfide.

Table 10.7. Butyl Body-mount Compound.

Exxon Butyl 268	100
HAF carbon black	45
MT carbon black	15
Paraffinic oil	20
Zinc oxide	5
CdDEDC[a]	2
MBTS	0.5
Sulfur	1

[a]Cadmium diethyldithiocarbamate.

difficult to provide a typical butyl body-mount compound, but the one listed in Table 10.7 can be considered representative of a 50 Shore A hardness vulcanizate.

Higher damping behavior of an elastomer can be associated with higher coefficients of friction between a rubber and a surface with a measurable degree of roughness or undulations.[20] This property of butyl has potential for improving the coefficient of friction of tire-tread materials against a variety of road surfaces.[21]

Chemical and Moisture Resistance

The essentially saturated hydrocarbon nature of butyl obviously imparts moisture resistance to compounded articles. This capability is utilized in applications such as electrical insulation and rubber sheeting for external use. This same hydrocarbon nature provides useful solubility characteristics that can be applied to a variety of protective and sealant applications. These useful solubility characteristics are again based upon the hydrocarbon nature of the elastomer backbone that is expressed as a solubility parameter of 7.8.[21] This value is similar to the solubility parameters of aliphatic and some cyclic hydrocarbons (7 to 8), but very different from the solubility parameters of more polar oxygenated solvents, ester-type plasticizers, vegetable oils, and synthetic hydraulic fluids (8.5 to 11.0).

Thus while butyl vulcanizates will be highly swollen by hydrocarbon solvents and oils, they are only slightly affected by oxygenated solvents and other polar liquids. This behavior is utilized in elastomeric seals for hydraulic systems using synthetic fluids, as Fig. 10.4 demonstrates.

The low degree of olefinic unsaturation in the saturated hydrocarbon backbone also imparts mineral acid resistance to butyl-rubber compositions. After 13 weeks immersion in 70% sulfuric acid, a butyl compound experiences little loss in tensile strength or elongation while a natural rubber or SBR will be highly degraded.

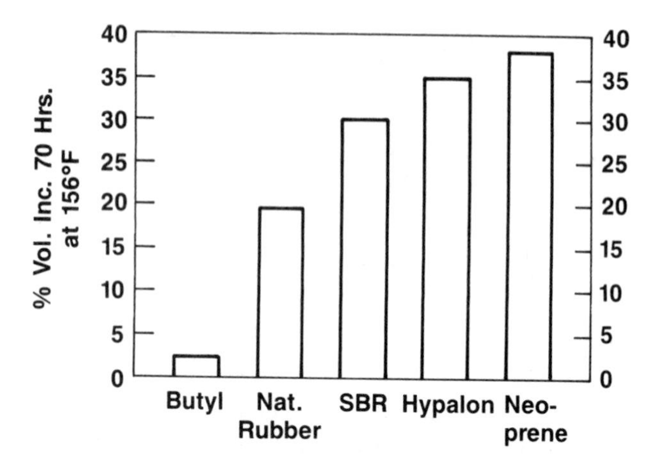

Fig. 10.4. Swelling of butyl and other elastomeric vulcanizates in phosphate ester fluids.

BUTYL VULCANIZATION

Regular butyl rubber is commercially vulcanized by three basic methods. These are (1) accelerated sulfur vulcanization, (2) crosslinking with dioxime and related dinitroso compounds, and (3) the polymethylol-phenol resin cure. The three methods will be briefly reviewed, but more detail can be found in reference 23.

Accelerated Sulfur Vulcanization

In common with more highly unsaturated rubbers, butyl may be crosslinked with sulfur, and activated by zinc oxide and organic accelerators. In contrast to the higher unsaturated varieties, however, adequate states of vulcanization can be obtained only with the very active thiuram and dithiocarbamate accelerators. Other less-active accelerators such as thiazole derivatives may be used as modifying agents to improve the safety of processing scorch. Most curative formulations include the ranges of ingredients shown in Table 10.8.

Table 10.8. Curative Formulations.

	Parts by wt.
Butyl elastomer	100
Zinc oxide	5
Sulfur	0.5–2.0
Thiuram or dithiocarbamate accelerator	1–3
Modifying thiazole accelerator	0.5–1

Thiurams with the structure

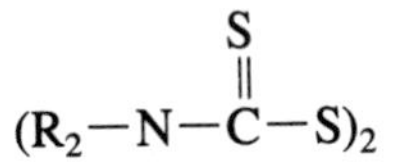

$$(R_2-N-\overset{\displaystyle S}{\overset{\|}{C}}-S)_2$$

and dithiocarbamates,

$$(R_2-N-\overset{\displaystyle S}{\overset{\|}{C}}-S)_xM$$

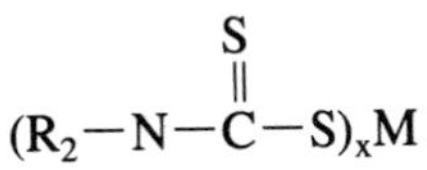

where M is a metallic element, provide the primary accelerating activity which promotes the most efficient use of sulfur. Thiazoles, generally mercaptobenzothiazoles,

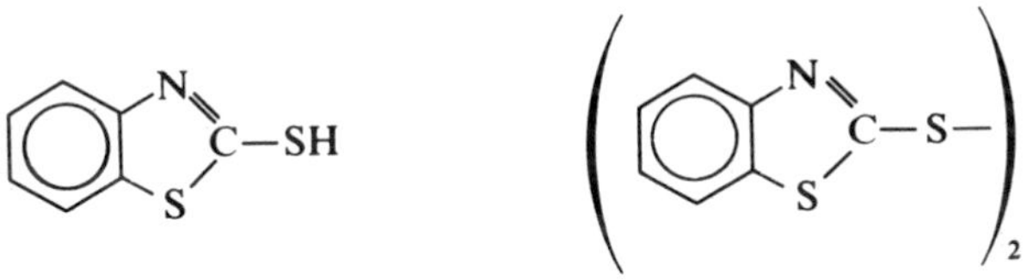

[2-mercaptobenzothiazole: benzene ring fused to a thiazole ring (N, S), C—SH] [dibenzothiazyl disulfide: (benzothiazole C—S—)$_2$]

reduce scorch during processing. High levels of dithiocarbamates and low levels of sulfur favor the formation of more stable monosulfidic crosslinks.[24] The use of sulfur donors in place of elemental sulfur also promotes simpler monosulfidic bonds.

Vulcanization temperatures may range from 275°F to 375°F, with temperature coefficients of vulcanization of 1.4 per 10°F for carbon black containing compounds.[25] This means that the vulcanization time, to a given state of cure, is multiplied by 1.4 for every 10°F decrease in temperature. Conversely, for every 10°F rise in temperature, the vulcanization time is divided by 1.4.

The Dioxime Cure

The crosslinking of butyl with *p*-quinone dioxime or *p*-quinone dioxime dibenzoate proceeds through an oxidation step that forms the active crosslinking agent, *p*-dinitroso-benzene, as shown in Fig. 10.5.

The use of PbO_2 as the oxidizing agent results in very rapid vulcanizations, which can produce room-temperature cures for cement applications. In dry rubber processing, the dioxime cure is largely used in butyl-rubber electrical insulation formulations as outlined in a preceding section, to provide a maximum of ozone resistance and moisture impermeability to the vulcanizate.

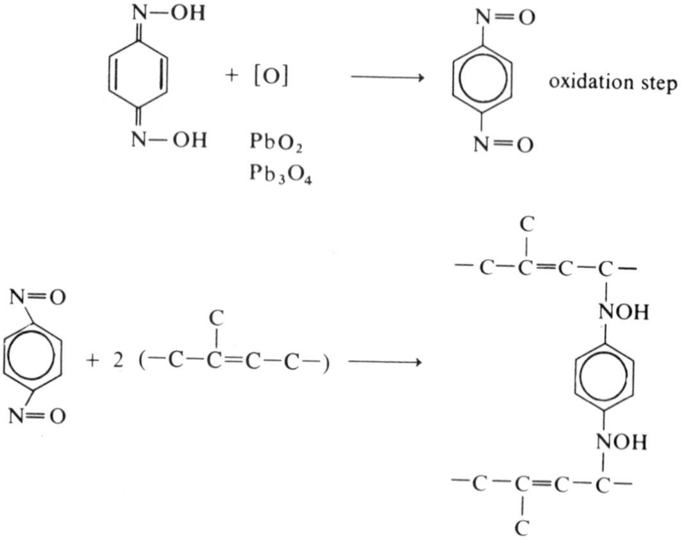

$$\text{HON}{=}C_6H_4{=}\text{NOH} + [O] \xrightarrow{PbO_2,\ Pb_3O_4} O{=}N{-}C_6H_4{-}N{=}O \quad \text{oxidation step}$$

$$O{=}N{-}C_6H_4{-}N{=}O + 2\,(-C-\overset{\displaystyle C}{\overset{|}{C}}=C-C-) \longrightarrow (-C-\overset{\displaystyle C}{\overset{|}{C}}=C-\underset{|}{C}-)\,NOH{-}C_6H_4{-}NOH\,(-C-\underset{\displaystyle C}{\underset{|}{C}}=C-\overset{|}{C}-)$$

Fig. 10.5. Dioxime cure of butyl rubber.

The Resin Cure

The crosslinking of butyl rubber (and other elastomers containing olefinic unsaturation) by this method is dependent upon the reactivity of the phenolmethylol groups of reactive phenol-formaldehyde resins:

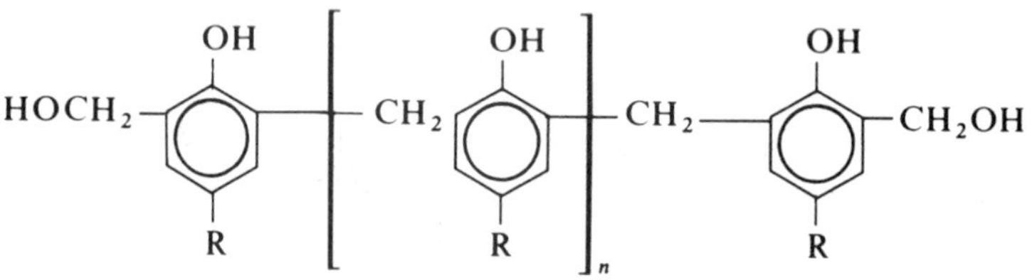

$$HOCH_2-C_6H_2(OH)(R)-\left[CH_2-C_6H_2(OH)(R)\right]_n-CH_2-C_6H_2(OH)(R)-CH_2OH$$

Two mechanisms involving the isoprenoid unit have been postulated. One involves reaction with allylic hydrogen through a methylene quinone intermediate,[26] and the other an actual bridging of the double bond.[27] The low levels of unsaturation of butyl require resin cure activation by halogen-containing materials such as $SnCl_2$ or halogen-containing elastomers such as neoprene.[15] A series of curves in Fig. 10.6 from reference 14 illustrates the activating effect of stannous chloride and the stability of the resultant crosslink to reversion upon prolonged heating. This feature of the resin cure is utilized in the fabrication of tire-curing bladders, as previously mentioned.

A more reactive resin cure system requiring no external activator is obtained

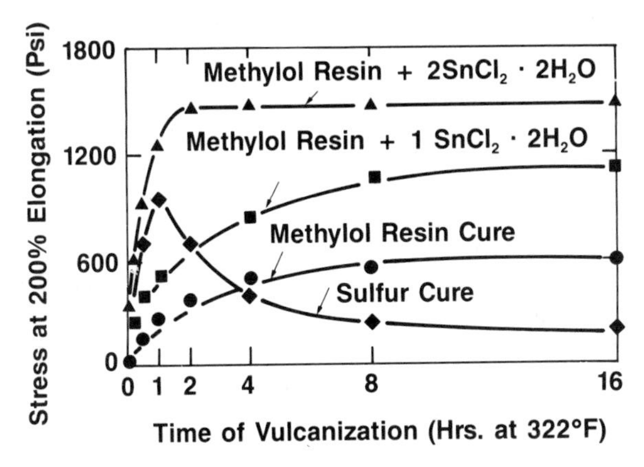

Fig. 10.6. The relative rates of cure and reversion of resin-and sulfur-cured butyl compounds. (Ref. 14)

if some of the hydroxyl groups of the methylol group are replaced by bromine. Such a resin commonly used is Schenectady Chemical's SP-1055. Examples of formulations were discussed for tire-curing bladders in the section "Thermal Stability."

HALOGENATED BUTYL

The concept of halogenation to provide more active functionality to the butyl molecule was first introduced by Goodrich researchers in 1954–56.[28–32] Their work emphasized the attributes of the brominated derivative. Goodrich commercialized a brominated butyl in 1954 (Hycar 2202) prepared from a bulk-batch halogenation, but withdrew it in 1969. In this same period, Exxon Researchers originated the chlorobutyl product concept.[33–36] Chlorobutyl was commercially introduced in 1961 by Exxon Chemical Company. Bromobutyl was again commercially introduced by Polysar Limited of Canada[37] in 1971 and by Exxon Chemical Co. in 1980.

The introduction of chlorine to the butyl molecule, in approximately 1/1 molar ratio of chloride to double bond, achieved a broadening of vulcanization latitude and rate, and enhanced covulcanization with general-purpose, high-unsaturation elastomers, while preserving the many unique attributes of the basic butyl molecule. Bromination in the same approximate molar ratio further enhanced cure properties and provided greater opportunity for increased covulcanization or adhesion, or both, to general-purpose elastomers. These property enhancements were vital to the major applications for halobutyl tubeless tire innerliners. Thus halogenated butyls contributed the required level of covulcanizability and adhesion to the highly unsaturated tire elastomers to carry the advantages of butyl inner tubes forward through the development of the radial tubeless tire.

Halogenation and Production Description

The reactions of elemental bromine and chlorine with the isoprene residue in butyl rubber can be quite complicated. The "dark" reaction of a solution of butyl in an inert diluent with these halogens can result in the incorporation of up to about 3 gram atoms of halogen per mole of unsaturation originally present in the polymer.[34,38,39] McNeil's study indicates that the overall halogenation occurs substantially as a series of consecutive reactions, each being slower than the preceding one.[39] The reported study by Van Tongerloo and Vukov[40] and Vukov[41], using the model compound 2,2,4,8,8 pentamethylnonene-4, is revealing as to the presence of possible structures.

Synthesis in the halogenated-butyls production facilities involves the "dark" reaction of a solution of butyl in hexane with elemental halogens at conventional process temperatures (40–60°C). The target is to produce a product in which no more than 1 halogen atom is introduced into the polymer per unsaturated site initially present, within the constraints of a final product weight percent halogen specification range.

Under the above conditions, the reaction with chlorine is very fast, probably completed in 15 seconds or less, even at the low molar concentration of reactants employed. The bromine reaction is considerably slower, perhaps about 5 times that of chlorination. In both cases, thorough mixing is a prerequisite to meet the synthesis targets (i.e., to avoid multiple halogenation at a particular site). That these reactions lead primarily to substituted products has been well documented.[37–41]

These fast reactions are presumed to occur via an ionic mechanism. The halogen molecules are polarized at the olefinic sites, and undergo heterolytic scission and consequent reaction. This ionic reaction can be depicted as follows:

1. Heterolysis and addition of halogen (X).

```
      |     |                            |     |
     -C-   -C-                          -C-   -C-
      |     |     |                      |     |     |     |     |
  ~~C-C-----C=C---C~~ + X2  →   ~~C-----C-----C-----C-----C~~ + X⊖
      |     |     |  |                   |     |    \ ⊕ /      |
     -C-                                -C-          X
      |                                  |

              (I)
```

There is interaction between postively charged carbon and added X up until

2. Proton abstraction by $X^{\ominus}$:

```
    |      |
   -C-    -C-
    |  |   |   |   |
 ~~C--C--C--C--C~~ + X⊖ →
    |  |   ⊕   |   |
   -C-      `-.X
    |

    |      |
   -C-    -C-
    |  |   |   |   |
 ~~C--C==C--C--C~~          (II) Thermodynamically most
    |          |   |              favored structure
   -C-         X
    |

        + HX

    |
   -C-    -C-
    |  |   ||  |   |
 ~~C--C--C--C--C~~          (III) Product found in highest
    |  |       |   |               concentration
   -C-         X
    |

    |      |
   -C-    -C-X
    |  |   |       |
 ~~C--C--C==C--C~~          (IV) Isomerized form of III
    |  |       |   |
   -C-
    |
```

Apparently the steric constraints imposed by the dimethyl-substituted carbon alpha to the in-chain charged carbon render III the kinetically favored structure by far. The structural features of commercial products can be accounted for to a level ≥95% by the summations of the concentrations of structures I–IV with III dominant.

Actual production is conducted in hexane solvent and the process flow chart is shown in Figure 10.7.[42] Butyl rubber in solution is treated with chlorine or bromine in a series of high-intensity mixing steps. Hydrogen chloride or hydrogen bromide is generated during the halogenation step, and must be neutralized, usually with a dilute aqueous caustic solution. After neutralization, the caustic aqueous phase is separated and removed and the halogenated cement is then stabilized and antioxidant is added to protect the halogenated product during the polymer recovery and finishing steps, which are much along similar lines as for butyl recovery. Commercial grades of halobutyl are summarized in Table 10.9 together with pertinent physical properties.

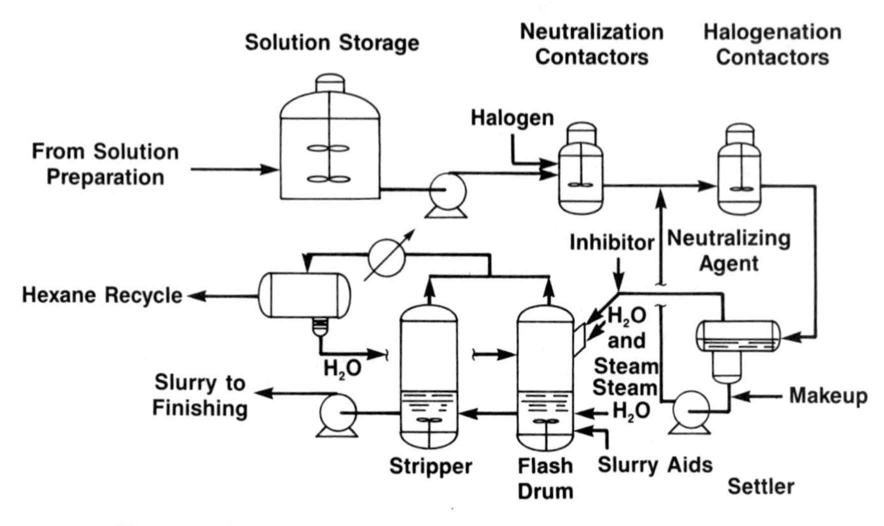

Fig. 10.7. Schematic flow plan of halogenated butyl rubber process.

Stabilization

As in all hydrocarbon polymers, the presence of an antioxidant is required to protect those elastomers during finishing, storage and compounding. These are chosen on the basis of cost effectiveness, discoloration tendencies, FDA approval, and compatibility with the overall process.

Table 10.9. Commercial Grades of Halobutyl.

	Typical Range for Mooney Viscosities		
Supplier and Typical Grade[a]	*M_L @ 100°C (1 + 8)*	*M_L @ 125°C (1 + 8)*	*Typical Wt. % Halogen*
Exxon Chlorobutyl 1065[b]	40–50	–	1.2
Exxon Chlorobutyl 1066	50–60	–	1.2
Exxon Chlorobutyl 1068	–	45–55	1.2
Polysar Chlorobutyl 1240	51–60		1.2
Polysar Chlorobutyl 1255	–	44–54	1.2
Exxon Bromobutyl 2222[c]		27–37	2.0
Exxon Bromobutyl 2233		31–40	2.0
Exxon Bromobutyl 2244		41–52	2.0
Exxon Bromobutyl 2255		41–52	1.9
Polysar Bromobutyl X2		41–52	2.1
Polysar Bromobutyl 2030		26–36	2.1

[a]All grades contain non-staining antioxidants at about 0.1 wt. %.
[b]Chlorobutyl contains Ca stearate to protect polymer from dehydrohalogenation.
[c]Bromobutyl contains Ca stearate and epoxidized soybean oil to protect polymer from dehydrohalogenation.

These reactive halogens having been implanted in the polymer, the principal stabilization problem is how to preserve them until they can display their main utility in crosslinking reactions. Unstabilized, both elastomers will undergo thermal dehydrohalogenation and simultaneous crosslinking,[34,43] the brominated product more readily than the chlorinated. The dehydrohalogenation is catalyzed by the evolved acid[43,44] as well as by Friedel Crafts metal halide catalysts (e.g., $AlCl_3$).[44]

Useful stabilization must meet numerous criteria. The stabilizer or package thereof must:

1. Prevent the accumulation of evolving HX and yield harmless by-products.
2. Not be itself highly reactive with allylic halogens.
3. Have process compatability.
4. Interact favorably, or at least not unfavorably, with crosslinking systems to be employed.
5. Not introduce a health hazard.
6. Be economical within the foregoing constraints.

Many materials will satisfy one or more of these demands, but very few all of them.

Experiment and experience have combined to teach that calcium stearate fulfills these objectives for chlorinated butyl.[45] Brominated butyl, being less stable and more reactive, requires a combination of calcium stearate and an epoxidized vegetable oil.[46]

Compounding Halobutyl

The following summarizes the influence of compounding ingredients on processing and vulcanizate properties of halobutyls:

1. *Carbon Black.* Carbon blacks affect the compound properties of halobutyl in a similar way as they affect the compound properties of other rubbers: particle size and structure determine the reinforcing power of the carbon black and hence the final properties of the halobutyl compounds.

   - Increasing reinforcing strength, for example, from GPF (N660) to FEF (N550) to HAF (N347) raises the compound viscosity, hardness, and cured modulus.
   - Cured modulus increases with the carbon black level up to 80 phr. Tensile strength goes through a maximum at 50–60 phr carbon black level.

2. *Mineral Fillers.* Mineral fillers vary not only in particle size but also in chemical composition. As a result, both cure behavior and physical prop-

erties of a bromobutyl compound are affected by the mineral filler used, although to a lesser extent than chlorobutyl compounds.

Generally the common mineral fillers may be used with halobutyl but highly alkaline ingredients and hygroscopic fillers should be avoided.

- *Clays* are semireinforcing. Acidic clays give very fast cures, therefore extra scorch retarders may be needed. Calcined clay is the preferred filler for pharmaceutical stopper compounds based on halobutyl.
- *Talc* is semireinforcing in halobutyl without a major effect on cure.
- *Hydrated silicas* even at moderate levels cause compound stiffness and slower cure rate, so their use should be restricted.
- *Silane-treated mineral fillers:* one way to enhance the interaction between polymer and silicates, and hence to improve compound properties, is to add small (1 phr) amounts of silanes. Particularly useful silanes are the mercapto- and amino-derivatives.

3. *Plasticizers.* Petroleum-based process oils are the most commonly used plasticizers for halobutyl. They improve mixing and processing, soften stocks, improve flexibility at low temperatures and reduce cost.

   Paraffinic/naphthenic oils are preferred for compatibility reasons. Other useful plasticizers are paraffin waxes and low-molecular-weight polyethylene. Adipates and sebacates improve flexibility at very low temperatures.
4. *Process Aids.*

- Struktol 40 MS and Mineral Rubber not only improve the processing characteristics of halobutyl compounds by improving filler dispersion, but they also enhance compatibility between halobutyl and highly unsaturated rubbers.
- Tackifying resins should be selected with care. Phenol-formaldehyde resins, even those where the reactive methylol groups have been deactivated, react with halobutyl especially bromobutyl, causing a decrease in scorch time, while partially aromatic resins such as Koresin have an intermediate effect on scorch in bromobutyl.
- *Stearates, stearic acid:* it should be noted that zinc stearate (which can also be formed via the zinc oxide and stearic acid reaction) is a strong dehydrohalogenation agent and a cure catalyst for halobutyl. Similar effects will be observed with other organic acids such as oleic acid or naphtenic acid. Alkaline stearates, on the other hand such as calcium stearate, have a retarding action on the halobutyl cure.

Anti-degradants. Amine-type antioxidants/antiozonants such as Flectol

H, mercaptobenzimidazole and especially *p*-phenylenediamines will react with halobutyl. They should preferably be added with the curatives, not in the masterbatch. Phenol derivative antioxidants are generally preferred.

Processing Halobutyl

The following recommended processing conditions are applicable to both chloro- and bromobutyl.

1. *Mixing*. The mixing is done in two stages. The first stage contains all the ingredients except for zinc oxide and accelerators. The batch weight should be 10–20% higher than that used for a comparable compound based on general-purpose rubbers. A typical mixing cycle and processing for a halobutyl innerliner compound are as follows:

First stage	0 min:	Halobutyl, carbon black, retarder.
	1.5 min:	Process aids, plasticizers, fillers, stearic aid . . . .
	3.5 min:	Dump at 120–140°C.

Higher dump temperatures could result in scorching.

Second stage	0 min:	Masterbatch + curatives.
	2 min:	Dump at 100°C.

Mill-mixing on a two-roll mill is best accomplished with a roll-speed ratio of 1.25/1 and roll temperatures of 40°C on the slow roll and 55°C on the fast roll.

The following sequence of addition is recommended: part of the rubber together with a small amount of a previous mix as a seed; 1/4 fillers plus retarder; remainder of polymer; rest of filler in small increments; plasticizers at the end; acceleration below 150°C.

2. *Calendering*. Feed preparation can be done either by mill or by extruder. Halobutyl follows the cooler roll; therefore a temperature differential of 10°C between calender rolls is recommended. Starting roll temperatures should be:

Cool roll	75–80°C
Warm roll	85–90°C

Normal calendering speeds for halobutyl compounds are between 25–30 meters/minute.

Rapid cooling of the calendered sheet is beneficial for optimal processibility (handling) and maximum tack retention.

3. *Extrusion.* Feed temperature should be 75–80°C while the temperature of the extrudate is around 100°C. During calendering and extrusion of halobutyl compounds, the most important problem is blister formation. The reason for this phenomenon is the low permeability of these polymers, which tend to retain entrapped air or moisture. Preventive action should be taken at all stages of the process, for example:

    - Ensure the stock is well-mixed in a full mixer to prevent porosity.
    - Avoid moisture at all stages.
    - Keep all rolling banks on mills and calender nips to a minimum.

4. *Molding.* Halobutyl can be formulated to have a fast-cure rate, good mold flow and mold release characteristics, and can, therefore, be molded into highly intricate designs with conventional molding equipment. Entrapped air can be removed by bumping of the press during the early part of the molding cycle.

    Halobutyl is also very well-suited for injection molding because of its easy flow and fast, reversion-resistant cures. Low-molecular-weight polymer grades may be required for optimum flow and good scorch safety.

CHLOROBUTYL VULCANIZATION AND APPLICATIONS

The presence of both olefinic unsaturation and reactive chlorine in chlorinated butyl provides for a great variety of vulcanization techniques. While conventional sulfur-accelerator curing is possible and useful, in this section attention will be confined to two general vulcanization techniques not available to regular butyl.

Zinc Oxide Cure and Modifications

Zinc oxide, preferably with some stearic acid, can function as the sole curing agent for chlorobutyl. After vulcanization, most of the chlorine originally present in the polymer can be extracted as zinc chloride. A proposed mechanism by Baldwin is based on formation of stable carbon-carbon crosslinks through a cationic polymerization route.[34] When zinc oxide is used as the vulcanizing reagent, the necessary initiating amounts of zinc chloride are likely formed as a result of thermal dissociation of some of the allylic chloride to yield hydrogen chloride. Subsequent reaction of the hydrogen chloride with zinc oxide then provides catalyst.

It is not likely that the propagation step proceeds very far, but for vulcanization purposes only one step is needed, particularly since both of the termination processes suggested result in the production of more catalyst.

The attainment of the full crosslinking potential of chlorobutyl by the ZnO system is relatively slow. This situation can be remedied by the inclusion of thiurams and thioureas into the curing recipe. It has been observed that chemical compounds with the grouping

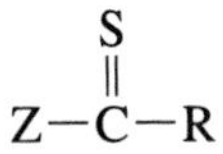

$$\mathrm{Z-\overset{\overset{\displaystyle S}{\|}}{C}-R}$$

(where Z is some type of activating group) will accelerate the ZnO cure of chlorobutyl. Examples are thiourea and tetramethyl thiuram disulfide.

The vulcanization of chlorobutyl with this type of accelerator can proceed via a mechanism similar to the mechanism of polychloroprene vulcanization with substituted thioureas, as proposed by Pariser.[47] The increase in cure rate and modulus as compared to the straight ZnO cure is obtained without sacrifice in vulcanizate stability.

Vulcanization through Bis-alkylation

The other unique and valuable curing method for chlorobutyl is that involving bis-alkylation reactions. This type of crosslinking reaction is perhaps best illustrated by a crosslinking with primary diamines, a vulcanization reaction that proceeds rapidly to yield good vulcanizates. This crosslinking reaction is believed to occur by the mechanism show below:

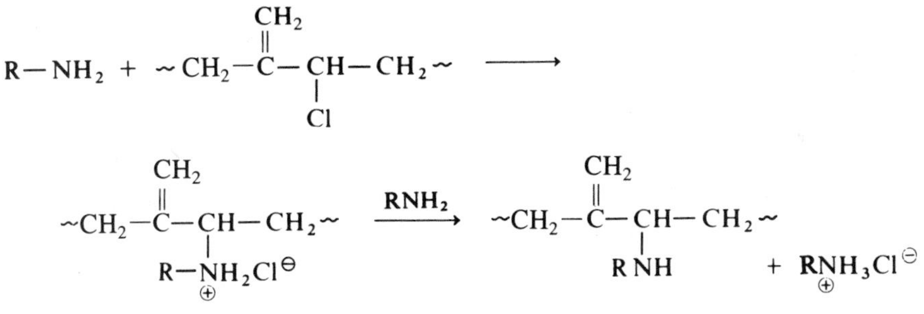

$$\mathrm{R-NH_2 + \sim CH_2-\overset{\overset{\displaystyle CH_2}{\|}}{C}-\underset{\underset{\displaystyle Cl}{|}}{CH}-CH_2\sim \longrightarrow}$$

$$\mathrm{\sim CH_2-\overset{\overset{\displaystyle CH_2}{\|}}{C}-\underset{\underset{\displaystyle R-\overset{\oplus}{N}H_2Cl^{\ominus}}{|}}{CH}-CH_2\sim \xrightarrow{RNH_2} \sim CH_2-\overset{\overset{\displaystyle CH_2}{\|}}{C}-\underset{\underset{\displaystyle RNH}{|}}{CH}-CH_2\sim + R\overset{\oplus}{N}H_3Cl^{\ominus}}$$

Obviously, in the presence of a diamine, reaction of both functional amino groups in the molecule with different polymer molecules results in crosslinking. Careful adjustment of the diamine concentration is required for development of the highest crosslink density. Too little could not possibly provide the maximum crosslinks, whereas too much would allow for reaction with only one of the amino groups per diamino molecule. In the presence of a hydrogen chloride scavenger, maximum modulus is developed when the ratio of NH_2/Cl is very nearly unity.

Extensions of this bis-alkylation vulcanization technique are numerous and in general any molecule having two active hydrogens can, under the proper catalytic conditions, crosslink the polymer. Typical examples are vulcanization with dihydroxy aromatics such as resorcinol and with dimercaptans.

Resin Cure

Both chlorobutyl and regular butyl are capable of vulcanization with heat-reactive phenolic resins, which are usually characterized by 6-9 weight-percent of methylol groups. Unlike conventional butyl, no promoter or catalyst other than zinc oxide is needed for efficient vulcanization, and a fast, tight cure is obtained with considerably less reagent.

Scorch Control

The modified zinc oxide cures are very fast and tend to be scorchy. As a general rule, acidic materials, such as channel blacks, activate the ZnO cure of chlorobutyl while basic materials hamper or retard it. The retarding effects of some alkaline materials, such as magnesium oxide, can be used in a very practical way to provide processing safety. The effects of the addition of magnesium oxide to the TMTDS-zinc oxide cure system on both cure rate and processing safety in carbon black compounds are shown in Table 10.10 based on work of Ziarnik.[48] It can be seen that the addition of 0.25 phr of MgO increases the margin against incipient vulcanization during processing at 126.5°C (260°F), as judged by Mooney scorch measurements, from 5 to 15 minutes, without greatly affecting tensile properties at the vulcanization temperature of 153°C (370°F). However, if the concentration is increased to 0.5 phr, the cure rate is depressed significantly.

As a scorch retarder, MgO is effective with all cure systems except the amine cure. In the latter case, it has the reverse effect since it prevents the hydrogen chloride, generated during crosslinking, from reacting with the curing agent. The choice of magnesium oxide type and concentration depends upon the type of compound used as well as the particular application involved.

Stability of Chlorobutyl Crosslinks

Since chlorobutyl has essentially the same structure as the butyl polymer, all the properties inherent to the butyl backbone are found in chlorobutyl rubber. These properties include low gas and moisture permeability, high hysteresis, good resistance to ozone and oxygen, resistance to flex fatigue, and chemical

Table 10.10. Effects of Magnesium Oxide Addition on Vulcanization.

Compound No.	1	2	3	4
TMTDS	1.0	1.0	1.0	0.5
Magnesium oxide	–	0.25	0.5	–
Mooney scorch measurements,[a] *time to 5 point rise at* 126.5°C:				
Minutes	5	15	30	8
Room temperature tensile properties,[a] *cured 45 minutes at* 153°C:				
Modulus at 300%, psi	2500	2370	1850	2360
Tensile strength, psi	2600	2720	2410	2580
Ultimate elongation, %	315	335	380	320

[a]Formulation (phr): Chlorobutyl—100, Antioxidant 2246—1, HAF Black—50, stearic acid—1, ZnO—5, curatives as indicated.

resistance. Additionally, chlorobutyl offers appreciably higher heat resistance than regular butyl cured with conventional sulfur vulcanization systems.

The chlorobutyl compound accelerated by ZnO/stearic acid displays a constantly increasing modulus or crosslink density with increasing vulcanization time. The problem of reversion does not exist in properly vulcanized chlorobutyl. Berger[49] has compared the behavior of regular butyl and TMTDS-ZnO-cured chlorobutyl with respect to stress relaxation under conditions of fixed strain, which is a common method to follow the degradation of crosslinked networks. Gum vulcanizates were used and the recipes and curing times were adjusted to give approximately equal crosslink densities for both materials. The specimens were held in air at a fixed elongation of 50% and data were taken as a function of time at various temperatures. The curves shown in Fig. 10.8 represent the ratio of the stress (S) at any time to that at 30 minutes (S_0) plotted as a function of time.

As the temperature increases, the relaxation of the butyl vulcanizates becomes more pronounced, as expected. The chlorobutyl vulcanizate by contrast shows very little stress relaxation over the time period shown. Only the 100°C data are shown for the chlorobutyl vulcanizate, since the behavior of this vulcanizate over the whole temperature range approximates that of sulfur-cured conventional butyl at the lowest test temperature. These data show the basic stability of the crosslinks formed by this cure system, and the stability of the vulcanizate toward oxidative scission reactions.

The combination of TMTDS with ZnO for curing chlorobutyl is effective in producing vulcanizates capable of withstanding temperatures up to 193°C (380°F). At that temperature, most vulcanizates, other than chlorobutyl, become excessively soft after a few hours of exposure, having lost all semblance

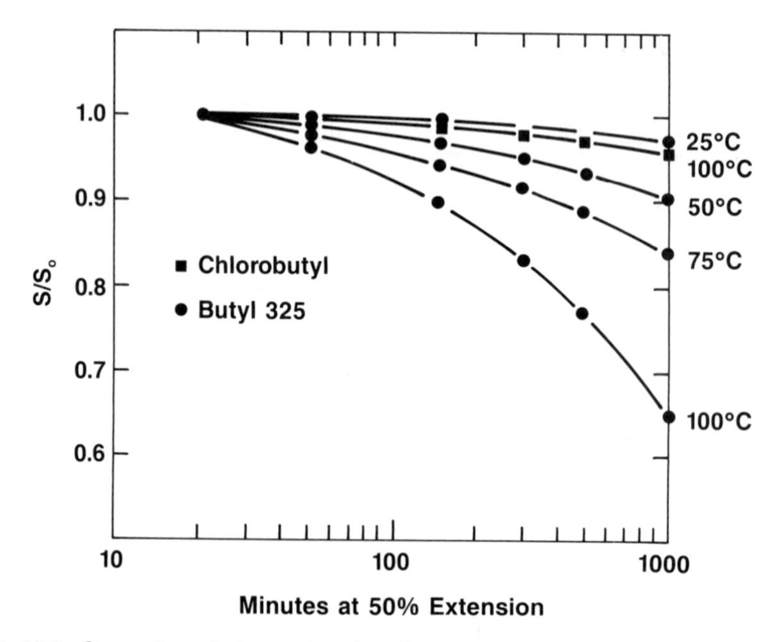

Fig. 10.8. Comparison of stress relaxation of gum vulcanizates: chlorobutyl and butyl.

of elastomeric behavior. Many chlorobutyl cure systems, other than the ZnO-TMTDS cure described above, can be chosen for heat resistance, such as straight zinc oxide, ZnO-dithiocarbamate, and resin cure systems. Dithiocarbamates such as the lead and zinc derivatives have been successfully used in chlorobutyl-zinc oxide systems.

Chlorobutyl Applications

Since its introduction in 1961, chlorobutyl has proven to be highly useful in many commercial rubber products. These products generally take advantage of desirable characteristics generic to butyl polymers, such as the resistance to environmental attack and low permeability to gases. In addition, chlorobutyl offers the superimposed advantages of cure versatility, highly heat-stable cross-links, and the ability to vulcanize in blends with highly unsaturated elastomers. Tire applications for halobutyl, chloro- and bromo- are illustrated in Fig. 10.9.

Innerliners for Tubeless Tires

The combination of low permeability, high heat resistance, excellent flex resistance, and ability to covulcanize with high unsaturation rubbers makes chlorobutyl particularly attractive for use in innerliners for tubeless tires.[50–53]. In Ta-

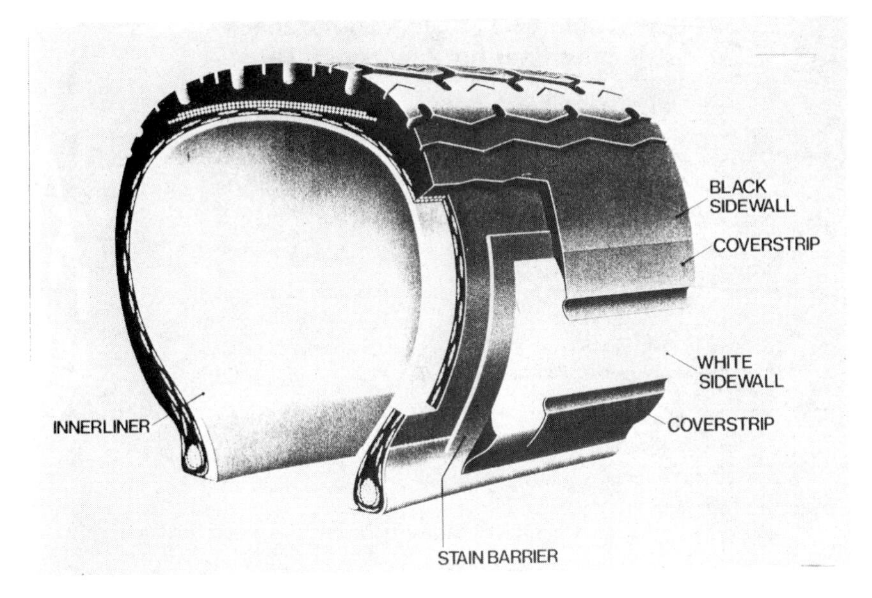

Fig. 10.9. Halobutyl in tire applications.

ble 10.11, a typical chlorobutyl innerliner formulation is presented. It is necessary to have 60–70 phr % chlorobutyl for good air barrier and heat resistance performance, while adequate building tack is provided by the inclusion of 20–30 phr of natural rubber. Cure systems based on Vultac No. 5 (alkylphenol disulfide) for black stocks and on thioureas for mineral-filled blends are widely used and preferred to realize the best balance of properties.

In a tubeless tire, the function of an innerliner is to provide an effective air barrier and minimize intracarcass pressure. Intracarcass pressure is a result of migration of air from the tire cavity into the cord area of the tire and the damming effect of the thick sidewall and tread exterior. The reduction in air (and oxygen) in the carcass serves to lessen oxidative degradation of the tire fabric and carcass rubbers. Intracarcass pressure and oxidation can weaken the body of the tire and effect tire failures through flex fatigue or loss of adhesion between components.

Tire performance data confirm the superiority of halobutyl innerliners for increasing tire durability. In a severe accelerated wheel test, shown in Table 10.12, a 60/40 chlorobutyl natural-rubber innerliner has increased mileage life by about 50% in a radial-ply-constructed tire. The data in Table 10.12 clearly illustrate that the excellent heat and flex resistance of chlorobutyl blends maintains the integrity and air barrier qualities of the liner, even under these severe operating conditions.

Table 10.11. Chlorobutyl Black Innerliner for Passenger Tires.

Exxon Chlorobutyl 1068	65
Natural rubber (No. 1 Smoked Sheet)	25
Whole tire reclaim	20
GPF Carbon Black	70
Paraffinic oil	12.5
Stearic acid	1
Amberol ST-317X	4
Maglite K	0.5
Zinc oxide	5.0
TMTMS[a]	0.25
Vultac 5 (Penwalt Co.)	0.7
MBTS	1.25
Permeability, $Q \times 10^3$ [b]	
Tested at 150°F	6.8
Adhesion to SBR-NR carcass stock	
Cured 45 minutes at 287°F	
Pulled at a rate of 2 in./min	
Tested at 75°F, lb/in.	48
Tested at 250°F, lb/in.	24

[a]Tetramethylthiuram monosulfide
[b]Q—Cubic feet of air (at 32°F and 29.92 in. Hg) diffusing through 0.001 in. of material under pressure differential of 1 psi/sq ft/day.

Tire Sidewall Components

Tire sidewall performance is critical from both an appearance and durability standpoint. Longer tire service life, particularly with the advent of belted bias and radial-ply tires, and increasing amounts of atmospheric degradants are contributing to a higher service severity. Tire manufacturers find it more difficult and costly to obtain the desired performance requirements with antiozonants. The substitution of chlorobutyl, either alone or in combination with ethylene-

Table 10.12. ASM Wheel Test.[a]

Innerliner	*Miles to Failure*	*Intracarcass Air Pressure*
SBR control	3129	16.0 lb
Chlorobutyl blend	4803	10.3 lb
	(average of five tests of each type of tire)	

[a]11.205″ diameter wheel, 120% of TRA load, 28 psi initial inflation pressure, 40 miles per hour.

propylene terpolymer, for a portion of the high unsaturation polymers commonly in use, offers a simple, economical means to upgrade the weathering and flex resistance of tire-sidewall components.

Heat-resistant Truck Inner Tubes

Chlorobutyl offers improved resistance to heat softening and "growth" while maintaining the desirable butyl property of excellent air retention in inner tubes. This feature is of particular importance in severe service conditions, for example, high-speed, heavy-load trucks, and buses for which tire service temperatures exceed 280°F and sometimes go as high as 300°F. After prolonged exposure, butyl tubes will soften, whereas chlorobutyl tubes will remain serviceable. A modified zinc oxide cure shown in the formulation below provides excellent thermal stability for a truck inner tube.

Exxon Chlorobutyl 1068	100
GPF Carbon Black	70
Paraffinic oil	25
Stearic acid	1
Zinc oxide	5
TMTDS	0.2
Maglite D	0.1
(Cure 15′ at 350°F)	

Other Applications

Chlorobutyl is used in many rubber articles in addition to tires and tubes. In these applications, the cure versatility of chlorobutyl and the stability of its vulcanizates are of particular importance. For example, chlorobutyl can be cured with nontoxic cure systems, such as zinc oxide with stearic acid, for use in products that will contact food. Most cure systems provide fast, reversion-resistant cures with chlorobutyl. This property facilitates the attainment of a uniform cure state in thick articles. Due to its exceptional heat resistance and compression-set properties, properly compounded chlorobutyl will give good service at temperatures up to 150°C (300°F).

Examples of typical applications are hose (steam, automotive), gaskets, conveyor belts, adhesives and sealants, tire-curing bags, tank linings, truck-cab mounts, aircraft engine mounts, rail pads, bridge bearing pads, pharmaceutical stoppers, and appliance parts.

BROMOBUTYL VULCANIZATION AND APPLICATIONS

Bromobutyl is very close in structure and properties to chlorobutyl. The approximately 2-weight-percent bromine present in bromobutyl is located on the isoprene units, in a similar way as the chlorine atoms in chlorobutyl, as indicated earlier. The main difference between bromo- and chlorobutyl is in the higher reactivity of the C—Br bond compared to that of C—Cl. This manifests itself in:

1. A higher versatility in vulcanization, i.e.:

   - Greater choice in curatives.
   - Generally faster cure.
   - Generally shorter scorch times.
   - Lower curative level needs.

2. A higher tendency for covulcanization with highly unsaturated rubbers.

In any particular application these differences will determine the preference for one halobutyl type or the other, taking also into account the slightly higher polymer cost of bromobutyl. Most cure systems used for chlorobutyl vulcanization can be applied to bromobutyl. Usually an adjustment for scorch is required. In addition, due to its higher reactivity and its more versatile chemistry, bromobutyl can be cured with some chemicals which are inactive or less active with chlorobutyl. Table 10.13 summarizes the most common bromobutyl cure systems together with their cure characteristics and application areas. It should be emphasized that, in contrast to highly unsaturated rubber-sulfur cures, bromobutyl cures (as with chlorobutyl) are generally accelerated by acids (and retarded by bases), and usually result in very stable crosslinks, i.e. C—C or C—S—C bonds.

Cure Systems Specific to Bromobutyl

Straight Sulfur Cure. Bromobutyl cures with sulfur as sole curative in the absence of zinc oxide. This cure, which is of more academic than practical interest, is retarded by phenolic antioxidants, pointing to a radical reaction mechanism. It also shows how efficiently bromobutyl uses the sulfur atom, which partially explains its enhanced co-vulcanizability with high unsaturation rubbers.

Zinc-free Cures. Zinc-free compounds are required in some special pharmaceutical closure applications. Bromobutyl is capable of vulcanizing without zinc oxide or any other zinc salt. Preferred curatives are diamines such as hexamethylene diamine carbamate (Diak No. 1).

Peroxide Cures. Although butyl polymers undergo molecular weight break-

Table 10.13. Summary of Most Common Bromobutyl Cure Systems.

Cure System		*Scorch Safety*	*Cure Rate*	*Other Characteristics*	*Applications*
Sulfur/MBTS	ZnO – 3 S – 0.5 MBTS – 1.5 Maglite D – 0.5 or Carbowax 4000 – 0.5 TMTDS – 0 to 0.25	Excellent	Moderate	Covulcanization with highly unsaturated rubbers.	Standard cure system. Tire innerliners. Blends with general-purpose rubbers.
Sulfur/sulfenamide	ZnO – 3 Sulfur – 0.5 Santocure NS – 1.0 TMTDS – 0.25	Excellent	Fast	Covulcanization with highly unsaturated rubbers.	Blends with general-purpose rubbers.
Dithiocarbamate	ZnO – 3 ZDEDC – 1.5 Maglite – 0.5	Fair	Very fast	Low compression set. Heat resistant.	Mechanical goods. Injection molding.
Resin	ZnO – 3 SP 1045 – 2	Good	Fast	Sulfur-free.	Clay-filled compounds. Pharmaceutical closures.
Peroxide	Dicup – 2 HVA 2 – 1	Fair	Moderate	Very-low compression set. Highly heat resistant.	High temperature and steam resistance.
Room-temperature cure	ZnO – 5 $SnCl_2$ – 2 $ZnCl_2$ – 2	Very scorchy	Very fast	Curatives should be premixed in Vistanex®.	Room-temperature curing sheeting, tank lining.
Zinc-free	Maglite D – 3 Diak No.1 – 1	Scorchy	Fast	Low compression set.	Special pharmaceutical closures, zinc free.

Cure rate: Very fast: <5 min at 160°C; fast: <15 min at 160°C; moderate: 15–30 min at 160°C.
Scorch T_3 135°C: Very scorchy: <5 min; scorchy 5 min; fair 10 min; good 10–15 min; excellent >15 min.

down under the influence of peroxides, chlorobutyl, and more so bromobutyl, are capable of crosslinking with peroxides; a good combination is dicumylperoxide and a bismaleimide (HVA-2) as coagent. However, further studies have shown that bromobutyl cures with HVA-2 alone without peroxide, giving low compression set and extremely high heat resistance.

Control of Cure Parameters

Bromobuytl is a very reactive rubber with a sensitive vulcanization chemistry. It requires, therefore, careful selection of type and level of compounding ingredients. Table 10.14 gives some general guidelines for adjustment of scorch, cure rate and cured modulus in 100% bromobutyl formulations. It says, for example, that Mooney scorch can be improved by adding a base, by lowering the sulfur and thiuram levels, increasing the MBTS level, replacing phenolic resins by hydrocarbon resins, lowering the carboxylic acid level, and avoiding amine-type antioxidants. The most classical scorch retarder for halobutyl is magnesium oxide. It has, however, some drawbacks: it affects both cure rate and cure state, and tends to lose its activity on storage. These drawbacks do not occur with polyethylene glycols (e.g., Carbowax), which are efficient bromobutyl scorch retarders due to their ether function. Further comments with regard to effects of other compounding ingredients on cure behavior are made in the section of this chapter dealing with compounding.

Cured Properties

Bromobuyl, like chlorobutyl, has several outstanding features related to both the polyisobutylene backbone and the cured network structure. The most important properties are summarized:

Permeability. Both chloro- and bromobutyl have outstanding low permeability to air and water vapor, unsurpassed among rubbers, as further illustrated in Table 10.15.

Heat Resistance. For several high-temperature applications, particularly those where low permeability is also required, bromobutyl is preferred. The highest heat resistance is obtained when using HVA-2 (bismaleimide) as a curative either with or without peroxide.

Resistance to Chemicals and Solvents. Bromobutyl is a nonpolar hydrocarbon rubber with very low unsaturation. It will, therefore, be swollen in the cured state in hydrocarbon solvents but will be resistant to polar solvents, much the same as chlorobutyl. Moreover, chemical attack will be minimal, i.e., attack by acids, bases, oxygen, and ozone. Except for special cases, bromobutyl compounds usually do not require addition of antioxidants/antiozonants.

Table 10.14. Compound Optimization in 100% Bromobutyl Formulations

Effect of Increased Amount ON / OF	Scorch Time	Rheometer Cure Rate	Rheometer Modulus
$M_{G}O$	↗	↘	↗
Carbowax 4000	↗	→	→
Sulfur	↘	→	↗
MBTS	↗	→	→
TMTDS	↘	↗	↗
Phenolic Resin	↘	→	→
Hydrocarbon	→	→	↘
Stearic Acid	↘	↗	↘
Amine Type Antioxidants	↘	↗	↗

Flex Resistance/Dynamic Properties. Bromobutyl has excellent flex resistance, both alone and in blends with other rubbers. Of course, the formulation should be adjusted according to the type of deformation involved in a particular application. For example, in a tire innerliner compound, lower modulus compounds will lead to higher flex resistance. Bromobutyl has the same vibration damping properties as the other rubbers of the butyl family.

Table 10.15. Air and Water Vapor Permeability Rates at 65°C (Index).

Typical Tire Innerliner Compound	*Air*	*Moisture*
All natural rubber	8.3	13.3
All SBR	6.8	11.0
60/40 halobutyl/NR	3.1	3.0
All halobutyl	1.0	1.0

Compatibility with Other Elastomers

Although bromobutyl and natural rubber have completely different backbone structures, different unsaturation (1.5 mole percent versus 100 mole percent), and different vulcanization chemistry, they can be blended in all proportions to attain the desired combination of properties. This is due to the versatility of the bromobutyl vulcanization. Not only can bromobutyl vulcanize independently of the NR sulfur cure, merely via crosslinking with zinc oxide, but it also uses sulfur very efficiently. Moreover, it has been shown that many cure systems result in chemical interaction between the two rubber phases, i.e., covulcanization. Recommended cure systems for optimal covulcanization are:

ZnO-3 / Sulfur-0.5 / MBTS-1.5 / TMTDS-0 to 0.25
ZnO-3 / Sulfur-0.5 / Santocure NS-1.0 / TMTDS-0 to 0.25
ZnO-3 / Sulfur-0.5 / BNDS*-1.5 / MBTS-1.5

Generally bromobutyl adheres better to polyisoprene than chlorobutyl does. Also, cured adhesion with other general-purpose rubbers such as polybutadiene and SBR is possible, but adhesion levels are lower than with polyisoprene.

Bromobutyl Applications

Tire Innerliners. Because of its low permeability to gas and moisture vapor and its ability to adhere to high unsaturation rubbers, bromobutyl is generally the preferred choice for 100% halobutyl innerliner formulations in tubeless tires. Over 80% of the total bromobutyl production is currently used in tire innerliners. Its use is mainly in truck tires, which require the highest quality because of their severe service conditions, but its use in passengers tires in 100% halobutyl compositions is also gaining importance and growing rapidly. A high-quality halobutyl innerliner maintains the tire inflation pressure better, with subsequent improvements in rolling resistance, tread abrasion, tire duration and, most importantly: safety.

Table 10.16 shows the considerable benefits in test results of inflation-pressure retention (IPR) and intracarcass pressure (ICP) when bromobutyl innerliners are used. The IPR time is defined as the time for inflation pressure to drop from 260 kPa to 165 kPa at 65°C, which is considered the minimum pressure below which the tire rolling resistance increases rapidly. The ICP test measures the equilibrium air pressure within the carcass plies of tubeless tires. The formulations shown in Table 10.17 are typical examples of:

*BNDS = Betanaphtholdisulfide.

Table 10.16. Effect of Innerliner Composition on Tire Pressure.

Innerliner type	*Inflation Pressure Retention, days*	*Intracarcass Pressure at 30°C, kPa*
Commercial NR/SBR	26	100
60% bromobutyl/40% (SBR + NR)	36	70
100% bromobutyl	54	25

1. A 100% bromobutyl innerliner suitable for truck and passenger tires.
2. A lower-quality (lower bromobutyl content) recipe that might be preferred for cost reasons in passenger-tire application.

Radial-tire Black Sidewalls. Critical properties for a black sidewall on radial passenger-car tires are dynamic ozone and flex resistance for the life of the tire. Also, retention of the black color (nonblooming, nondiscoloring, nonstaining properties) and serviceable adhesion to adjoining general-purpose rubber components are essential.[54]

Common natural rubber/polybutadiene sidewall compounds depend on antiozonants such as *p*-phenylenediamines to obtain the required ozone resistance. These molecules, however, are generally discoloring, staining and toxic. Compounds based on blends of bromobutyl (at approximately 45 parts) with EPDM (approximately 10 parts) and natural rubber do not require the use of these protective agents, and therefore avoid the drawbacks of general-purpose rubber sidewall compounds. Indeed, they are nonstaining and nondiscoloring since the

Table 10.17. Composition, Two Qualities of Bromobutyl Innerliners.

100% Bromobutyl Innerliner Compound		*60% Bromobutyl Innerliner Compound*	
Exxon Bromobutyl 2255	100	Exxon Bromobutyl 2255	60
GPF (N.660)	50	Nat. Rubber #20	25
Flexon 871 oil	8	SBR 1712	20.5
Stearic acid	2	HAF (N.330)	40
Maglite D	0.5	Whiting	40
Struktol 40 MS	7	Flexon 876 oil	10
ZnO	3	Stearic acid	1
Sulfur	0.5	Maglite D	0.5
MBTS	1.5	ZnO	3
		Sulfur	0.5
		Santocure NS	1.5
		TMTDS	0.25

protection is inherent in the polymer system. An example of a bromobutyl-based sidewall tri-blend compound follows:

Exxon Bromobutyl 2244	45
Vistalon 6505	10
SMR 5 NR	45
HAF (N-339)	25
SRF (N-774)	25
Flexon 641 oil	12
Stearic acid	1
Escorez 1102	5
Zinc oxide	3
Sulfur	0.25
MBTS	1.5
Betanaphtholdisulfide*	0.5

(An alternate cure system is: zinc oxide, 3; sulfur, 0.25; Santocure NS, 1.25; TMTDS, 0.25.)

The above type of compound exhibits excellent dynamic ozone resistance. It exceeds 900 hours in 100 pphm ozone, while flexing 0–25% at 30 cycles per minute at 38°C without cracking. (Paraphenylenediamine-protected general-purpose rubber compounds start to crack in less than 48 hours in this test.) The bromobutyl compound also possesses excellent flex fatigue resistance and retains this performance very well after aging.

Pharmaceutical Closures. Butyl and halobutyl rubbers are extensively used in the pharmaceutical closure field because of their inherent properties:

- Low permeability.
- Resistance to heat, UV and ozone.
- Chemical/biological inertness.
- "Pure", nontoxic vulcanization systems.

While butyl still requires sulfur for vulcanization, halobutyl can be cured with sulfurless cure systems, resulting in very low extractables. A preferred cure system is based on zinc oxide combined with 1.5–2.5 phr phenolformaldehyde resin.

Bromobutyl has a further advantage of being vulcanizable without zinc, which is useful for some special closure types. A diamine such as Diak No. 1 is suitable for vulcanization in this case. Calcined clay is the preferred filler, but it

*BNDS from Pennwalt Co.

can be partially replaced by talc, whiting, or even carbon black. For improved hot tear resistance, silica can be added, but only in low amounts, since it has a cure-retarding effect. All the ingredients should be of the purest grade possible.

An example of a sulfurless zinc-free formulation is given:

Exxon Bromobutyl 2244	100
Whitetex No. 2	60
Primol 355 (white oil)	5
Polyethylene AC 617 A	3
Paraffin wax	2
Vanfre AP-2	2
Stearic acid	1
DIAK No. 1	1

This compound cured for 10 minutes at 170°C gives a Shore A hardness of 45, a tensile strength of 6.0 MPa and a compression set of 16% (22 hrs at 70°C).

Heat-resistant Conveyor Belt

Several types of rubber are used for heat-resistant conveyor-belt cover and skim compounds: resin-cured butyl, chlorobutyl, EPDM and SBR. Bromobutyl gives the best compromise between heat resistance and adhesion to textile. Moreover, it generally cures faster than the other rubbers just listed. The following is a typical bromobutyl belt-cover compound:

Exxon Bromobutyl 2244	100
FEF carbon black	45
Maglite D	1
Carbowax 4000	1.5
Stearic acid	1
MBT	1
ZnO	3
Dicup 40C	2
HVA-2	1

Cured 35 minutes at 160°C, it exhibited the properties shown in Table 10.18.

Table 10.18. Bromobutyl Belt-cover Compound.

	Original	*Heat-aged 7 days at 150°C*
Hardness, Shore A	59	64
Tensile strength	11.2 MPa	9.1 MPa
Elongation at break	250%	200%

Other Applications. Other applications are possible with bromobutyl, taking advantage of the wide spectrum of properties of this polymer. Examples are:

- Chemical-resistant tank linings.
- Curable contact cements for rubber-to-rubber adhesion.
- Applications such as sheeting, tank lining, adhesives, where room temperature cures are required.

ACKNOWLEDGMENT

The authors wish to especially acknowledge the guidance and contributions of Goodyear Medalist, Dr. F. P. Baldwin, retiree from Exxon Chemical Company. Dr. Baldwin's wealth of knowledge and experience was of invaluable assistance in the preparation of the section on halobutyl.

REFERENCES

1. R. M. Thomas and O. C. Slotterbeck, U.S. Pat. 2,243,658 (May 27, 1941).
2. R. M. Thomas and W. J. Sparks, U.S. Pat. 2,356,128 (Aug. 22, 1944).
3. W. J Sparks and R. M. Thomas, U.S. Pat. 2,356,129 (Aug. 22, 1944).
4. R. M. Thomas, I. E. Lightbown, W. J. Sparks, P. K. Frolich and E. V. Murphree, *Ind. Eng. Chem.* **32,** 1283 (1940).
5. R. M. Thomas and W. J. Sparks, *Synthetic Rubber*, G. S. Whitby, ed. John Wiley & Sons, New York, 1963, ch. 24.
6. S. G. Gallo, H. K. Weise, and J. F. Nelson, *Ind. Eng. Chem.* **40,** 1277 (1948).
7. J. Rehner, Jr., and Priscilla Grey, *Ind. Eng. Chem.* **17,** 367 (1945).
8. J. Rehner, Jr., *Ind. Eng. Chem.* **36,** 46 (1944).
9. K. W. Powers, U.S. Pat. 3,562,804 (Feb. 9, 1971): G. W. Burton, C. P. O'Farrell, *J. Elastomers & Plastics* **9,** 94 (Jan. 1974).
10. D. A. Paterson, *Adhesives Age* **12** (8), 25 (Aug. 1969).
11. A. M. Gessler, *Rubber Age* **74,** 59 (1953).
12. A. M. Gessler and F. P. Ford, *Rubber Age* **74,** 243 (1953).
13. A. M. Gessler and J. Rehner, Jr., *Rubber Age* **77,** 875 (1955).
14. G. J. Van Amerongen, *J. Applied Phys.* **17,** 972, (1946); *J. Polym. Sci.* **5,** 307 (1950).
15. P. O. Tawney, J. R. Little, and P. Viohl, *Rubber Age* **83,** 101 (1958).
16. J. V. Fusco, S. B. Robison, and A. L. Miller, U.S. Pat. 3,093,613 (Jun. 11, 1963).
17. D. J. Buckley and S. B. Robison, *J. Polym. Sci.* **19,** 145 (1956).
18. D. F. Kruse and R. C. Edwards, *SAE Transactions* **77,** 1868 (1968).
19. R. C. Puydak and R. S. Auda, *SAE Transactions* **76,** 817 (1967).
20. D. Tabor, G. G. Giles, and B. F. Sabey, *Engineering* **186,** 838 (1958).
21. R. L. Zapp, *Revue Generale du Caoutchouc* **40,** 265 (1963).
22. C. J. Sheehan and A. L. Bisio, *Rubber Chem. Technol.* **39,** 149 (1966).
23. W. C. Smith, *Vulcanization of Elastomers*, G. Alliger and I. J. Sjothum, eds., Rheinhold, New York, 1964, ch. 7.
24. C. J. Jankowski, K. W. Powers, and R. L. Zapp, *Rubber Age* **87,** 833 (Aug. 1960).
25. J. G. Martin and R. F. Neu, *Rubber Age* **86,** 826 (Feb. 1960).

26. S. Vander Meer, *Rec. Trav. Chim.* **65,** 149, 157 (1944).
27. A. Groth, *Kunstoffe* **31,** 345 (1941)
28. R. A. Crawford and R. T. Morrissey, U.S. Pat. 2,631,984 (Mar. 17, 1953).
29. R. T. Morrissey, *Rubber World* **138,** 725 (1955).
30. R. T. Morrissey, *Ind. Eng. Chem.* **47,** 1562 (1955).
31. R. T. Morrissey, U.S. Pat. 2,816,098 (Dec. 10, 1957).
32. V. L. Hallenbeck, U.S. Pat. 2,804,448 (Aug. 27, 1957).
33. F. P. Baldwin and R. M. Thomas, U.S. Pat. 2,964,489 (Dec. 13, 1960).
34. F. P. Baldwin, D. J. Buckley, I. Kuntz, and S. B. Robison, *Rubber and Plastics Age* **42,** 500 (1961).
35. B. R. Tegge, F. P. Baldwin, and G. E. Serniuk, U.S. Patent 3,023,191 (Feb. 27, 1962).
36. J. A. Johnson, Jr., and E. D. Luallin, U.S. Patent 3,275,349 (Jun. 21, 1966).
37. J. Walker, R. H. Jones and G. Feniak, Philadelphia Rubber Group Technical Meeting, Oct. 22, 1972.
38. F. P. Baldwin and I. Kuntz, *Kirk-Othmen Encyclopedia of Chemical Technology*, 2d Suppl., A. Standen, ed., Interscience, New York, 1960, p. 716.
39. I. G. McNeil, *Polymer* **4,** 15 (1963).
40. A. Van Tonoerloo and R. Vukov, *Proc. Int. Rubber Conf.*, Venice (1979), p. 70.
41. R. Vukov, *Rubber Chem. Technol.* **57,** 275 (1984).
42. F. P. Baldwin and R. Schatz, *Encyclopedia of Chem. Tech.*, 3d Ed., John Wiley and Sons, New York, 1979, Vol. 8, p. 470.
43. P. Hous, unpublished data, Exxon Chemical Company.
44. V. M. Goncharov and I. P. Cherenyuk, *Khim. Tekhnol. Polim.* **3,** 13 (1974).
45. L. T. Eby, D. C. Cottle, and T. Lemiska, U.S. Pat. 2,958,667 (1960).
46. R. T. Morrissey and H. W. Weiss, U.S. Pat. 2,833,734 (May 6, 1958).
47. R. Pariser, *Kunstoffe*, **50,** 623 (1960).
48. G. J. Ziarnik, *Rubber Chem. Technol.* **35,** 467 (1962).
49. M. Berger, *J. Appl. Polym. Sci.*, **5,** 322 (1961).
50. J. V. Fusco and R. H. Dudley, Paper No. 27, "New Compounds for Air Barriers in Tubeless Tires," presented before Rubber Division, ACS, New York, Sept. 1960.
51. J. V. Fusco and R. H. Dudley, *Rubber World* **144,** 67 (Aug. 1961).
52. D. M. Coddington, *Rubber Chem. Technol.* **52,** 905 (1979).
53. D. M. Coddington, "Tire Inflation Pressure Loss, Its Causes and Effects", presented to Fifth Australian Rubber Technology Convention, Canberra, Australia, Oct. 1–3, 1980.
54. D. G. Young, E. N. Kresge, and A. J. Wallace, *Rubber Chem. Technol.* **55,** (2) (May–Jun. 1982).

11

NITRILE AND POLYACRYLIC RUBBERS

DONALD A. SEIL AND FRED R. WOLF
The BFGoodrich Company
Elastomers & Latex Division
Avon Lake, Ohio

INTRODUCTION

Nitrile-butadiene and polyacrylic elastomers fall into the class of special-purpose, oil-resistant rubbers. Nitrile elastomers offer a broad balance of low temperature, oil, fuel and solvent resistance as related to *acrylonitrile* content. These characteristics, combined with their good abrasion and water-resistant qualities, make them suitable for use in a wide variety of applications with heat-resistant requirements to 149°C (300°F).

Conventional unsaturated diene-type nitrile rubbers tend to embrittle (or case-harden) when exposed to temperatures above 149°C due to (1) oxidative cross-linking or (2) crosslinking by oils containing sulfur-bearing additives. Acrylic elastomers contain little or no residual unsaturation and therefore are more resistant to these effects. In general, the acrylics exhibit 55°C (100°F) higher service-temperature capability than the nitriles and are specifically recommended for those applications requiring heat and oil resistance in the 149°C to 204°C (400°F) temperature range.

Nitrile rubbers carry the ASTM designation of NBR,[1] XNBR[2] and are covered by ASTM D2000/SAEJ200 BF, BG, BK, CH classifications. Acrylic rubbers carry the ASTM designation of ACM,[3] ANM,[4] or EA[5] and are covered by ASTM D2000/SAEJ200 DH, DF classifications.

(1) Acrylonitrile-butadiene.
(2) Carboxylic-acrylonitrile-butadiene.
(3) Copolymers of ethyl acrylate or other acrylates with a cure monomer.
(4) Copolymers of ethyl or other acrylates and acrylonitrile.
(5) Ethylene/acrylic.

NITRILE RUBBER

History

The first literature reference to nitrile rubber is found in a French patent issued in 1931, covering the polymerization of butadiene and acrylonitrile. The first commercial nitrile rubber was produced in 1935 by I. G. Farbenindustrie in Germany and was called "Perbunan." The first commercial nitrile rubber was produced in the U.S. by the BFGoodrich Company in 1939. Other manufacturers such as Goodyear Tire and Rubber Company, Firestone Tire and Rubber Company, and the Standard Oil Company were soon producing various grades of nitrile rubber. At the present time, there are 10 major producers of nitrile rubber worldwide, including four in the United States; these producers and their corresponding trade names are listed in Table 11.1.

Chemistry

Nitrile rubber is defined as a copolymer of a diene and an unsaturated nitrile. The majority of the nitrile elastomers produced today are copolymers of acrylonitrile and butadiene. The basic polymer reaction may be shown as follows:

$$\underbrace{CH_2 = CH - CH = CH_2}_{\text{butadiene}} + \underbrace{CH_2 = \underset{\displaystyle CN}{\underset{|}{CH}}}_{\text{acrylonitrile}}$$

$$(CH_2 - CH = CH - CH_2 - CH_2 - \underset{\displaystyle CN}{\underset{|}{CH}})$$

copolymer unit

(depending on the ratio of acrylonitrile to butadiene)

Table 11.1. Primary Producers of Nitrile Rubber with Trade Names.

Producer	*Trade Name*
Copolymer Rubber and Chemical Corp.	Nysen
BFGoodrich Chemical Company	Hycar®
Goodyear Tire and Rubber Company	Chemigum
Uniroyal Chemical Division of Uniroyal, Inc.	Paracril
B.P. Chemicals Ltd. (UK)	Breon
Huels Mexicanos, S.A. (Mexico)	Humex
Japan Synthetic Rubber Co. (Japan)	JSR
Mobay (Germany)	Perbunan
Nippon Zeon Company (Japan)	NIPOL
Polysar, Ltd. (Canada)	Krynac

Manufacture

In addition to the monomers, a basic polymerization recipe would contain water, emulsifier, modifier, electrolytes, catalyst, activator, shortstop, and stabilizers. The basic steps involved in the manufacture of nitrile rubber are as follows:

The monomers are emulsified in water, a free radical-generating catalyst is added, and the mixture is agitated while a constant temperature is maintained. After the desired degree of polymerization is reached, a shortstop and stabilizers are added; residual monomers are removed; the latex is concentrated, coagulated, washed, dewatered; and the resulting crumbs are dried and compacted to form bales. Both batch and continuous methods of manufacture are in use today.

Nitrile rubbers are available in several forms such as sheets, crumb, powders and liquid.

Properties of Nitrile Rubber

Oil resistance is the most important property of nitrile rubber. Nitrile rubber is available in several grades of oil resistance based on the acrylonitrile content of the polymer. Available grades range from 15% to 50% acrylonitrile content.

Many of the properties of nitrile rubber are directly related to the proportion of acrylonitrile in the rubber. The most important property trends related to the acrylonitrile content are shown in Table 11.2.

Table 11.3 illustrates some of the properties one can expect of a typical nitrile rubber compound when only the ratio of acrylonitrile to butadiene content of the polymer is varied. Other important characteristics which can affect the properties of nitrile rubber, more particularly the processing properties are Mooney

Table 11.2. Property Trends Influenced by the Percent of Acrylonitrile in a Nitrile Rubber.

	% Acrylonitrile *15 → 50*
Oil resistance improves	→
Fuel resistance increases	→
Tensile strength increases	→
Hardness increases	→
Abrasion resistance improves	→
Gas impermeability increases	→
Heat resistance increases	→
Low-temperature flexibility improves	←
Resilience improves	←
Plasticizer compatibility increases	←
Compression set decreases	←

Table 11.3. Typical Properties of Nitrile Rubber Related to Acrylonitrile Content.

% Acrylonitrile	*41*	*33*	*29*	*21*
Original Properties ASTM D412				
100% Modulus, MPa (psi)	2.0 (300)	2.3 (340)	2.3 (340)	2.4 (355)
300% Modulus, MPa (psi)	12.0 (1750)	12.0 (1750)	11.7 (1710)	14.6 (2130)
Tensile Strength, MPa (psi)	22.0 (3200)	20.9 (3040)	20.8 (3030)	21.0 (3055)
Elongation, %	380	500	490	390
Hardness, points	66	64	62	62
Compression Set, Plied Discs, ASTM D395, Method B				
70 Hrs @ 125°C	34	34	39	38
Low-Temp. Brittleness, ASTM D2137				
°C	−20	−36	−40	−45
Oil-Resistance, Volume-Change Percent, 70 Hrs @ 100°C				
ASTM No. 1	−1	+1	+3	+4
ASTM No. 3	−10	+17	+32	+49
Fuel-Resistance, Volume-Change Percent, 70 Hrs @ 23°C				
ASTM Reference Fuel B	+29	+39	+54	+76
ASTM Reference Fuel C	+48	+66	+88	+133
Distilled-H_2O-Resistance, Volume-Change Percent, 70 Hrs @ 23°C				
% Change	+11	+10	+8	+5

plasticity and gel content. Typical Mooney viscosity ranges of nitrile rubbers are 30 to 95.

Gel content generally ranges from 0% to over 80%. The type of gel present also affects the properties. Permanent gel or crosslinking generally results in lower tensile strength and lower-percent elongation and abrasion resistance in cured compounds, but in calendered or extruded goods, lack of nerve or reduced shrinkage is valuable. Nonpermanent gel, which generally can be temporarily milled out of a compound, results in increased green strength and reduced cold flow of uncured compounds, but also results in reduced surface tack and poor mold flow characteristics.

A modification of the nitrile rubber molecule, i.e., the addition of carboxyl groups, results in a rubber with the usual oil resistant characteristics of nitrile rubber, but with much improved abrasion resistance. Blends of poly(vinyl chloride) and nitrile rubber also are produced either by blending the two latices or by physically blending the polymers. The resulting products have much improved ozone resistance and improved flame-retardant properties.

Uses of Nitrile Rubber

Because of its oil-and fuel-resistant characteristics, nitrile rubber finds its greatest market in applications where these characteristics are necessary.

There are many applications in the hose industry where oil, fuel, chemicals and solutions are transported. Typical hose constructions requiring a nitrile-rubber tube or liner are automotive, marine, aircraft fuel lines, and bulk-fuel transfer hose; oil-tanker, barge-loading, and unloading hose; chemical transfer hose, industrial air hose and food handling, dairy and creamery hose. Other automotive applications include emission hose and tubing, shaft seals, bushings, gaskets, carburetor parts, fuel-pump diaphragms and oil-filter gaskets.

Another large user of nitrile rubber is the oil-drilling industry in such items as blowout preventors, down-hole packers, drill-pipe protectors, pump-piston elements and rotary drilling hose. Powder and particulate forms of nitrile rubber are especially useful in cements, adhesives, binders for cork gaskets and brake linings. Another interesting application for nitrile rubber is in plastics modification to improve impact strength and flexibility.

Compounding

Nitrile rubber can be compounded to obtain a broad range of properties. Basically, nitrile rubbers are compounded in much the same fashion as natural rubber or the styrene-butadiene rubbers. Polymer selection is important to obtain the best balance of oil resistance and of low-temperature flexibility. The higher acrylonitrile content polymers have the highest oil and fuel resistance, but poorer low temperature properties. The lower acrylonitrile content polymers have good low temperature properties at some sacrifice in oil and fuel resistance. Table 11.4 will serve as a guide in choosing the optimum polymer to meet the end use requirements.

As with natural rubber and styrene-butadiene rubber, zinc oxide at a 3- to 5-phr level and stearic acid at a 1- to 2-part level are added for proper activation. A good antioxidant is usually added in all nitrile compounds to improve stability. Proper choice of antioxidant depends on the requirements such as high-heat resistance, extraction resistance, or staining characteristics. Here the antioxidant suppliers' literature should be consulted for the most suitable antioxidant.

Nitrile rubber is not inherently ozone-resistant; therefore protection must be added to the compound to achieve a desired degree of ozone-resistance. Again, the suppliers' literature will be helpful in choosing the best antiozanant. Protective waxes, usually at the 1- to 2-part level, are generally added to the compound in addition to the ozone inhibitors. Ozone resistance can also be obtained

Table 11.4. Typical Effect of Fillers on Stress-Strain Properties.

Filler to a Nominal 65 Durometer, A	*Tensile Strength MPa (psi)*	*Elongation %*	*Modulus at 300% Elongation, MPa (psi)*
SAF, N110	23.4 (3410)	420	15.2 (2210)
FEF, N550	18.6 (2700)	490	12.1 (1730)
SRF, N774	17.0 (2470)	550	10.4 (1520)
MT, N990	13.5 (1960)	780	5.2 (760)
Hard clay	13.0 (1900)	600	8.2 (1200)
Hydrated silica	20.5 (2980)	720	4.2 (610)

through blending techniques with other materials such as poly(vinyl chloride), chlorosulfonated polyethylene, chlorinated polyethylene, epichlorohydrin and ethylene-propylene terpolymers.

Reinforcing fillers are necessary in order to achieve optimum properties with nitrile rubbers. Carbon black is the most widely used filler and the nitrile rubbers respond to the full range of available carbon blacks from the largest to the smallest particle sizes and range of structures. Nonblack applications will require the use of reinforcing silicas of various types, calcium carbonates, hard clays, talc and other pigments. Table 11.4 illustrates the effects of typical fillers on the stress-strain properties of a 33% acrylonitrile-content polymer.

Plasticizers are generally used in nitrile-rubber compounds to improve processing and low-temperature properties. Typically they are ester types, aromatic oils, and polar derivatives and can be extractible or nonextractible depending upon the end-use requirements.

Vulcanization can be achieved with sulfur, sulfur-donor, or peroxide systems. Table 11.5 illustrates some typical vulcanization systems in general use. Tables 11.6, 11.7 and 11.8 represent typical compounds designed for automotive-hose tubes, oil-well-packer elements, and general-purpose molded goods.

Mixing and Processing

All of the commercially available nitrile rubbers can be mixed either on a two-roll mill or with internal mixing equipment. Compounds can be designed to be easily extruded, calendered, or molded with injection, compression, or transfer techniques.

The proper choice of polymer Mooney viscosity determines how well a compound will process. The low-Mooney-viscosity polymers lend themselves to injection molding and calendered friction compounds. The medium-Mooney-viscosity polymers will generally transfer- and compression-mold and provide

Table 11.5. Typical Vulcanization Systems.

System	*Level*	*Characteristics*
Sulfur MBTS	1.5 parts 1.5 parts	Low cost, general purpose, slow curing
Sulfur TMTD	0.3 parts 3.0 parts	Fast-curing, low compression set, blooms
Sulfur TETD TMTD CBTS	0.5 parts 1.0 parts 1.0 parts 1.0 parts	Faster curing, lower compression set, nonblooming
DTDM OTOS TMTM	1.0 parts 2.0 parts 1.5 parts	EV system, excellent heat aging
Dicumyl Peroxide (40%)	4.0 parts	General-purpose Peroxide system, low compression set

Table 11.6. Automotive Fuel-Hose Compound

Hycar® 1092-80[a]	100.0
Zinc oxide	5.0
Stearic acid	1.0
Antioxidant	1.5
Hard clay	30.0
N774 Black	90.0
Plasticizer	15.0
Sulfur	1.5
Accelerator	1.5
	245.5

[a]Typically, 80 Mooney viscosity, 33% Acrylonitrile.

Expected properties:

Tensile strength, psi	2000
Elongation, %	300
Hardness, points	70

good calendering characteristics. The high-Mooney-viscosity polymers provide excellent extruded compounds, particularly when high "green" strength is required. Blends of polymers with different Mooney-viscosity ranges often provide the solution to many processing difficulties.

Latest Developments

With the increase in automobile under-hood temperatures and increasing demands of the energy-recovery markets, two new developments should be noted:

Table 11.7. Oil-well-Packer Element.

Hycar® 1051[a]	100.0
Zinc oxide	5.0
Stearic acid	1.0
N550 Black	70.0
Antioxidant	2.0
Plasticizer	5.0
Sulfur	0.4
MBTS	3.0
ZDMC	1.5
	187.9

[a]Typically, 75 Mooney, 41% acrylonitrile.

Expected properties:	
Tensile strength, psi	2940
Elongation, %	370
Hardness, points	78

Table 11.8. General-purpose Molding Compound.

Hycar® 1052[a]	100.0
Zinc oxide	5.0
Stearic acid	1.0
Antioxidant	1.5
N774 Black	65.0
Plasticizer	15.0
Sulfur	1.5
MBTS	1.5
	190.5

[a]Typically 50 Mooney, 33% acrylonitrile.

Expected Properties:	
Tensile strength, psi	2700
Elongation, %	550
Hardness points	65

1. Nitrile rubbers containing "bound" antioxidants, that is, antioxidants attached to the polymer chain, have appeared on the market. These antioxidants are less fugitive and are less likely to be soluble in fuels or oils and are less volatile, thereby improving dry-heat resistance.
2. Hydrogenated nitrile rubbers containing little or no unsaturation have recently appeared. These nitrile rubbers show promise of better heat resistance and resistance to oxidized gasoline as well as improved resistance to the harsh environments found in deep, sour wells.

POLYACRYLIC RUBBERS

History

Unmodified acrylic rubbers were developed in the early 1940's by the BFGoodrich Company. The U.S. Department of Agriculture's Eastern Regional Research Laboratory and the University of Akron's Government Laboratories pioneered in the development of modified ''Lactoprene EV'' and ''Lactoprene BN'' acrylic elastomers. The first commercial products of both types were manufactured and marketed by BFGoodrich Chemical in 1948. These were identified as Hycar PA and Hycar PA-21. It was found that Hycar PA-21 offered the greatest advantage and development work was concentrated on this polymer, which was subsequently identified as Hycar 4021.

Since then, significant advances have been made in the low-temperature capability and cure technology of acrylic elastomers by the BFGoodrich Chemical Group and subsequent suppliers.

Suppliers

Worldwide suppliers of acrylic elastomers are presented in Table 11.9. Product literature is available upon request.

Composition

Acrylic elastomers are copolymers based on two major components: the backbone (95–99%) and the reactive cure site (1–5%) shown in Fig. 11.1. Reactivity of the cure site monomer governs the cure behavior of the polymer.

Method of Manufacture

Conventional acrylic elastomers are commonly emulsion-polymerized to form a latex, or are polymerized in suspension. The solid product is recovered by

Table 11.9. World Suppliers of Acrylic Elastomers.

Company	*Country*	*Trade Name*
BFGoodrich Chemical Group	USA	Hycar®
American Cyanamid Company	USA	Cyanacryl®
E. I. DuPont DeNemours & Co., Inc.	USA	Vamac®
Enichem	Italy	Europrene AR
Nippon Zeon Co., Ltd. (NZ)	Japan	Nipol
TOA Paint Co., Ltd.	Japan	Acron
Nippon Oil Seal-NOK	Japan	Noxtite
Japan Synthetic Rubber Co., Ltd.	Japan	JSR AR

Conventional ACM (Hycar, Cyanacryl)

Backbone

(1) Alkyl

$$\begin{array}{l} (\mathrm{CH_2-CH})_x \\ \quad\quad\ \ | \\ \quad\quad\ \mathrm{C=O} \\ \quad\quad\ \ | \\ \quad\quad\ \mathrm{O-C}_n\mathrm{H}_{2n+1} \end{array}$$

e.g., if $n = 2$: ethyl acrylate
e.g., if $n = 4$: butyl acrylate

and (2) Alkoxy

$$\begin{array}{l} (\mathrm{CH_2-CH})_x \\ \quad\quad\ \ | \\ \quad\quad\ \mathrm{C=O} \\ \quad\quad\ \ | \\ \quad\quad\ \mathrm{O-C}_n\mathrm{H}_2n\mathrm{O-C}_m\mathrm{H}_{2m+1} \end{array}$$

e.g., if $n = 2$ and $m = 1$: methoxy ethyl acrylate
e.g., if $n = 2$ and $m = 2$: ethoxy ethyl acrylate

Cure Site

Chlorine

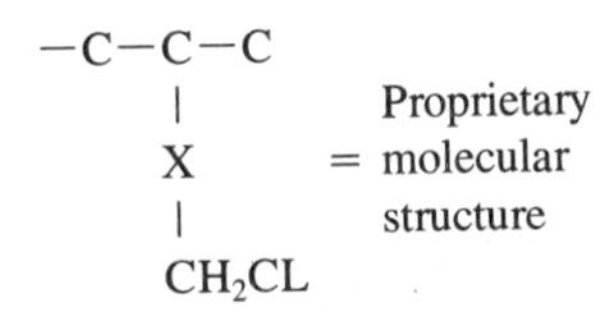

Ethylene/Acrylic (Vamac)

Backbone

(1) Ethylene

$$(-\mathrm{CH_2-CH_2}-)_x$$

(2) Methyl Acrylate

$$\begin{array}{l} (-\mathrm{CH-CH_2}-)_y \\ \quad\ \ | \\ \quad\ \mathrm{C=O} \\ \quad\ \ | \\ \quad\ \mathrm{OCH_3} \end{array}$$

Cure Site

Carboxylic Acid

$$\begin{array}{l} (-\mathrm{R}-)_z \\ \quad\ | \\ \quad \mathrm{C=O} \\ \quad\ | \\ \quad \mathrm{OH} \end{array}$$

Fig. 11.1. Major components, acrylic elastomers.

conventional coagulation, washing procedures, and dried by hot air, vacuum, or extrusion techniques. The ethylene/acrylic type is solution-polymerized under pressure. However, its method of recovery is not revealed in the literature.

Raw Polymer Characteristics

Conventional acrylic polymers have a specific gravity of 1.10 and are off-white in color. Current commercial types are manufactured in a 25-55 Mooney (ML − 1 + 4′/100°C) viscosity range. All are normally supplied in solid, slab form. Some special solution grades are also available in both solid and crumb form. The ethylene/acrylic polymer has a specific gravity of 1.03, is pale-gray in color, and supplied in a 16 ± 3 Mooney (ML − 1 + 4/100°C) viscosity range.

Black and nonblack masterbatch versions are also available. All are supplied in slab form only. The shelf stability of acrylic rubbers is excellent (> 12 months) when stored in their original packages under normal room-temperature, dry conditions.

Physical Characteristics

As a class, acrylic elastomers are inherently resistant to:

1. Temperatures from −40 to 204°C (400°F).
2. Oils at elevated temperatures.
3. Oxidation.
4. Ozone.
5. Aliphatic hydrocarbons.
6. Sunlight (UV) discoloration.

Typical vulcanizate properties are as follows:

Tensile Strength (MPa/psi)	5–17/700–2500
Elongation (%)	100–600
Hardness (Duro A)	40–90

In addition, the acrylics exhibit good damping qualities, are not highly corrosive to steel, and can be compounded to provide excellent flex life, fair electrical-insulation resistance and some degree of flame resistance. The more polar conventional acrylics can also be compounded to provide a high degree of electrical conductivity. The ethylene/acrylic type also provides improved abrasion resistance, but tends to lose this advantage after oil-aging as a result of its high swell characteristics.

Ethyl-acrylate-base polymers provide the best overall balance of processing characteristics, physical properties, and heat and oil resistance, but possess limited low-temperature resistance. The newer, improved low-temperature versions exhibit some sacrifice in this balance. The choice of polymer is primarily dependent on the balance of low-temperature/oil-swell resistance required for a given application.

Heat, Fluid, Low-temperature Resistance

The high-temperature service capability of acrylic elastomers is rated at >70 hr maximum/190–204°C short term, intermittent; and >1000 hr maximum/163–177°C long term, continuous, in dry heat. Short-term exposure above 204°C is also possible. The ultimate mode of heat-aging failure for all acrylic elastomers is embrittlement.

Acrylic elastomers are highly resistant to petroleum base lubricating oils, transmission fluids and greases, including the sulfur-bearing, extreme-pressure gear lubricants. In addition, they are also quite resistant to the newer, longer-service SF and synthetic ester-type lubricants. The use of butyl acrylate and ethylene/acrylic-based polymers is limited to applications that can tolerate high-oil-swell characteristics. Acrylic elastomers are not recommended for service in aromatic hydrocarbons, e.g., gasoline. Most conventional acrylic polymers are also not suitable for continuous use in hot water or steam, or in water-soluble materials such as methanol or ethylene glycol. However, the ethylene/acrylic type is recommended for service in hot water and in ethylene glycol media.

Current commercial types are supplied in a −14°C to −40°C low-temperature range as characterized by differential thermal analysis, glass-transition temperature (DTA T_g).

Applications

Historically, conventional acrylic elastomers have been successfully utilized in a wide variety of critical automotive seal applications. These include automatic-transmission seals, valve-stem seals, crankshaft seals, pinion seals and oil-pan seals. The newer, more versatile types are also gaining rapid acceptance in other mechanical-goods applications such as hose, tubing, electrical-cable jacketing, rolls, and belting.

Typical Compounds

The basic physical property parameters of representative commercially available acrylic elastomers in standard black reinforced, 80 ± 5 Durometer A hard test compounds are presented in Table 11.10. Compound recommendations for specific applications can be obtained from the polymer suppliers.

General Compounding

Cure Systems. Because acrylic elastomers have a saturated backbone, crosslinking is accomplished via incorporation of copolymerized reactive cure sites. The nature of this cure site varies in commercial products, and therefore different cure systems have been developed for specific types. Thus acrylic elastomers from different suppliers are not generally directly interchangeable in a given recipe. Conventional domestic acrylics respond to soap/amine, activated thiol, soap/sulfur or sulfur donor, lead/thiourea, diamine and trithiocyanuric acid (TCY) cure systems. The ethylene/acrylic type responds to diamine and peroxide cures. Since the majority of acrylic cure systems are basic in nature,

Table 11.10. Typical Properties, Commercial Standard Black, Reinforced Acrylic Elastomers.

Recipe	*1*	*2*	*3*	*4*
Hycar® 4051	100	–	–	–
Hycar® 4052	–	100	–	–
Hycar® 4054	–	–	100	–
Vamac® B-124	–	–	–	124
TE-80 Process Aid[a]	2	2	2	–
Stearic acid	1	1	1	–
N539 FEF-LS Black	65	80	80	–
N774 SRF-HM Black	–	–	–	55
Sodium stearate	4	4	3	–
Poly-Disp. T(KD)D-75[b]	2	2	4	–
Methylene dianiline	–	–	–	1.25
Diphenylguanidine				4
	174	189	190	184.25
Mold-cure time (min.)	4	4	4	20
Mold-cure temp. (°C)	190	190	190	177
All oven-post-cured 4 hr/177°C.				
Ultimate Properties—ASTM D412				
Stress at 100% (MPa/psi)	10.54/1530	9.65/1400	7.23/1050	7.03/1020
Tensile strength (MPa/psi)	15.23/2210	11.51/1670	8.96/1300	14.68/2130
Elongation (%)	140	120	120	250
Hardness (Duro. A)	76	77	75	78
Compression Set, 70 hr/150°C—ASTM D395				
Solid (%)	20	28	34	15
Low-temperature Flexibility—ASTM D1053				
T100 (°C)	−14.5	−30.5	−37.5	−29
Low-temperature Brittleness—ASTM D2137				
Pass (°C)	−10	−21	−28	−24
Air Test Tube—ASTM D865				
70 Hrs	204°C	190°C	204°C	204°C
Tensile strength (MPa/psi)	10.40/1510	9.37/1360	7.10/1030	8.27/1200
Tensile change (%)	−32	−19	−21	−44
Elongation (%)	100	80	60	80
Elongation change (%)	−29	−33	−50	−68
Hardness (Duro. A)	88	87	90	88
Hardness change (pts)	+12	+10	+15	+10
180° bend	Pass	Pass	Pass	Pass
ASTM Oil No. 3, 70 hr/150°C—ASTM D471				
Volume change (%)	+15	+20	+60	+55

Note: BFGoodrich Chemical Group Data
[a]Technical Processing, Inc.
[b]Wyrough & Loser, Inc.

they are retarded by acids and accelerated by bases. Specific cure-system recommendations should be obtained directly from the polymer suppliers.

Despite significant advancements in acrylic-cure technology, all current, state-of-the-art acrylic elastomers require a relatively long cure cycle or must be subsequently post-cured (tempered) in a circulating hot-air environment to realize optimum compression-set resistance.

Reinforcing Agents. Acrylic elastomers do not provide high gum strength when cured. Reinforcing agents are required to develop useful properties. Carbon black reinforcing agents provide the best overall balance of vulcanizate properties. Of the many available grades, N550FEF, N539 FEF-LS, and N774 SRF-HM are the most popular.

The use of mineral reinforcing agents is primarily limited to electrical-resistant and color-coded applications. Aluminum silicate or silica types are commonly utilized, either alone or in combination. Amino or vinyl silane coupling agents are also sometimes employed to gain improved vulcanizate properties. Organic colorants are generally recommended since inorganic metallic-oxide color pigments tend to have an adverse effect on the cure and heat-aging characteristics of silica-reinforced compounds.

Synthetic graphite is also utilized in conjunction with carbon black and/or mineral reinforcing agents to promote improved surface lubricity characteristics in, for example, rotary-shaft-seal applications. Neutral-to-high-pH-alkaline pigments are recommended since acidic types will tend to retard the basic cure mechanism of most acrylic elastomers.

Plasticizers. Plasticizers are used in acrylic-rubber compounds as process aids or to gain improved low-temperature resistance. However, type and amount is limited by their volatility and extraction characteristics, as related to post-cure conditions and service requirements. Low-volatility ester and polyester plasticizers are generally utilized.

Process Aids. Lubricating agents are essential to promote the release characteristics of acrylic compounds. Stearic acid is commonly used, usually in combination with commercial process aids, which provide both external (release) and internal (viscosity-reducing) lubricating qualities. However, it is cautioned that high lubricant levels can interfere with mold knitting and metal-bonding properties.

Antioxidants. Although acrylic elastomers are highly resistant to oxidation, certain antioxidants can promote marginally improved dry-heat resistance. Low-volatility diphenylamines have been found useful for conventional acrylics, while hindered phenols are suggested for the ethylene/acrylic type. The use of antioxidant is specifically recommended in the ethylene/acrylic gum polymer

Table 11.11. Banbury vs. Mill-mixing.

Banbury	Mill
Start: slow speed, full-cooling water.	Start: room-temperature rolls, full-cooling water.
Master	
1. Charge polymer.	1. Band polymer.
2. Add filler, lubricant, antioxidant.	2. Adjust nip to provide rolling bank.
3. Add process aid, plasticizer.	3. Add lubricant, filler, antioxidant. Open nip as required to retain rolling bank.
4. Bump and sweep.	4. Add process aid.
5. Dump Time 3–6′. Dump temp 100–150°C. Dip cool.	5. Add plasticizer. Dip cool (optional).
Finish	6. Add curatives. Dip cool.
6. Charge ½ master, curatives ½ master. Dump at 100–121°C. Dip cool.	

On completion of mixing, the stocks should be given 6 cuts from each side plus 6 refining passes on the mill before stripping.

and for all nonblack reinforced compounds. Additional general compounding information can be obtained from the polymer suppliers.

General Processing

Mixing. Acrylic compounds can be Banbury- or mill-mixed. However, internal Banbury mixing is preferred and generally utilized by the industry. Typical Banbury and mill-mixing procedures are as shown in Table 11.11.

A two-pass Banbury-mix procedure is most common, but a one-pass mix is also possible on relatively nonscorchy compounds. An "upside-down" mixing procedure is sometimes desirable for use with highly loaded, dry compounds. Because acrylic elastomers are somewhat thermoplastic in nature and tend to lose shear resistance fairly rapidly on mixing, it is recommended that reinforcing agents be incorporated very early in the mixing cycle to obtain good dispersion. Maximum cooling is also essential, to maintain polymer integrity during the initial phase of mixing.

Extrusion/Calendering. Most acrylic compounds extrude well enough for standard mold-preparation. However, for finish goods, special compounding may be required to obtain satisfactory green strength, size, and finish characteristics. In general, the use of high-structure reinforcing agents along with increased process aid is usually effective. Compound extension through the use of softener is also sometimes necessary. The range of die temperatures is 65–107°C.

Calendering compounds are designed much the same as for finished extrusions. However, additional lubricant may be required to promote good release characteristics. The range of calender-roll temperatures is 27–107°C.

Compound Storage Stability. Typical finished acrylic compounds provide reasonably good shelf-stability characteristics in the range of one week to several weeks when stored under normal room-temperature, dry conditions. Shelf life is primarily dependent on the activity of the cure system employed. Storage under refrigerated (4°C), low-humidity conditions is also an effective means of extending shelf life. In practice, aged compounds are usually mill-freshened prior to subsequent processing.

Vulcanization. Acrylic elastomers lend themselves to all common cure processes. They are commonly compression-, transfer- and injection-molded, but can be steam-cured as well. Typical cure cycles (in minutes) are as follows:

Compression mold	3′/190°C to 1′/204°C
Transfer mold	12′/163°C to 4′/177°C
Injection mold	3′/190°C to 1′/204°C
Open-steam cure	30–90′/163°C

In addition, acrylic elastomers respond to most continuous vulcanization (CV) cure techniques. Special compounding may be required for the low viscosity types to gain adequate backpressure in compression molding operations.

As indicated previously, a post-cure is usually necessary to develop optimum compression-set resistance. Oven post-cure cycles in the range of 4–8 hr/177°C are commonly utilized.

More comprehensive processing recommendations are available from the polymer suppliers.

Bonding Characteristics

A variety of commercially available solvent-base, curing-type adhesives have been found to provide excellent bonds of acrylic compounds to metals and other substrates. Bonding is carried out during the vulcanization process. Recommended substrate preparation and application procedures are provided by the manufacturer.

Solution Characteristics

Certain acrylic elastomers also lend themselves to solvent-solution coatings and adhesive applications. These dissolve readily in common industrial solvents such as methyl ethyl ketone, acetone, ethyl acetate, and toluene. Polymer concentrations in the 30-40% total solids range are possible.

Blends

The blend capability of acrylic elastomers is limited. However, certain conventional types can be utilized to reduce the cost of diamine curable fluorocarbon (FKM) compounds or to improve the heat-resistance of epichlorohydrin (CO/ECO) compounds. In addition, modified low-T_g versions are believed to have potential as high-temperature resistant-impact improvers for poly(vinyl chloride), polycarbonate, and poly(phenylene oxide) plastic materials. Although not reported, the ethylene/acrylic type may also have interesting blend possibilities.

Future Developments

Although significant advances have been made in acrylic-cure technology, continued research is aimed at the development of more highly reactive polymers and/or cure systems to completely eliminate the necessity of post-cure for short-cycle molded-goods applications.

REFERENCES

1. John P. Morrill, ''Nitrile and Polyacrylate Rubbers,'' in *Rubber Technology*, 2d Ed., M. Morton, ed., Van Nostrand Reinhold, New York, 1973.
2. John P. Morrill, ''Nitrile Elastomers,'' in *Vanderbilt Rubber Handbook*, 12th Ed., R. T. Vanderbilt Co., Norwalk, CT, 1978.
3. Carl H. Lufter, ''The Vulcanization of Nitrile Rubber,'' in *Vulcanization of Elastomers*, G. Alleger and I. J. Sjothun, eds., Reinhold, New York, 1969.
4. F. R. Wolf and R. D. DeMarco, ''Polyacrylic Rubber,'' in *Vanderbilt Rubber Handbook*, 12th Ed., R. T. Vanderbilt Co., Norwalk, CT, 1978.
5. J. F. Hagman, ''Ethylene/Acrylic Elastomers,'' in *Vanderbilt Rubber Handbook*, 12th Ed., R. T. Vanderbilt Co., Norwalk, CT, 1978.
6. R. D. DeMarco, ''New Generation Polyacrylate Elastomers,'' *Rubber Chem. and Technol.* **52** (1979).
7. R. J. Mayers and D. A. Seil, ''Effect of Long Term Fluid Immersions on Properties of Polyacrylic and Ethylene/Acrylic Elastomers,'' Paper, ACS Rubber Division, Oct. 1981.
8. P. H. Starmer and F. R. Wolf, ''Acrylic Elastomers,'' in *Encyclopedia of Polymer Science and Engineering*, 2d Ed., John Wiley & Sons, New York, 1984.
9. J. R. Hagman and J. W. Craley, ''Ethylene-Acrylic Elastomers,'' in *Encyclopedia of Polymer Science and Engineering*, 2d Ed., John Wiley & Sons, New York, 1984.
10. Bulletin HPA-1, ''Hycar Acrylic High Heat, Oil Resistant Elastomers,'' BFGoodrich Chemical Group, Cleveland, 1984.

12

NEOPRENE AND HYPALON

Part I

Neoprene

RALPH S. GRAFF
E. I. du Pont de Nemours & Company
Polymer Products Department
Wilmington, Delaware

HISTORY

On December 30, 1925, at an organic chemistry symposium of the American Chemical Society in Rochester, New York, Father Nieuwland of Notre Dame University reported on his research on acetylene gas,[1,2] and it was there that Dr. Elmer K. Bolton of the Du Pont Company became interested in Nieuwland's work. Out of their discussions came an alliance and unique research effort between Notre Dame and a group of Du Pont scientists including Wallace Carothers, Arnold Collins, William Calcott, and A. S. Carter. For his contributions in the field of acetylene and chloroprene chemistry, Arnold Collins later received the Charles Goodyear Medal from the Rubber Division.[3]

On November 2, 1931, a historic meeting of the Akron Rubber Group took place. Three Du Pont chemists discussed various aspects of the development of a new polychloroprene polymer that they called "Duprene."[2,3] Some time later, this 2-chloro-1,3-butadiene synthetic rubber was named *neoprene*. This discussion generated intense interest among the research people in attendance at the Akron meeting, but the commercial people paid little attention. They couldn't foresee much market potential for a synthetic rubber the price of which was $1.05 a pound at a time when natural rubber was selling for 3¢ to 5¢ per pound. It took both managerial and technical courage to launch neoprene commercially but it became the first widely used synthetic rubber.

A general summary of the development of 2-chloro-butadiene polymers has been published in the *Encyclopedia of Polymer Science and Technology*,[4] one of the most comprehensive treatises of this kind ever printed. It includes background information on the theoretical aspects, both polymerization and polymer characteristics, types, compounding, processing and uses.

COMMERCIAL POLYCHLOROPRENE

After over 50 years of development and manufacture, the beginning of 1985 finds polychloroprene made by at least eight manufacturers, namely:

1. E. I. du Pont de Nemours & Company (plants in USA and Northern Ireland)
2. Showa Neoprene K.K. (Japan, operated by the Du Pont Company and Showa Denko K.K.)
3. Denka Chemical Corporation (USA)
4. Bayer-Baypren (Germany)
5. Denki Kagaku-Denka (Japan)
6. Distugil-Butaclor (France)
7. Toya Soda-Skyprene (Japan)
8. U.S.S.R.

Each company uses emulsion polymerization and makes a number of variations of the product. Varieties are marketed both as solid polymers and as latices, but not all of them are readily available in all trading areas. The solid types vary over a wide viscosity range. In some instances, either nonstaining or staining antioxidants are included at the time of manufacture.

Lists of available neoprenes vary. In late 1984, the seven free-world manufacturers of neoprene were marketing the polymers shown in Table 12.1. The chronological development of these polychloroprene variants is a classic story of necessity and creativity. These variations were developed to satisfy the ever changing needs of the rubber industry and its customers, including greater raw material stability, processing reliability and uniformity, economy, minimum waste, and upgraded performance properties to meet new demands in end products.

GENERAL-PURPOSE SOLID TYPES

The solid types may conveniently be divided into the more widely used general-purpose types on the one hand and the specialty types on the other. The general-purpose types are available in three families—G, W, and T (Du Pont grades)—with variations within each family. They are used in a great variety of elastomer products because they offer a broad range of physical properties and adaptability to current processing conditions, thus allowing the user to formulate for his or her specific requirements. The characteristics of the raw polymer and the vulcanizates of each family are noted in general terms in Table 12.2.

Table 12.1. Commercial Polychloroprenes.

Du Pont Neoprene	*Showa Neoprene KK*	*Denka Chemical Corporation Neoprene*	*Bayer Baypren*	*Denki Kagaku Denka*	*Distugil Butaclor*	*Toya Soda Skyprene*
AC	AC	–	321, 331	TA-85, 95	MA41H, K	G-41
AD	AD	–	320, 330	A-70, A-120	MA40R, S	G-40
ADG	ADG	–	–	–	MA40T	–
AF	–	–	–	–	–	–
AG	–	–	–	–	–	–
FB	–	–	–	–	–	–
GN	–	S5	–	–	SC22	–
GNA	–	S5S	–	PM-40	SC22	–
GRT	GRT	S3	610	–	SC10	R-10
–	–	–	–	PT-60	SC11	–
GS	GS	–	710	PM-40NS	SC22	R-22
GW	–	–	–	OCR-42	X2211	–
W	W	M1	210, 220	M40, M41	MC30A, B	B-30
W-MI	–	M1.1	211	M30, M31	MC31	B-31
WB	WB	–	214	EM-40	ME20	Y-20E
WD	–	–	130	DCR-30	–	–
WHV	WHV	M2	230	M-120	MH30	Y-30
WHV-100	WHV-100	M2.7	230	M-100	MH31	Y-31
WHV-A	–	–	–	–	–	–
–	WHV-200	–	–	–	–	–
WK	WK	–	124	ES-70	–	–
WRT	WRT	M3.5	110	S-40V	MC10	B-5
WX	WX	–	112	S-40	MC20	B-10
–	WX-K	–	–	–	–	–
–	WX-KT	–	–	–	–	–
–	WK-J	–	–	–	–	–
TW	TW	EM1	215	MT-40	DE302	E-33
TW-100	TW-100	–	235	MT-100	DE305	–
TRT	TRT	–	115	ES-40	DE102	–
–	–	–	–	–	MC122	–
–	–	–	–	–	MC322	–
–	–	–	–	–	MC323	–

Table 12.2. Family Characteristics of the Neoprenes.

G Types	*W Types*	*T Types*
	Raw Polymers	
Limited storage stability	Excellent storage stability	Excellent storage stability
Peptizable to varying degrees	Not peptizable	Least nerve
Fast curing but safe processing		Best extrusion, calendering performance
No accelerators necessary		
	Vulcanizates	
Best tear strength	Best compression-set resistance	Properties similar to W types
Best flex	Best heat-aging	
Best resilience		

The G, W, and T types of neoprene are available in many variations. Table 12.3 makes qualitative comparisons between types for the raw polymer, uncured compound, and the vulcanizate. By selecting the most important properties required for an end-use, the compounder can choose the appropriate type(s) for compound development.

PROPERTIES

Most neoprene compositions are made by selecting the base elastomer, adding a variety of ingredients, and then processing and vulcanizing. The choice of the ingredients used affects the physical properties of the finished product. Because of the abundance of applicable ingredients, a wide variety of compounds of different compositions and properties are available. The introduction of new synthetic rubbers has vastly increased the number of possible permutations and added to the complexity of designing with an elastomer.

Consider the property called *oil resistance*. When an elastomeric product is exposed to a liquid hydrocarbon, several reactions take place. The most obvious, and readily documented, is a change in dimensions. Less apparent, but frequently of much greater importance than swelling, is a deterioration of the original physical properties. If dimensional stability is functionally necessary, as in O-rings or tubing, swelling would spell failure. In many applications, however, retention of original mechanical characteristics such as high tensile strength, resilience, and abrasion resistance in the swollen state is vital to long service life. For practical purposes, then, oil resistance is the ability of an elastomeric part to retain its utility in spite of contact with oil.[5]

Table 12.3. Relative Property Comparisons of Types of Neoprene. Rating: O—Outstanding; A—Acceptable; L—Limited or not appropriate

Properties	GW	GN	GNA	GRT	FE[a]	W-MI	W	WHV-100	WHV	WRT	WD	WK	WB	TW	TW-100	TRT
Raw Polymers																
Crystallization resistance	A+	A	A	A+	L	L	L	L	L	O	O	O	L	L	L	O
Extendable	A	A	A	A	L	A	A	O−	O	A	O	O−	A	A	O−	A
Peptizable	A	O	O	O	O	L	L	L	L	L	L	L	L	L	L	L
Viscosity[b]	Lo	Lo	Lo	Lo	[a]	Lo	Lo+	M	H	Lo	H	M	Lo	Lo	M	Lo
Stability	A	A	A	A	A	O	O	O	O	O	O	O	O	O	O	O
Uncured Compound																
Cure rate	O	O	O	O	O	A+	A+	A+	A+	A	A	A	A+	A+	A+	A
Green strength	A	A−	A−	A−	L	A+	A+	O	O	A−	O	O−	A−	A	O	A−
Processability	A	A	A	A	A	A	A	A+	A+	A	A+	O	O	O	O	O
Vulcanizate																
Compression-set resistance	A	L	L	L	L	O	O	O	O	O	O	O	O	O	O	O
Good flex	O	O	O	O	L	A	A	A	A	A	A	A	A−	A	A	A
Good tear	O	O	O	O	L	A	A	A	A	A	A	A	A−	A	A	A
Heat-aging resistance	A+	A−	A−	A−	A−	O	O	O	O	O	O	O	O	O	O	O
Low-temperature properties	A+	A	A	O−	A	A	A	A	A	O	O	O	A	A	A	O

[a]Fluid polymer at 50°C, (122°F). [b]H = high; M = medium; Lo = low.

Originally, neoprene was selected as a replacement for natural rubber because its superior resistance to swelling in oil justified its premium price. During the 50 plus years of its commercial existence, successful performance has resulted more from its unique and balanced combination of properties than for any single property alone.

Properly compounded, neoprene resists sun, ozone, and weathering while remaining tough and durable. It is practical in cold environments to −25°C (−13°F) and in heat to 93°C (199°F). Specifically formulated compositions permit service at temperatures as low as −55°C (67°F). The flame resistance of neoprene products can be enhanced by special compounding.

CRYSTALLIZATION

Crystallinity is an inherent property of all polychloroprene rubbers[6,7] but some of the neoprenes crystallize more readily and to a greater extent than others. An understanding of this predilection is essential to the understanding of polychloroprene technology.

Neoprene, as well as natural rubber, butyl, and other linear elastomers, becomes highly oriented upon being stretched and shows X-ray diffraction patterns that indicate varying degrees of crystallinity. The crystalline structure of neoprene is indicated by a sharp peak in diffraction intensity at the Bragg angle of 19°30′. As crystallization develops, a small decrease in volume occurs, and stressed specimens tend to relax and elongate in the direction of the stress. Crystallization does not take place at high temperatures because these forces (orientation) are overcome by vigorous molecular motion. Crystallization rate slows down at low temperatures because thermal stiffening inhibits movement, and the incidence of favorable alignment is reduced. Anything that aids mobility of molecules at low temperatures actually encourages crystallization at those temperatures. Crystallization is a completely reversible phenomenon. Warming a crystallized specimen to a temperature above that at which the crystallites were formed destroys them, for example, and increasing molecular motion by vibrating or flexing causes decrystallization. When neoprene is decrystallized, it returns to its original softness and flexibility and recovers from any relaxed condition that may have developed as a result of crystallization under stress.

Crystallization, which is time-dependent, should not be confused with thermal stiffening and embrittlement effects, which occur at very low temperatures and are a function of temperature only. Neoprene, as do all elastomers, becomes progressively stiffer as it is cooled. This type of stiffening becomes evident just as soon as thermal equilibrium is established. An unplasticized neoprene compound, for example, will have an ASTM D-749 brittleness temperature of about −40°C for both crystallized and uncrystallized vulcanizates.

The practical aspects of crystallization are seen in the hardness of the raw polymer in the bag and the "quick grab" and high cohesive strength of a neo-

Table 12.4. Room-temperature Crystallization Rates (measurements made on freshly-milled polymer).

Relative Time	*Neoprene*
Hours	AD, AC, ADG
Up to 3 days	W, TW
Up to 7 days	FB, GN, GNA, WX
One or two weeks	AF
More than two weeks	GW, GRT, WRT, WD, AG, WB, WK, TRT

prene adhesive. Other evidences are the stiffening of an unvulcanized compound that has been in shelf storage for a week or two and the hardening of a vulcanized part with time at moderately low temperatures.

It would be a good idea to show a comparison of the relative crystallization rates of the various neoprenes. Dry neoprenes are produced with widely different crystallization rates so that crystallization can be avoided if it might be a liability in a product or exploited if it might be an advantage. Uncompounded polymer crystallizes faster than its vulcanizate. The relative crystallization rates of the neoprenes are presented in Table 12.4. This table is based upon room-temperature observations important to adhesives; it should be noted, however, that maximum crystallization occurs at $-10°C$.

SPECIAL-PURPOSE SOLID TYPES

The special-purpose types of neoprene—AC, AD, AF, AG, and FB—are designed primarily for a particular application (e.g., adhesives, coatings, sealants). Neoprene AC, AD, and ADG have a high degree of crystallinity, whereas Neoprene AF has a low degree of crystallinity. None contains a staining type of stabilizer.

Toluene is a satisfactory solvent for dissolving AC or AD. With AF, some polar solvent should be used with toluene to increase polymer-solution stability.

Neoprene AC is a sol polymer that crystallizes rapidly. The viscosities of AC and its solutions change very little during storage. Low-viscosity grades readily go into solution without milling.

Neoprene AD is a readily crystallizing sol polymer with solubility characteristics similar to those of AC. Color stability and solution stability are superior to those of AC, even when stored in metal containers.

Neoprene AF, which contains small amounts of carboxyl functionality, is essentially a noncrystallizing polymer designed for one-part adhesives. Films from AF adhesives rapidly develop high room temperature or elevated bond strengths. Adhesive solutions made with AF are resistant to phasing.

Neoprene AG, a chloroprene polymer of high gel content, exhibits thixotropy when dissolved or dispersed in solvents. It is very suited for high solids and high-viscosity mastics, which require easy extrudability and resistance to slump. Low-viscosity adhesives prepared from AG not only spray more easily than those based on other neoprene polymers but offer the performance properties required for many contact adhesive applications. Generally speaking, AG can be formulated in the same fashion as AC and AD.

Neoprene FB is a soft, partly crystalline solid at room temperature, but its viscosity drops rapidly as temperature is increased and it becomes fluid at 50°C. It is used in adhesives and caulks and as a nonvolatile, vulcanizable plasticizer for other neoprenes. It dissolves readily in aromatic hydrocarbons, chlorinated hydrocarbons, or low-molecular-weight ketones and esters.

COMPOUNDING DRY NEOPRENES

As indicated elsewhere, the key to the all-around performance of neoprene products in a variety of services lies in the words "properly compounded." Once the service requirements have been documented, the compounder is free to take over, but with his freedom of choice subject to the limitations of process requirements and economics, of course.

Compound development logically begins with the selection of the type of neoprene that best provides the required physical properties within the applicable cost and processing limits. The response of all general-purpose types to compounding ingredients is similar. Compounding of the G types, however, differs from that of the W and T types in three ways:

1. They can be softened by the addition of chemical peptizers.
2. They soften under mechanical shear and may therefore need less plasticizer to prepare compounds in a workable viscosity range.
3. They can be used with metal oxides alone; organic accelerators are not required.

By peptizing, the compounder can achieve workable compound viscosity at high filler loading without the reduction in vulcanizate hardness and strength that usually occurs when additional plasticizers are used to lower compound viscosity.

Minimum requirements for a practical compound include antioxidant, metallic oxide, filler or reinforcing agent, processing aid, and vulcanizing system. Optional ingredients may include antiozonants, retarders, extenders, plastics or resins, other elastomers, and blowing agents.

CURING SYSTEMS

Curatives for neoprene are limited by the ingenuity of the compounders, the processing and end user requirements, and economics. A comprehensive "Summary of Curing System" is tabulated elsewhere.[8]

Metal oxides are an essential part of the curing system. They regulate scorch, cure rate, and cure state; in addition, they serve as acid acceptors for trace amounts of hydrogen chloride that are released from the polymer during processing, curing, and vulcanizate aging. Combinations of magnesium oxide and zinc oxide produce the best balance of processibility, cure rate, and vulcanizate performance. With the G types, the relationship between these properties is markedly affected by the amounts of MgO and ZnO used. Similar effects are noted in the W and T types, but to a much lesser degree because of the pronounced influence of the accelerator.

For improved resistance to swell and deterioration by water, an oxide of lead may be used in place of the magnesia/zinc oxide combination. Red lead (Pb_3O_4) is the preferred oxide. Using lead oxide results in a sacrifice in colorability, tensile strength, compression set resistance, and heat resistance compared to the characteristics offered by MgO/ZnO combination. It is more expensive usually.

The oxides of other metals are either impractical to use or exert no appreciable influence on the curing characteristics. Calcium oxide is sometimes used as a desiccant to facilitate vacuum-extrusion and, with Permalux acceleration, to control scorch. Iron oxide (Fe_2O_3) and titanium dioxide (TiO_2) are often used for their tinctorial properties in colored stocks.

The grade of magnesia used in neoprene compounds is important to the processing and properties of the finished product. The grade of magnesia used has two characteristics: (1) It is precipitated (not ground) and calcined after precipitation, and (2) it is very active, having a high ratio of surface to volume.

Types of zinc oxide that have fine particle size and high surface area are preferred. This ingredient should be added late in the mix cycle to avoid scorch, with appropriate care exercised to insure good dispersion. Incomplete dispersion results in localized overcure and inferior physical properties. Because of the importance of good dispersion, of the protection of magnesia from the effect of moisture contamination, and of the economics of reduced mixing cycles, predispersed forms of the various metallic oxides are available from many suppliers.

For practical cure rate and vulcanizate properties, an organic accelerator is required in addition to metal oxides for the W and T types of neoprene. Amines, phenols, sulfenamides, thiazoles, thiurams, thioureas, guanidines, and sulfur are the common accelerators and/or curing agents. The curing mechanism for organic curing agents other than thioureas was reviewed by Kovacic.[9] Sulfur vulcanization as described in *The Neoprenes*[6] is probably similar to that of natural rubber or of copolymers containing butadiene. Thiourea accelerators, particularly ethylene thiourea, have for many years been the accelerators of choice for a broad range of applications. Characteristics of thiourea acceleration, information on the potential toxicity of ethylene thiourea, and suggested handling precautions for it, as well as alternatives to ETU, are covered in detail in the literature elsewhere.[10] Today, ETU is the single most popular accelerator.

Antioxidants.

Antioxidants are essential in all neoprene compounds for good aging vulcanizates. Those commonly used in other elastomers usually behave similarly in neoprene.

Processing Aids

Processing aids include lubricants, tackifiers, and agents for controlling viscosity and nerve. Stearic acid, microcrystalline waxes, and low-molecular-weight polyethylenes make good lubricants. Hydrogenated rosin esters and coumarone-indene resins are preferred as tackifiers. Viscosity modification may be achieved in either direction. Dithiocarbamates and guanidines peptize (soften) the G types. Increased viscosity is best obtained by blending the neoprene with a high-molecular-weight neoprene WHV. Decreased viscosity is achieved by using plasticizers or blending with the low-molecular-weight neoprene FB. Nerve may be reduced for smooth extrusion and calendering by blending with neoprene WB, employing high structure carbon black or hard clay as fillers.

Fillers

The effects of carbon blacks and mineral fillers on the processing and vulcanizate properties of neoprene are generally similar to their effects on other elastomers, with the possible exception that reinforcing fillers are of less importance. Like natural rubber and other polymers with pronounced crystallization tendencies, neoprene gum vulcanizates have high tensile strength. The need for reinforcement, therefore, is less than for such elastomers as SBR, butyl rubber, and nitrile rubber. Practical vulcanizates are obtained for the majority of uses by filling (''loading'') with the soft, relatively nonreinforcing thermal type of carbon blacks (MT and FT) or with clays and fine-particle calcium carbonate. These fillers are not only relatively inexpensive but afford extra economy because they may be added in greater amounts than more reinforcing fillers for a given hardness. If the superior physical properties ordinarily associated with higher tensile strength are required, SRF blacks are usually used. Carbon blacks of finer particle size—either furnace or channel black—are used as a rule only when the ultimate in certain properties such as abrasion and tear resistance is required.

Of the fillers other than carbon black, clays are usually preferred for good processing and overall vulcanizate quality. Whitings, especially coarse ones, provide poor vulcanizates that weather badly. Whitings of finer particle size, however, are used to provide vulcanizates with superior resistance to heat aging. Hydrated calcium silicate and precipitated silicon-dioxide fillers of fine particle size give neoprene vulcanizates with remarkably high tensile strength, hot tear resistance, and abrasion resistance, but with impaired dynamic properties.

Plasticizers

The plasticizers and softeners most often used in neoprene are low-cost petroleum derivatives. Amounts ranging from 10 to 20% by weight of the filler loading are usually required for processing reasons alone. The W types require more than the G types. When the total amount of petroleum plasticizer fails to exceed 20 to 25 parts per 100 of neoprene, naphthenic oils may be used. These have the advantage over aromatic oils of not darkening light-colored vulcanizates or staining contacting surfaces. When plasticizing oils are used in greater amounts in order to produce very soft vulcanizates or to accommodate the high amounts of filler used in low-cost compounds, aromatic oils are recommended to insure compatibility.

Petroleum plasticizers seldom improve the flexibility of a vulcanizate at low temperature. Dioctyl sebacate is excellent for this purpose, although many other organic chemicals, mostly high-molecular-weight esters, are widely used.

PROCESSING DRY NEOPRENES

The methods for processing neoprene reflect the principles common to the processing of most elastomers, but some peculiarities bear mention.

Mixing

With few exceptions, neoprene compounds should be mixed at as low a temperature and with as short a cycle as possible. This handling minimizes the danger of scorch and is especially important with the G types, which are likely to give soft and sticky stocks on over-mastication. Omission of both zinc oxide and accelerators until late in the mixing cycle is also necessary to avoid scorching. Magnesia should be added early, if possible, with the neoprene.

Calendering and Extrusion

Calendering of neoprene stocks demands more critical control than other processing operations, particularly of the G types, which are very temperature sensitive and perform best below 70°C. The W types are usually calendered with slightly hotter rolls because they release more easily and require a higher temperature to avoid shrinkage. The three states of neoprene (elastic, granular, and plastic) are most apparent in calendering, especially with stocks of low loading. In the elastic phase, the sheet will shrink and become rough on release but may be fairly smooth on the roll. The granular phase is self-descriptive, and the sheet is relatively free from nerve. In the plastic phase, it is glossy smooth, weak, and nerve-free. The temperature at which phase changes occur varies with the type of neoprene, the phase change from elastic to granular ranging from 60°C in the case of type WRT to 70°C for type GN. The plastic phase exists at about 93°C for all neoprenes.

For frictioning, the plasticizable, crystallization-resistant neoprene GRT is recommended. Best results are obtained using prewarmed fabric, a top roll temperature of 93°C or above, a center roll at 65°C to 82°C, and a bottom roll at room temperature to 65°C, depending on the weave.

Most neoprene compounds extrude best with a cool barrel and screw, a warm head, and a hot die.

These are the major variations in the processing of neoprene that differ from the normal routines followed with other elastomers. Detailed guidelines are presented in supplier literature.[11-15]

APPLICATIONS

Why and where is neoprene in use? This multipurpose synthetic rubber has over half a century of proven performance in thousands of applications. Some of the most important are in adhesives, the transportation industry, the energy industry, the construction industry, wire and cable, hose, belting, and consumer products.

Adhesives

Neoprene is prominent among elastomers for adhesives because of its combination of polarity and crystallinity. The polarity gives a greater versatility in bonding a wide range of substrates, and the crystallinity gives improved strength. There are hundreds of different kinds of neoprene-based adhesives available for use by the fabricator of finished goods. Manufacturers of shoes, aircraft, automobiles, furniture, building products, and industrial components rate neoprene adhesives as the most versatile of materials for joining.

Neoprene-based adhesives are available both in fluid and dry-film form. The fluid types are classed as (1) solvent, (2) latex, or (3) 100%-solids compositions.

Solvent adhesives consist of the neoprene polymer and suitable compounding ingredients dissolved in an organic solvent or combination of solvents (toluene, ethyl acetate, naphtha, methyl ethyl ketone). *Latex adhesives* are composed of particles of neoprene and compounding ingredients dispersed in water. Both solvent and latex types are normally supplied in ready-to-use "one-part" adhesive systems. (Included here are contact-bond adhesives—the "workhorses" of industry). When specified, "two-part" adhesive systems can also be produced to meet specific end-use requirements. Solvent and latex neoprene adhesives achieve their bonds through evaporation of the fluid and subsequent *crystallization and curing* of the elastomeric residue.

Solid adhesives, on the other hand, are based on fluid neoprene polymers and contain neither solvent nor water. Adhesives of this type are normally the choice in specialty applications where fluidity is required, yet where volatile loss or shrinkage cannot be tolerated. They possess little "green" strength and usually require application of either heat or catalytic agent to develop an adhesive bond.

Dry-film adhesives based on neoprene are 100%-solids materials normally supplied in the form of a tape or sheeting. This type of adhesive is compounded so that it softens when heated. Cooling then resolidifies the material and forms the adhesive bond.

Neoprene adhesives are made in a range of consistencies, i.e., from very thin liquids through heavy-bodied "putties" to dry films. These adhesives can be controlled to offer a wide range of properties, for example, tack, "open tack time," storage stability, bond-development, flexibility and strength, heat and cold resistance, and all of the basic characteristics of polychloroprene polymers. The most significant ingredient in a neoprene adhesive, apart from the polymer, is the resin. Various types of resin are used in order to enhance specific adhesion, improve cohesive strength and hot bond strength, or to impart better tack retention. Resins are usually compounded at a level of 25 to 60 phr and are rarely added above 100 phr. The most commonly used resin types are alkyl phenolic resins, terpene phenolic resins, hydrogenated resin and rosin esters, coumarone indene resins, and hydrocarbon resins such as poly(x-methyl styrene). Specific resin recommendations and compounding for adhesives are found in the supplier literature.[16]

Transportation

In the automotive field, neoprene is the base elastomer for a variety of components such as V-belts; timing belts; blown sponge gaskets for door, deck, and trunk; spark plug boots; power brake bellows; radiator hoses; steering and suspension joint seals; tire sidewalls; ignition wire jackets; and many other items. In aviation, neoprene is used in mountings, wire and cable jackets, gaskets, seals, de-icers, etc. In railroading, it is used in track mountings, car body mountings, air brake hoses, flexible car connectors, freight-car interior linings, journal box lubricators, and so forth. All of these uses require physical strength, compression set, resilience, fluid, weather, and temperature degradation resistance over a wide range of destructive service conditions for many years.

Energy Industry

A durable oil-resistant neoprene rubber has a multitude of uses in the exploration, production, and distribution of petroleum, such as packers, pipeline "pigs," seals, gaskets, hose, coated fabrics, wire and cable, and so forth.

Construction Industry

This application arena, while not new, is now gaining in popularity. Outstanding items are extruded window-wall sealing gaskets, calendered sheeting for membrane waterproofing, flashing and roof covering, highway joint seals, bridge mounting pads, soil pipe gaskets, sound and vibration isolation pads, etc. Neoprene-modified asphalt is gaining acceptance in roads, parking lots, and airport runways. Proven serviceability and economic acceptability account for this growth.

Wire and Cable

Practically the oldest use of neoprene is as a jacket for electrical conductors (low and high voltages). Its superior resistance to abrasion, oil, flame, weather, and aging over natural rubber accounts for its broad acceptance. Billions of feet of neoprene-covered telephone dropwire have been installed in the United States alone since 1946. In addition, billions of more feet of neoprene-jacketed appliance cord, industrial power and control cable, illumination, and miscellaneous secondary cable have been placed in service in the air, underground, and underwater over the past fifty-odd years.

Hose

Neoprene hose has been made almost continuously since 1932. Polychloroprene is used in the cover, cushion, and tube compounds. All types of hose are involved.

Belts

Maximum durability in service climates requiring a range of temperature and flex resistance makes neoprene the polymer of choice for belting. Applications include V-belts, power-transmission belts, conveyor belts, and escalator hand rails.

Consumer Products

Sponge shoe soles made of closed-cell neoprene combine cushioned comfort with long wear. Comfortable mattresses made of neoprene foam have served aboard ships for over 30 years. Neoprene latex foam has played a critical role in mattress and cushioning fire safety, offering an added margin of improved flame-protection and low smoke for institutional furnishings.

NEOPRENE LATEX

Neoprene latices are aqueous, colloidal dispersions of polychloroprene or of copolymers of chloroprene and other monomers such as methacrylic acid or 2,3-dichloro-1,3-butadiene. They are available in both anionic and nonionic surfactant systems. The theory, compounding, processing, and applications of these latices have been described in detail by J. C. Carl[17] in 1962 and P. R. Johnson[18] in 1976. Table 12.5 is a listing of types available from various manufacturers.

A variety of articles, particularly those that are relatively thin or of complicated shape, are more easily made from latex than from dry neoprene. An example would be neoprene household and industrial gloves, which are readily made from latex. Also, the production of neoprene balloons from dry neoprene would be difficult. Other major applications include latex-based adhesives, protective coatings, binder for cellulose and other fibers, and elasticizing additive for concrete, mortar, and asphalt.

As with dry neoprene, the neoprene latex must be compounded with other materials to be converted into useful products.[18,19] Some of these materials are mandatory for optimum performance, whereas others are optional. In some cases, a more expensive type of neoprene latex may actually provide greater value-in-use to the user because it can accommodate more additives or because its inherent strength eliminates the need for curing.

End-use performance, processing, and economics usually dictate the choice of neoprene latex type. Included in processing are the cost of storage and handling, coagulation and drying rates, curing time, and production yield. Usually cost can be minimized by maintaining low compound viscosity at the highest practical solids content.

In view of those factors, the user should determine whether the lowest priced, high-solids latices (latex 654 or latex 671A) will provide the necessary performance. Moving from either of these into any other neoprene latex type will usually result in higher in-use costs. These higher costs are justified only by processing and performance requirements that are unsatisfied by 654 or 671A types. The characteristics of many neoprene latices are detailed in Table 12.6.

COMPOUNDING NEOPRENE LATICES

All neoprene latex compounds must contain a metal oxide and an effective antioxidant. Metal oxides function in several ways: They serve to vulcanize the polymer; they make it more resistant to aging, heat, and weathering; and they act as acid acceptors. Zinc oxide has proven to be the most desirable of these oxides from both processing and film-property standpoints. The presence of a good antioxidant in neoprene is as important as a metal oxide. Both are necessary to bring out the inherently good aging properties of the neoprene and, as

Table 12.5. Types of Neoprene Latex.

Du Pont Neoprene	*Showa Neoprene KK*	*Denka Chemical Corporation Neoprene*	*Bayer Baypren Latex*	*Denki Kagaku Denka*	*Distugil Butaclor*	*Toya Soda Skyprene*
60						
115						
400	400					
571						
622						
654						
671A						
735A						
750						
842A	842A			LM-50		LA502
–	650					
–	601A			LM-60		
–	572			LA-50		
–	950			LK-50		
–	635					
–	736					
–	–	L300				
–	–	L345				
–	–	L360				
			B			
			T			
			MKB			
			SK			
			GK			
			4R			
					L540	
					L632	

a general rule, it is recommended that neoprene latex compounds contain at least 5 phr of zinc oxide and 2 phr of an antioxidant.

Accelerators are used to increase the rate of cure and to enhance the physical properties of neoprene. The addition of thiocarbanilide to neoprene latex produces polymers with high modulus. Where tensile strength is of primary importance, the use of both tetraethylthiuram disulfide and a water solution of sodium dibutyldithiocarbamate is recommended. This combination imparts outstandingly high tensile strength with little increase in modulus. Other accelerators commonly used in neoprene latexes are zinc dibutyldithiocarbamate and di-o-tolylguanidine.

Mineral oils and light process oils are used as plasticizers and to improve the "hand" in neoprene latex. Petroleum-based plasticizers have been found most effective as crystallization inhibitors, and ester-type plasticizers are employed to improve the low-temperature serviceability of the product. Three to 10 phr are usually used.

Fillers such as clay, whiting, titanium dioxide, carbon black, hydrated alumina, and fine silicas can be used to impart specific properties to the neoprene or to act as low-cost diluents. Loadings vary from as little as 10 phr in dipped goods to several hundred parts in adhesives and coatings.

Anionic, nonionic, and amphoteric surfactants are used as both shear and chemical stabilizers in the preparation of neoprene latex compounds. Usually 0.1 to 1 phr is sufficient in most applications to produce the necessary processing stability.

Thickeners are more often needed with neoprene latex compounds than with natural-rubber latex because the viscosity of neoprene latex at a given solids content is less than that of natural rubber latex. Both natural and synthetic gums are excellent thickeners for neoprene latex. If it is necessary to decrease the viscosity of neoprene latex compositions, they can be diluted with deionized or distilled water.

PROCESSING NEOPRENE LATICES

The manufacture of articles from both neoprene and natural rubber latex is similar, and the processes used with natural rubber are adaptable in most instances for use with neoprene. Exceptions exist, however, the most important being that of cure. Neoprene has the advantage of not precuring or aftercuring, but in order to obtain the optimum physical properties, neoprene films must be cured at temperatures above 100°C and preferably in the range of 120 to 140°C.

The general techniques suggested for good rubber latex compounding procedures also hold for neoprene latex. Fine-particle-size dispersions and emulsions, compounding pH above 10.5, and the use of deionized or distilled water in preparing all ingredients are all recommended for processing uniformity.

Table 12.6. Comparison Chart of the Neoprene Latices.

Neoprene Latex Type	*Co-Monomer*	*Emulsifier*	*Emulsifier Class*	*Chlorine Content, %*	*pH at 25°C [77°F] (Typical Values)*	*Standard Solids, %*	*Distinguishing Features*	*Major Applications*
60	Sulfur	Sodium salts of disproportionated rosin acids	Anionic	37.5	11.0	59	Develops high wet gel strength	Foam
115	Methacrylic acid	Poly(vinyl alcohol)	Nonionic	36	7	47	Carboxylated polymer with outstanding mechanical and chemical stability	Adhesives
400	2,3-dichloro-1,3 butadiene	Potassium salts of disproportionated rosin acids	Anionic	48	12.5	50	Maximum chlorine content; rapid crystallizing; outstanding ozone and weather resistance	Coatings
571	Sulfur	Principally sodium salts of rosin acids	Anionic	37.5	12.0	50	High-strength cured films combined with low permanent set	Adhesives Coatings Dipped goods
622	–	Potassium salts of disproportionated rosin acids	Anionic	38	12.5	61	High solids, low viscosity, high gel polymer	Foam
654	–	Potassium salts of disproportionated rosin acids	Anionic	38	12.0	59	High solids, low viscosity, low gel polymer	Adhesives Binders Coatings Foam

Neoprene Latex Type	*Co-Monomer*	*Emulsifier*	*Emulsifier Class*	*Chlorine Content, %*	*pH at 25°C [77°F] (Typical Values)*	*Standard Solids, %*	*Distinguishing Features*	*Major Applications*
671A	–	Potassium salts of disproportionated rosin acids	Anionic	38	12.5	59	High solids, low viscosity, medium-high gel polymer	Adhesives Binders Coatings Dipped goods Elasticized asphalt Elasticized concrete Foam
735A	–	Sodium salts of disproportionated rosin acids	Anionic	38.5	12.0	45	Designed for wet-end addition to fibrous slurries	Binders Elasticized asphalt
750	2,3-dichloro-1,3 butadiene	Potassium salts of disproportionated rosin acids	Anionic	40	12.5	50	High wet gel strength; low-modulus films having a very slow crystallization rate	Adhesives Dipped goods
842A	–	Principally sodium salts of rosin acids	Anionic	37.5	12.0	50	Medium-strength cured films having a slow crystallization rate	Adhesives Binders Coatings Dipped goods

PRODUCT SAFETY

Over the 50 years plus of the commercial existence of neoprene, the Du Pont Company has not been aware of any unusual health hazards associated with the solid polymers of neoprene except AH and FB, which contain chemicals that may cause irritation and allergic skin reaction. For all the solid polymers, however, routine industrial hygiene practices are recommended during handling and processing to avoid such conditions as dust build-up or static charges.

Neoprene latices similarly have a very low order of oral toxicity. Since most are strongly alkaline, however, they may cause burns if they come in contact with eyes or skin. Volatile organic materials in neoprene latices include chloroprene monomer, toluene, and butadiene.

Compounding ingredients used with neoprene to prepare finished products may present hazards in handling and use. Before proceeding with any compounding work, consult and follow label directions and handling precautions from suppliers of all ingredients.

For additional detailed information on handling precautions and first aid treatment, consult the referenced literature.[21-23] Major neoprene types are on the approved list of the FDA (Food and Drug Administration of the U.S. Department of Health, Education, and Welfare).[24]

Du Pont is familiar with the inventory-reporting provisions of the Toxic Substances Control Act and the inventory-reporting regulations of the U.S. Environmental Protection Agency. It has reported, and will continue to report, all reportable chemical substances that Du Pont currently manufactures or may manufacture in the future.

REFERENCES

1. J. A. Niewland, W. S. Calcott, F. B. Downing, and A. S. Carter, ''Acetylene Polymers and Their Derivatives, I—The Controlled Polymerization of Acetylene.'' *J. Am. Chem. Soc.* **53,** 4197 (1931).
2. W. H. Carothers, J. Williams, A. M. Collins, and J. E. Kirby, ''Acetylene Polymers and Their Derivatives, II—A New Synthetic Rubber: Chloroprene and Its Polymers.'' *J. Am. Chem. Soc.* **53,** 4203 (1931).
3. A. M. Collins, ''The Discovery of Polychloroprene.'' *Rubber Chem.* **46,** G45–G52 (1973).
4. C. A. Stewart, Jr., T. Takeshita, and M. L. Coleman, ''2-Chloro-butadiene Polymers.'' *Encyclopedia of Polymer Science and Technology*, John Wiley & Sons, New York, 1985, vol. 4.
5. ''The Language of Rubber,'' Du Pont Report, Jan. 1963. Revised 1986.
6. R. M. Murrary and D. C. Thompson, *The Neoprenes*, Du Pont, 1963.
7. ''First and Second Order Transitions in Neoprene,'' Du Pont Report BL-373, October, 1961.
8. ''Summary of Curing Systems for Neoprene,'' Du Pont Report NP-330.1 (R1).
9. P. Kovacic. *Ind. Eng. Chem.* **47,** 1090 (1955).
10. ''NA-22F, An Accelerator for Polychloroprene Rubber,'' Du Pont Report NP-730 NA22F.
11. S. W. Schmitt and G. Newton, ''Mixing of Neoprene,'' Du Pont Report NP-420.1.
12. S. W. Schmitt, ''Extrusion of Neoprene,'' Du Pont Report NP-430.1.

13. S. W. Schmitt, ''Calendering Compounds of Neoprene,'' Du Pont Report NP-440.1.
14. S. W. Schmitt, ''Molding Neoprene,'' Du Pont Report NP-450.1.
15. S. W. Schmitt, ''Vulcanization Methods For Neoprene,'' Du Pont Report NP-460.1.
16. D. G. Coe, ''Neoprene Solvent-Based Adhesives,'' Du Pont Report -100.1 (R1).
17. J. C. Carl, *Neoprene Latex*, Du Pont, 1962.
18. P. R. Johnson, ''Polychloroprene Rubber,'' *Rubber Chem. Technol.* **49,** no. 3, 1976.
19. C. H. Gelbert, ''Compounding Neoprene Latex for Colloidal Properties,'' Du Pont Report NL-200.1.
20. C. H. Gelbert, ''Basic Compounding of Neoprene Latex,'' Du Pont Report NL-310.1.
21. ''Toxicity and Handling Guidelines for Neoprene,'' Du Pont Report NP-110.1 (Nov. 1985).
22. ''Toxicity and Handling Guidelines for Neoprene Latexes,'' Du Pont Report NL-110.1 (Nov. 1985).
23. ''Transfer and Storage of Neoprene Latexes,'' Du Pont Report NL-120.1.
24. ''FDA Status of Du Pont Neoprene Solid Polymers and Latexes,'' Du Pont Report NP-120.1 (Mar. 1984).

Part II

HYPALON®

GERALD A. BASEDEN
E. I. du Pont de Nemours & Company
Polymer Products Department
Wilmington, Delaware

HISTORICAL BACKGROUND

The development of HYPALON was the result of a long-term program at Du Pont that originated in the wartime need of the U.S. Government for new synthetic rubbers. This program was aimed at deriving a vulcanizable rubber from polyethylene, partly because such a rubber might reasonably be expected to exhibit many of the chemical and electrical characteristics of polyethylene, and partly because it was recognized that polyethylene was likely to continue to be a low-cost starting material for the foreseeable future.

The initial outcome of the work was the discovery that a rubbery product could be obtained by chlorinating polyethylene in solution. Although this rubber was found to be vulcanizable with peroxides, the curing process was difficult because the vulcanizing peroxides now used in the rubber industry were not then available. To give a CPE-like elastomer with improved curing characteristics, Du Pont developed a modification of the solution chlorination process that permitted the simultaneous chlorination and chlorosulfonation of polyethylene. The product, chlorosulfonated polyethylene, was an elastomer, similar to CPE, but now readily curable with sulfur-containing curatives.

The two products, chlorinated polyethylene and chlorosulfonated polyethylene, were both commercialized in 1951 as HYPALON S-1 and HYPALON S-2, respectively. HYPALON S-1 was soon withdrawn in favor of HYPALON S-2[25,26] which was subsequently renamed HYPALON 20. The latter is still commercially available and has proven to be the first of a family of chlorosulfonated polyethylene synthetic rubbers.

DESCRIPTION

HYPALON synthetic rubber is made at plants in Beaumont, Texas, and at Maydown in North Ireland. The process for its manufacture involves the simultaneous chlorination and chlorosulfonation of polyethylene in solution. HYPALON, as a chlorosulfonated polyethylene, is described as a CSM rubber, according to ASTM D1418. A simplified form of its chemical structure is shown:[27]

$$-CH_2-\underset{\displaystyle Cl}{\underset{|}{CH}}-CH_2-CH_2-CH_2-\underset{\displaystyle SO_2Cl}{\underset{|}{CH}}-CH_2-$$

The rubberiness of HYPALON is derived from the natural flexibility of the polyethylene chain in the absence of crystallinity; the introduction of the chlorine atoms along the polyethylene chains provides sufficient molecular irregularity to prevent crystallization in the relaxed state.

The crystallinity of the polyethylene chain can also be eliminated by an alternative method of introducing molecular irregularity: by copolymerizing the ethylene with propylene. The process is used in the manufacture of EPDM polyolefin elastomers.

Gum vulcanizates of noncrystalline grades of EPDM and CSM are compared in Table 12.7. CSM vulcanizates are seen to have better oil and flame resistance; and, as indicated by the substantial difference in their tensile strengths, better all-around mechanical toughness. The chlorine atoms on the polyethylene backbone not only provide elastomeric properties, but also give useful improvement in oil resistance and flame resistance. The sulfonyl chloride groups provide crosslinking sites for the nonperoxide curing processes.

The sulfur content of most grades of HYPALON is maintained at 1% by weight on the polymer. The chlorine content varies between the different grades, as shown in Table 12.8, and has a significant effect on the oil resistance of the polymer.

Many other properties of compounds and vulcanizates of HYPALON are affected by chlorine content, besides oil resistance. At low-chlorine levels, the

Table 12.7. S-cured Elastomers with a Saturated Hydrocarbon Backbone.

	EPDM	*CSM*
O_2, O_3 resistance	Excellent	Excellent
Color stability	Excellent	Excellent
Gum strength, psi	500	5000
Oil resistance	Poor	Fair–good
Heat resistance	Very good	Very good
Flame resistance	Poor	Good
Low-temperature properties	Excellent	Fair–excellent

Table 12.8. Typical Raw-polymer Properties of Available Grades of HYPALON

Grade	*Chlorine Content, %*	*Sulfur Content, %*	*Mooney Viscosity ML(1 + 4) at 100°C*	*Specific Gravity 25/4°C*
20	29	1.4	28	1.14
30	43	1.1	30	1.27
LD-999	35	1.0	30	1.18
40S	35	1.0	45	1.18
40	35	1.0	56	1.18
4085	36	1.0	97	1.20
623	24	1.0	21	1.08
45	24	1.0	37	1.08
48S	43	1.0	62	1.26
48	43	1.0	78	1.26

polymers retain some polyethylene-like characteristics: they are harder and stiffer because they are partially crystalline; they also show good electrical properties, good heat resistance, and good low-temperature flexibility. With increasing chlorine, the compounds and vulcanizates get increasingly rubbery but then at higher chlorine levels again become stiffer, because of increasing glass transition temperature. The high-chlorine grades are characterized by their excellent oil resistance and good flame resistance.

AVAILABLE GRADES

The raw-polymer properties of the available grades of HYPALON are shown in Table 12.8. All grades are produced in the form of irregular creamy-white chips with excellent storage stability and no odor. All grades show certain properties that are characteristic of CSM elastomers in general: their vulcanizates are tough, colorable, weather-resistant, and show excellent resistance to attack by oxygen and ozone.

HYPALON 20, formerly used for the manufacture of molded and extruded goods, is now used mainly in coatings applications to give curable rubbery films with good low-temperature flexibility. Its principal coatings applications are for roof coatings and the manufacture of tarpaulins, but it is also used in caulks and adhesives.

HYPALON 30 is also used mostly for coatings. It gives solutions of much lower viscosity than HYPALON 20. Its higher chlorine content gives its coatings a dryer, slicker feel than ones from HYPALON 20, and a better oil and flame resistance. Low-temperature properties are not as good as those for HYPALON 20, however.

LD-999 can be considered a low-viscosity analog of HYPALON 40 and is used mostly in the extrusion of wire and cable jackets. It is also used as a blending elastomer to reduce the viscosity of hose compounds from HYPALON 48 and 48S.

HYPALON 40 and 40S are 35% chlorine, medium-viscosity elastomers having good heat and oil resistance. Compounds from these grades have good extrusion characteristics. Major applications are in hose construction, wire and cable coverings, industrial rolls, and automotive spark-plug boots.

HYPALON 4085 is similar to HYPALON 40, but is useful where higher green strength or better vulcanizate properties are required.

HYPALON 45 and its low-viscosity analog *HYPALON 623* are semicrystalline polymers that have useful physical properties in the uncured state. Their main uses are in pond liners and in roofing membranes. A smaller but growing application is in vulcanized wire covering with very good electrical properties, excellent heat-aging resistance, and good low-temperature flexibility.

HYPALON 48 and its lower-viscosity analog *HYPALON 48S* have a high-chlorine content and show excellent oil resistance, including a very low permeability to FREON® refrigerants. Their principal use is in the manufacture of automotive air-conditioner hose. HYPALON 48S is also used as a coatings polymer; it gives a higher viscosity solution than HYPALON 30, but it has better resistance to low-temperature brittleness while providing the same low-friction property to the dry film.

COMPOUNDING INGREDIENTS

A typical vulcanizable compound is made up of the components listed in Table 12.9. Components of the curing system are considered to be both the acid acceptor and its activator as well as the vulcanizing agent and its accelerator. These will be described first because the curing of HYPALON involves substantially different chemistry from other elastomers.[28–32]

Table 12.9. A Typical CSM Compound.

Ingredient	*Function*	*phr*
1. HYPALON 40 (35% Cl)	Polymer	100
2. Magnesia	Acid acceptor	4
3. Pentaerythritol	Activator for 2	3
4. Hard clay	Filler	80
5. Aromatic oil	Plasticizer	25
6. Paraffin wax	Process aid	3
7. Sulfur	Vulcanizing agent	1
8. Tetramethylthiuram disulfide	Accelerator for 7	2

Acid Acceptor/Activator

The choice of acid acceptor will depend on the vulcanizing agent used and the functional requirements of the vulcanizate. The main functions of an acid acceptor are to act as a heat stabilizer, to absorb acid by-products of the curing reaction, and to maintain sufficient alkalinity to allow effective curing reactions to proceed. Note that zinc oxide is excluded from this list because its reaction product, zinc chloride, causes polymer degradation on heat-aging or natural weathering. For the same reason, other zinc-containing additives should also be excluded.

The activator increases the effectiveness of the acid acceptor apparently by solubilizing it in the polymer. Adding an activator allows a substantial reduction in the amount of acid acceptor needed and this, in addition to a cost saving, leads to lower-viscosity, safer processing stocks.

Fillers

Carbon black is the preferred filler for CSM vulcanizates because it gives best reinforcement of physical properties and best resistance to chemical degradation, to compression set, and to water absorption. SRF carbon black gives a good balance of properties and is widely used as a general-purpose filler. Weathering of CSM with as little as three parts of carbon black as a protective pigment is outstanding.[33]

Mineral fillers are often used to take advantage of CSM's nondiscoloring characteristics. Best heat resistance is obtained with whiting or blanc fixe; best electrical properties and very good water resistance are obtained with calcined clay; and hydrated silicas and alumina are preferred for improved flammability performance. When compounded with suitable protective pigments, nonblack CSM compositions show excellent long-term weathering performance.[33]

Plasticizers

Aromatic petroleum oils are widely used as plasticizers primarily because of their low cost. Ester plasticizers give compounds that are light in color, with good low-temperature properties, while liquid chlorinated paraffins give vulcanizates with good flammability performance and weatherability. Polymeric plasticizers are less volatile than the other types and are preferred for heat-resistant compounds.

Processing Aids

Stearic acid and stearates are effective release agents but should not be used in litharge-containing compounds in which they reduce scorch safety. Zinc stear-

Table 12.10. The Ionic Curing Process.

HYPALON 45	100
Magnesia	5
Carbon black	100
Process aids	1.6
Stress/Strain Properties of 30 mil (0.76mm) Calendered Sheeting	
Original	
100% Modulus, MPa (psi)	5.0 (725)
Tensile strength, MPa (psi)	7.6 (1100)
Elongation at break, %	260
Aged 250 hrs. Weather–Ometer	
100% modulus, MPa (psi)	7.6 (1100)
Tensile strength, MPa (psi)	12.4 (1800)
Elongation at break, %	310

ate should be avoided in compounds designed for heat resistance or weatherability. Stearic acid should also be avoided in maleimide cures because of cure retardation.

Paraffin wax, poly(ethylene glycol) and low-molecular-weight polyethylene, separately or in combination, are all effective process aids for CSM that do not affect scorch safety in litharge-containing compounds, or the effectiveness of maleimide cures.

Curatives

Two types of curing processes are available with CSM elastomers and both are effective regardless of the level of chlorine in the elastomer.

Ionic Cures. Ionic cures of CSM are possible when the acid acceptor is a divalent metal oxide. Ionic cures are characterized by their ability to proceed at low temperatures, by the fact that they are accelerated by moisture, and by the development of a high modulus. Ionic crosslinking is responsible for the limited bin stability of CSM compounds when they contain divalent metal oxide acid acceptors and are stored under humid conditions.

Table 12.10 shows a compound in which significant ionic cure of the HYPALON by the magnesia has taken place as a result of exposure to the warmth and humidity of a Weather-Ometer. This ability of HYPALON to cure under mild ambient conditions is useful for both pond liners and roofing membranes; it allows a sheeting that is readily bondable to itself, by means of solvent- or heat-welding, to be put in place and then to slowly cure as a result of exposure to ambient conditions.

Covalent Cures. CSM is also capable of undergoing covalent crosslinking, and most applications for CSM make use of this type of curing. Three different systems are commonly used: sulfur cures, which give the widest choice of compounding ingredients; peroxide cures, which are nondiscoloring and give good heat and compression-set resistance; and maleimide cures, which also give good heat and compression-set resistance, but which are less sensitive to the type of compounding ingredient. Covalent curing systems rely on a crosslinking agent and one or more accelerators for their curing action.

Sulfur Cures. A recipe for a typical sulfur cure is shown in Table 12.9. Sulfur-based cures are very versatile and are practical for all types of molding and extrusion processes. They allow high states of cure to be achieved even in low-cost, highly extended compounds.

Vulcanizates with properties of both covalent and ionic crosslinking result from the low-temperature moisture curing of a compound such as shown in Table 12.11. Practical cures can also be achieved in hot water. Although the cure rate is slow, it is very convenient since it allows wire insulation to cure while the wire is on the reel. Typically, insulation is cured in this way by standing the reels in a steam room or immersing them in a tank of hot water. For a comparison, it would take about 20 hours in a press at 100°C to achieve the state of cure obtained with 2 hours in steam at 100°C.

Peroxide Cures. CSM is readily cured with a peroxide. Frequently, a cure promotor such as triallyl cyanurate is added to improve the effectiveness of the peroxide. An acid acceptor must also be added. Compounds to be cured with peroxides can contain low levels (e.g., up to 40 phr) of fully-saturated plasticizers; higher levels will cause progressive cure retardation. Chlorinated paraf-

Table 12.11. Low-temperature Moisture Cure of CSM.

HYPALON 40	100
Magnesia	10
Pentaerythritol	3
Mineral fillers	100
Plasticizers	35
TETRONE® A	2

Stress/Strain Properties[a] of 0.64mm (.025 in) Slabs

Cured in Steam, 2 hrs/100°C (212°F)

100% modulus	MPa (psi)	8.2 (1190)
Tensile strength	MPa (psi)	16.7 (2420)
Elongation at break	%	400

[a]Samples tested after drying in a desiccator.

fins are best; and ester plasticizers are less satisfactory, but adequate. Aromatic process oils will drastically retard the cure.

Maleimide Cures. Du Pont's HVA-2, (N, N′-*m*-phenylenedimaleimide) is a primary curing agent for HYPALON. Maleimide cures also require the presence of calcium hydroxide as an acid acceptor and the antioxidant VANOX AT (a butyraldehyde-amine condensation product) as an accelerator. Maleimide cures are very safe and characteristically produce good water resistance and excellent resistance to compression set and to heat aging. Unlike peroxide cures, maleimide cures are relatively insensitive to type and proportion of plasticizer. Although the maleimide cure system is deactivated by moisture, it is effective in high-pressure steam; some cure retardation may be seen in low-pressure steam cures, however.

Special Additives

Nickel dibutyl dithiocarbamate (NBC), a dark-green powder, is widely used with CSM to improve the heat resistance of black or dark-colored compounds. Up to 3 parts of NBC are added in sulfur-cured compounds based on mixed metal oxides. The addition of NBC reduces scorch safety. Also, NBC is not effective in peroxide cures.

HVA-2 is an effective secondary cure accelerator for CSM that can improve both the state of cure and the electrical properties of the vulcanizate.

Other additives include low-melting thermoplastic polymers such as PVC copolymers and high styrene resins, which may be added to CSM to prevent contact-sticking and distortion of wire covering while it is being cured on the reel in steam or hot water.

MIXING

All elastomers will soften as their temperature rises while being worked on processing equipment. This softening effect is somewhat greater for CSM than for most other elastomers. However, it is entirely thermal and reversible so that a CSM stock that is warmed up on a mill and then allowed to cool back to room temperature will regain its former stiffness. This means that CSM compounds can be reworked without significant change in processing behavior, as long as scorch is not a factor.

Mill-mixing

CSM compounds are easily mixed on conventional mills. Because of longer heat exposure, mill-mixed compounds are likely to be scorchier than those mixed in internal mixers. Cold water should be used on the mill except when processing the more thermoplastic grades, such as HYPALON 45 and 623, which require a mill temperature of around 70°C. A typical mill-mixing time for a normal CSM batch would be about 25 minutes, depending on the type and amount of fillers and plasticizers used.

Internal Mixing

Internal mixing is the most cost-efficient volume system for producing well-mixed CSM compounds. Short mix cycles are preferable due to the need to minimize the heat history of the batch.

The preferred mix procedures generally use upside-down mixing without pre-mastication of the polymer, and as high a ram pressure as practical. Load factors varying from 0.55 to 0.80 are normal, with the higher load factors being used for the lower-viscosity stocks. Typical ram-down times will be $1\frac{1}{2}$ to $3\frac{1}{2}$ minutes.

Production compounds can generally be made on a single-pass mix cycle, but if processing safety is marginal, a two-pass mix procedure may be needed. Dump temperature should be around 121°C (250°F) if no sulfur curatives are included, and around 107°C (225°F) when sulfur curatives are added.

PROCESSING

Molding

CSM compounds are successfully processed by compression, transfer, or injection molding. In these processes, where reproducibility of flow of the compound is important, steps should be taken to compound for good bin stability, to mix so as to minimize scorch, to store the mixed compound under cool, dry conditions, and to rotate the inventory so that compounds to be molded have a similar storage history.

Due to the marked thermoplasticity of CSM compounds, multicavity transfer and injection molds require carefully balanced runner systems and similar flow restrictions for the gates, so that uniform filling of all cavities is achieved.

Extrusion

CSM compounds can be processed in standard hot- or cold-feed rubber extruders. The compounds are normally fed in strip form. In general, CSM compounds require little work to attain physical uniformity and very-low-compression-ratio screws are usually adequate. Die L/D ratios of 1/1 are common, with stock temperatures of around 120°C at the die.

Extruded parts can be cured in zero-pressure steam, hot-air ovens, LCM, fluidized bed, CV, and by microwave. Curing with ionizing radiation is also effective for CSM.

Calendering

Substantial amounts of CSM are used in calendering operations. Compounds from the intermediate chlorine grades, for example, HYPALON 40, are calendered at 60–100°C (140–212°F) onto a fabric with, normally, uneven roll speeds on the calender and usually with the top roll slightly hotter than the middle roll. The maximum practical gauge for these sheets is about 40 mils (1.02 mm) for a single pass. The calendered sheets are normally cured in an autoclave while wrapped on a drum with a curing liner. Adhesion between the plies is usually excellent.

Compounds from the low-chlorine CSM grades, as represented by HYPALON 45 and the lower-viscosity HYPALON 623, are calendered without curatives for use in pond liners and roofing membranes. Because of their crystallinity and scorch safety, these compounds are calendered at higher temperatures, 100–150°C (212–302°F), with the easier-processing compounds from HYPALON 623 being calendered at the lower end of this temperature range.

PROPERTIES AND PERFORMANCE

CSM is classified as a specialty synthetic rubber, with a "DE" rating in the "heat-resistance/oil-resistance" classifications of ASTM D2000 and SAE J200. The applications for CSM depend only in part on its combination of oil and heat resistance. CSM is much more resistant to corrosive or oxidizing chemicals including ozone, than are neoprene and nitrile rubbers; it is tougher than silicones and EPDM rubbers; and it has electrical properties that are not as good as EDPM's but exceed those of many other elastomers. CSM has excellent radiation resistance, making it well-suited for wire and cable coverings in nuclear applications. Further, it is color-stable and highly weather-resistant.

Oil-resistance and Low-temperature Properties

The oil-resistance and low-temperature properties of CSM vulcanizates depend mainly on the chlorine content of the CSM used, as shown in Table 12.12. Low-temperature properties can additionally be controlled by the choice of plasticizer in the compound.

Heat Resistance[34]

CSM elastomers, as a class, have good-to-excellent heat resistance, depending mainly on the choice of compounding ingredients, although within this class of elastomer the low-chlorine grades tend to have slightly better heat resistance than the high-chlorine grades. The thermal limits of typical CSM vulcanizates are shown in Figure 12.1.

Table 12.12. Effect of Chlorine Content on Low-temperature Properties and Oil Resistance.

	% Chlorine	*Glass-transition Temperature, °C*	*% Swell*[a]	*Brittleness Temperature, °C*[b]
HYPALON 45	24	−30	86	−65
HYPALON 40	36	−20	38	−60
HYPALON 48	43	0	13	−14

[a]Typical vulcanizate, aged 70 hrs at 121°C in ASTM Oil No. 3.
[b]Same vulcanizate, tested by ASTM D746.

Flame Resistance

Best flame resistance is obtained with the high-chlorine grades of CSM. Chlorinated plasticizers, brominated resins, and antimony trioxide synergist are commonly used additives. Mineral fillers, excluding carbonates, are preferred. Hydrated alumina is frequently used as the major filler where smoke suppression is important.

Water Resistance

The use of magnesia acid acceptor is the major cause of water sensitivity in CSM vulcanizates. Replacing magnesia with litharge in black stocks with a

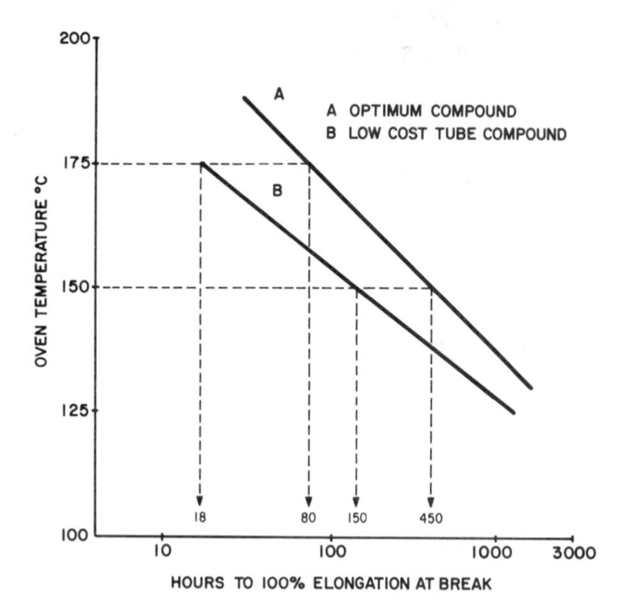

Fig. 12.1. The thermal limits for practical performance of CSM vulcanizates.

sulfur cure, or with dibasic lead phthalate in all other systems, will result in improved water resistance. Where the presence of lead is undesirable, synthetic hydrotalcite will provide significantly better water resistance than magnesia. Water resistance is also affected by the choice of filler: carbon black is the best, followed by nonblack fillers with calcined clay preferred.

Weather Resistance[33]

When suitably compounded with protective pigments, CSM vulcanizates and the unvulcanized crystalline grades are highly weather-resistant; they are also colorable with good color stability. CSM compounds, provided they are free of nutritive compounding ingredients, are also resistant to mildew growth.

Oxygen/Ozone Resistance

CSM vulcanizates are highly resistant to attack by ozone and do not need the addition of antiozonants in their compounds. They also are very resistant to oxidation; however, the addition of an antioxidant will improve performance under severe heat-aging conditions.

Electrical Properties[35]

Best insulation resistance is obtained with the low-chlorine grades of CSM. Good electrical properties may also require formulating for good water resistance in compounds containing little or no carbon black. Compounds can be formulated so that combustion products are low in both smoke and corrosive gases.

Chemical Resistance

CSM vulcanizates are very resistant to degradation caused by exposure to corrosive chemicals. Litharge is the preferred acid acceptor for vulcanizates for generally good chemical resistance although for certain highly concentrated chemicals such as 66° Baumé sulfuric acid, or concentrated nitric acid, magnesia will give better service. The best general-purpose filler for chemical-resistant stocks is blanc fixe or barytes.

Resistance to solvents is limited by the resistance of the vulcanizates to swelling in those solvents rather than to any chemical degradation caused by the solvents. Table 12.13 shows a guide that will indicate the likely performance of vulcanizates based on HYPALON 40 in a variety of chemical media. The guide refers to room-temperature exposure; higher temperatures of exposure may cause worse deterioration.

Table 12.13. CSM Chemical Resistance Guide.

Chemical	*Rating*[a]
Acids, mineral	A
Acids, organic	A
Alcohols	A
Aldehydes	C
40% Formaldehyde	A
Alkalis, caustic	A
Aniline	B
Aromatic solvents	C
ASTM Oil No. 1	A
ASTM Oil No. 3	B
ASTM Reference Fuel A	A
ASTM Reference Fuel C	C
Asphalt	B
Chlorinated solvents	C
Esters	C
Ethers	B–C
Gasoline	B
Hydrogen peroxide	A
Inorganic salts	A
Ketones	B–C
Triethanolamine	A
Water	A

[a]For room temperature exposure, using vulcanizates based on HYPALON 40:
A—little or no effect.
B—minor to moderate effect.
C—severe effect.

Dynamic Properties

The resilience of CSM vulcanizates is generally in the 55–75% range for grades having medium- and low-chlorine content. High-chlorine grades are more highly damped and the resilience of their vulcanizates are generally too low at room temperature to measure on a Yerzley Oscillograph. The dynamic mechanical properties of HYPALON 20 and HYPALON 40 are described in greater detail in references 36 and 37.

APPLICATIONS

Industrial Uses

Two areas in which CSM has proven property advantages are in hose and in electrical applications. In the hose industry, CSM is particularly useful because of its oil and chemical resistance, its outstanding ozone and weather resistance, and because of the frequent need for bright colors.

CSM is used in the electrical industry as a protective jacketing or sheathing on top of other elastomers with superior insulating properties. CSM also finds considerable usage as an integral or "one-shot" insulation/jacket for wiring carrying up to 600 volts. The combination of insulation resistance and long-term resistance to the deteriorating effects of ozone, weather, oil, and heat make CSM a clear choice for high-quality wire applications.

Automotive Uses

In the auto industry, high-volume CSM usage includes hose, tubing, and electrical applications. In addition, low permeability to moisture and refrigerants make high-chlorine CSM an ideal polymer for air-conditioning hoses.

Construction Uses

Considerable usage of CSM in reservoir-liner construction has resulted from a combination of excellent weather resistance and the ease of fabrication of uncured liners made of CSM. Over 500 million square feet of CSM liners have been installed, with an exceptionally low rate of failure due to the liner's easy and reliable seaming characteristics.

For many of the same reasons, CSM use is also well-established in single-ply roofing membranes. Here, the flame resistance and moisture resistance of CSM is very desirable and the fact that weather-resistant membranes can be made in heat-reflective light colors helps reduce the cost of air conditioning.

In an alternative construction method, colorable weather resistant roof surfaces of HYPALON are prepared by roller coating HYPALON paint over a neoprene base.

Miscellaneous Applications

Other applications for CSM include roll covers for use in strong acids, and in applications where corona discharge produces ozone and electrical stress; CSM is also used for tank linings, industrial maintenance coatings, and extruded sponge. CSM compositions are binders for cork in automotive gaskets, and for magnetic fillers in refrigerator gaskets.

PRODUCT SAFETY

When recommended handling procedures are followed, HYPALON synthetic rubber polymers and the products derived from them present no health hazards of which the Du Pont Company is aware. HYPALON synthetic rubber may contain a small amount of carbon tetrachloride (CCl_4) and a lesser amount of chloroform ($CHCl_3$) as residues from its manufacturing process. Both are regulated as air contaminants in the United States under the Occupational Safety and Health Act (OSHA); refer to 29 C.F.R. 1910, 1000. When large quantities of raw polymer are stored or processed, it is advisable for the protection of personnel to provide adequate ventilation to keep employee exposure below the regulated levels.

Additional information is available concerning these and other potential health hazards that could arise in the handling of HYPALON synthetic rubber, its compounding ingredients, thermal decomposition products, or its disposal as waste.[38]

REFERENCES

25. R. R. Warner, *Rubber Age*, May 1952.
26. R. E. Brooks, D. E. Strain, and A. McAlevy, *India Rubber World*, Mar. 1953.
27. A. Nersasian and D. E. Anderson, *J. Appl. Poly. Sci.* **4,** 74 (1960).
28. M. A. Smook, I. D. Roche, W. B. Clark, and O. G. Youngquist, *India Rubber World*, April 1953.
29. W. F. Busse and M. A. Smook, *India Rubber World*, June 1953.
30. J. T. Maynard and P. R. Johnson, *Rubber Chem. Technol.* **36,** 963 (1963).
31. P. R. Johnson, I. C. Kogan, and F. J. Rizzo, "Crosslinking Chlorosulfonated Polyethylene with Ammonia," Du Pont HYPALON® Report, Oct. 1964.
32. I. C. Dupuis, "Selecting a Curing System," Du Pont HYPALON® Report HP-320.1.
33. P. A. Peffer, Jr., and R. R. Radcliff, *Rubber World*, Oct. 1960.
34. I. C. Dupuis, SAE Booklet 730541, 1973.
35. J. R. Hoover, A. L. Moran, and K. H. Whitlock, *Elastomerics*, March 1982.
36. F. Kurath, E. Passaglia, and R. Pariser, *J. Appl. Poly Sci.* **1,** 150 (1959).
37. R. W. Fillers and N. W. Tschoegl, *Trans. Soc. Rheol.* **21,** 51 (1977).
38. "Toxicity and Handling Guidelines for HYPALON®," Du Pont Report HP-110.1 (Nov. 1985).

13
SILICONE RUBBER

J. C. CAPRINO AND R. F. MACANDER
General Electric Company
Silicone Products Division
Waterford, New York

INTRODUCTION

Silicone rubber sales have had a spectacular growth in the 40 years since the product was first introduced commercially. Even in the last 10 years, silicone rubber sales have done very well in contrast to most organic rubbers. This growth has taken place not only in the United States, but especially in Japan and Western Europe. The United States now accounts for 50% of the total sales of 100 million pounds.[1,2] The increase in sales is due to two factors: first, the unique properties of silicone rubber relative to organic elastomers over a wide temperature range (−150°F to 600°F); and second, the cost of silicone rubber, which has improved relative to the organic elastomers.[3]

The properties of silicone rubber are due to the unusual molecular structure of the polymer, which consists of a backbone of silicon atoms (approximately 7000) alternating with oxygen atoms. The silicon-oxygen-silicon linkage in silicones is similar to the linkage in quartz and glass. Although the silicones are not as heat-resistant as quartz, they have superior heat resistance compared to other elastomers.

One reason for this greater heat resistance of silicone rubber can be found in the bond energy of the silicon-oxygen bond. Values for the silicon-oxygen bond range from 88–117 kcal/mole while values for the carbon-carbon single bond range from 83–85 kcal/mole.[4] While the bond energy is helpful in understanding the heat resistance of silicone rubber, a more thorough understanding of the chemical behavior of the bond would require a look at the three components of the bond, ie., covalent-, polar- and double-bond contribution.[5] We only have space here to say that whatever the potential reactivity of this bond, it is shielded by outward-pointing and unusually mobile methyl groups.

The mobility of the methyl groups as well as the larger volume of a silicon atom relative to a carbon atom results in a large free space for the dimethylsiloxane unit and limits the close approach of neighboring molecules. The distance between molecules and the nonpolar nature of the methyl groups contributes to the low glass-transition temperature for siloxanes. It also contributes to the high permeability of silicones to gases and to the high compressibility of silicones. Further, the inorganic backbone is responsible for the fungus resistance of silicones and their lack of appeal to rodents.

The methyl groups usually make up 99.9% of the side groups off the backbone. The methyl groups contain only primary hydrogens rather than the secondary, tertiary, allylic or benzylic hydrogens of some elastomers. The primary hydrogens are less susceptible to oxidative attack; thus silicone rubber retains its properties longer at high temperatures. Finally, silicone rubber contains no carbon-carbon double bonds in the backbone. These double bonds are susceptible to ozone and ultraviolet light, especially at higher temperatures, which lead to breaking of the backbone.

The result is that a longer service life can be expected at elevated temperatures as shown in Table 13.1.

In addition to the long life at 400°F, silicone rubber has a higher tensile strength than most organic elastomers at 400°F even though the tensile strength is lower than most organic rubbers at room temperature. Young's modulus, or the ratio of stress to strain, of low-temperature silicone rubber shows very little change down to −100°F, and rises to only 10,000 psi at −150°F.

Silicone rubber performs unusually well when used as a gasket or O-ring in sealing applications. Over the entire temperature range of −120°F to +500°F, no available elastomer can match its low compression set.

Many types of wire and cable are insulated with silicone rubber, mainly because its excellent electrical properties are maintained at elevated temperatures. Even when the insulation is exposed to a direct flame, it burns to a nonconducting ash that continues to function as insulation in a suitably designed cable.

The ozone and corona resistances of silicone rubber are outstanding, approaching that of mica. These properties are important in many electrical applications, and in exposure to outdoor weathering.

Table 13.1. Estimated Service Life of Silicone Rubber.

Temperature	Service Life
90°C (194°F)	40 years
121°C (250°F)	10–20 years
150°C (300°F)	5–10 years
200°C (392°F)	2–5 years
250°C (482°F)	3 months
315°C (600°F)	2 weeks

Many samples of elastomeric silicones have been exposed to outdoor weathering for 15 years with no significant loss of physical properties. This demonstrates unique resistance to temperature extremes, sunlight, water, and ozone and other gases. The rubber has good resistance to the low concentrations of acids, bases, and salts normally found in surface water. It has been estimated that a silicone elastomer will last in excess of 30 years under weathering conditions that would cause most organic rubbers to fail within a few years.

Silicone rubber is odorless, tasteless, and nontoxic. When properly fabricated, it does not stain, corrode, or in any way deteriorate materials with which it comes in contact. Consequently, it has found application in gas masks, food and medical-grade tubing, and even in surgical implants in the human body.

SYNTHESIS OF SILICONE POLYMERS

Frederick Stanley Kipping (1863–1949) is considered the founder of silicone chemistry.[6] He not only made many chlorosilanes, but he hydrolyzed them to silicones. Many of those which now interest us he considered polymeric nuisances. However, Kipping's contributions were remarkable considering the time in which he worked.

Dr. J. F. Hyde, of the Corning Glass Works, impregnated glass fibers with diphenyl silicone resins in 1937 to make tape for electrical insulation. He realized this combination would have considerably more heat stability than the materials then in common use. Dr. Eugene G. Rochow, at the General Electric Research Laboratories, began work on the methyl silicones in 1938, using the Grignard Chemistry developed by Kipping. He made several important discoveries and useful products but soon realized that the Grignard route was unsuitable for the commercial production of silicones. In 1940, he discovered the direct process for methylchlorosilanes.[7] This process made the methylchlorosilanes available at a reasonable cost and it made the silicone industry of today possible.

The silicon used in this process is made by the reduction of SiO_2 that in the free or combined form makes up about 60% of the earth's crust. The abundance of SiO_2 is one reason for the favorable trend in the cost of silicone rubber relative to the organics. The direct process involves the reaction of methyl chloride with silicon in the presence of a copper catalyst. Dimethyldichlorosilane is the major product, but many other products are produced, some of which are shown in Table 13.2; these are separated by large distillation towers.

Both the methyltrichlorosilane and trimethylmonochlorosilane must be reduced from percent levels to low ppm levels either by distillation or subsequent purification steps. The trimethylchlorosilane is a chainstopper and affects the molecular weight of the silicone polymer. The methyltrichlorosilane causes branching or crosslinking of the desired linear silicone polymer, which affects

Table 13.2. Major Products of the Direct Process.

Compound	*BP (°C/760 mm)*
Dimethyldichlorosilane	70
Methyltrichlorosilane	66
Trimethylchlorosilane	57
Methyldichlorosilane	40
Dimethylchlorosilane	35
Trichlorosilane	32

processing. After purification, the dimethyldichlorosilane is then hydrolyzed as shown below to a mixture of linear polymers and low-molecular-weight cyclics.

$$Me_2SiCl_2 + H_2O \rightarrow HO(Me_2SiO)_nH + (Me_2SiO)_m + HCl$$

The hydrolyzate was originally polymerized to high molecular weight with acid catalyst. Today the cyclics are usually removed as an additional purification step. Dr. J. F. Hyde of the Dow-Corning Corp. discovered in 1946 that a superior silicone rubber polymer is produced when solid alkalis are added in small amounts to the cyclics.[8] The polymer normally used in dimethylsilicone rubber varies from 300,000 to 800,000 in molecular weight. This corresponds to a viscosity of 4×10^6 to 140×10^6 cps, or 4000–11,000 silicon atoms per molecule. There are also two monofunctional groups per molecule as chainstopper, or 180–500 ppm chainstopper. One can then see how relatively small variations in the level of chainstopper can lead to large variations in molecular weight and even larger variations in viscosity.

Another discovery of importance was made in 1943 by Wright and Oliver, who found that a superior rubber could be made by crosslinking with benzoyl peroxide.[9] This was followed in 1944 by the work of J. Marsden who discovered an impurity in dimethyldichlorosilane which increased the crosslinking.[10] This impurity was shown to be methylvinyldichlorosilane. Today most dimethylsilicone polymers and compounds contain this vinyl source. When the latter is present at less than 0.5%, the resulting polymers make more efficient use of peroxides and thus require less for cure. The cured rubber has less tendency to revert and thus has lower compression set. J. C. Caprino and R. J. Prochaska were the first to introduce unsaturated groups at the ends of the silicone polymer chain. They patented this composition and its use in a low-compression-set rubber.[11]

Dimethyl silicone rubber tends to become stiff below −60°F. However, the low temperature flexibility may be improved by substitution of phenyl ($-C_6H_5$) groups for some of the methyl groups attached to the silicon atoms in the poly-

Table 13.3. Types of Silicone Polymers.

Classification	*Uses*	*Basic Polymer Units*
MQ	The first silicone polymer	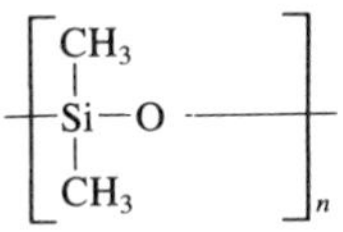 $\left[-\mathrm{Si}(CH_3)(CH_3)-O- \right]_n$
MPQ	The first low-temperature copolymer	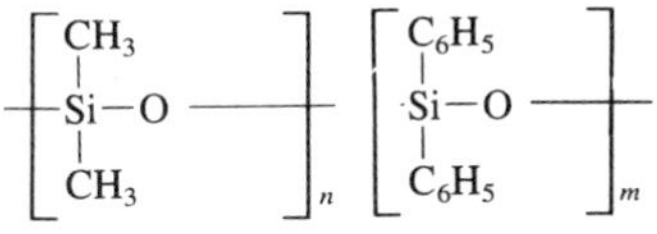 $\left[-\mathrm{Si}(CH_3)(CH_3)-O- \right]_n \left[-\mathrm{Si}(C_6H_5)(C_6H_5)-O- \right]_m$
MVQ	Low-compression-set copolymer	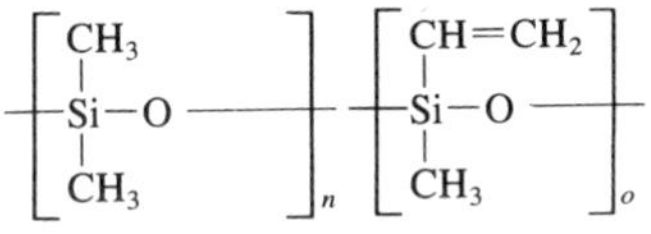 $\left[-\mathrm{Si}(CH_3)(CH_3)-O- \right]_n \left[-\mathrm{Si}(CH{=}CH_2)(CH_3)-O- \right]_o$
MPVQ	The first high-strength, low temperature terpolymer	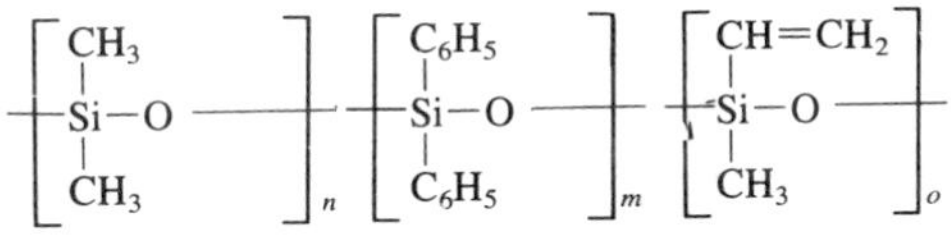 $\left[-\mathrm{Si}(CH_3)(CH_3)-O- \right]_n \left[-\mathrm{Si}(C_6H_5)(C_6H_5)-O- \right]_m \left[-\mathrm{Si}(CH{=}CH_2)(CH_3)-O- \right]_o$
FVQ	Oil-resistant polymer	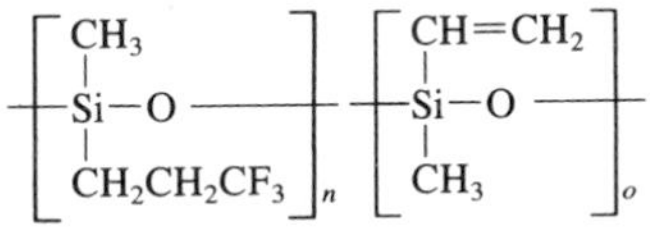 $\left[-\mathrm{Si}(CH_3)(CH_2CH_2CF_3)-O- \right]_n \left[-\mathrm{Si}(CH{=}CH_2)(CH_3)-O- \right]_o$

mer chain as shown in Table 13.3. Replacement of only 5% of the methyl groups by phenyl groups will lower the crystallization temperature and extend the useful service temperature range to below −130°F.

The dimethyl silicone rubbers show good performance in acetone and diesters but undergo up to 200% swell in aliphatic and aromatic hydrocarbons. The replacement of one methyl group on each silicon atom with a polar group such as trifluoropropyl ($-CH_2CH_2CF_3$), as shown in Table 13.3, reduces the swelling in aliphatic and aromatic hydrocarbons to less than 25% but increases swelling in acetone and diesters. Fig. 13.1 compares neoprene rubber, methyl silicone rubber and fluorocarbon rubber in various solvents. Fluorosilicone rubber behaves similarly to fluorocarbon rubber in these solvents. In solvents with a higher solubility parameter such as methanol or ethanol, the swelling of fluorosilicones drops down again to less than 25%.

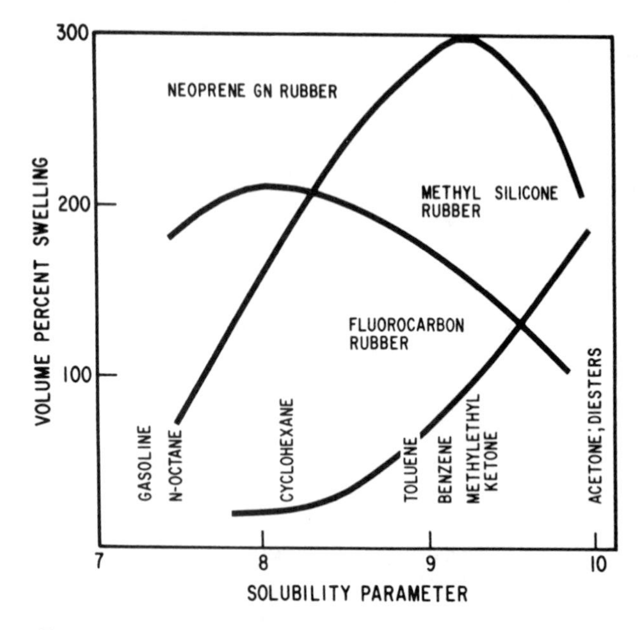

Fig. 13.1. Variation of volume swell with solubility parameter.

VULCANIZATION

Silicone rubber compounds are normally heat-cured in the presence of one of the organic peroxides shown in Table 13.4. Some other peroxides are used to a limited extent; however, the first four in the table are the most important ones.

In the case of the cure of a dimethyl polymer with a diaroyl peroxide, the cure mechanism shown in Fig. 13.2 has been fairly well established.[12, 13]

It is apparent that these reactions could produce no more than 1 mole of chemical crosslinks per mole of peroxide. The hydrogen abstraction reaction has been estimated at 50% efficient, and the ethylene bridge formation at 40% efficient. And, indeed, independent work has shown that the actual crosslink yield in an unfilled polymer is 0.1 to 0.3 moles of chemical crosslinks per mole of diaroyl peroxide.[14] The above ratio decreases as the level of diaroyl peroxide is increased. This corresponds to a chemical crosslink density of 0.4 to 0.7 moles of crosslinks per 1000 moles Si in the polymer. It is not practical to increase the crosslink concentration much further by raising the peroxide levels.

The cure mechanism for the methyl vinyl siloxy containing copolymers shown in Fig. 13.3 seems to account for available experimental observations.[12, 15, 16]

This mechanism predicts more than 1 mole of crosslinks per mole of peroxide, and not more than 1 mole of crosslinks per mole of vinyl groups.

In the case of unfilled methyl vinyl siloxy containing copolymers, actual

Table 13.4. Organic Peroxides Used for Silicone Rubber Vulcanization.

Peroxide	*Temperature, °F, for Half-life = 1 minute*
Bis(2,4-dichlorobenzoyl) peroxide	234
Di-benzoyl peroxide	271
Di-cumyl peroxide	340
2,5-dimethyl-2,5-*bis* (*t*-butyl peroxy) hexane	354
Di-tertiary butyl peroxide	379

crosslink yields have also been measured.[14] Data on a polymer with Vi/Si = 0.0026 are shown in Table 13.5.[14] Efficiencies are high, and these results are consistent with the "trimethylene bridge" cure mechanism.

2,5-dimethyl-2,5-*bis*(t-butyl peroxy) hexane, dicumyl peroxide, and di-*t*-butyl peroxide are usually considered to be "vinyl-specific" in that they will give good cures only with vinyl containing polymer. The two diaroyl peroxides, benzoyl and *bis*(2,4-dichlorobenzoyl), will cure both vinyl and nonvinyl containing gums. This difference is illustrated in Fig. 13.4, again using the polymer with Vi/Si = 0.0026.[14] Di-tertiary butyl peroxide seems to be truly "vinyl specific" in that the chemical crosslink density is constant in the range 0.2 to

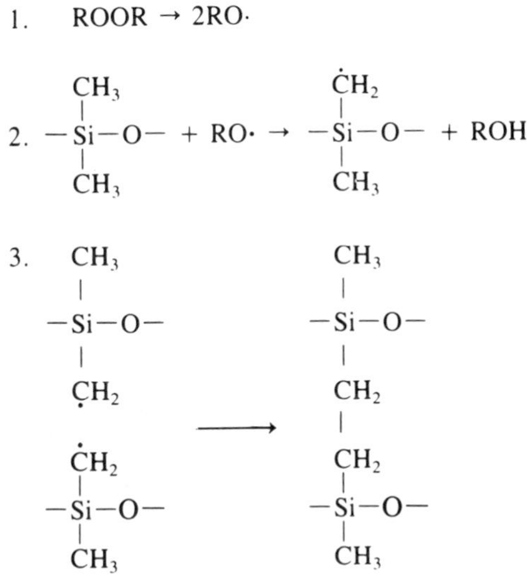

1. $ROOR \rightarrow 2RO\cdot$

2. $-Si(CH_3)_2-O- + RO\cdot \rightarrow -Si(\dot{C}H_2)(CH_3)-O- + ROH$

3. $-Si(CH_3)(\dot{C}H_2)-O- + -Si(CH_3)(\dot{C}H_2)-O- \longrightarrow -Si(CH_3)-O-$ / CH_2-CH_2 / $-Si(CH_3)-O-$ (two Si linked through $-CH_2-CH_2-$)

Fig. 13.2. Peroxide cure mechanism for methyl siloxanes.

```
1.   ROOR → 2RO ·

2.   CH3                       CH3                  CH3
     |                         |                    |
    —Si—O— + RO· →           —Si—O—               —Si—O—
     |                         |                    |
     CH                       · CH                  CH3
     ||                        |                 ————————→
     CH2                       CH2OR

     CH3                       CH3
     |                         |
    —Si—O—                   —Si—O—
     |                         |
     CH2                       CH2 + RO ·
     |                         |
     CH2OR                     CH2
                ————————→      |
    ·CH2                       CH2
     |                         |
    —Si—O—                   —Si—O—
     |                         |
     CH3                       CH3
```

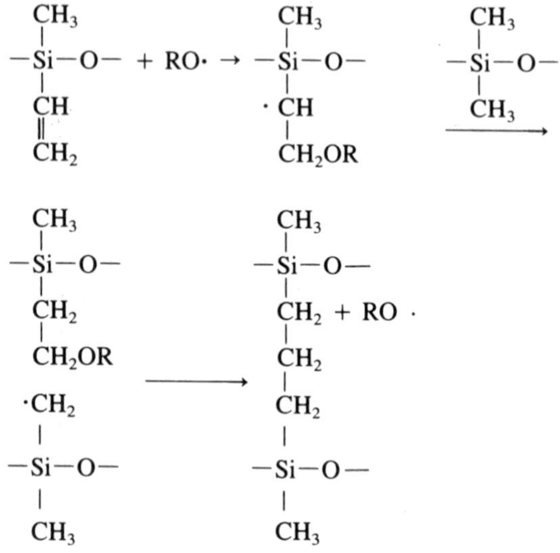

3. Repetition of Step 2 followed eventually by termination.

Fig. 13.3. Peroxide cure mechanism for methyl vinyl siloxanes.

6% peroxide. However, the crosslink density is dependent upon peroxide concentration in the case of the *bis*(2,4-dichlorobenzoyl) peroxide.

Figure 13.5, drawn from data in reference 14, gives more insight into the effect of vinyl on vulcanizate properties. When the vinyl level is zero, using 1.95×10^{-5} moles of *bis*(2,4-dichlorobenzoyl) peroxide/g polymer, one obtains 0.44 moles crosslinks/1000 moles Si in the polymer. This can be increased to 0.55 moles crosslinks per 1000 moles Si in the polymer by doubling the peroxide concentration, or it can be increased to 1.5 moles crosslinks per 1000 moles Si in the polymer with the use of a 0.2 mole % methyl vinyl siloxy copolymer. It is therefore possible to reach a higher level of crosslinking by

Table 13.5. Crosslinking Efficiency. Polydimethylcomethylvinylsiloxane (Vi/Si = 0.0026)

Peroxide	*% Peroxide (Optimum Level)*	*Moles Chemical Crosslinks per 1000 moles Si*	*Moles Chemical Crosslinks per mole Peroxide*	*Moles Chemical Crosslinks per mole Vinyl*
2,5-dimethyl-2,5-*bis* (*t*-butyl peroxy) hexane	0.315	2.4	3.2	0.9
Dicumyl peroxide	0.315	2.4	3.0	0.9
Di-*t*-butyl peroxide	0.21	3.3	2.4	0.9

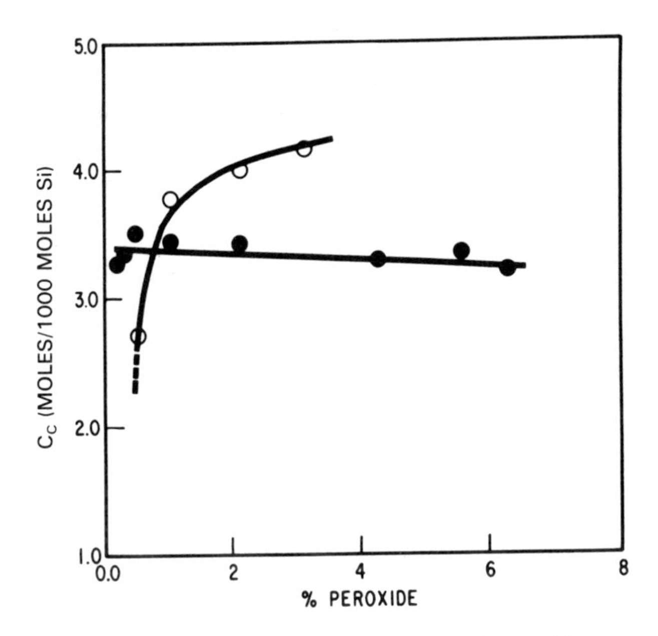

Fig. 13.4. Variation of chemical crosslink density with peroxide concentration.

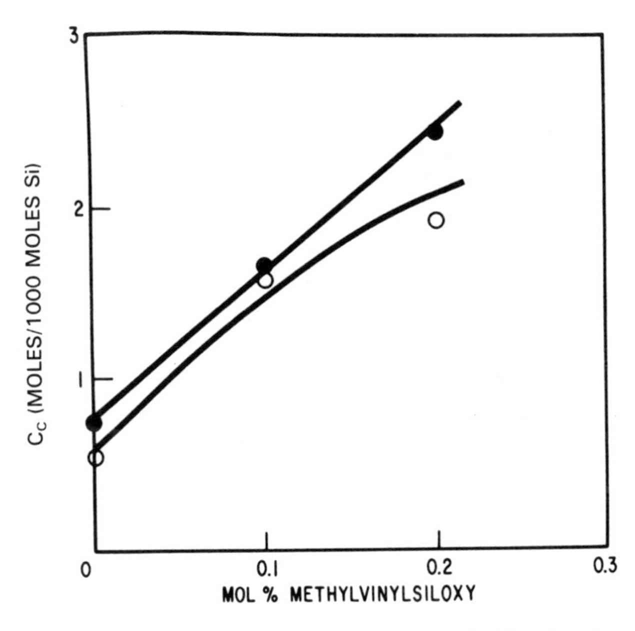

Fig. 13.5 Variation of chemical crosslink density with vinyl level and concentration of *bis*(2,4-dichlorobenzoyl) peroxide.

using a vinyl containing siloxane polymer. The higher level of crosslinking results in less tendency of the cured rubber to creep. Of equal importance, crosslinking can be accomplished with lower levels of acid-forming peroxides and peroxides that don't lead to acidic decomposition products. This last process in turn leads to less tendency to revert in curing or during post-bake processing. By optimizing the vinyl level and the type and level of peroxide catalyst, the compression set at elevated temperatures can be minimized.

Vulcanization rate is conveniently studied by means of the Monsanto Rheometer. This instrument, described elsewhere in this book, provides a continuous measurement of complex dynamic shear modulus while a rubber is being cured in a mold under heat and pressure. This study is accomplished by measurement of the torque on a conical disk rotor that is embedded in the rubber, and is being sinusoidally oscillated through a small arc. The torque is a linear function of crosslink density as determined by swelling measurements though the proportionality constant varies with the stock. The torque readings can, therefore, be considered as relative crosslink densities.

From a rheograph, it is possible to get an idea of how the compound will flow in the mold before cure starts (minimum viscosity and scorch time), cure rate (and consequently time to various degrees of cure), and a measure of the final crosslink density. If a series of these curves are run at different temperatures with the various peroxides, the data may be used to estimate hot-mold residence times required to effect a given degree of cure at a given temperature with a given peroxide. Figs. 13.6, 13.7, and 13.8 may be used to make such estimates for the vulcanization of a high-tear-strength methyl vinyl compound. For example, with *bis*(2,4-dichlorobenzoyl) peroxide at 0.65 phr, and at a temperature of 230°F, the induction period is 2 minutes. The hot-mold-residence time is 4 minutes for 90% cure, and 8 minutes for full cure. A technically satisfactory cure (in terms of durometer, tensile, and elongation) is obtained by introduction of 90% of the crosslinks; however, a full cure is required to obtain the lowest compression set. In order to obtain full cure in 8 minutes, the mold temperature should be 230°F for *bis*(2,4-dichlorobenzoyl) peroxide, 270°F for benzoyl peroxide, 255°F for dicumyl peroxide, and 365°F for 2,5-dimethyl-2,5-*bis*(*t*-butyl peroxy) hexane.

COMPOUNDING INGREDIENTS

A typical silicone rubber formulation contains a silicone polymer, reinforcing and (or) extending fillers, process aids or softeners to plasticize and retard crepe-aging, special additives (e.g., heat-ageing and flame retardant additives), color pigments, and one or more peroxide curing agents.

Silicone Polymers

Pure silicone rubber polymers, differing from one another in polymer type and molecular weight, are available from the basic suppliers. These polymers or gums are described in Table 13.3.

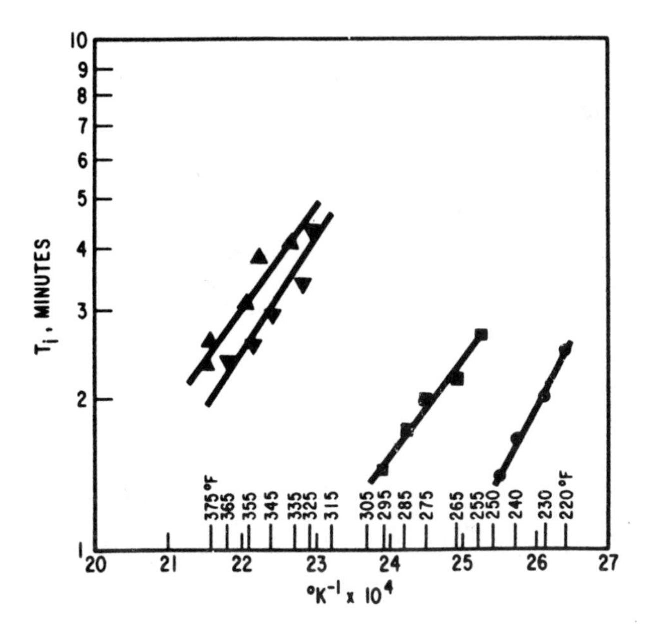

Fig. 13.6. Induction times for cure with various peroxides.

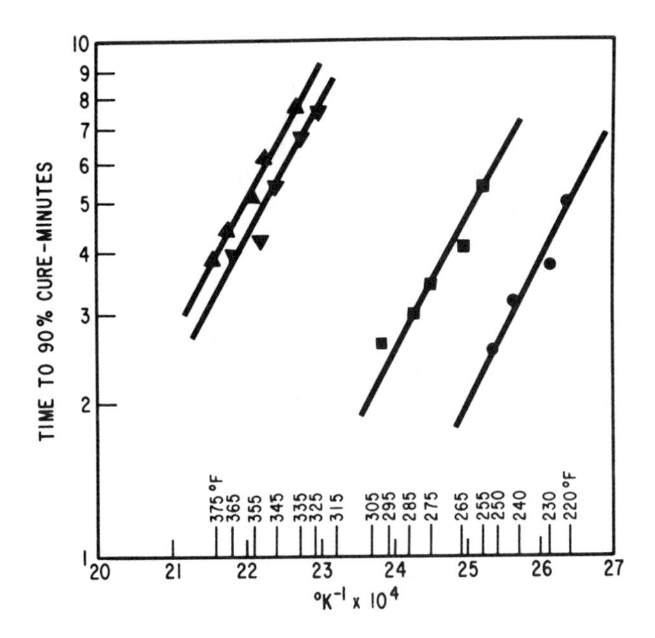

Fig. 13.7. Time to 90% cure with various peroxides.

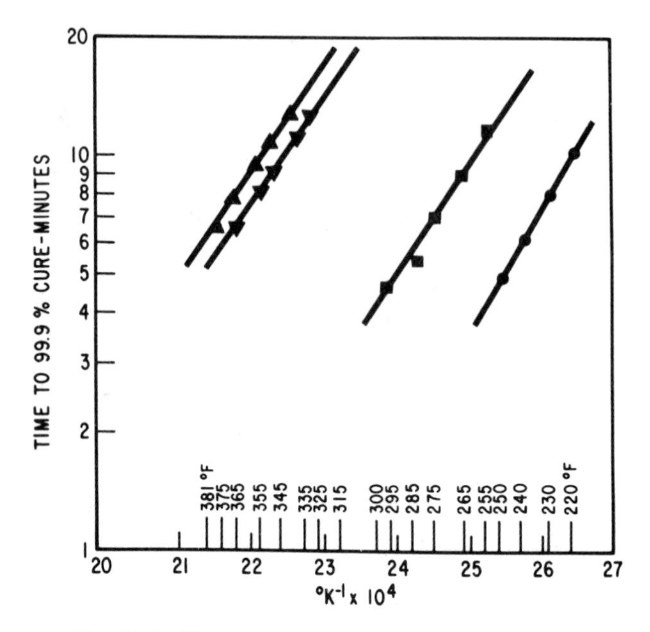

Fig. 13.8. Time to full cure with various peroxides.

Silicone Rubber Elastomers

Although pure polymer may be used, it is generally easier and more economical for the rubber fabricator to compound from silicone-reinforced gums or bases. Over the years, silicone suppliers have developed a wide variety of silicone rubber compounds to meet special requirements and specifications. However, the most recent offerings have been in bases which are mixtures of pure polymer, of process aids, and of reinforcing silica fillers that have been specially processed. For example General Electric offers its Silplus® Elastomeric System and Dow Corning provides its Silastic® Compounding System Elastomers. Both of these product lines are made up of blendable bases plus selected additives for enhanced properties such as heat-ageing, flame retardancy, and oil resistance as well as processing aids. The other major suppliers such as Rhone-Poulenc and SWS Silicones also offer a variety of bases and compounds.

Reinforcing Fillers

The fumed process silicas reinforce silicone polymer to a greater extent than any other filler. Due to the high purity of the filler, the rubbers containing it have excellent insulating properties, especially under wet conditions.

Precipitated silicas impart moderately high reinforcement, as shown in Table 13.6. Some precipitated filler may give high water absorption due to residual

Table 13.6. Reinforcing Fillers for Silicone Rubber.

Filler[a]	*Type*	*Particle Size, Mean Diameter (millimicrons)*	*Surface Area (square meters per gram)*	*Specific Gravity*	*Supplier*	*Reinforcement Produced in Pure Silicone Gum*	
						Tensile Strength Range, psi	*Elongation Range, %*
CAB-O-SIL HS-5	Fumed silica	10	300–350	2.20	Cabot Corporation CAB-O-SIL Div. P.O. Box 188 Tuscola, IL 61953	600–1800	200–1000
CAB-O-SIL MS-7	Fumed silica	15	175–200	2.20	Cabot Corporation CAB-O-SIL Div. P.O. Box 188 Tuscola, IL 61953	600–1200	200–800
AEROSIL 300	Fumed silica	7	270–330	2.20	Degussa Corporation Route 46 at Hollister Rd. Teterboro, NJ 07608	600–1800	200–1000
AEROSIL 200	Fumed silica	12	175–225	2.20	Degussa Corporation Route 46 at Hollister Rd. Teterboro, NJ 07608	600–1200	200–800
FK160	Precipitated silica	18	160	2.20	Degussa Corporation Route 46 at Hollister Rd. Teterboro, NJ 07608	600–1100	200–800
HiSil	Precipitated silica	20	140–160	2.20	PPG Industries 1 Gateway Center Pittsburgh, PA 15272	600–1100	200–800

[a]The filler product designations are trademarks of the companies listed in the Supplier column.

salts. If so, the wet electrical properties will not be very good. However, in recent years some precipitated silicas have been placed on the market with very-low-salt content and provide better electrical properties.

Semireinforcing or Extending Fillers

The extenders (Table 13.7) are important for use in compounds containing reinforcing fillers in order to obtain an optimum balance of physical properties, cost, and processibility.

Ground silica and calcined kaolin do not provide significant reinforcement. As a consequence, they can be added to a reinforced gum or compound in relatively large quantities in order to reduce pound-volume cost. These extenders are satisfactory in either mechanical- or electrical-grade rubber.

The reinforcement obtained with calcined diatomaceous silica is greater, though quite modest, than that obtained with any other extender. Therefore, as an extender, it is not as useful as ground silica. However, it is used in electrical stocks, low-compression-set stocks, and in general mechanical stocks to reduce tack and modify handling properties.

Calcium carbonate and zirconium silicate are special-purpose extenders, used mainly in pastes that are coated on fabrics via solvent dispersion.

Zinc oxide is used as a colorant, and as a plasticizer. It imparts tack and adhesive properties to a compound.

Additives

Organic colors, and many inorganic colors, have adverse effects on the heat-aging of silicone rubber. The inorganic pigments listed in Table 13.8 have been found suitable for use. Usually 0.5 to 2 parts per 100 parts of compound are sufficient for tinting purposes. It is often desirable to masterbatch color pigments in order to get good dispersion and close color matches.

Red iron oxide is used as a color pigment, and as a heat-ageing additive; 2 to 4 parts per 100 parts of gum will give improved heat stability at 600°F.

Process aids are used with highly reinforcing silica fillers. These have a softening or plasticizing effect, and they retard the "crepe-ageing" or "structuring" or "pseudocure" of the raw compound that occurs due to the high reactivity of the reinforcing filler with the silicone polymer.

Curing Agents

In commercial practice, it has been found that none of the six commonly used peroxides is a universal curing agent. The three aroyl peroxides (Table 13.9) may be considered general-purpose in that they will cure both nonvinyl and vinyl-containing polymers. However, no one of them is suitable in all types of

Table 13.7. Semireinforcing or Extending Fillers for Silicone Rubber.

Filler[a]	Type	Particle Size, Mean Diameter (microns)	Surface Area (square meters per gram)	Specific Gravity	Supplier	Reinforcement Produced in Pure Silicone Gum: Tensile Strength Range, psi	Reinforcement Produced in Pure Silicone Gum: Elongation Range, %
BLANC ROUGE		1–5	5	2.65	Illinois Mineral Company 2035 Washington Ave. Cairo, IL 62914	100–400	200–300
CELITE SUPER FLOSS	Flux calcined diatomaceous silica	1–5	5	2.30	Johns-Manville Filtration & Minerals Div. Ken-Caryl Ranch Denver, CO 80217	400–800	75–200
CELITE 350	Calcined diatomaceous silica	1–5	5	2.15	Johns-Manville Filtration & Minerals Div. Ken-Caryl Ranch Denver, CO 80217	400–800	75–200
IRON OXIDE RP-3097	Iron Oxide	1		4.80	Pfizer, Inc. 235 E. 42nd St. New York, NY 10017	200–500	100–300
IRON OXIDE RY-2196	Iron Oxide	1		4.95	Pfizer, Inc. 235 E. 42nd St. New York, NY. 10017	200–500	100–300
MIN-U-SIL	Ground silica				Pennsylvania Glass Sand Corp. 300 Penn Center Blvd. Pittsburgh, PA 15235	100–400	200–300
5		5	5	2.65			
10		10	5	2.65			
15		15	5	2.65			

[a]The Filler product designations are trademarks of the companies listed in the Supplier column.

(cont.)

Table 13.7. (*Continued*)

Filler[a]	*Type*	*Particle Size, Mean Diameter (microns)*	*Surface Area (square meters per gram)*	*Specific Gravity*	*Supplier*	*Reinforcement Produced in Pure Silicone Gum*	
						Tensile Strength Range, psi	*Elongation Range, %*
SUPERPAX	Zirconium silicate			4.50	TAM Ceramics, Inc. 4511 Hyde Park Blvd. Niagara Falls, NY 14305	400–600	100–300
TITANOX RA	Titanium dioxide	3		4.2	NL Industries, Inc. Titanium Pigments Div. 100 Chevalier Ave. South Amboy, NJ 08879	200–500	300–400
WHITETEX CLAY	Calcined kaolin	1–5	5	2.55	Freeport Minerals Co. 200 Park Ave. New York, NY 10017	400–800	75–200
WITCARB R	Precipitated calcium carbonate	.03–.05	32	2.65	WITCO Chemical Co. 520 Madison Ave. New York, NY 10022	400–600	100–300
ALBACAR 5970	Precipitated calcium carbonate	1–4	8	2.71	Pfizer Inc. 235 E. 42nd St. New York, NY 10017	400–600	100–300
ZINC OXIDE XX-78	Zinc oxide	.3	3.0	5.6	Gulf & Western Ind. New Jersey Zinc Co. One Commerce Place Nashville, TN 37239	200–500	100–300

[a]The Filler product designations are trademarks of the companies listed in the Supplier column.

Table 13.8. Color Pigments for Silicone Rubber.

Reds	Red (RY-2196)[a]
	Red (RO-3097)[a]
	Red (F-5893)[g]
	Maroon (F-5891)[g]
	Dark Red (F-5892)[g]
Greens	Chromium Oxide Green (X-1134)[b]
	Chromium Oxide Green (G-6099)[a]
	Yellow Green (F-5688)[g]
	Blue Green (F-5687)[g]
	Turquoise (F-5686)[g]
Blues	Cobalt Aluminum Blue (F-6279)[g]
	Medium Blue (F-5274)[g]
	Dark Blue (F-6279)[g]
	Violet Blue (F-5273)[g]
	Ultra Marine Blue[i]
Oranges	Mapico Tan #20[c]
	Orange Red (F-5894)[g]
	Orange (F-5895)[g]
	Light Orange (F-5896)[g]
Whites	Titanium Dioxide (TITANOX RA)[d]
	Titanium Dioxide (TITANOX ALO)[d]
Yellows	Cadmolith Yellow[b,e]
	Cadmium Yellow (F-5897)[g]
	Lemon Yellow (F-5512)[g]
Buffs	Dark Buff (F-6115)[g]
	Buff (F-2967)[g]
Blacks	Thermax Carbon Black[f]
	Black Iron Oxide (Drakenfeld 10395)[h]
Browns	Light Yellow Brown (F-6109)[g]
	Medium Brown (F-6111)[g]
	Red Brown (F-6112)[g]

[a]Pfizer Inc., 235 E. 42d St., New York, NY 10017
[b]Ciba-Geigy, 875 Providence Highway, Dedham, MA 02026
[c]Cities Service Corp., Columbian Chemicals Co., 3200 W. Market St., Akron, OH 44313
[d]NL Industries, Inc., 100 Chevalier Ave., South Amboy, NJ 08879
[e]Glidden Pigments Group, Div. of SCM Corp., 3901 Glidden Rd., Baltimore, MD 21226
[f]R. T. Vanderbilt Co., Inc., 30 Winfield St., Norwalk, CT 06855
[g]Ferro Corp., 4150 E. 56th St., Cleveland, OH 44101
[h]Hercules Inc., Coatings & Specialty Product Dept., Washington, PA 15301
[i]Whittaker, Clark & Daniels, Inc., 1000 Coolidge St., South Plainfield, NJ 07080

Table 13.9. Peroxide Curing Agents for Silicone Rubber: General Purpose.

Peroxide	Commercial Designation	Form	Assay (%)	Cure Temperature (°F)	Decomposition Products	Uses
Bis(2,4-dichloro-benzoyl) perox-ide	CADOX TS-50[a]	Paste	50	220–250	Nonvolatile	Hot-air vulcanization
	LUPERCO CST[b]	Paste	50		Acidic	Continuous-steam vulcanization
						Autoclave
						Thick-section molding
						Low compression set
Benzoyl peroxide	CADOX BSG-50[a]	Paste	50	240–270	Volatile	Continuous-steam vulcanization
	LUPERCO AST[b]	Paste	50		Acidic	Autoclave
	CADOX 99[a] (200 mesh)	Powder	99			Tower coating
						Low compression set
						Thin-section molding
Tertiary butyl perbenzoate	*t*-Butyl Perbenzoate[b]	Liquid	95	290–310	Volatile Acidic	Usually with other peroxides; sponge; generally for high-temperature activation

[a]Noury Chemical, Burt, NY, 14028. CADOX is a trademark of Noury Chemical.
[b]Lucidol Division, Pennwalt Corp., 1740 Military Rd., Buffalo, NY 14240. LUPERCO is a trademark of Lucidol Division, Pennwalt Corp.

fabrication procedures. The three dialkyl peroxides (Table 13.10) are "vinyl-specific" since they will give good cures only with vinyl-containing polymers.

Bis(2,4-dichlorobenzoyl) peroxide normally requires a curing temperature of about 220–250°F. Its decomposition products (2,4-dichlorobenzoic acid and 2,4-dichlorobenzene) volatilize relatively slowly at commercial curing temperatures; and, consequently, compounds containing it may be cured without external pressure, provided that air removal and forming have been done (e.g., by extrusion or calendering) prior to the application of heat. In fact, one of the main uses of this peroxide involves hot-air vulcanization of extrusions in a few seconds at temperature of 600–800°F.

Although this peroxide may be used for molding, it has some undesirable features. Since it starts crosslinking at an appreciable rate as low as 200°F, thin sections may start to gel before flow and air removal are complete. In addition, a thick section must be carefully programmed through a post-vulcanization oven bake cycle in order to remove acidic decomposition products without degrading the interior of the part. The peroxide may be used for steam cures in autoclaves and in continuous-steam vulcanizers. However, benzoyl peroxide is more desirable, particularly in the latter case, because of its higher cure temperature (240–270°F). With benzoyl peroxide, there is less tendency for the compound to scorch in the dies when extruding at high speed into a continuous-steam vulcanizer that is operating at 100–200 pounds of steam.

Due to the volatility of its decomposition products (benzoic acid and benzene), external pressure is required to prevent porosity when curing with benzoyl peroxide. The only exception is in the very thin sections involved in tower-coating fabrics from solvent dispersions of silicone rubber. In this case, benzoyl peroxide is nearly always used because it has long shelf life in the dip tanks, and because it is not volatilized from the rubber during the solvent-removal operation prior to cure.

Low-compression-set rubber may be produced with either of the diaroyl peroxides, provided that the polymer contains vinyl groups and that the acidic decomposition products are removed by an oven post-bake.

Unlike the aroyl peroxides, the so-called "vinyl-specific" peroxides (Table 13.10) may be employed to vulcanize stocks containing carbon black. If dicumyl peroxide is used, these stocks can be hot-air-vulcanized.

All three "vinyl-specific" peroxides are good for thick-section molding, with dicumyl being less preferable due to a slight tendency to "air-inhibit" in the same manner as benzoyl peroxide. In addition, dicumyl peroxide has somewhat less volatile decomposition products (acetophenone and $\alpha,\alpha,$-dimethylbenzyl alcohol) than do the other two. This means that external pressure during cure is less important, but still required; and that longer oven post-bakes are needed for thick sections than in the case of the other two peroxides. Just as with the diaroyl peroxides, optimum physical properties require relatively close control of dicumyl peroxide concentration.

Table 13.10. Peroxide Curing Agents for Silicone Rubber: Vinyl-specific.

Peroxide	*Commercial Designation*	*Form*	*Assay (%)*	*Cure Temperature (°F)*	*Decomposition Products*	*Uses*
Dicumyl peroxide	DICUP R[a]	Solid	95	300–320	Fairly volatile	Thick-section molding
	DICUP 40C[a]	Powder	40			Low compression set
						Carbon black
2,5-dimethyl-2,5-*bis*(*t*-butyl per-oxy) Hexane	VAROX[b]	Power	50	330–350	Volatile	Thick-section molding
	LUPERSOL 101[c]	Liquid	95			Low compression set
	LUPERCO 101XL[c]	Powder	50			Carbon black
Di-tertiary butyl peroxide	Di-*t*-butyl perox-ide[c,d]	Liquid	97	340–360	Volatile	Thick-section molding
	CW-2015[e]	Powder	20			Low compression set
						Carbon black

[a]Hercules, Inc., 910 Market St., Wilmington, DE 19894. DICUP is a trademark of Hercules, Inc.
[b]R. T. Vanderbilt Co., 30 Winfield St., Norwalk, CT 06855. VAROX is a trademark of R. T. Vanderbilt Co.
[c]Lucidol Div., Pennwalt Corp., 1740 Military Road, Buffalo, NY 14240. LUPERSOL is a trademark of Lucidol Div., Pennwalt Corp.
[d]Shell Chemical Corp., P.O. Box 2463, Houston, TX 77001.
[e]Harwick Chemical Corp., 60 S. Sieberling St., Akron, OH 44305.

Due to their nonacidic decomposition products, the "vinyl-specific" peroxides require less oven post-bake after vulcanization; and, in addition, close-stepped program post-bakes are not necessary. Lower compression sets are obtained with these peroxides than with the general-purpose curing agents.

During vulcanization with di-*t*-butyl peroxide, external pressure is especially important due to the extreme volatility of the peroxide and its decomposition products (acetone and methane). The state of cure is determined primarily by the vinyl content of the polymer, and not by peroxide concentration; air inhibition is absent; prevulcanization or scorch is seldom a problem; and the innocuous decomposition products can be removed by short, high-temperature oven post-bakes. This peroxide also produces a rubber with the best overall balance of properties. Its main deficiency lies in its extreme volatility. A stock must be molded with external pressure shortly after the peroxide has been added.

Although it does not produce rubber with quite as good a balance of properties, 2,5-dimethyl-2,5-*bis*(*t*-butyl peroxy) hexane is quite similar to di-*t*-butyl peroxide in its performance. It has the advantage of lower vapor pressure at room temperature. In fact, it can be added to a compound 1 to 2 months before vulcanization. However, external pressure is still required during cure due to the volatility of the decomposition products.

COMPOUNDING

The basic suppliers have developed a very extensive body of compounding information that has resulted in compounds, bases and formulations with which the fabricator can meet well over a hundred industrial and military specifications for silicone rubber. For details the fabricator should contact one of the basic suppliers listed below. The ranges of physical properties typically obtained are shown in Table 13.11.

Basic suppliers of silicone rubber:

General Electric Co., Silicone Products Div.,
260 Hudson River Road, Waterford, NY 12188.
(518) 237-3330.

Dow Corning Corp.,
2200 W. Salzburg Rd., P.O. Box 1767,
Midland, Michigan, 48640.
(517) 496-4000.

Rhone-Poulenc Chemical Co., Chemicals Div.,
P.O. Box 125, Monmount Junction, NJ 08852.
(201) 297-0100.

SWS Silicones Corp., Adrian, MI 49221. (517) 263-5711.

Table 13.11. Typical Physical Properties of Selected Formulations.

Property	*Low Compression Set*	*High-strength Methyl*	*Addition Cure*	*Extreme High Temperature*	*Extreme Low Temperature*	*High Strength, Extreme Low Temperature*	*Solvent Resistant*	*Wire and Cable Insulation*		
Hardness, Shore A	60	50	55	46	50	63	50	60	68	72
Tensile strength, psi	925	1300	1300	1000	900	1500	1000	900	1400	900
Elongation, %	200	650	700	430	400	700	220	300	500	220
Tear strength, Die B pi	70	200	250	95	95	200	75			
% compression set (22 hrs/350°F)	14	30	40	14	20	42	50			
Brittle pt (°F)	−85	−90	−90	−85	−150	−150	−90			
Oil resistance, % volume change:										
7 days @ 160°F Skydrol 500A				+10	+25		+30			
70 hr @ 300°F ASTM 1	+7			+6	+10	+10	+1			
70 hr @ 300°F ASTM 3	+40				+90	+90	+5			
Heat-aging (48 hr/600°F)										
Hardness, Shore A				56						
Tensile strength, psi				400						
Elongation, %				194						
Volume resistivity, ohm-cm								5×10^{15}		1×10^{15}
Electric strength, vpm								650		650
Dielectric constant, 60 cps								3.15		3.4
Power factor, 60 cps								0.001		0.005

Mixing

Silicone rubber may be compounded in conventional equipment, such as Doughmixers, Banburys and two-roll mills. A Banbury works well with a dry, nonsticky stock, but is undesirable with compounds that require a relatively long heat cycle and become too tacky to unload easily.

While a two-roll mill is capable of compounding silicone rubbers, the nature of the silica fillers creates an undesirable work atmosphere. However, two-roll mills are excellent for coloring, catalyzing, and preforming some of the firmer stocks.

FABRICATING

Freshening

Freshening has historically been the first fabrication step. This is a remilling operation, to reverse the "crepe-hardening" or "structuring" that has taken place since the compound was made. "Structuring" occurs more quickly and proceeds farther if the compound has been aged at higher temperatures. It is caused by the formation of hydrogen bonds between the hydroxyl groups of the filler and the hydroxyl groups or oxygen atoms of the polymer. Two factors have reduced the amount of structuring that occurs in the current silicone compounds: first, the hydroxyls on the polymer are kept to a low level; second, the hydroxyls on the filler are reduced by treating the filler with silanes or siloxane process aids.[17] However, freshening is still often carried out by the fabricator because the material has been in inventory at a high temperature or a long time. It is also freshened when the fabricator catalyzes it.

Before the mill is loaded, the roll clearance should be set fairly loose. Usually the silicone compound will band on the slow roll first (crumbling and lacing can be, but usually are not, encountered). As the milling continues, the nip should be gradually tightened. In the case of a hard compound, the original roll settings will be somewhat tighter for a short time. Then the mill is set fairly loose and gradually tightened as above. This procedure, combined with the proper amount of stock on the mill (slight bank), minimizes air entrapment during further milling and blending. After a smooth sheet has formed on the fast roll, compounding ingredients may be added if desired.

Milling is continued until the stock reaches the desired consistency. Under-freshened stock will flow poorly upon molding, and will form parts with a rough surface upon extrusion. Over-freshened compound loses green strength, and becomes sticky and hard to handle. Stock must not be allowed to warm up above 100–130°F after a curing agent has been added.

At this point, the stock is freshened and ready for use in one of the following fabrication procedures. Additional information on fabrication can be found in Lynch's excellent book.[18]

Molding

Compression and transfer molding are the most widely used methods for molding silicone rubber parts. However, large-volume applications, in the automotive industry for spark-plug boots and in the health-care industry for catheters, are resulting in large growth for injection molding. Compression and transfer moldings are run at pressures of 800–3000 psi and temperatures of 220 to 370°F. Three mold variables that must be controlled are temperature, speed of mold closing, and pressure. Temperature is influenced mainly by the choice of peroxide curing agent. If the temperature of the mold increases after repeated moldings, the temperature of the platens may have to be decreased to prevent scorching. The speed of mold closing has to be adjusted to allow complete filling of the mold and escape of all air, again without scorching. The pressure has to be sufficient to prevent thick flash but low enough to avoid excessive wear and tear on the mold. The molded silicone piece will vary in size if the pressure is varied, because of the great compressibility of silicones. To optimize the use of material in compression molding, the preform must be shaped and sized to minimize wasted flash.

The mold shrinkage of silicone rubber is about 2–4% and is affected by a number of factors, but primarily that the linear thermal expansion of silicone rubber is 17–20 times that of steel and about 2 times that of organic rubber. The shrinkage is therefore largely dependent on the temperature of the molding. It should be noted that silicone compounds with higher loadings of filler have lower shrinkage. Mold shrinkage is augmented by the release of volatiles during the cure and post-bake stages. This release of volatiles has been reduced by the basic silicone suppliers stripping the polymer prior to compounding.

Injection molding involves pressures of 5000 to 20,000 psi, and temperatures of 370 to 485°F. Injection times are approximately 3 to 10 seconds, while molding time is in the range of 25 to 90 seconds. This method relative to compression molding gives less flash, better properties, and greater uniformity. Units costing $60,000–$350,000 are being used.[18] Blow molding has also been used on some simple parts. In this process, heavy-walled parts are usually made because of the variation in wall thickness.

Prehardened stainless steel is recommended for the molds with chromium plating also used where a high finish is desired and where the undercuts are minimal. Assistance in mold design is available from the basic silicone rubber suppliers. Silicone mold-release agents are unsuitable for use with silicone rubber. Household-detergent water solutions of 0.5–2.0% are recommended for spray or brush applications to the mold. A thin layer is preferred to avoid build-up on the mold.

Extrusion

Gaskets, tubing, tape, wire and cable, seals, rods, channels, and hose in a variety of shapes and sizes may be extruded. The equipment is similar to that used with organic rubber. When silicone rubber is extruded, however, the low green strength and the decomposition temperature of the peroxide curing agent must be considered.

Typical extruders have screws with a length-to-diameter ratio of 10/1 to 12/1 although shorter screws are sometimes used. A single flight screw is usually used but sometimes a second flight is added near the discharge to minimize the pulsating of the extrusion. The screw has a compression ratio of 2/1 for harder stocks, and up to 4/1 for stickier stocks, or when less porosity or close tolerances are needed. The compression ratio is preferably provided by a variable-pitch screw, keeping the flight depths constant from feed to discharge. If the flight depth is reduced, there is a tendency in the longer barrels to build excessive heat. Because of the abrasive nature of silica fillers, especially the larger particle sizes, both the barrel and screws should be built of abrasion-resistant alloys. However, many of the organic rubber houses are successfully extruding tubing and wire and cable using unhardened equipment. Both the barrel and screw should also be water-cooled to prevent scorch.

To prevent feeding problems, a roller feed is usually used. The roller is fed from a "hat" (coiled strips of compound) purchased from a silicone supplier or fed from strips removed from an in-house mill. The roller should have a higher speed than the screw. The breaker plate should contain at least two screens (40–200 mesh) to ensure clean extrusions. A fine mesh is backed by a coarser backing screen. This assembly increases the back pressure that ensures air removal and better dimensional control. The finer screens will give cleaner extrusions, but will slow production and will have to be changed more often. Usually the screens are not cleaned but are discarded after removal.

A front flange assembly is used to hold and center both the die and the mandrel (pin) when extruding tubing. Fine adjustments are made with adjusting screws on either the die or the pin. The die should be made from prehardened stainless steel and designed to produce smooth flow with no dead spaces to hold up material that may start to cure. The pin should be drilled so that low-pressure air can be used to support the tubing and keep it round. The die opening is often different from both the shape and size of the extrusion because of differential flow and die swell. These differences can be compensated for to some extent by reducing the thickness of the lasts where the die dimensions are smaller. Die swell is also greater if the silicone polymer is branched or if the extrusion is speeded up. More heavily loaded compounds have less die swell. The extrusion can be considerably reduced in size by stretching it before cure. Again, the shrinkage that takes place on curing and post-bake must be taken into account.

In forming wire and cable insulation, the extruder must be fitted with a crosshead. Reinforced hose may also be made by feeding tubing reinforced with high-temperature fiber or wire through a crosshead and extruding a second layer of silicone rubber over it.

Several methods of vulcanization are available for extrusions. Pictures or diagrams of these are shown in Lynch's book.[18] Hot-air vulcanization is very common and the catalyst is *bis*-(2,4-dichlorobenzoyl) peroxide. Other catalysts are too volatile and cause bubbling in the extrudate and lead to undercure due to loss of catalyst. Two types of hot-air vulcanization are possible: vertical and horizontal. In horizontal, the extrudate is laid on an endless belt and passes through an oven 10 to 30 feet long. The oven is usually at 600–800°F although the first zone may be cooler to minimize loss of catalyst. Air turbulence should be present inside the oven to improve heat transfer and to drive off the volatile by-products. A vertical hot-air vulcanizer with a variable-speed drum at the top may also be used. Cure usually takes place only on the up-cycle, and speeds up to 60 feet per minute are obtained.

Although hot-air vulcanization can be used to cure wire and cable insulation, continuous steam vulcanization is usually used because it is much faster with speeds up to 1200 feet per minute. Benzoyl peroxide is preferred to 2,4-dichlorobenzoyl peroxide to avoid scorching. Continuous-steam vulcanization is more expensive, longer, and more complicated. Other less common types of vulcanization are steam autoclave vulcanization, hot-liquid vulcanization and fluidized-bed vulcanization.

Calendering

Continuous thin sheets of unsupported silicone rubber are produced on a calender. One can also coat a reinforcing fabric on one side with a 3-roll calender or on both sides with a 4-roll calender. The resulting product gets its strength from the fabric and its flexibility, and good electrical properties and moisture resistance from the silicone rubber. Silicone rubber should be processed at slower speeds than organic rubber. 0.2 to 2 feet per minute is best for start-up until even release from the rolls is assured and running speeds are generally around 5 to 10 feet per minute.

Silicone rubber is often calendered onto fabrics such as glass, nylon, aromatic nylons, polyester, and cotton. A variety of weaves are used. Woven fiber glass has the best combination of properties and is the most commonly used with silicone rubber. The high-temperature properties of aromatic nylons such as Kevlar or Nomex are making them increasingly popular.

A typical 4-roll calendering set up is shown in Figure 13.9. An extruder or a ram feeds silicone rubber to the first and fourth rolls. These rolls are slower and warmer than the center rolls. Since silicone rubber moves to the faster and cooler rolls, it is transferred to the center rolls. The center rolls then transfer the rubber to the fabric. While the outside rolls may be warmed to facilitate transfer, the temperature must be kept below 130°F in order to prevent scorching. If the silicone rubber is unsupported, the fabric is replaced with a liner. Polyester film, holland cloth, polyethylene and release-coated paper are commonly used. The liner is stripped off after cure or just before use if the sheet is to be used green. A major use for green sheet is to cut it into strips to build

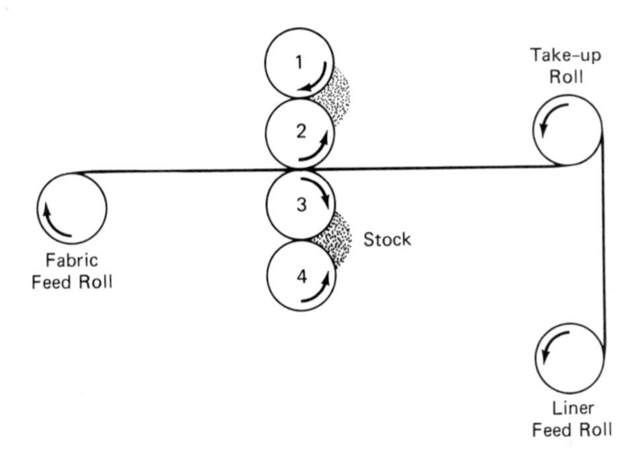

Fig. 13.9. Schematic diagram of set-up for calendering supported sheet.

hoses or to manufacture electrical tape. A 3-roll calender is actually more common than a 4-roll calender. In that case, roll 4 in Fig. 13.9 is eliminated.

Calendering sheet can be vulcanized in a steam autoclave provided the roll is first pressure-taped. The thickness of the roll must be controlled and preferably heated inside and out to achieve complete cure. The sheet can also be cured or semicured by eliminating the liner feed, and bypassing the supported or unsupported sheet over a hot drum placed between the calender and the take-up roll.

Dispersion Coating of Fabric

This technique of fabric coating permits thinner coatings and provides more thorough penetration of the fabric than does calendering. A thin coat from a 5% to 15% silicone-rubber dispersion will improve the strength and flex life of glass cloth, and provide a good "anchor coat" for calendering. Excellent high-temperature electrical insulating materials, diaphragms, and gaskets can be made from glass cloth with thicker coatings. Silicone rubber can also be dispersion-coated on organic fabrics, and then used in many applications such as aircraft seals, radome covers, and general-purpose control diaphragms.

The basic manufacturers will supply silicone rubber dispersions of soft, readily dispersed pastes that have been especially designed for cloth coating. However, any silicone rubber compound can be dispersed in solvent and probably most dispersions are made at the fabricator's plant.

Xylene, toluene and mineral spirits are common solvents, except for the fluorosilicones for which the solvent must be a ketone such as methyl ethyl or methyl isobutyl ketone. Chlorinated solvents and solvents containing antioxidants, rust inhibitors, and similar additives should be avoided since they interact with the peroxide vulcanizing agent.

The coating pastes are readily dispersed with a propeller mixer. Compounds containing reinforcing fillers should be freshened, sheeted off thin, cut into small pieces, and soaked overnight in just enough solvent to cover the compound. The mixture should then be stirred with the propeller mixer until uniform. The remaining solvent is then added, with mixing, in small portions. The dispersion should be filtered through an 80- to 150-mesh screen (depending upon consistency) before use.

Benzoyl peroxide is normally used for curing because it has a high-enough decomposition temperature and a low-enough vapor pressure to permit the use of heat to remove solvent after coating. It is best to add crystalline benzoyl peroxide to the dispersion in the form of a 5% solution in toluene or xylene. Preventing overheating during or after peroxide addition is essential.

Dip-and-flow coating is used for priming and for applying thin coatings (for example, coating a 4-mil electrical-grade glass cloth to 10 mils overall thickness). Thickness is controlled by coating speed and by dispersion solids concentration (ordinarily 5% to 25%).

Usually, the uncoated cloth enters the dip tank at a 45° angle, passes under an idler roll at the bottom of the tank, and then up into a vertical oven or coating tower. Excess dispersion flows off the cloth between the tank and the tower. The coating tower is heated by hot air, and, ideally, is divided into three zones. Solvent is removed in the first zone at 150–175°F. Vulcanization takes place in the second zone at 300–400°F in the case of glass cloth, and at 250–300°F in the case of organic fabrics. When glass cloth is being coated, the third zone is maintained at 480–600°F. This last temperature range removes final traces of volatiles, including peroxide decomposition products. Maximum bond to the cloth and optimum electrical properties are developed in the third zone.

Dispersion coating is somewhat faster than calendering, with speeds of 10–20 feet per minute common.

Dip-and-knife coating is similar to dip and flow. However, thicker dispersions (35% solids is typical) are used, and a knife or rod is placed between the dip tank and the tower. The coated fabric is pulled past the knife (a second knife can be placed on the opposite side of the fabric), which wipes off excess dispersion. Thickness is controlled by dispersion solids concentration and knife position relative to the fabrics. Heavier coatings can be obtained by this method than by dip and flow.

In *reverse-roll coating*, the knife is replaced by a roll which rotates in opposition to the direction of cloth movement. This often improves penetration.

Heavy-duty Hose

The production of hose on a continuous basis was described earlier in the section on extrusions. Heavy-duty hose for autos, trucks, aircraft and industrial use requires different techniques.

Uncured and semicured, unsupported and fabric-supported silicone rubber sheets and tape may be fabricated into ducts and heavy-duty hose. This is done by wrapping a hollow mandrel, a collapsible mandrel, a core made with a low-temperature-melting alloy or a foundry sand core. Aluminum mandrels are widely used, but release of the finished part is often a problem. The mandrel should be sprayed or brushed with a dilute aqueous solution of household detergent or dusted with mica or talc.

The aluminum mandrel works best for straight sections of hose. Silicone-rubber sheets are wrapped around the mandrel to the desired thickness. It is easier to get a tight, wrinkle-free construction if the mandrel is turned on a lathe. Smooth liner for hose may be made by butt-wrapping uncured tape or sheet on the mandrel, or by using extruded and cured tubing.

For more complex shapes with curves, a core made from a low-temperature-melting alloy or from foundry sand and resin is often used. The cores are spiral-wrapped with fabric-reinforced tape cut on a 45° bias. Bias-cut tapes are more stretchy and give the hose maximum flexibility.

After the hose has been built, it should be pressure-taped with wet cotton or wet nylon. This is necessary to prevent sponging during cure, and to provide maximum ply adhesion.

Curing is usually accomplished by running steam into the hollow mandrel, or by placing the wrapped core in a steam autoclave. Curing time and temperature depend upon the peroxide used, and the heat capacity of the hose assembly. Following cure, the pressure tape should be removed, and the mandrel stripped, while the assembly is still warm. The release agent should be washed off before the part is oven post-baked. If a core of a low-temperature-melting alloy has been used, the alloy is melted to free the cured hose. If a core of foundry sand has been used, the brittle core is broken up to release the cured hose.

Bonding

Silicone rubber can be bonded to many materials, including iron, nickel, copper, zinc, aluminum, titanium, various steels, ceramics, glass, masonry, many plastics, organic and inorganic fabrics, vulcanized and unvulcanized silicone rubber and other elastomers.

In all cases, it is essential to clean thoroughly the surface to be bonded. Metal surfaces containing loose scale, oxides, other salts, and embedded dirt should be sandblasted, sandpapered or acid-etched. Metal surfaces should be cleaned with a solvent such as methylene chloride or trichloroethylene. Some surfaces, such as plastics and vulcanized rubbers should be roughened with abrasives. It helps to wash the surface with acetone before application of the primer. Primers are applied in dilute solution by brushing, dipping or spraying. If the bond is inadequate, the layer of primer may be too thick or the primer solution may be too concentrated. One may also try several primers to find the one best suited to the particular materials being bonded.

The typical primer for silicone rubber has two types of reactive sites. The first reactive site reacts with the surface hydroxyls of the metals, glass, or masonry. The reactive specie is usually an alkoxy silane (e.g., SiOMe) which hydrolyzes to a hydroxy silane that can then form either a hydrogen bond or a covalent bond with the surface hydroxyls. A catalyst to speed this reaction may also be present. The second reactive site in the primer is some type of unsaturation to react with the silicone rubber. The hydrolysis of the alkoxysilanes usually takes place within 60 minutes of the time the primer is applied. The reaction with the surface hydroxyls takes place while the primer is drying and also during vulcanization. The reaction of the primer unsaturation with the silicone rubber takes place during vulcanization and is catalyzed by peroxide.

Bonding Unvulcanized Silicone Rubber. Curing agent should be added to the freshened compound. After preforming, the stock should be carefully laid on the freshly primed surface. Vulcanization must be done under heat and pressure at temperatures appropriate to the peroxide used. Approximately 15–30 minutes at 330–350°F is usually sufficient when using dicumyl peroxide or VAROX, and 15–30 minutes at 260–280°F, when using powdered benzoyl peroxide. A stepped oven post-bake is usually required to develop optimum bond strength.

The self-bonding compounds may be bonded to many metals in the same manner, except that no primer is required. Unlike bonding to primed surfaces, this technique is insensitive to part geometry and to degree of compound flow in the mold during bonding.

Bonding Vulcanized Silicone Rubber. Cured silicone rubber may be bonded to a primed surface with a 10–40 mil interlayer of heat-curing silicone-rubber adhesive. Vulcanization is conducted under heat and pressure. The pressure must not be so great as to squeeze out the adhesive. This bonding can also be accomplished under pressure at room temperature with a room-temperature-vulcanizing silicone rubber adhesive. Both techniques can be used for splicing, or for bonding cured silicone rubber to itself.

The self-bonding compounds are excellent for bonding cured silicone rubber to itself, and are recommended for splicing. These bonds will usually be stronger than those obtained with adhesives.

Post-baking

With trimming or deflashing, oven post-baking is often the final step in the fabrication of silicone-rubber parts. Post-baking removes volatile materials such as low-molecular-weight silicones and peroxide decomposition products. Removal of the low-molecular-weight silicones results in less shrinkage later and in low extractables. Removal of the peroxide decomposition products improves reversion resistance, compression set, electrical properties, chemical resistance, and the bond to other substrates.

Ovens should have forced-air circulation with no dead-air pockets. Air flow should be sufficient to keep the oven atmosphere outside the explosion limits of the volatiles from the rubber. Oven vents to outside the building should be checked periodically to be sure they are not plugging up and reducing air flow. Temperature control should be $\pm 10°F$ and the oven should go up to 500–600°F. The oven should be equipped with a temperature-limit switch, and a safety switch should be provided to turn off the heat if the blower stops.

The rubber charge should be on stainless steel trays designed to provide maximum contact with the circulating air. The parts must *not* be in contact with one another, which would slow the removal of volatiles and lead to parts sticking together.

For some applications, no post-bake is necessary. However, it is generally desirable to post-bake at least a few hours at 300–400°F. Usually, longer cures to at least 50°F above the service temperature are required. Thicker sections should be step-cured, especially if the aroyl peroxides have been used. Also, 2,4-dichlorobenzoic acid and benzoic acid will cause internal reversion if the part containing them is not step-cured.

There are no set curing schedules that will cover every situation that a fabricator may encounter. In each case, the optimum curing schedule must be worked out. The size and shape of the part, number of parts in the oven, and the air flow through the oven have major influences on the curing cycle. A possible schedule for a thicker part is 3 hours at 300°F followed by 4 hours at 350°F followed by 10 hours at 400°F.

LIQUID SILICONE RUBBER COMPOUNDS

The silicone rubber polymers we have dealt with up to now have a common formula with the polymers in liquid silicone rubbers, i.e.,

$$X-\left[\begin{matrix} CH_3 \\ | \\ Si-O \\ | \\ CH_3 \end{matrix}\right]_n-\begin{matrix} CH_3 \\ | \\ Si-X \\ | \\ CH_3 \end{matrix}$$

However, while $n = 200–1500$ for the liquid silicone rubbers, it equals 3000–11,000 for the conventional silicone rubbers. For liquid silicone rubber the chainstopper X depends on the cure system and is typically hydroxy, vinyl or methoxy. For the conventional silicone rubbers, the chainstopper X is typically methyl or vinyl. The liquid silicone rubber compounds have historically been called *room-temperature-vulcanizing* (RTV) rubber, but the group does contain some compounds which are vulcanized at high temperature. Indeed the high-temperature-addition-cure liquid systems have a combination of easy processing, fast cure, and excellent physical properties. Good growth is anticipated in this area as experience is gained by the processors.

Liquid silicone rubber is not quite as old as the higher-consistency silicone rubber, but the first references for the former appeared about 30 years ago. Since then, both the technology and the commercial use have developed to the point at which their market rivals the market for conventional silicone rubber. The production of liquid silicone rubbers requires careful quality control if the material is not to cure prematurely or too slowly. Thus they are sold as "ready-to-use" products. The manufacture and packaging is still done mainly by the basic silicone suppliers although the last 5 years have seen some custom formulators introduce standard and specialty products.

Suitable compounding ingredients include the silicone polymer, crosslinker, and catalyst as well as the usual reinforcing and extending fillers, color pigments, heat age additives, adhesion promotors, stabilizers, thickeners, plasticizers, and other additives.

Liquid silicone rubbers come in both one- and two-component systems. Chemically they can be subdivided by cure type into the three categories below.

Condensation Cure: One-component

These are called one-component systems because both the catalyst and crosslinking agent are incorporated in the base compound at the time of manufacture.

These one-component systems are dependent on moisture for curing. They are generally made by compounding a silanol-stopped silicone polymer with an excess of crosslinker. A reaction takes place during compounding, endstopping the polymer with the crosslinker.

Upon exposure to atmospheric moisture, further reaction takes place leading to cure of the rubber. Both steps are shown below.

(1) Reactions during compounding:

$$\mathrm{HO}\!\left(\begin{array}{c}\mathrm{Me}\\|\\\mathrm{Si{-}O}\\|\\\mathrm{Me}\end{array}\right)_{\!n}\!\mathrm{H} + \mathrm{MeSi(Y)_3} \rightarrow \mathrm{Me}\begin{array}{c}\mathrm{Y}\\|\\\mathrm{Si}\\|\\\mathrm{Y}\end{array}\!\mathrm{{-}O}\!\left(\begin{array}{c}\mathrm{Me}\\|\\\mathrm{Si{-}O}\\|\\\mathrm{Me}\end{array}\right)_{\!n}\!\begin{array}{c}\mathrm{Y}\\|\\\mathrm{Si}\\|\\\mathrm{Y}\end{array}\!\mathrm{{-}Me}$$

(2) Reactions during cure:

$$2\ -\overset{|}{\underset{|}{\mathrm{Si}}}\mathrm{Y} + 2\mathrm{H_2O} \rightarrow 2\ -\overset{|}{\underset{|}{\mathrm{Si}}}\mathrm{OH} \rightarrow -\overset{|}{\underset{|}{\mathrm{Si}}}-\mathrm{O}-\overset{|}{\underset{|}{\mathrm{Si}}}-$$

$$2\ -\overset{|}{\underset{|}{\mathrm{Si}}}\mathrm{Y} + \mathrm{H_2O} \rightarrow -\overset{|}{\underset{|}{\mathrm{Si}}}\mathrm{OH} + -\overset{|}{\underset{|}{\mathrm{Si}}}\mathrm{Y} \rightarrow -\overset{|}{\underset{|}{\mathrm{Si}}}-\mathrm{O}-\overset{|}{\underset{|}{\mathrm{Si}}}-$$

Vulcanization will occur first at the surface, and then progress inward with the diffusion of moisture into the rubber.

These rubbers are classified as acidic, basic, or neutral, depending on the nature of the byproduct hydrolyzed during the cure. Acids released are acetic[19] and 2-ethylhexanoic.[20] Bases released are cyclohexylamine and butylamine.[21] Neutral groups used are aldoxime,[22] ketoxime,[23] alkoxy,[24] enolate,[25] aminoxy,[26] and amido.[27]

Organo-tin and organo-titatium catalysts are used in small amounts. These products make excellent adhesive sealants, formed-in-place gaskets and can be used to form films from solvent dispersion.

Condensation Cure: Two-component

These are called two-component systems because the crosslinker and (or) catalyst must be added to the base compound just before use. They are generally made by using a silanol chainstopped polymer. The crosslinker is usually an alkoxy silane or oligomers thereof (e.g., ethyl o-silicate) and the catalyst is an organic tin compound such as dibutyltindiacetate, dioctyltinmaleate[28] or dibutyl tin dilaurate.[29] The catalyst level is generally much higher in the two-component systems than in the one-component. Applications include molds for plastic parts, coatings, adhesives, therapeutic gels, and encapsulants. This system is used where deep-section cures are needed although aeration to remove the ethanol byproduct is desirable for complete cure. The cure can be independent of moisture although moisture can accelerate the cure:

$$6\ -\overset{|}{\underset{|}{\mathrm{Si}}}\mathrm{OH} + (\mathrm{EtO})_3\mathrm{SiOSi(OEt)}_3 \overset{\mathrm{Sn}}{\rightarrow} (-\overset{|}{\underset{|}{\mathrm{Si}}}\mathrm{O})_3\mathrm{SiOSi(O}\overset{|}{\underset{|}{\mathrm{Si}}}-)_3$$

Addition Cure

These are usually two-component systems although one-component is possible if the catalyst is inhibited sufficiently. In this case, the inhibitor must be volatilized, or decomposed by heat in order to start the cure. These systems are made by adding a polyfunctional silicon hydride crosslinker to a vinyl containing silicone polymer. Vinyl on the end of the polymer is more reactive than vinyl on the chain. The catalyst is usually a complex of platinum although palladium, rhodium and ruthenium are also used:

$$-\overset{|}{\underset{|}{Si}} - H + CH_2{=}CH\ \overset{|}{\underset{|}{Si}}- \xrightarrow{Pt} -\overset{|}{\underset{|}{Si}} - CH_2 - CH_2 - \overset{|}{\underset{|}{Si}}-$$

The reaction is exothermic, but the concentration of reactive groups is too low to cause much of a temperature rise. This system is used where deep section cures are needed in a confined space since there are no volatile by-products. They have excellent resistance to compression set and to reversion. Applications include flexible molds, dip-coating, and the potting and encapsulation of electrical and electronic components.

The cure rate and physical properties of all the liquid silicone rubber compounds can be varied over a rather wide range by suitable choice of polymer type and molecular weight, reinforcing and extending fillers, pigment, cross-linker, and type and concentration of catalyst.

REFERENCES

1. *Chemical Week*, Jan. 28, 1981, pp. 26–27.
2. *Rubber & Plastics News*, May 21, 1984, p. 11.
3. *Chemical Economics Handbook—SRI International*, Dec. 1980.
4. W. Noll, *Chemistry and Technology of Silicones*, 2 Ed., Academic Press, New York, 1968, p. 305.
5. Ibid, pp. 306–309.
6. H. A. Liebhafsky, *Silicones Under the Monogram*, John Wiley & Sons, New York, 1978, p. 67.
7. Ibid, p. 105.
8. J. F. Hyde, U.S. Pat. 2,490,357 (Dec. 6, 1949).
9. J. G. E. Wright and C. S. Oliver, U.S. Pat. 2,448,565 (Sept. 7, 1948).
10. J. Marsden, U.S. Pat. 2,445,794 (Jul. 27, 1948).
11. J. C. Caprino and R. J. Prochaska, U.S. Pat. 3,096,303 (Jul. 2, 1963).
12. S. W. Kantor, 130th Meeting, Am. Chem. Soc., Sept. 1956.
13. F. M. Lewis, *High Polymer Series*, Vol. 23, *Polymer Chemistry of Synthetic Elastomers*, Part 2, Chap. 8, J. P. Kennedy and E. G. M. Tornquist, eds., John Wiley & Sons, New York, 1969.
14. W. J. Bobear, *Rubber Chem. Technol.* **40,** 1560 (1967).
15. G. Alliger, and I. J. Sjothun, eds., *Vulcanization of Elastomers*, Reinhold, New York, 1964, p. 370.
16. M. L. Dunham, D. L. Bailey, and R. Y. Mixer, *Ind. Eng. Chem.* **49,** 1373 (1957).
17. B. B. Boonstra, H. Cochrane, and E. M. Dannenberg, ''Reinforcement of Silicone Rubber by Particulate Silica,'' *Rubber Chem. Technol.* **48,** 558 (1975).
18. W. Lynch, *Handbook of Silicone Rubber Fabrication*, Van Nostrand Reinhold, New York, 1978.
19. Brit. Pat. 835,790; Belg. Pats. 569,320 and 577,012; U.S. Pats. 3,032,532, 3,035,016, 3,133,891, and 3,382,205.

20. U.S. Pat. 4,247,445.
21. Ger. Pats. 1,120,690, and 1,255,924; U.S. Pats. 3,032,528 and 3,291,772.
22. Fr. Pat. 1,432,799.
23. Belg. Pats. 614,394 and 637,096; U.S. Pats. 3,184,427 and 3,189,576.
24. Ger. Pats. 1,169,127 and 2,713,110; U.S. Pats. 3,294,739, 3,334,067, and 4,100,129.
25. Ger. Pats. 1,248,049, 2,335,569, and 2,827,293.
26. U.S. Pat. 3,359,237.
27. Belg. Pat. 659,254; Ger. Pat. 1,224,039; Neth. Pat. 6,501,494.
28. J. C. Weis, in *Progress of Rubber Technology*, Vol. 46, Chap. 3, ''Silicone Rubber,'' Elsevier, U.K., 1984, p. 91.
29. U.S. Pats. 2,843,555 and 3,127,363.

14

FLUOROCARBON ELASTOMERS*

HERMAN SCHROEDER (*Ret.*)
E. I. du Pont de Nemours & Company
Polymer Products Department
Wilmington, Delaware

INTRODUCTION

The fluoroelastomers are highly specialized products that as a class show the best resistance of all rubbers to the attack of heat, chemicals, and solvents. This chapter covers the technology of the fluorocarbon elastomers based on carbon chains containing fluorine, usually hydrogen, and occasionally other substituents. Fluorosilicones, the other major fluoroelastomer, are described in Chapter 13, on silicones. Over the 12 years since the last edition, understanding and application of these products has increased substantially as the result of the development of new types possessing better processibility and greatly improved compression set and steam resistance. There are now five manufacturers (Table 14.1). By 1983, world-wide usage (excluding the U.S.S.R.) surpassed 7 million pounds worth over $80,000,000, but their value-in-use in terms of critical problems solved is far greater.

HISTORY

The fluorocarbon elastomers are all offshoots of Plunkett's 1938 discovery[1] of the inertness of polytetrafluoroethylene, later marketed under the trademark TEFLON® by Du Pont. The Du Pont scientists pursued their new lead and soon found other interesting fluoroolefin polymers and copolymers. Copolymers of

*I would like to thank my friend and colleague, Dr. Arthur Nersasian, for his valued advice and help in preparation of this chapter.

vinylidene fluoride (VF_2) were observed to be leathery[2] while elastomeric products were found in the copolymers of olefins with tetrafluoroethylene (TFE)[3] and with 3,3,3-trifluoropropene-1.[4]

In the press of the post-World War II industrial boom, these developments were put aside until, in the early 1950's, the emerging aerospace industry found it had a critical need for an elastomer with better heat and fuel resistance to be used in seals and hose for military jet engines. This stimulated increased interest in the earlier fluoroolefin polymerizations, which soon generated elastomeric products. First was a copolymer of vinylidene fluoride (VF_2) with chlorotrifluoroethylene (CTFE),[5] later called Kel-F.® This was followed very quickly by a better copolymer made with VF_2 and hexafluoropropene (HFP)[6] and then an even more thermally stable and more solvent resistant terpolymer VF_2/HFP/TFE.[7] These products now known as VITON® A (copolymer) and B (terpolymer) were soon joined by FLUOREL,® and later by DAI-EL® and TECNOFLON® brands of the same products.[16]

The first samples of VITON® were enthusiastically welcomed by the Air Force Laboratory in Dayton, Ohio, and quickly applied to key problems in jet engines. This generated a need for large amounts of product, which was met by the installation of manufacturing facilities in 1957.

Following a spectacular success in military planes, these VF_2/HFP polymers gradually moved into civilian applications in the petroleum and process industries, where their high value-in-use became apparent. As business grew, however, their poor processibility and certain other product deficiencies became obvious. These were met with a series of developments. First came delayed-action amine-type curing agents that improved processibility.[8] Next was a major advance in high-temperature compression-set resistance based on a cure system which involved a quaternary phosphonium accelerator and bisphenol AF.[9] This system was used in the then novel VITON® E-60 and related FLUOREL® types. Finally, products with good steam resistance were obtained through use of a peroxide-curable structure in which free radical systems attack a reactive bromine atom attached to the chain.[10]

The fluorocarbon rubbers, though very expensive and elastomers of last resort, have been of critical importance in solving urgent problems in the aerospace, automotive, chemical and petroleum industries. Their success has led to the development of related products often superior in selected special uses. These include copolymers of pentafluoropropene with VF_2 and with VF_2/TFE[11] and the rediscovered TFE/propylene copolymer which was found to have excellent steam resistance.[12] Outstanding was the discovery of the first elastomeric perfluorocarbon, a TFE/perfluoro(methyl vinyl ether) copolymer, the most stable of all known elastomers.[13] It has found extensive use in solving critical problems in aggressive environments.

COMMERCIAL FLUOROCARBON RUBBERS

Manufacturing Processes

The major VF_2/HFP and VF_2/HFP/TFE elastomers are usually prepared by radical polymerization in emulsion using peroxy compounds such as ammonium persulfate as initiator, occasionally in redox systems with or without chain-transfer agents such as carbon tetrachloride, alkyl esters, or halogen salts. A fluorinated soap may be used but is not required. The normally continuous polymerization is followed by centrifugation or coagulation and isolation of dry polymer by standard manufacturing techniques. Product is sold in the form of pellets, chips, strips, and slabs, which are translucent, white, or slightly colored depending on the presence of compounding agents.

The other fluoroolefin copolymers are prepared by similar processes. In all cases, great care must be exercised because of the potential toxicity and explosiveness of some of the monomers or monomer mixtures, particularly TFE and HFP. Monomers must be pure since the polymerization of fluoroolefins is inhibited by a wide variety of substances.

A list of the manufacturers is shown in Table 14.1.

Commerical Types

With the rapid expansion of the fluorocarbon elastomers into a very wide spectrum of uses requiring diverse properties, each manufacturer sought to develop

Table 14.1. Manufacturers.

Company	*Trademark*
E. I. du Pont de Nemours & Co. 1) Polymer Products Dept. 2) Finishes and Fabricated Products Dept. Wilmington, DE 19898	VITON® KALREZ®
3M Company Commercial Division, 3M Center St. Paul, MN 55144	FLUOREL® KEL-F®
Montedison, USA 1114 Avenue of the Americas New York, NY 10036	TECNOFLON®
Daikin Industries Ltd. Shinjuku Sumitomo Blvd. 6-1, 2 Chome, Nishi Shinjuku Shinjuku-ku, Tokyo, Japan	DAI-EL®
Xenox Inc. P.O. Box 79773 Houston, TX 77279	AFLAS®

specialties to fit. As a result, there has been a proliferation of products. They are listed in Table 14.2 by categories since space limitations do not permit a complete description. For details on the properties, applications and handling of these very special products, it is wise to contact the manufacturers.

RAW-GUM FLUOROCARBON RUBBERS

The fluorocarbon rubbers are transparent to translucent gums ranging in molecular weight from 5000 for a waxy specialty like VITON LM; through 100,000 to 200,000 for the mainline types from all manufacturers; to over 200,000 for the KEL-F products, which are very tough. The vinylidene fluoride copolymers are soluble in certain ketones and esters, the tetrafluoroethylene/propylene copolymer in halogenated solvents, whereas a perfluorinated structure like that in KALREZ parts is almost completely insoluble. Excepting KEL-F and KALREZ, the products are all designed to process readily in standard rubber equipment such as mills, internal mixers, extruders, and calenders depending upon the particular type. They can be compounded with appropriate fillers, curatives, plasticizers, and so on, and then converted to cured articles by normal techniques. Since they are unreactive, they cure relatively slowly and a high-temperature post-cure is generally necessary. The raw polymers are very stable thermally but have no useful tensile or mechanical properties.

COMPOUNDING

A typical fluorocarbon rubber formulation contains curing ingredients, acid acceptors, fillers, and processing aids, plus occasional release agents, colors, and so forth with details dependent on the proposed application.

Curatives

In general, curing is effected with agents that remove hydrogen fluoride to generate a cure site that reacts with a diamine,[14] or a bisphenol,[9] or with organic peroxides[10] that promote a radical cure by hydrogen or bromine extraction. Preferred amines are blocked diamines such as hexamethylene diamine carbamate (Diak No. 1) or bis(cinnamylidene) hexamethylenediamine (Diak No. 3). Preferred phenols are hydroquinone and the bisphenols such as 4,4′-isopropylidene bisphenol or the corresponding hexafluoro derivative, bisphenol AF. The preferred accelerator is a quaternary phosphonium salt, benzyltriphenylphosphonium chloride. VITON curative masterbatches containing bisphenol AF are sold by Du Pont as Curative 30 and by Montedison as TECNOFLON MI. Masterbatches containing quaternary phosphonium salt are sold by Du Pont as VITON Curative 20, by 3M as FLUOREL FC-2172, and by Montedison as

Table 14.2. Commercial Fluorocarbon Rubbers.

Type Name	*Mooney Visc. ML 1 + 10 100°C (212°F)*	*Sp. G.*	*Uses*	*Comments*
Vinylidene Fluoride/Hexafluoropropylene (FKM) Copolymers $(CH_2CF_2)_x(CF_2\text{-}CF(CF_3))_y$				
VITON LM	2000 cps (100°)	1.72	Processing aid	Semiliquid rubber
VITON C10	7–17	1.82	Caulks, sealants, coatings	
DAI-EL G-101	–	1.76	Processing aid	Semiliquid rubber
VITON A-35	32–42	1.80	Easy molding	Low-viscosity VITON A
TECNOFLON-FOR-LHF	60	1.79	Low-hardness compounds	Plasticized (contains curative)
FLUOREL 2145 (2220)	23 at 121°C	1.81	Blending, (solutions)	Low viscosity
FLUOREL FC 2175	51 at 121°C	1.81	Binders and adhesives	FDA-approved
FLUOREL FC 2230	37 at 121°C	1.81	Blending polymer	FDA-approved
FLUOREL FC 2178	120 at 121°C	1.82	High-strength uses	
VITON A	59–71	1.82	General purpose	Pellet form
VITON A-HV	147–173 (at 121°C)	1.82	High-strength uses	Pellet form
TECNOFLON NMLB	60–79	1.81	General-purpose types	Amine-curable
TECNOFLON NML	60–70	1.81	General-purpose types	Amine-curable
TECNOFLON NMB	80–90	1.81	Metal adhesion	Amine-curable
TECNOFLON NM	80–90	1.81	General purpose	Amine-curable
TECNOFLON NH	130 at 121°C	1.81	High-strength uses	Amine-curable
Special Polymers with Low Compression Set				
VITON E 45	45–55	1.82	Blend polymer for mold flow and demolding tear	Base for E-430
VITON E 60	60–70	1.82	As above and also O-rings	Base for E-60C
VITON E 60C	50–65	1.81	High-temperature comp-set resistance	Contains curatives
VITON E 430	38–52	1.81	Molding polymer	Contains curatives
FLUOREL FC 2170	30 at 121°C	1.81	O-rings, diaphragms	Contains curatives

Type Name	Mooney Visc. ML 1 + 10 100°C (212°F)	Sp. G.	Uses	Comments
Vinylidene Fluoride/Hexafluoropropylene (FKM) Copolymers $(CH_2CF_2)_x(CF_2\text{-}CF(CF_3))_y$				
Special Polymers with Low Compression Set (cont.)				
FLUOREL FC 2173	28 at 121°C	1.80	High flow, molding	Contains curatives
FLUOREL FC 2174	38 at 121°C	1.80	Good mold release	Contains curatives
FLUOREL FC 2177	32 at 121°C	1.81	Good metal adhesion	Contains curatives
FLUOREL FC 2176	30 at 121°C	1.81	All-purpose grade	Contains curatives
FLUOREL FC 2180	40 at 121°C	1.80	O-rings	Contains curatives
FLUOREL FC 2181	43 at 121°C	1.81	Valve seats, O-rings	Contains curatives
FLUOREL FC 2179	80 at 121°C	1.81	Down-well packers	Contains curatives
FLUOREL FC 2152	50 at 121°C	1.80	Fast cure, molded goods	Contains curatives
FLUOREL FC 2182 (2120)	28 at 121°C	1.80	Fuel hose, (extrusions)	Contains curatives
TECNOFLON FOR-45, 45BI	50–60	1.80	Injection molding, fast cure	NML/B, NML with curatives
TECNOFLON FOR-70, 70BI	70–80	1.80	General molding	NM/B with curatives
TECNOFLON FOR-45, C, CI	50–65	1.80	Hot tear resistance	Contains curatives
TECNOFLON FOR-60K, KI	50–60	1.80	Shaft seals	Contains curatives
TECNOFLON FOR-50E	45–55	1.80	High flow, extrusions	Contains curatives
DAI-EL 701	60	1.81	Static applications	Contains curatives
DAI-EL 751	57	1.81	High adhesive strength	Contains curatives
DAI-EL 702	60	1.81	Dynamic applications	Contains curatives
DAI-EL 704	50	1.81	Good extrudability	Contains curatives
DAI-EL 755	50	1.81	Roll processability	Contains curatives

(cont.)

Table 14.2. (*Continued*)

Type Name	*Mooney Visc. ML 1 + 10 100°C (212°F)*	*Sp. G.*	*Uses*	*Comments*
Vinylidene Fluoride/Hexafluoropropylene/Tetrafluoroethylene (FKM) Terpolymers $(CF_2\text{-}CH_2)_x(CF_2\text{-}CF(CF_3))_y(CF_2CF_2)_z$ **and other high-fluorine types**				
VITON B-50	50–60	1.86	Coatings, fuel hose	Blending polymer
VITON B	66–82	1.86	General purpose	General-purpose type
VITON B-70	60–80	1.77	Automotive applications	Low-temperature uses
VITON B-910	50–65	1.85	Gaskets and seals	Contains curatives
FLUOREL FC 2350	55 at 121°C	1.86	Fabric composites	Contains curatives
FLUOREL FC 2481	85 at 121°C	1.87	Molded goods	General-purpose type
FLUOREL FC 2330	30 at 121°C	1.86	Moldings and tubing	Contains curatives
TECNOFLON TH	70–90	1.91	Fluid resistance	High-fluorinate type
TECNOFLON TN-50	55–65	1.86	Moldings, gaskets and shaft seals	Lower-viscosity TN
TECNOFLON TN	115–130	1.86		
TECNOFLON FOR-THF	60–80		Fluid resistance	TH with curatives
TECHNOFLON FOR-TF50	55–65		Flue ducts and expansion joints, moldings	Lower-viscosity TN + curatives
TECHNOFLON FOR-TF	100–110	1.84		TN with curatives
DAI-EL G-201	55	1.84	Blending polymer	Amine cure
DAI-EL G-501	110	1.84	General purpose	Amine cure
DAI-EL G-602	90	1.84	Seals	Contains curatives
DAI-EL G-603	60	1.90	Low compression set	Contains curatives
Peroxide-curable Types				
DAI-EL® 801	75	1.81	Food industry	Flex resistance
DAI-EL 901	105	1.90	Food industry	Alcohol- and solvent-resistant
DAI-EL 902	55	1.90	Good mechanical properties	
FLUOREL FC 2460	60 at 121°C	1.80	O-rings, molded goods	Covulcanizable
FLUOREL FC 2690	90 at 121°C	1.87	Diesel liner, oil-field seals	Low solvent swell

Type Name	*Mooney Visc. ML 1 + 10 100°C (212°F)*	*Sp. G.*	*Uses*	*Comments*
		Peroxide-curable Types (cont.)		
FLUOREL FC 2480	80 at 121°C	1.80	General purpose	Steam resistant
VITON GLT	80–100 at 121°C	1.78	Seals, hose, etc.	Flexible at lowest temperatures
VITON GF	55–70	1.91	Fluid-resistant applications, fuel systems	Highest fluorine content
VITON VTR-4730	80–100 at 121°C	1.86	Fuel systems, tubing	For extrusions
		Tetrafluoroethylene/Propylene Copolymer: $(CF_2-CF_2)_x(CH_2CH(CH_3))_y$		
AFLAS 100	85	1.55	General purpose	
AFLAS 150E	70	1.55	Extrusions and moldings	Slab form
AFLAS 150P	130	1.55	General purpose	All resistant to polar solvents
AFLAS 100H	-	1.55	High-pressure extrusion resistance	
		Vinylidene Fluoride/Chlorotrifluoroethylene Copolymer: $(CF_2CH_2)_x(CF_2CFCl)_y$		
KEL-F 3700	127 at 170°C	1.85	Resists oxidizing acids	
		Perfluorocarbon Rubbers (Tetrafluoroethylene Copolymers) $(CF_2CF_2)_x(CF_2CF(OCF_3))_y(CF_2CF(X))_z$		
KALREZ perfluoroelastomer parts	Sold only as parts		Seals, diaphragms, O-rings, hose sheet	Highest temperature and solvent resistance

Table 14.3. Cure System.

Characteristics	*Diamine*	*Bisphenol*	*Peroxide*
Scorch safety	P–F[a]	G–E	G–E
Balance of fast cure and scorch safety	P	E	E
Mold release	G	E	F
Ability to single-pass Banbury mix	No or risky	Yes	Yes
Adhesion to metal	E	G	G
Tensile strength	G–E	F–E	G–E
Compression-set resistance	F	E	G
Steam, acid resistance	F	G	E

[a]E = excellent; G = good; F = fair; P = poor.

TECNOFLON M2. Relative performance of these systems is summarized in Table 14.3.[15]

Some properties are more dependent upon the amount of curative than type, e.g., demolding tear. Scorch in bisphenol cures depends upon the ratio of accelerator to curative (see Table 14.12).

Acid-acceptor Systems

Acid acceptors are necessary in compounded fluorocarbon elastomers since they serve to neutralize hydrogen fluoride generated during the cure or on extended ageing at high temperature.

The materials most commonly used are shown in Table 14.4. Note that it is important that low-activity magnesium oxide *not* be used in bisphenol cures and also that high-activity magnesium oxide *not* be used in diamine cures.

Table 14.4. Acid Acceptors.

Acid Acceptor	*Usage*
Magnesium oxide (MgO)—low activity	General-purpose diamine cures.
Magnesium oxide (MgO)—high activity	General-purpose bisphenol cures.
Litharge (PbO)	Steam and acid resistance in all cures.
Zinc oxide/Basic lead phosphate (ZnO/Dyphos)	Low compound viscosity in bisphenol stocks.
Calcium oxide (CaO)	Added to minimize fissuring; can aid metal adhesion.
Calcium hydroxide ($Ca(OH)_2$)	General purpose with MgO.

Table 14.5. Fillers for Fluorocarbon Rubbers.

Filler	*Comments*
MT Black (N908)	Best general-purpose filler; excellent compression set and heat-ageing.
Austin Black (coal fines)	Better high-temperature compression-set resistance than MT, but less reinforcing and poorer in processing and tensile strength/elongation.
SRF Black	High strength, high modulus compounds; aggravates mold-sticking in peroxide cures.
Blanc Fixe ($BaSO_4$)	Best compression set of nonblack fillers; neutral filler good for colors; poorer tensile strength than MT Black.
Nyad 400 (fibrous $CaSiO_3$)	General-purpose mineral filler, neutral and good for control stocks; tensile comparable with MT.
Ti-Pure® R-960 (TiO_2)	Good for light-colored compounds; good tensile but poorer heat-ageing than other fillers.
Red Iron Oxide	Used at 5–10 phr with other neutral mineral fillers for red-brown compounds.
Graphite Powder, or TEFLON® Powder	Combined at 10–15 phr with other fillers to improve wear resistance.
Celite 350	General-purpose neutral filler; good tensile strength.

Fillers

Various carbon blacks and mineral fillers are all used in fluorocarbon elastomers to confer special properties as shown in Table 14-5.

PROCESSING AIDS PLASTICIZERS RELEASE AGENTS

The processing characteristics of fluorocarbon rubber compounds can be improved. Particularly effective aids are the high-molecular-weight hydrocarbon esters such as dioctyl phthalate (DOP), pentaerythritol stearate (PET), carnauba wax, sulfones, and low-molecular-weight polyethylene (AC-617A, 1702). They insure smoother extrusions, and better die definition and mold release. Polyethylene should not be used in peroxide cures because it aggravates mold-sticking. Stocks can be softened with esters, but such additives reduce high-temperature stability since esters are far less stable than fluorocarbons. Many other process aids are commercially available and very effective, including Du Pont's VPA No. 1 and VPA No. 2. The low-viscosity FKM rubbers are also often used to facilitate processing without compromising properties.

PROCESSING AND CURING

Fluorocarbon elastomers must be processed with care. Equipment must be clean and free of oil and grease. The usual contaminants such as sulfur, water, and other polymers must not be present. They either interfere with the cure or compromise fluorocarbon elastomer properties. This need for cleanliness is especially important in Banbury mixing, where inadvertent contamination is more likely.

Mill Mixing. This should be done on as cool a mill as possible with chilled cooling water to assure shearing characteristics for good dispersion and minimum scorch. Raw polymer bands readily. Once the band is formed and cut a few times, compounding should be started immediately. For fast mixing, compounding ingredients should be preblended. This is particularly important for preventing mill-roll sticking, which commonly occurs if magnesium oxide is added separately. If polymer blends are needed, the higher-viscosity fluorocarbon rubber should be banded first, followed by the lower-viscosity variety to assure a uniform blend.

Internal Mixing. Most fluorocarbon compounds cured with bisphenol or peroxide cure systems can be mixed in "one-pass" in an internal mixer. Diamine cures are too scorchy for the Banbury. Properly handled with ample cooling water, compounds present less of a potential scorch problem in a Banbury than on large mills. The mix cycle should be kept to 3–4 minutes and the batch dropped when the temperature indicator is in the range 93–104°C (200–220°F).

Extruding. Extrusion of VITON® and FLUOREL® fluorocarbon rubbers is widely practiced to produce finished goods and preforms. In general, lower-viscosity grades will extrude more easily. Compounds with diamine-cure systems, particularly Diak No. 3, will generally extrude well at temperatures generated by the shear heat from the screw and need only moderate heat added. Bisphenol compounds generally require more heat. The head and die should be heated to 110°C (230°F) and 127°C (260°F) respectively. The very-high-fluorine products such as VITON GF also require heating to assure smooth extrudates.

Calendering

Most of the FKM group of fluoroelastomers (VF_2/HFP copolymers and terpolymers) can be calendered. Quality and economy depend upon many factors, particularly viscosity at the time of calendering. Mixed stocks should be used promptly or stored cool (<18°C [65°F]) and great care taken to exclude moisture. Following careful warming to 27°C ± 6°C (81°F ± 10°F) stock should be continuously strip-fed across the calender roll to maintain a uniform rolling bank (1–1.5 inches, 2.5–3.8 cm). Typical roll temperatures for calendering VITON E-60C or related FLUOREL types are:

Top roll	85°C ± 3.5°C (185° ± 5°F)
Middle roll	74°C ± 3.5°C (165° ± 5°F)
Bottom roll	Cool
Speed	7–10 yd or meters/min.

Molding

Most fluoroelastomer compounds can be molded by compression, transfer, or injection processes. Higher-viscosity polymers are best-suited for compression molding and may even be required, depending on the design of the mold (vs. lower-viscosity analogs) for elimination of trapped gases. VITON E-45 is probably as low a viscosity fluoroelastomer as should be considered. Transfer and injection molding generally require polymers of lower viscosity than does compression molding. Also, for transfer- and, especially, injection-molding processes, good scorch safety will be more important than in compression molding.

Injection molding is becoming very important. Injection-molding systems require proper selection of elastomeric compound, molding conditions, and mold design to insure the right combination of good flow, low scorch, and optimal physical properties such as the best (or lowest) compression set for O-rings. Scorch safety is needed to allow flow through parts and throughout the mold.

For all these processes, products such as VITON E-45, E-60, B-50, FLUOREL FC 2152, 2176, 2181 and related TECNOFLON and DAI-EL types are recommended. Mold-release agents are normally needed; products such as AQUAREX L®, silicone oils and polyethylene emulsions are effective. Shrinkage is caused by relatively low quantities of reinforcing filler and loss of volatiles generated during post-cure. Molds must be designed to correct for shrinkage to provide specification parts.

Curing

Although fluorocarbon elastomers are inert and slow to react, cures can be effected with surprising speed with appropriate formulations, details for which should be obtained from the respective manufacturers.

In general, full compounds are subjected to a two-stage cure cycle to achieve the best balance of vulcanizate properties. The initial cure (pre-cure) in most cases requires pressure and is usually carried out in a hydraulic press or steam autoclave at temperatures up to 204°C (400°F). Pre-cure times may be as short as 1–5 minutes or as long as 2–3 hours (steam). This is followed by a long oven cure at atmospheric pressure.[17] An oven-cure cycle of 15 hours at 232°C (450°F) is required to optimize tensile and compression set[18] properties, though a cycle as short as 10 hours may be sufficient. Higher-temperature oven-cure

Table 14.6. Oven-cure Time vs. Physical Properties.

	Percent of 24-hour Properties[a]		
Oven-cure Time (hr)	*100% Modulus*	*Tensile Strength*	*Compression Set*[b]
	VITON® E-60C		
0	68	83	42
5	91	87	83
10	95	102	125
15	102	101	107
24	100	100	100
	VITON B-910		
0	73	73	43
5	93	93	100
10	99	92	104
15	100	98	100
24	100	100	100

[a]24 hr at 232°C is standard cure.
[b]70 hr at 200°C, Dash 214 O-rings.

cycles can improve tensile strength and elongation at break for VITON E60C and compression-set resistance for VITON E60C and VITON B-910 or related FLUOREL products. In most cases, a cycle of 10–12 hours will develop over 90% of maximum physical properties (Table 14.6). Properties suitable for special applications are obtained by careful selection of cure system and product type.

Processing of Other Fluorocarbon Rubbers

The vinylidene fluoride/chlorotrifluoroethylene copolymer KEL-F 3700, though very tough, can be formulated with ingredients on a two-roll rubber mill at 77–88°C (170–190°F). Curing is effected with amine (Diak No. 1) or peroxides with metal salts or oxides as acid acceptors. The tetrafluoroethylene-propylene copolymer AFLAS can be processed like the other fluorocarbon rubbers using similar fillers. Curing is effected with peroxide systems. The perfluoroelastomer in KALREZ parts is very difficult to process and is sold only as finished parts or stock.

PHYSICAL PROPERTIES OF CURED FLUOROCARBON RUBBERS

The fluorocarbon rubbers are exceptionally stable, excelling all rubbers in overall resistance to combinations of heat, light, ozone, solvents, and aggressive

Table 14.7. Elastomer Comparison ASTM D2000-SAEJ200 Classification.

Type	*Heat-ageing Temp., °C (70 hr)*[a]	*Volume Swell, % 10 hr/130°C in ASTM No. 3 Oil*
Nitrile[b]	100	10, 40 or 60
Polyacrylic[b]	130	0 or 60
Silicone	200 or 225	120 or 80
Fluorosilicone	200	10
Fluorocarbon (FKM)	250	10

[a]Tensile change ± 30%, elongation change −50%, hardness change ± 15 points.
[b]Varying acrylonitrile content or acrylate content.

chemicals. They also have good high-temperature compression-set resistance and low-temperature flexibility. Many products have been developed to give the particular combinations of properties required for the varied highly sophisticated applications for which very special formulations have been developed. Because it is not possible in a chapter of this length to cover the many details, only broad principles will be indicated. In Table 14.7, heat and oil resistance of some fluorocarbon rubbers are compared with a few other rubbers.[18, 19]

In Table 14.8 the fluoroelastomers are compared with one another. These comparisons are approximations representing a variety of compounds and are intended only to point out salient features and differences between types.

Some specific properties for an important FKM-type elastomer are presented in Table 14.9. The three compounds are based on VITON A and represent a standard formulation and those designed for high elongation or for high modulus.

Other physical properties of a typical FKM stock (FLUOREL 2146) containing 15 parts Thermax, 20 parts calcium oxide, and 1.2 parts Diak No. 1 cured 30 min/149°C and post-cured 24 hr/204°C are shown in Table 14.10.

Heat Resistance

FKM fluoroelastomer vulcanizates are considered serviceable almost indefinitely when exposed continuously up to 200°C. Estimated service life at other temperatures is shown in Table 14.11.

The effects of heat ageing on typical FKM vulcanizates is illustrated in Table 14.12. These properties are illustrative of those obtainable with the VF_2/HFP and VF_2/HFP/TFE copolymers in general.

Compression-set Resistance

The most important market for fluorocarbon elastomers is in seals and O-rings, for which compression-set resistance has proved to be the best criterion of ser-

Table 14.8. Selected Properties of Fluoroelastomer Vulcanizates.

	VITON A *FLUOREL 2140*	*VITON E-60C* *FLUOREL 2170*	*VITON B* *FLUOREL 2481*	*VITON GF* *FLUOREL 2690*	*VITON GLT*	*KEL-F 3700*	*KALREZ Perfluoro-elastomer*	*SILASTIC LS 53*	*AFLAS*
			Stress Strain 24°C (75°F)						
100% modulus, MPa (psi)	← ← ←	← ← ←	4.14–6.9 (600–1000)	→ → →	→ →	2.1–5.5 (300–800)	0.7–1.4 (100–200)		2.1–18 at 50%
Tensile, psi, MPa (psi)	← ← ←	← ← ←	12.1–20.7 (1750–3000)	→ → →	→ →	→ →	→ →	8.3 (1200)	17.2 (2500)
E_B, %	← ← ←	← ← ←	100–300	→ → →	→ →	200–300	100–200	220	55–400
Hardness, durometer	← ← ←	← ← ←	60–95	→ → →	55–90	53–85	65–95	50	55–75
			Heat Resistance						
<50% elong. in 1 wk at °C	285	285	295	295	285	250	315	260	280
Continuous service Temp. °C	200	200	210	210	200	175	250+		200
			Low-Temp. Properties, °C						
Brittle point	−30	−30	−42	−45	−59	−40	−40	−66	−46
Clash-Berg, T-10000	−14	−14	−12	−5	−23	-	−3	−60	−2
TR_{10}	−17	−17	−13	−1	−30	−15	−3	-	3
			Compression Set						
(Method B), % 200°C (70 hrs)	20–70	10–20	20–70	30–40	20–40	40	63–80	60	30

	VITON A	VITON E-60C	VITON B	VITON GF					
	FLUOREL 2140	*FLUOREL 2170*	*FLUOREL 2481*	*FLUOREL 2690*	*VITON GLT*	*KEL-F 3700*	*KALREZ Perfluoro-elastomer*	*SILASTIC LS 53*	*AFLAS*
Fluid Resistance, Volume % days at 75°F									
ASTM ref. Fuel A	0	0	0			2		15	Fair
ASTM ref. Fuel B	2	2	1			16	1	28	
ASTM ref. Fuel C	5	5	4	3	7				
Diester aircraft lubricant (Stouffer 7700)	20	20	15	10	22				
ASTM Oil, No. 3	3	3	2	1	3	9			
Cyclohexane	4	4	4				<1	16	13
Benzene	22	22	12			55	3	23	31
Toluene	8	8	6	4		50	<1	22	
Butyl acetate	200	200	190				3	160	
Ethyl acetate	280	280	250				3		88
Carbon tetrachloride	1	1	1			30	4	20	86
Acetone	200	200	175				2	180	50
Methyl ethyl ketone	240	240	210				<1	220	
Tetrahydrofurane	200	190	200				<1		
Methanol	40	40	40						
Ethanol	6	6	6			6	0	5	
Nitrobenzene	24	24	20				<1		
Water	2	2	2			15			

Table 14.9. Physical Properties of FKM Vulcanizates.

Formulated for:	*High Elongation Low Modulus*	*Standard 75 ± 5 Duro. A*	*High Modulus*
M_{100} MPa	2.07	6.07	17.2
(psi)	(300)	(880)	(2500)
T_B, MPa	14.5	13.8	19.3
(psi)	(2100)	(2000)	(2800)
E_B, %	450	180	120
Hd, Durometer A	59	77	95
E_B, % at 177°C (350°F)	210	80	40
Compression set, Method B 70 hr at 200°C (392°F)	38	22	32

Table 14.10. Additional Physical Properties of FKM Vulcanizate.

Tensile strength, MPa (psi)	12.1 (1750)
Elongation, %	195
Hardness, Shore A	73
Tear strength (ASTM D-624-54, B) lb/in.	180
Bashore resilience	5
Abrasion resistance Tabor, H-22, 1000 g at 1000 rev, wt. loss g	0.031
Low-temperature Properties	
Brittle point (ASTM D746-52T)	
0.075″ thickness, °C (°F)	−35 (−30)
0.025″ thickness, °C (°F)	−40 (−40)
Gehman stiffness (ASTM D-1053-52T), °C (°F)	
T_2	− 8 (+17)
T_5	−14 (+ 7)
T_{10}	−16 (+ 3)
T_{100}	−24 (−11)
Electrical Properties	
D.C. resistivity	
at 50% rel. humidity, ohm-cm;	2×10^{13}
at 90% rel. humidity, ohm-cm;	1.5×10^{13}
Dielectric strength (short time), volts/mil (20 mil specimen)	630
Dielectric constant, 100 cps, 25°C	11.4
Dissipation factor, 100 cps, 25°C	0.0125

Table 14.11. Service Life vs. Temperature.

Limit, hr of Service	*Temperature, °C (°F)*
>3000	230 (356)
1000	260 (410)
240	290 (464)
48	315 (509)

vice performance. As a result of the discovery of the phosphonium chloride accelerator system for bisphenol AF and other phenol cures, excellent results are readily obtained with VITON® E-60C and FLUOREL® 2170 and related products. Such products are now available in a variety of viscosities, both in precompounded form and uncompounded, for which accelerator masterbatches are offered. Effects of variation in accelerator system on compression set and some other properties are shown in Table 14.13.

Compression-set resistance is also strongly affected by the filler. MT blacks provide the best overall balance of tensile properties and heat and compression-set resistance. Only Austin Black and Koch MDC coal provide better compression set, but at a sacrifice in mechanical properties and processibility.

Low-temperature Flexibility

The commercial FKM elastomers have brittle points of about −25°C to −40°C, which are not easily improved by compounding. Plasticizers tend to harm heat stability and accelerate deterioration on ageing since they are less stable than the fluoropolymers. Peroxide-curable fluorocarbon rubbers can be blended with fluorosilicones to improve low-temperature flexibility, but at a sacrifice in high-temperature stability and solvent resistance. One novel copolymer, VITON GLT, is inherently superior at low temperatures with a brittle point of −51°C (−59°F) and Clash-Berg stiffness (T − 10,000) of −31°C (−24°F). By contrast, a typical FKM vulcanizate has a T − 10,000 of −17°C (+1°F). Another new product, Tecnoflon® FOR, is said to have a stable fluorinated amide plasticizer to improve low-temperature hardness, brittle point, and compression set without sacrificing physical properties.[20] Low-temperature properties of typical FKM elastomers are shown in Table 14.14.

COMPOUNDING FOR SPECIFIC PROPERTIES

Fluids and Chemical Resistance

FKM fluorocarbon elastomer vulcanizates are very resistant to hydrocarbons, chlorinated solvents and mineral acids. They swell excessively in many polar

Table 14.12. Heat Resistance of FKM Fluoroelastomers.

	A-12	*B-12*
VITON A	100	-
VITON B	-	100
Magnesia	15	15
MT Carbon Black	20	20
Diak No. 3	2	3
Pressure: min/°C	30/163	30/163
Oven posture: hr/°C	24/204	24/204

	Original		*100 days at 232°C (450°F)*		*20 days at 260°C (500°F)*		*2 days at 316°C (600°F)*		*1 day at 343°C (650°F)*	
	A-12	*B-12*	*A-12*	*B-12*	*A-12*	*B-12*	*A-12*	*B-12*	*A-12*	*B-12*
Tensile strength, MPa	15.0	15.5	6.90	4.31	8.62	3.79	7.24	3.45		3.97
psi	2175	2250	1000	625	1250	550	1050	500	Brittle	575
Elongation at break, %	470	410	160	480	100	400	60	240		15
Hardness, Duro A	68	74	87	75	94	83	91	83	99	91
Weight loss, %	-	-	-	-	-	-	18	11	36	22

Table 14.13. Compression Set with Cure System Variation[a]

	a[b]	*b*	*c*	*d*	*e*	*f*	*g*	*h*
VITON Curative 20[c]	1.8	0.75	1.5	2.25	2.25	3	2.25	0.75
VITON Curative 30[d]	4.0	2.0	2.0	2.0	3.00	3	6.0	6.0
Mooney Scorch (MS) at 121°C (250°F)								
Minimum value	35	39	37	40	37	41	37	38
5 pt. rise, min.	1 pt 45 min	1 pt 45 min	17	9	15	9	1 pt 45 min	2 pts 45 min
Cure: Press-minutes at 177°C (350°F)								
	10	10	10	10	10	10	10	30
Oven	<			24 hrs at 232°C (450°F)				>
Compression Set (Method B), %—1 × 0.139 in. (25.4 × 3.5 mm) O-rings:								
70 hr at 24°C (75°F)	9	30	23	23	10	14	5	4
70 hr at 200°C (392°F)	16	59	37	39	22	24	13	10
336 hr at 200°C (392°F)	37	73	67	65	45	50	32	25
70 hr at 232°C (450°F)	37	76	64	65	50	50	34	30
Stress-strain								
100% modulus, MPa	4.1	1.9	2.2	2.1	3.1	3.8	6.4	5.3
psi	600	275	325	300	450	550	925	775
Tens. strength, MPa	9.8	5.7	7.8	7.6	9.0	9.3	9.7	9.7
psi	1425	825	1125	1100	1300	1350	1400	1400
Elongation	220	430	390	420	310	280	150	160
Hardness, Duro A	70	64	63	64	67	68	74	71

[a]All the compounds contain VITON E-60 (100 parts), MT carbon black (30 parts), Maglite D (3 parts) and calcium hydroxide (6 parts).
[b]Roughly equivalent to VITON E-60C. [c]33% organophosphonium salt. [d]50% dihydroxy aromatic compound.

Table 14.14. Low-temperature Properties.

	VITON E-60C FLUOREL FC 2170	*VITON B-910 FLUOREL FC 2350*	*VITON B-70*	*VITON GLT*
Brittle point °C (°F)[a]	−25 to −30 (−13 to −22)	−35 to −40 (−30 to −40)	−35 to −40 (−30 to −40)	−51 (−59)
Clash-Berg, °C (°F) at 69 MPa (10000 psi)	−16 (+2)	−13 (+9)	−19 (−3)	−31 (−24)

[a]These values are often difficult to reproduce.

solvents such as ketones, some esters and ethers. They are attacked by amines, alkali, and some acids, e.g. hot anhydrous HF and chlorosulfonic acid. In general, stability and solvent resistance are a function of fluorine content, as shown in Table 14.15.[19,20,21,22]

Other than type, the main determinant of chemical resistance is the metal oxide used in the compound. For example, VITON B compounded with magnesium oxide swells 61% in volume in red fuming nitric acid, but only 45% when compounded with zinc oxide-Dyphos, and 24% with litharge. General-purpose FLUOREL 2141 with magnesium oxide swells 110% in concentrated hydrochloric acid at 158°C while a litharge compound will swell only 2%. KALREZ and AFLAS compounds are markedly superior to the FKM elastomers in resistance to amines and strong alkali, AFLAS because it lacks the activated hydrogen of the vinylidene fluoride FKM polymers and KALREZ because it contains no hydrogen.

Adhesion

The adhesion of FKM elastomers to metals is important for many applications. Good adhesion during molding and curing is obtained if the metal surface is properly prepared and a suitable adhesive primer used.[24,25] Recommended ma-

Table 14.15. Effect of Fluorine Content on Solvent Swell.

		Percent Swell	
FKM Copolymer	*% Fluorine*	*Benzene/21°C*	*Skydrol B/121°C*
VF_2/HFP	65	20	171 (at 100°C)
VF_2/HFP/TFE	67	15	127
VF_2/HFP/TFE/CSM[a]	69	7–8	45
[b]TFE/PFMVE/CSM[a]	71	3	10

[a]Cure-site monomer.
[b]Perfluoroelastomer.

terials are Chemlok 607 or Chemosil 511 and Thixon XAV-273/66 or a 50/50 mix of Chemlok 607 with Dow Corning Z6020. Best adhesion is obtained with Diak No. 1 cures, but at a sacrifice in compression set. Bisphenol curing systems are often very successfully used when calcium oxide or hydroxide (2 phr) is added to a stock containing low-activity magnesium (15–17 phr).

Steam Resistance

Peroxide cures of FKM compounds give markedly better steam resistance than diphenol or diamine cures. Steam resistance also increases with fluorine content in FKM stocks. The TFE/propylene copolymer AFLAS and particularly the perfluoroelastomer KALREZ surpass the FKM rubbers in this respect.

Resistance to Automotive Fuels

A combination of automotive regulations, higher under-hood and under-body temperatures, and the use of more highly aromatic, nonleaded gasoline has focused attention on the FKM elastomers in automotive applications. As a result, they are used increasingly in automotive fuel hose because of superior resistance to ''sour'' gasoline (fuel containing peroxides), to gasoline/alcohol mixtures, and to permeation.[26] A comparison of the swelling of selected solvent-resistant elastomers is given in Table 14.16.

When compared with other fuel-hose-rubber materials (epichlorohydrin rubber and various NBR rubbers), fluorocarbon rubbers have exhibited outstanding resistance to attack by ''sour'' (peroxide-containing) gasoline and the lowest-volume swell. Epichlorohydrin copolymer reverts toward uncured product (becomes soft, devulcanized) and the various nitrile rubbers embrittle.[27a] Similar studies of permeation by fuels have shown a fluorocarbon elastomer hose com-

Table 14.16. Swelling of Elastomers in Fuel Blends.

Fuels	*Fuel Composition (Volume %)*				
Gasoline (42% aromatic)	100	85	75	85	75
Methanol	–	15	25	–	–
Ethanol	–	–	–	15	25
Rubber	*Equilibrium Volume Increase % at 54°C*				
VITON AHV	10.2	28.6	34.9	19.7	21.1
VITON B	10.2	22.9	25.6	17.3	18.3
VITON GF	7.5	13.6	14.6	12.6	13.1
Fluorosilicone (FK)	16.4	25.3	26.6	23.1	23.0
Nitrile rubber (NBR)	40.8	90.5	95.8	62.4	66.6
Epichlorohydrin (ECO)	42.4	92.6	98.1	75.6	78.5

pound to be superior to all hydrocarbon elastomer materials tested, with permeation rates often over 100 times better than for unfluorinated rubbers.[27b]

Resistance to Exhaust Gases

The largest existing application for fluorocarbon rubbers in energy-related industries is flue-duct expansion joints used in desulfurization systems for coal-fired plants. The fluorocarbon elastomer is needed to resist the high temperatures and the wet acidic flue gas streams that together cause metal or less chemically resistant elastomeric expansion joints to corrode and/or deteriorate.[20]

While all FKM terpolymers are more effective in this application than the copolymers, the peroxide-curable high-fluorine terpolymers exhibit the best resistance. Peroxide cures are more effective than bisphenol cures, which in turn surpass diamine cures.[28]

Coatings and Sealants

FKM elastomers are often needed in coatings and sealants for which the lower-viscosity types such as VITON® C-10, VITON A-35, and FLUOREL 2145 are usually preferred. Typical solvents used are methyl ethyl ketone, ethyl acetate, methyl isobutyl ketone, and amyl acetate and related ketones and esters. A typical sealant formulation based on a 50/50 VITON C-10/LM mix also contains magnesia (5 phr), MT carbon black (15 phr), FS-1265 fluorosilicone oil (10 phr); all then is mixed with 56 phr of methyl ethyl ketone and Epon H-2 curing agent (2 phr). This product has a useful storage life of one week at 24°C (75°F).[29] For coating applications, the elastomer is mill-mixed with magnesia (15 phr), MT black (20 phr), and diamine curing agent (ca. 1–1.5 phr), and all is dissolved in 200 phr of methyl ethyl ketone. Such a product has a seven-day shelf life. After two weeks at 24°C (75°F), it has a 100% modulus of 4.1 MPa (600 psi), tensile strength of 6.2 MPa (900 psi) and elongation of 940%.

VF_2/CTFE (75/25 vol.%) KEL-F 3700

Though hampered in its commercial development by its toughness, this product continues to perform very well in the presence of strong oxidizing acids, and under such conditions has demonstrated superior resistance to flex-cracking under dynamic loads. It is also used as a binder in solid-propellant systems. Typical compounds are cured with peroxide or Diak No. 1 (HMDA carbamate) as shown in Table 14.17.

TETRAFLUOROETHYLENE/PROPYLENE COPOLYMER—AFLAS

This product has a different chemical-resistance profile and better electrical resistance properties as compared with the FKM rubbers and is of about equal

Table 14.17. Properties of KEL-F Vulcanizates.

	A	B
KEL-F® 3700	100	100
Zinc oxide	10	10
Dyphos	10	10
Luperco 101 XL	1	-
Diak No. 1	-	3
TAIC (Diak No. 7)	6	-
Press cure: minutes/temp.	15/177°C(350°F)	30/177°C(350°F)
Oven cure: hours/temp.	24/232°C(2450°F)	16/177°C(350°F)

	A		B	
Properties	*Orig.*	*Aged 30 days at 204°C (400°F)*	*Orig.*	*Aged 30 days at 204°C (400°F)*
Tensile str., Mpa	27.7	26.2	16.6	15.5
psi	4000	3800	2400	2250
Elongation, %	220	300	350	330
100% modulus, MPa	5.5	5.4	2.1	3.0
psi	800	750	300	400
Hardness, Duro A	65	67	53	61
Compression Set, Method B				
70 hr/200°C (392°F)	41		62	
Chemical Resistance, Volume Swell %				
20% Fuming H_2SO_4, 7 Days at 24°C	3		8	
38% HCl, 3 Days at 70°C	40		54	
ASTM No. 3 Oil, 7 Days at 149°C	9		10	

thermal stability. The points of better resistance are in exposure to steam, amines and amine corrosion inhibitors, wet sour gas and oil, phosphate ester hydraulic fluids, glycol brake fluids, and some other polar systems plus high- and low-pH environments. It is not as resistant to many hydrocarbons, chlorinated solvents, and ethers. Because of these properties, the major areas of usage are oil-field applications, chemical processing, automotive, aerospace, and industrial; plus on wire and cable exposed to chemically aggressive media.[30,31] A typical formulation is shown in Table 14.18.

KALREZ PERFLUOROELASTOMER PARTS

KALREZ parts are made from a perfluoroelastomer that is in a sense a rubbery derivative of polytetrafluoroethylene plastic. In aggressive fluid or gaseous en-

Table 14.18. Properties of TFE/P Compounds.

	A	*B*
AFLAS® 150P	100	100
TAIC (75% dispersion)[a]	7.5	7.5
Vul-Cup 4015E	2.5	2.5
Sodium stearate	2.0	2.0
MT Black (N990)	28	–
FEF Black (N550)	–	25
Vulcanizate Properties		
Hardness, Shore A	70	83
Elongation, %	160	110
Tensile strength, MPa	17.2	17.9
psi	2500	2600
Electrical Properties		
Volume resistivity (ohms cm, 500 watts dc)		
at 21°C		3.0×10^{16}
at 200°C		1.7×10^{13}
Dielectric constant, 60 hz		
at 21°C		2.5
at 150°C		3.0

[a] Triallylisocyanurate.

vironments, no other elastomer can equal the overall performance of KALREZ parts. These products combine the resilience and sealing force of an elastomer with chemical inertness and thermal stability similar to polytetrafluoroethylene resins. They resist attack by nearly all chemical reagents including ethers, ketones, esters, amines, oxidizers, fuels, acid, and alkalis. They provide long-term service in virtually all chemical and petrochemical streams, even where corrosive additives cause other elastomers to swell or degrade. They are not suitable for use with molten potassium or sodium metals.

KALREZ parts retain electric properties in long-term service as high as 288°C (550°F) and in intermittent service up to 316°C (600°F). They are usually reliable at temperatures as much as 83°C (150°F) higher than those sustainable by parts made from even other fluoroelastomers. Consequently, KALREZ parts usually out perform other elastomeric sealing material in difficult environments.

The polymer is extremely expensive and in fabrication requires unique tooling plus difficult and complex manufacturing techniques. Consequently, it is offered only in the form of finished parts, including O-rings, tubing, rods, sheeting, and so on, plus custom-designed parts. These parts are used extensively in oil exploration and processing, in the chemical industry, and in scientific instrumentation.[31, 32]

SUMMARY OF APPLICATIONS

The fluorocarbon rubbers are widely used in critical applications demanding their unmatched heat and fluid resistance. Increasingly, they are also supplanting more conventional materials because they offer exceptional value-in-use, along with a long trouble-free lifetime that insures the soundness of product warranties. The most important uses are as O-rings, shaft seals, gaskets, fuel hose, valve-stem seals and flue-duct expansion joints. They also enjoy significant usage as binders for military fuels (in propellants), as processing aids for low-density polyethylene, in oil-field applications, and in coatings and sealants. For the most extreme environments—deep oil exploration, exotic chemical exposure, and so forth—KALREZ perfluoroelastomer parts, though very expensive, are required—nothing else works. Fields of application for fluorocarbon rubbers, particularly the FKM type, are:

1. Typical industrial uses:

   Valve seals, O-rings and special configurations
   Valve and pump linings
   Gaskets—in refineries and chemical plants
   V-ring packings
   Expansion joints
   Hose (for chemical resistance)—rubber-lined or rubber-covered
   Wire/cable cover—in steel mills and nuclear power plants
   Rolls—100% fluorocarbon rubber or laminated to other elastomers

2. Automotive applications:

   Valve-stem seals
   Shaft seals
   Transmission seals
   Fuel-handling systems:
   - inject or nozzle seals
   - in-tank pump coupler hose
   - carburetor-pump cups, needle valves, diaphragms
   - fuel shut-off valves
   - fuel hose or fuel-hose liner

3. Aerospace:

   O-ring seals in jet engines
   Hydraulic systems
   Lubricating systems
   Fuel systems

4. Oil Well:

Drill bit seals
Packers
V-ring packers
Valve seals
Blow-out preventers

5. Other:

Oil-suction and delivery hose
Truck chemical-transfer hose
Tank linings
Chimney linings
Seals for scientific instruments

PRODUCT SAFETY

When the recommended handling procedures are followed, fluoroelastomer polymers and products based on them in themselves present no health hazards of which we are aware. As with many polymers, minute quantities of residual gases that might be harmful diffuse from uncured FKM even at ambient temperatures; therefore, all containers should be opened and the FKM used only in well-ventilated areas to avoid employee exposure.

There are potential hazards that result from the use of certain compounding ingredients or from high-temperature processing and service conditions. Before proceeding with any compounding or processing work, one should consult and follow label directions and handling precautions from suppliers of all ingredients.

A complete review of handling precautions together with health and hazard-related information and disposal recommendations for FKM and related chemicals is given in reference 33.

REFERENCES

1. R. J. Plunkett (assignor to E. I. du Pont de Nemours & Co.), U.S. Pat. 2,230,654 (1941).
2. T. A. Ford (to E. I. du Pont de Nemours & Co.), U.S. Pat. 2,458,054 (1949).
3. W. E. Hanford and J. R. Roland (to E. I. du Pont de Nemours and Co.), U.S. 2,468,664 (1949).
4. H. E. Schroeder (to E. I. du Pont de Nemours & Co.), U.S. Pat. 2,484,530 (1949).
5. M. E. Conroy et al., *Rubber Age* **76,** 543 (1955); C. B. Griffis and J. Montermoso, *Rubber Age* **77,** 559 (1955).
6. S. Dixon, D. R. Rexford and J. S. Rugg, *Ind. Eng. Chem* **49,** 1687 (1957); D. R. Rexford (to E. I. du Pont de Nemours and Co.), U.S. Pat. 3,051,677 (1962).

7. J. R. Pailthorp and H. E. Schroeder (to E. I. du Pont de Nemours & Co.), U.S. Pat. 2,968,649 (1961).
8. L. E. Robb (to 3M Co.), U.S. Pat. 3,029,227 (1962); A. L. Moran (to E. I. du Pont de Nemours & Co.), U.S. Pat. 2,951,832 (1960).
9. A. L. Moran and D. B. Pattison, *Rubber World* **103,** 37 (1971); W. W. Schmiegel, *Kautsch und Gummi* **31,** 137 (1978).
10. D. Apotheker and P. J. Krusic (to E. I. du Pont de Nemours and Co.), U.S. Pat. 4,214,060 (1980); D. Apotheker et al., *Rubber Chem. Technol.* **55,** 1004 (1982).
11. D. Sianesi et al., (to Montedison SpA), U.S. Pats. 3,331,823 and 3,335,106 (1967).
12. W. R. Brasen and C. S. Cleaver (to E. I. du Pont de Nemours and Co.), U.S. Pat. 3,467,635 (1969).
13. G. F. Gallagher (to E. I. du Pont de Nemours & Co.), U.S. Pat. 3,069,401 (1962); E. K. Gladding and R. Sullivan (to E. I. du Pont de Nemours & Co.), U.S. Pat. 3,546,186 (1971); D. B. Pattison (to E. I. du Pont de Nemours & Co.), U.S. Pat. 3,876,654 (1975).
14. J. F. Smith and G. T. Perkins, *Proc. Inst. Rubber Conf.* Preprints 575 (1959).
15. J. G. Bauerle, *VITON® Technical Notes*, July 1983.
16. R. G. Arnold, A. L. Barney and D. C. Thompson, *Rubber Chem. Technol.* **46,** 619 (1973).
17. J. D. MacLachlan, "Effect of Oven Post-Cure Cycles on Vulcanizate Properties." Du Pont Technical Bulletin VT-440.1.
18. R. E. Knox and A. Nersasian, Paper No. 770867 presented at Society of Auto. Eng. Meeting, Detroit, MI, 1977.
19. J. D. MacLachlan, "Fluorocarbon Elastomers: A Technical Review," Du Pont Technical Bulletin VT-020.884.
20. S. Geri, C. Lagana and R. Grossman, "A New Easy-Processing Fluoroelastomer," Montedison Paper No. 30.
21. L. D. Albin and R. R. Campbell, "Current Trends in Fluoroelastomer Development, 3M Technical Bulletin, see also P. F. Tuckner and J. L. Kosmala, "Fluorocarbon Elastomers," Chicago Group Seminar, March 1984.
22. J. D. Eddy and J. G. Bauerle, Du Pont Technical Bulletin VT-250.GF—VITON® GF.
23. A. L. Moran, "Compounding with VITON® Curative Masterbatches," Du Pont Technical Bulletin VT-310.1.
24. D. A. Stivers, "Fluorocarbon Rubbers," *Vanderbilt Rubber Handbook*, R. T. Vanderbilt, Co., Norwalk, CT, 1978.
25. E. T. Hackett, Jr., "Adhering VITON® to Metal During Vulcanizates," Du Pont Technical Bulletin VT-450-1; see also Du Pont Technical Bulletin VC-240.C10.
26. A. Nersasian, *Elastomerics*, Oct. 1980.
27. A. Nersasian, Paper No. 790659; and J. D. MacLachlan, Paper No. 790657, *SAE Passenger Car Meeting*, Detroit, 1979.
28. F. J. Rizzo, "Corrosion Resistant Expansion Joints," Du Pont Technical Bulletin C-VT-650.353.
29. VITON® Technical Bulletin VT-240.C10.
30. M. J. Watkins, *Corrosion Inhibitors*, Energy Rubber Group, Houston, 1984.
31. T. W. Ray and C. E. Ivey, Paper No. 68, International Corrosion Forum, New Orleans, 1984.
32. A. L. Barney, G. H. Kalb, and A. A. Khan, *Rubber Chem. Technol.* **44,** 660 (1971); S. M. Ogintz, Preprint NO77-AM-5A-1, ASLE Meeting, Montreal, 1977.
33. "Handling Precautions for VITON® and Related Chemicals, Du Pont Report VT-100.1.

15

POLYURETHANE ELASTOMERS

C. S. SCHOLLENBERGER
Hudson, Ohio

BACKGROUND

Polyurethanes are a discovery and development of O. Bayer and his chemists in Germany, first in the Main Scientific Laboratory of I. G. Farbenindustrie beginning about 1937, and later, Farbenfabriken Bayer. Since then they have been studied, developed, and improved worldwide in many laboratories and have become a resounding commercial success.

They are extremely interesting polymers from many different standpoints. For example, they have multiple fabrication forms comprising the liquid, gum, and thermoplastic resin states. The processing options for these materials include casting, centrifugal molding and injection for the liquids; and milling, Banbury mixing, calendering, extruding, injection/compression/transfer/centrifugal (powder) molding, and solution applications for the gums and resins. Finished urethane products appear in the solid, microcellular or cellular, thermoset and thermoplastic, flexible, elastic, and rigid states. The applications of the polyurethanes seem without number. They include molded and extruded goods, insulating and comfort foams, binders, adhesives, caulks, films, coatings, jackets, encapsulants, prostheses, and so on. Important technical aspects of polyurethanes include raw materials, polymerization (melt, solution, suspension), chemical and physical structure (affecting mechanical properties, morphology, processibility, and environmental stability), molecular weight, environmental stabilization, compounding, processing, and so forth.

SCOPE AND CONTENT

It can be appreciated that full treatment of all of the polyurethane material types, as interesting and important as each is, would be far beyond the scope of this chapter; so the content has been limited to the consideration of solid polyurethane elastomers. A special feature of the chapter is the attention it gives to their chemical and physical aspects. This emphasis might be considered somewhat amiss in a volume dealing with technology, but it is necessary for several reasons.

First, there is no other synthetic polymer system in commercial use today that involves such varied and complex chemistry, considering both main and side reactions, as do the polyurethanes. Second, the liquid-process polyurethane fabricator, for example, runs an individual in situ chemical polymerization reaction for every separate part manufactured, regardless of its size. Consistent, high-quality parts are of the essence in the fabricator's operation, and there are several important variables that must be accommodated to achieve this. These include monomer purity, reactant mixing ratio, catalyst level, reaction temperature, and environmental conditions (e.g., humidity), which can all significantly influence the quality of the parts produced.

No blending away of substandard polymer products with standard material is possible here: parts are either acceptable and useful, or unacceptable and scrap. So, it is clearly quite important that the liquid process fabricator understand and practice the proper chemistry in all operations. Although there are fewer polyurethane material (reactive prepolymer, gum, and resin) suppliers than fabricators, it goes without further comment that the knowledge and appreciation of urethane chemistry are as important to this group as to the liquid process polyurethane fabricator.

Finally, as if the foregoing chemical knowledge requirement were not enough, we have come to recognize that physical structure (morphology) is a very important contributor to polyurethane properties and performance, although this is much less the case with the commercial millable gum polyurethanes which have been modelled after amorphous hydrocarbon gum rubbers. Therefore a knowledge and appreciation of the morphological contribution to polyurethane properties is also needed to understand and best exploit these polymers.

The foregoing subjects having been discussed, typical examples of the various solid polyurethane elastomer systems will be provided to demonstrate the mechanics of their preparation and use. Typical property levels for each type will be presented, and polyurethane elastomer compounding and environmental stabilization will be discussed.

OVERVIEW

Rubber-like elastomer systems that have achieved commercial importance through the years are largely amorphous, low glass transition temperature (T_g), gum-like masses of polymer chains whose interchain attractive forces are relatively weak and delocalized (randomly distributed) along the polymer chain lengths. To render them useful for mechanical applications, such polymer chains must be joined in lateral fashion by covalent chemical crosslinks (CC) in the process known as *vulcanization* or *curing*. O. Bayer and his chemists, and then other laboratories worldwide, subscribed to and applied the classical (CC) vulcanization concept in explaining and exploiting the excellent mechanical properties of the polyurethanes.[1,2,31] However, it was subsequently recognized that very high level, useful mechanical properties can be obtained in amorphous, low T_g polyurethane elastomer systems which are essentially linear-structured, thermoplastic, soluble, and therefore devoid of CC.[3a,3b,4,94] This phenomenon was attributed to tiepoints among the linear polyurethane chains that are reversible with heat or solvation, and the term ''virtual crosslink'' (VC), i.e., ''crosslinked in effect but not in fact,'' was applied to such tiepoints.[3a,3b] VC are a consequence of polymer chain structure and organization and will be discussed in more detail later in this chapter.

It can be imagined that CC polyurethanes of appropriate VC structure could have significant property contributions from both CC and VC; this fact has been demonstrated.[5a,5b,6] Even so, it seems clear that at service temperature the dominant contribution to most high-level polyurethane elastomer mechanical properties (e.g., tensile strength, tear strength, abrasion resistance, modulus, hardness, and so forth)—when these are realized—is the VC contribution.[3a,3b,5a,5b,81]

Polyurethanes with strong VC are not strengthened by the incorporation of the customary reinforcing agents (e.g., carbon black) used for conventional rubbers, but do respond strongly to polymer chemical-structure changes that further enhance their VC. The latter practice has been referred to as ''chemical'' or ''internal'' compounding and is the principal method of polyurethane mechanical-property regulation. So we see yet another aspect of polyurethanes that sets their technology apart from that of the conventional hydrocarbon elastomers.

Another notable feature of polyurethanes is the thermal dissociation tendency of the urethane group—the linkage generated during polyurethane formation that builds the polymer chains. This dissociation is a temperature-dependent, chemical equilibrium process whose degree is minor at room temperature, where the equilibrium constant is very large, and at use temperatures. With increasing and high temperatures, however, as more urethane groups dissociate to form isocyanate and hydroxyl terminal groups, the polymer chains shorten, their viscosities drop, and the equilibrium constant decreases.[7]

URETHANE CHEMISTRY

The chemical basis of the polyurethanes is the isocyanate group and its ability to react with the hydroxyl group to form the urethane linkage. A multitude of isocyanates has been prepared and described in the literature.[13,14] Eq. 1 shows the reaction of a monoisocyanate, phenylisocyanate, with a monohydroxy compound, ethyl alcohol, and the thermal reversibility of this equilibrium reaction.

Eq. 1. Urethane Formation

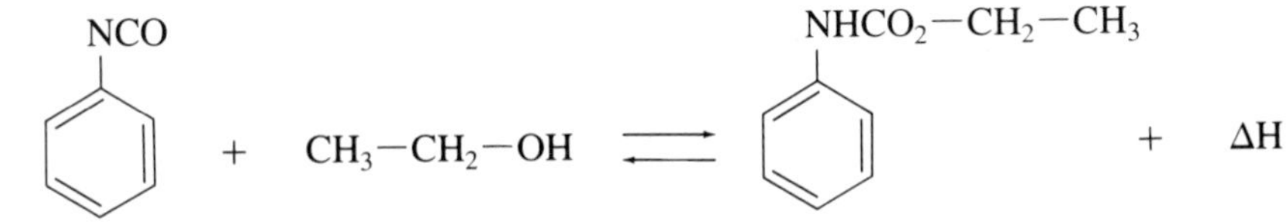

$$C_6H_5{-}NCO + CH_3{-}CH_2{-}OH \rightleftharpoons C_6H_5{-}NHCO_2{-}CH_2{-}CH_3 + \Delta H$$

phenylisocyanate ethyl alcohol ethylphenyl urethane

Urethane formation is a very rapid, exothermal reaction.[7,8,23,92] Aromatic-structured isocyanates such as phenylisocyanate react more rapidly than aliphatic isocyanates such as, e.g., cyclohexylisocyanate. Both primary and secondary hydroxyls are utilized, but not tertiary hydroxyls, the latter producing unstable urethanes that decompose into olefins.

The isocyanate group also reacts with a host of other active hydrogen compounds, i.e., those whose structures contain at least one hydrogen atom that is capable of being displaced by alkali metals,[1] or by the Zerewittinoff reagent, methylmagnesium iodide.[9] The more important of these other active hydrogen reactions appear in Eqs. 2 to 4.

Eq. 2. Urea Formation

(A)

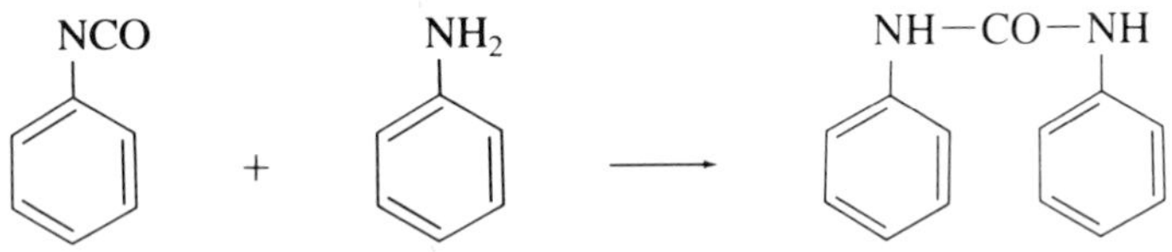

$$C_6H_5{-}NCO + C_6H_5{-}NH_2 \longrightarrow C_6H_5{-}NH{-}CO{-}NH{-}C_6H_5$$

phenylisocyanate aniline sym. diphenyl urea

The above reaction is also quite exothermal and even faster than urethane formation. Aliphatic amines, such as cyclohexylamine, being more basic, react faster than aromatic amines do. Both primary and secondary amines are useful reactants in urethane systems.

(B)

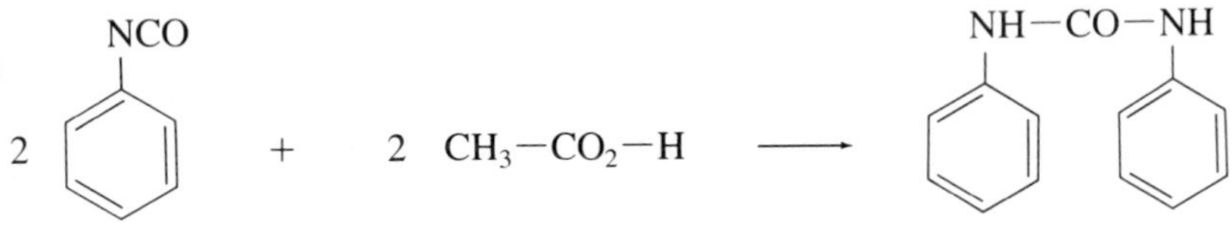

$$2\ C_6H_5{-}NCO + 2\ CH_3{-}CO_2{-}H \longrightarrow C_6H_5{-}NH{-}CO{-}NH{-}C_6H_5$$

phenylisocyanate acetic acid sym. diphenyl urea

$$+ \ (CH_3{-}CO)_2O + CO_2$$

acetic anhydride

(C)

$$2\ C_6H_5NCO + H_2O \longrightarrow C_6H_5NH{-}CO{-}NHC_6H_5 + CO_2$$

phenylisocyanate water sym. diphenyl urea

Eqs. 2B and 2C show that two other reactions of isocyanates—(B) with carboxylic acids, and (C) with water—also yield urea structures. (C) shows that adventitious water can be a bona fide reaction component/problem in polyurethane formation.

Eq. 3. Allophanate Formation

$$C_6H_5NCO + C_6H_5NHCO_2{-}CH_2{-}CH_3 \underset{\Delta}{\rightleftharpoons} C_6H_5NH{-}CO{-}N(C_6H_5){-}CO_2{-}CH_2{-}CH_3$$

phenylisocyanate ethylphenyl urethane ethyl-α,γ-diphenyl allophanate

Isocyanates react sluggishly with urethanes to form allophanates. The allophanate link is thermally unstable, tending to dissociate at ~106°C.[10,11,12]

Eq. 4. Biuret Formation

$$C_6H_5NCO + C_6H_5NH{-}CO{-}NHC_6H_5 \underset{\Delta}{\rightleftharpoons} C_6H_5NH{-}CO{-}N(C_6H_5){-}CO{-}NHC_6H_5$$

phenylisocyanate sym. diphenyl urea 1,3,5-triphenyl biuret

Isocyanates react somewhat faster with ureas than with urethanes to form biurets. The biuret link is also thermally labile, tending to dissociate at 130–145°C.[11,12]

In similar fashion, other active hydrogen groups including —SH (sulfhydryl), $-CONH_2$ (carbonamide), —CONHR (substituted carbonamide), and $-CSNH_2$ (thioamide), $-SO_3H$ (sulphonic) react with isocyanates.

The isocyanate group can undergo other reactions besides those with active hydrogen atoms; one need only refer to review articles on isocyanates and their chemistry to be thoroughly impressed regarding their number.[15,16,17] But those of the greatest recognized importance in polyurethanes involve self-additions as seen in Eqs. 5 and 6.

Eq.5. Isocyanate Dimerization (Uretidinedione Formation)

2 phenylisocyanate ⇌ (Δ) phenyl uretidinedione

Only aromatic isocyanate groups form dimer structures, since aliphatic isocyanates prefer to trimerize. Isocyanate dimers are thermally labile and revert to the original isocyanate on heating as seen in Eq. 5. They are reactive per se with some active hydrogen compounds.[12,23]

Eq. 6. Isocyanate Trimerization (Isocyanurate Formation)

3 phenylisocyanate → 1,3,5-triphenylisocyanurate

Both aromatic and aliphatic isocyanate groups form trimers (Eq. 6) which, in contrast with the dimer structure, are notably stable to heat.[67]

Eq. 7. Carbodiimide Chemistry

(A) Formation:

2 phenylisocyanate →(Δ) diphenyl carbodiimide + CO_2

Strongly heated isocyanate (and urethane reaction mixtures) can yield structures called *carbodiimides*, with attendant evolution of carbon dioxide gas, according to the reaction of Eq. 7A[18,23]. The isocyanate dimer structure (uretidinedione, Eq. 5) may be an intermediate in this reaction. The carbodiimides

are reactive materials and important in urethane chemistry, adding active hydrogen compounds such as carboxylic acids, isocyanates, and so forth.
(B) Reaction with Carboxylic Acids:[20,21]

N=C=N (diphenyl carbodiimide) + CH_3-CO_2-H (acetic acid) ⟶ NH−CO−N−$COCH_3$ (N-acetyl diphenyl urea)

diphenyl carbodiimide   acetic acid   N-acetyl diphenyl urea

(C) Reaction with Isocyanates:[21]

N=C=N (diphenyl carbodiimide) + NCO (phenylisocyanate) ⇌ (Δ) phenyl uretidinedione phenyl imine

diphenyl carbodiimide   phenylisocyanate

CO, N, N, C, ‖, N

phenyl uretidinedione phenyl imine

The chemistry of Eq. 7B is exploited in the stabilization of poly(ester-urethanes) against hydrolysis,[19a,19b] whereas Eq. 7C is utilized in transforming solid, crystalline diisocyanates to the liquid state at room temperature, thus aiding in the polyurethane manufacturing process.[22]

Isocyanate reactions including the foregoing, as ready as they are, nevertheless respond strongly to even low levels of a variety of catalysts including metal compounds, acids, and bases. And considerable specificity is apparent among the catalysts as to the particular isocyanate reactions they promote. Urethane catalysis is well-covered in the literature.[23]

POLYURETHANE COMPONENTS

Urethane polymer formation requires di- and polyfunctional reactants to enable the building of large linear chains and three-dimensional networks. The actual components used to prepare polyurethane elastomers are usually difunctional, and the classes are listed in Table 15.1. However, the formulations used to make some polyurethane systems such as foams, reaction-injection molding (RIM) products, coatings, caulks, adhesives, and so on, often include Table-15.1 component types of functionality greater than 2. These other systems, excepting RIM, will not be discussed in this chapter. Almost all polyurethane components are room-temperature liquids or low-melting solids for process advantages such as transport, metering, and mixing.

The specific natures of R, R′, and R″ in Table 15.1 become apparent in the typical examples of Tables 15.2–15.4 (see Structure columns), and also apply wherever they appear in subsequent equations, e.g., those dealing with polyurethane chain formation and chemical structure, and so forth.

Diisocyanates

The diisocyanate component may be any of a multitude of relatively small molecules.[13, 14] Table 15.2 lists the types of diisocyanates (with specific examples) which have achieved commercial importance in the polyurethane industry.

Polyols

The polyol component of polyurethane elastomers is usually a macroglycol, of molecular weight about 500 to 4000 (usually 1000 to 2000), whose bifunctionality permits the formation of long, strong, linear chains. Table 15.3 lists the types (with specific examples) of macroglycols that have gained commercial importance in the polyurethane industry. In some elastomer formulations, the polyol may be a triol or even of higher functionality.

Chain Extender (Crosslinker)

The chain extender component of polyurethane elastomers is a relatively small, usually difunctional molecule, about the same size as the diisocyanate. In the chain extending reaction with isocyanate, its small size results in structures rich in urethane groups (Eq. 1) and/or urea groups (Eq. 2A), both of which have the dual ability to form VC through hydrogen bonding, and CC or chain branching through allophanate formation (Eq. 3) and biuret formation (Eq. 4) by further reaction with isocyanate. Table 15.4 lists some types of chain extenders, with specific examples, which have achieved commercial importance.

POLYURETHANE CHAIN FORMATION AND CHEMICAL STRUCTURE

Now, with the necessary understanding of urethane chemistry and polyurethane components, we will consider how the components are combined chemically to produce polyurethane elastomer chains.

Table 15.1. Polyurethane Components.

Component	*General Structure*
Diisocyanate	OCN—R—NCO
Macroglycol (polyol)	HO—R′—OH
Chain extender (crosslinker)	HO (or NH_2)—R″—OH(or NH_2)

Table 15.2 Some Commercial Diisocyanates Important in Urethane Elastomers.

Abbreviation	*Common Chemical Name*	*Type*	*Structure*
MDI	Methylene *bis*-(4-phenylisocyanate) (diphenylmethane-*p*, *p*′-diisocyanate)	Aromatic	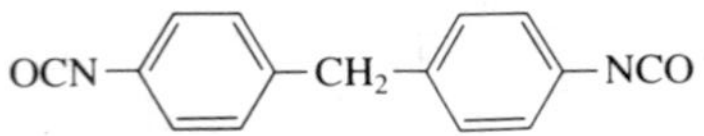 OCN—(benzene ring)—CH_2—(benzene ring)—NCO
80/20 TDI	2,4-tolylene diisocyanate (80% wt.)	Aromatic	CH_3, NCO, NCO (benzene ring)
	2,6-tolylene diisocyanate (20% wt.)	Aromatic	CH_3, OCN, NCO (benzene ring)
H_{12} MDI	Methylene *bis*(4-cyclohexylisocyanate)	Cyclo- aliphatic	OCN—(S)—CH_2—(S)—NCO
IPDI	3-isocyanatomethyl-3,5,5-trimethyl-cyclohexyl isocyanate (isophorone diisocyanate)	Cyclo- aliphatic	CH_3, CH_3, (S), NCO, CH_3, CH_2—NCO
HDI	1,6-hexane diisocyanate (hexamethylene diisocyanate)	Open chain-aliphatic	OCN—$(CH_2)_6$—NCO

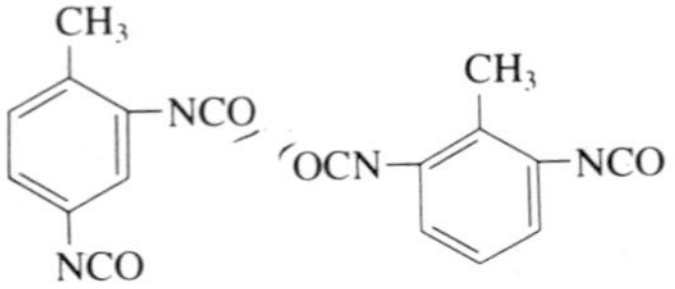

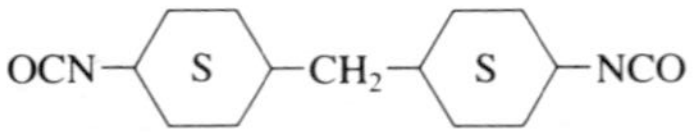

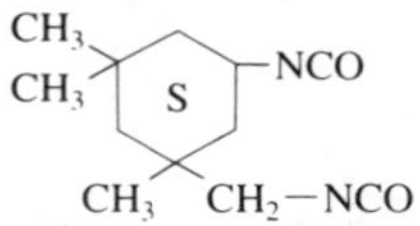

Table 15.3. Some Commercial Polyols Important in Urethane Elastomers.

Abbreviation	*Common Chemical Name*	*Type*	*Structure*
PTAd	Poly(tetramethylene adipate) glycol	Polyester	$HO[(CH_2)_4OCO(CH_2)_4COO]_n\text{-}(CH_2)_4\text{-}OH$
PCL	Poly(ϵ-caprolactone) glycol	Polylactone (polyester)	$H\text{-}[O(CH_2)_5CO]_xORO\text{-}[CO(CH_2)_5O]_y\text{-}H$
PHC	Poly(hexamethylene carbonate) glycol	Polycarbonate (polyester)	$HO\text{-}[(CH_2)_6OCOO]_n\text{-}(CH_2)_6OH$
PTMO	Poly(oxytetramethylene) glycol	Polyether	$HO[(CH_2)_4O]_nH$
PPG	Poly(1,2-oxypropylene) glycol	Polyether	$HO\text{-}[CH(CH_3)-CH_2-O]_nCH_2-CH(CH_3)\text{-}OH$
PBDG	Poly(butadiene) glycol	Polyhydro-carbon	$HO\text{-}(CH_2-CH=CH-CH_2)_x\text{-}OH$[a]

[a]Although only the 1, 4-polybutadiene structure is shown, it is understood that there is a certain proportion of 1,2 units $\left(\begin{array}{l}-CH_2-CH- \\ \qquad\quad | \\ \qquad\; CH=CH_2\end{array}\right)$ in the chain.

Table 15.4. Some Commercial Chain Extenders Important in Urethane Elastomers.

Abbreviation	*Common Chemical Name*	*Type*	*Structure*
1,4-BDO	1,4-butanediol	Aliphatic, open chain glycol	$HO \leftarrow CH_2 \rightarrow_4 OH$
CHDM	1,4-cyclohexanedimethanol	Cycloaliphatic glycol	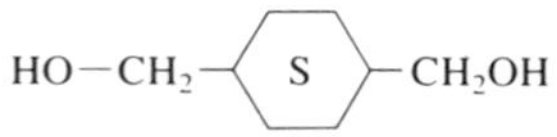 $HO-CH_2-\langle S \rangle-CH_2OH$
HQEE	1,4-*bis*(2-hydroxyethoxy) benzene	Aromatic-aliphatic glycol	 $HO \leftarrow CH_2 \rightarrow_2 O-\langle C_6H_4 \rangle-O \leftarrow CH_2 \rightarrow_2 OH$
MOCA	4,4′-methylene-*bis*(2-chloroaniline)	Aromatic diamine	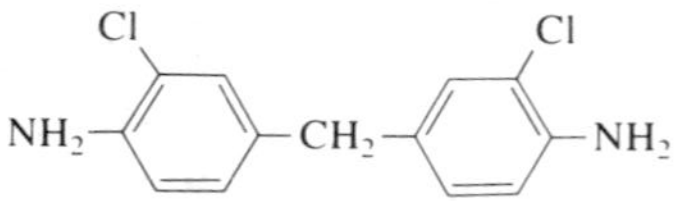 $NH_2-\langle C_6H_3Cl \rangle-CH_2-\langle C_6H_3Cl \rangle-NH_2$
P74OM	Trimethylene glycol di (*p*-aminobenzoate)	Aromatic diamine	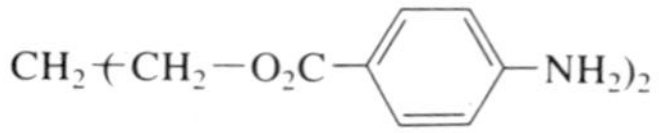 $CH_2 \leftarrow CH_2-O_2C-\langle C_6H_4 \rangle-NH_2)_2$

POLYMERIZATION

Basically, there are two commercial methods of polyurethane polymerization—the "prepolymer" (two step) process and the "one-shot" process. The former involves the preparation of a low-molecular-weight, isocyanate-terminated prepolymer (Eq. 8A), then its chain extension to a high polymer (Eq. 8B).

Eq. 8. Prepolymer Process

(A) $n\,\mathrm{OCN{-}R{-}NCO} + \mathrm{HO{-}R'{-}OH} \longrightarrow$

diisocyanate   macroglycol

$$\mathrm{OCN{-}R}\!\left(\underline{\mathrm{NHCO{-}O{-}R'{-}OCONH{-}R}}\right)\!\mathrm{NCO} + (n-2)\,\mathrm{OCN{-}R{-}NCO}$$

prepolymer   unreacted diisocyanate

(B) $\downarrow$ $(n-1)$ $\mathrm{HO{-}R''{-}OH}$

chain extender

$$\left[\left(\mathrm{NHCO{-}O{-}R'{-}OCONHR}\right)\!\left(\underline{\underline{\mathrm{NHCOO{-}R''{-}OCONH}}}\mathrm{{-}R}\right)_{n-1}\right]_x$$

polyurethane

In the first step (Eq. 8A), the macroglycol unites with excess of the diisocyanate through urethane link formation in an end-capping reaction to produce linear chains that terminate in isocyanate groups, remain relatively low in molecular weight, and retain low melt viscosity, thus enabling liquid processing of the prepolymer. The diisocyanate-macroglycol chain segments (singly underlined) comprise the urethane-sparse "soft-segment" structure in polyurethane chains.

In the second step (Eq. 8B), the terminal isocyanate groups of the prepolymer react with the added chain extender through urethane link formation to couple the prepolymer molecules and produce a high molecular weight polyurethane elastomer. In the process, a new structure, the diisocyanate-chain extender product (urethane-rich "hard segment," doubly underlined) has been built into the polyurethane chains to alternate with the soft segments.

At this point, we will see how certain methods of "chemical (internal)" compounding can be applied in polyurethanes by structure regulation. The prepolymer of Eq. 8A can be made longer by reducing the molar reactant ratio of diisocyanate/macroglycol charged until the prepolymer becomes high in molecular weight. The effect is to lengthen the soft segment and increase its overall concentration in the final product of Eq. 8B, tending to soften the polyurethane (if crystallization is not a consequence).

It can also be seen that in the extreme case of a 1/1 molar ratio of diisocyanate/macroglycol, the formulation cannot accommodate (react with) *any* added chain extender; and the linear polyurethane produced is simply a very high molecular weight soft segment, or one type of a millable polyurethane gum, having the structure shown in Eq. 8A, but differing in that one end of the product bears an unreacted hydroxyl group instead of an isocyanate group. In due course, the other terminal isocyanate group would likely hydrolyze to the amino group due to reaction with atmospheric moisture.

If the prepolymer of Eq. 8A contains free diisocyanate, which can be achieved by raising the diisocyanate/macroglycol molar ratio to $>2/1$ in the initial charge, or by subsequent supplemental diisocyanate addition to the preformed prepolymer, a "quasi-prepolymer" results whose reaction with equivalent amounts of chain extender (Eq. 8B) produces polyurethane elastomers that are harder and higher-modulus due to their increased, urethane-rich hard-segment content.

In Eq. 8B, when the isocyanate content of the prepolymer and the active hydrogen content (from, e.g., hydroxyl, amino groups) of the chain extender are charged to be equivalent, or when the latter is a bit greater, *linear* polyurethane thermoplastic resins or millable gums result, which can be processed as thermoplastics or as gum rubbers, respectively. In effect, the process has produced linear polyurethane chains that can be segmented, i.e., of $-(AB)_n-$ structure to any desired degree. But if somewhat less than an equivalent amount of the chain extender is charged with the prepolymer, then the excess prepolymer isocyanate can ultimately react with polymer-chain urethane groups to produce chemical crosslinks (CC) and/or branches in the product via allophanate linkages (Eq. 3), or biuret linkages (Eq. 4) when the chain extender contained amino groups or the system contains water. Such regulation of component stoichiometry allows the production of the self-curing, castable/injectable, liquid processing urethane systems that are widely used by urethane fabricators.

Eq. 9. One-shot Process

$$\underset{\text{diisocyanate}}{n\ \mathrm{OCN{-}R{-}NCO}} + \underset{\text{macroglycol}}{\mathrm{HO{-}R'{-}OH}} + \underset{\text{chain extender}}{(n-1)\ \mathrm{HO{-}R''{-}OH}} \rightarrow$$

$$\underset{\text{polyurethane}}{\left[-\!\left(\underline{\mathrm{NHCOO{-}R'{-}OCONH{-}R}}\right)\!-\!\left(\underline{\underline{\mathrm{NHCOO{-}R''{-}OCONH{-}R}}}\right)_{n-1}- \right]_x}$$

This is a single-step process in which all of the polyurethane components are mixed together at one time. Here the polymerization proceeds to completion, again yielding a high polymer with alternating soft and hard segments. And as in Eq. 8B, the stoichiometric balance of isocyanate/hydroxyl (or, e.g., amino) groups charged determines whether the product will be a linear-structured, thermoplastic urethane elastomer (NCO/OH $\lesssim 1.0$) as pictured, or an allophanate and/or biuret branched and/or crosslinked polyurethane elastomer (NCO/OH > 1.0).

The one-shot process is used to make thermoplastic polyurethane elastomer resins, RIM (reaction injection molding) products, foams, and so forth. As in the prepolymer process, polymer hardness and modulus can be increased by increasing the diisocyanate and (balancing) chain-extender levels charged in the polymerization, which, in effect, produces more and longer diisocyanate-chain extender hard segments in the polyurethane.

Chemical Crosslinking

The reader has already seen that polyurethane chemical crosslinking can be part of the polymerization process in the liquid processing systems (casting, RIM), accomplished by proper adjustment of the polymerization recipe. But in the case of the millable gum and thermoplastic resin urethane elastomers, conventional rubber-processing and curing techniques (mill/Banbury mixing, compression/injection molding) are applied. The structure of the chemical crosslinks involved is considered next.

Eq. 10. Chemical Crosslinks

(A) Via Isocyanate-Active Hydrogen Atom Reactions:

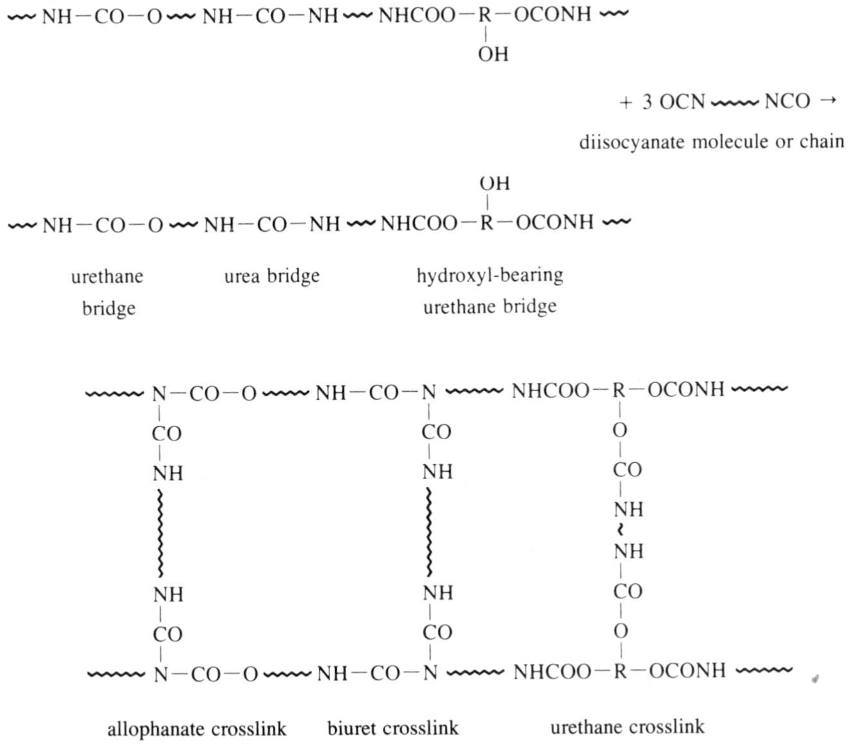

```
~NH—CO—O~NH—CO—NH~NHCOO—R—OCONH~
                            |
                            OH

                                  + 3 OCN~~NCO →
                          diisocyanate molecule or chain

                            OH
                            |
~NH—CO—O~NH—CO—NH~NHCOO—R—OCONH~

 urethane     urea bridge     hydroxyl-bearing
  bridge                      urethane bridge

~~N—CO—O~~NH—CO—N~~NHCOO—R—OCONH~~
  |               |              |
  CO              CO             O
  |               |              |
  NH              NH             CO
  ≀               ≀              |
  ≀               ≀              NH
  ≀               ≀              ≀
  ≀               ≀              NH
  NH              NH             |
  |               |              CO
  CO              CO             |
  |               |              O
  |               |              |
~~N—CO—O~~NH—CO—N~~NHCOO—R—OCONH~~

allophanate crosslink   biuret crosslink   urethane crosslink
```

Equation 10A compares three common types of chemical crosslinks possible in polyurethanes by isocyanate-active hydrogen atom reactions. Equations 3, 4, and 1, respectively, are involved.

(B) Via Free-radical Reactions (Peroxide):

$$\begin{matrix} \sim CH_2 \sim \\ \sim CH_2 \sim \end{matrix} \xrightarrow{\text{peroxide}} \begin{matrix} \sim CH \sim \\ \cdot \\ \sim \dot{C}H \sim \end{matrix} \rightarrow \begin{matrix} \sim CH \sim \\ | \\ \sim CH \sim \end{matrix}$$

methylene positions in polyurethane chains — free radical positions in polyurethane chains — carbon-carbon crosslinked polyurethane chains

Polyurethane chains can also be chemically crosslinked by the generation, then coupling of free-radical positions in the polymer chains as pictured in Eq. 10B. Added organic peroxide has been used to crosslink poly(ester-urethane) millable gums,[24,25,26,27,29,30] thermoplastic polyurethane elastomers,[5,28] and castable/injectable liquid polyurethane systems.[28] Millable gum cures via peroxide can be rapid.[29] Some sites suggested as preferred for free-radical formation in polyurethane chains are the α-methylenic positions in the adipyl moities of adipate-based polymers[29,30] and the methylene group between the two phenyl rings in the diurethane bridge structure of MDI-based polymers.[31] Carbon-carbon crosslinks enhance polyurethane network stability since they are not subject to the thermal dissociation characteristic of allophanate, biuret, and urethane crosslinks.[24,67]

(C) Via Free-radical Reactions (Sulfur-accelerator):

$$\begin{matrix} \sim\sim\sim \\ R-C-CH=CH_2 \\ | \\ H \\ \\ H \\ | \\ R-C-CH=CH_2 \\ \sim\sim\sim \end{matrix} \quad + \text{ S/accelerator} \rightarrow \quad \begin{matrix} \sim\sim\sim \\ R-C-CH=CH_2 \\ | \\ (S)_x \\ | \\ R-C-CH=CH_2 \\ \sim\sim\sim \end{matrix}$$

polyurethane millable gum chains — sulfur-crosslinked polyurethane millable gum chains

Finally, the sulfur-accelerator type of free-radical curing system employed to vulcanize natural rubber and conventional unsaturated synthetic rubbers can also be used to crosslink special-structured polyurethane millable gums.[32,33,34,35,36] Such polymers are rendered responsive to sulfur-accelerator cure by building occasional special cure sites into the polymer chains which are more reactive in the free radical cure system. The preferred active structure is an ethylenically unsaturated appendage on the polyurethane chain that contains at least one methylene hydrogen atom on the carbon atom adjacent to the double bond.[36] Such structures are readily incorporated in the polyurethane chains via ethylenically unsaturated glycols, e.g., glyceryl-α-allyl ether, trimethylolpropane monoallyl ether, and so forth, as chain extender[32,36,37,38,39] or as macroglycol polymerization initiator.[38,39] The cure proceeds according to Eq. 10C if analogy to the same cure of conventional hydrocarbon rubbers holds, as it likely does.

Due to their linear structure and their concomitant solubility, thermoplasticity, and so forth, the thermoplastic polyurethane elastomer resins have provided special advantages in the scientific study and understanding of all polyurethane elastomers, and these findings have been applied to help understand the CC urethane elastomers.

POLYURETHANE CHAIN ORGANIZATION AND PHYSICAL STRUCTURE

The urethane and, if present, the urea linkages in polyurethane chains provide ample opportunity for interchain hydrogen bonding. This bonding occurs to different degrees between, for example, the urethane/urea hydrogen atoms of one chain and the urethane, urea, and ester carbonyl groups, or the ether oxygen atoms of adjacent chains.[40,46]

In the urethane millable gum polymers, which usually have relatively low concentrations of urethane groups, hydrogen bonding is distributed randomly along the polymer chains. This results in *delocalized* interchain associations, and thus gum polymers that can be processed and chemically crosslinked (CC) like rubber to yield strong, useful CC elastomers.

On the other hand, the thermoplastic resins and the most useful castable polyurethane elastomers are formulated to have higher urethane concentrations and longer or better-structured hard segments. In these types, their polymerization produces, as already shown, polymer primary chains which consist of alternating, urethane-sparse, low-T_g soft segments (diisocyanate-macroglycol linear reaction product) and urethane-rich, high-T_g hard segments (diisocyanate-chain extender linear reaction products). These primary chains tend to "virtually crosslink" and "virtually chain-extend" with one another, principally through the association (aggregation) of their hard segments by hydrogen bonding of their urethane groups, the association of aromatic π electrons, and so forth, to produce a giant "virtual network" of polymer chains, and thus elasticity in the product. These virtual linkages are relatively labile and are reversible with heat and solvation.

In polyurethane elastomer chains, urethane-urethane association through hydrogen bonding, with the attendant ordering (aggregation) of the urethane units thereby,[40] and the same greatly magnified process in *segmented* polyurethane

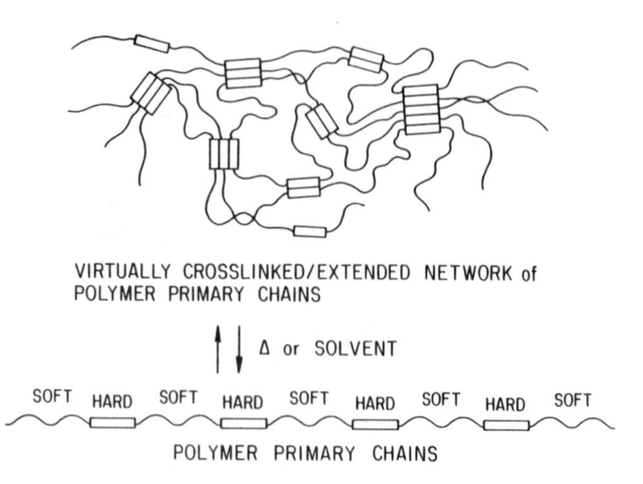

Fig. 15.1. Thermoplastic urethane elastomer chain structure and organization. (Reproduced from C. S. Schollenberger and K. Dinbergs, "Thermoplastic Polyurethane Elastomer Molecular Weight-Property Relations" in *Advances in Urethane Science and Technology*, Vol. 7, Technomic Publ. Co., Lancaster, PA, 1979, pp. 1–34.)

elastomer chains,[41] would seem to account for the elastic behavior of the thermoplastic polyurethane elastomer resins. The same phenomenon is also a very strong contributor to the properties of some other polyurethane elastomer forms, including the castable urethanes and the injectable RIM systems, in instances in which their structures can produce such VC among the polymer chains.

The foregoing type of chain organization is depicted schematically in Fig. 15.1,[5a, 5b] which although specifically representing the TPU system, applies as well to CC polyurethane systems such as the casting and RIM types. The difference, of course, is that in the latter two types the CC tiepoints, while allowing VC network disruption and reformation, can be stable enough to prevent or significantly impede disruption of the coexisting CC network.

The return to the VC state after VC thermal disruption has been shown to be a morphology-related, time-dependent, phase-separation phenomenon that is concurrent with, and explains, the redevelopment of polymer mechanical properties on cooling and ageing heated polyurethane samples.[42] Comparisons of VC network disruption in VC vs. CC-VC polyurethanes, as by heat, show that the CC of the latter system impede but do not prevent the redevelopment of VC on cooling and ageing.[5a, 5b, 6]

The hard-segment aggregates (domains) in polyurethane elastomers have been studied from several standpoints. In unmanipulated (i.e., unoriented, unannealed) samples of many useful polyurethane elastomer compositions, the hard-segment aggregates do not register well, if at all, as crystalline regions by wide-angle X-ray diffraction.[41,42,43,44] Inadequate crystallite size has been proposed to explain this, or the difficulty in achieving a good crystalline lattice due to the impeding action of hydrogen bonding on segment mobility.[5a,5b,45] Even so, small-angle X-ray scattering clearly shows the presence, size, and separation of hard-segment aggregates in relaxed polyurethane elastomer samples.[6,47,48,49,50] And thermal analysis shows their presence, thermal disruption, and reformation on cooling.[45] The term, "paracrystalline"[51,52] has been applied to the semicrystalline state of aggregated hard segments in unmanipulated, *N*-unsubstituted polyurethane elastomer samples.[47]

However, the annealing and orientation of polyurethane elastomer samples can develop and improve their crystallinity, and increase the ability to detect and to study both hard- and soft-segment domains by wide-angle X-ray diffraction. Such treatment has led to much useful information about the crystalline character of the domains, including hard-segment packing, hydrogen bonding, conformation, density, size, and so forth.[53,54]

Both the structure and the size of hard segments are factors in their effectiveness as VC in polyurethanes. Good component symmetry, compactness, and rigidity, and long segments are favorable factors.[79] But it has been shown that hydrogen bonding is *not* a prerequisite, since certain *N*-substituted thermoplastic poly(ether-urethane) elastomers whose urethane groups bear no hydrogen, precluding urethane hydrogen bonding, nevertheless exhibit elasticity and good strength.[55] In this type of polyurethane composition, whose structure is shown in Fig. 15.2, the hard segments again associate to form aggregates (VC) in the soft-segment matrix, but *these* hard-segment aggregates have considerable crystallinity,[55,56,57,58] possibly due to the absence of hydrogen bonding to impede segment ordering within the aggregates. Studies have indicated that hard segments of only one unit length (MW 228) can produce effective VC in urethane elastomers.[55,57] Thus one now sees that crystallinity alone, as well as hydrogen bonding, can produce VC.

In addition, the VC principle has been extended to include glassy (amorphous) hard-segment aggregates (VC) as in the segmented, tri-block A-B-A

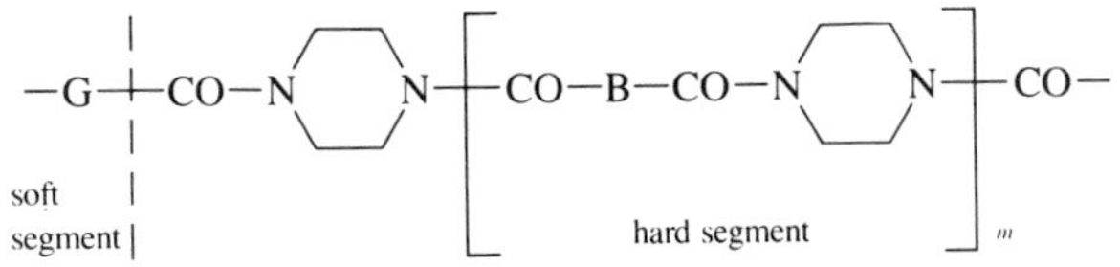

$$-G-CO-N\bigcirc N-[CO-B-CO-N\bigcirc N]_m-CO-$$

soft segment | hard segment

$$G = (OCH_2CH_2CH_2CH_2-)_X-O-$$
$$B = -O-CH_2-CH_2-CH_2-CH_2-O-$$

Fig. 15.2. Nonhydrogen bonding thermoplastic urethane elastomer structure. (From H. L. Harrell, Jr., "Segmented Polyurethanes: Properties as a Function of Segment Size and Distribution," *Macromolecules*, **2** [6] [Nov.–Dec. 1969]. Copyright 1969 American Chemical Society.)

(e.g., polystyrene-polybutadiene-polystyrene) type of thermoplastic elastomers, and to the ionomers in which the VC are characteristically ionic (electrostatic) in nature.

POLYURETHANE PROPERTIES

In view of the several polyurethane elastomer types and the many compositions of each of these that have been or are marketed commercially, it is preferable that polymer properties be discussed in terms of type ranges rather than as data for specific polymers. This approach is used in the present chapter in presenting the tyical physical property information of Table 15.5.[59]

The abrasion resistance of polyurethanes has been recognized as one of their outstanding attributes. Laboratory tests show polyurethanes to abrade only about one-third as readily as natural rubber, whereas in actual field applications the urethanes often abrade only about one-tenth as much, according to one source. However, this reference also points out that such performance is observed at

Table 15.5. Typical Properties of Urethane Elastomers.[59]

Property	*Cast Elastomers*		*Millable Gums*[f]	*Thermoplastic Elastomers*[g]
	Polyester[d]	*Polyether*[e]		
Hardness, Shore	A65–D70	A75–D70	A62–95	A83–D60
Tensile, MPa	41.4–31.0	31.0–48.3	20.7–37.9	41.4–34.5
Elongation, %	650–350	600–300	450–500	500–400
100% modulus, MPa	1.72–11.0	4.83–25.51		4.14–17.93
300% modulus, MPa	4.14–20.68	7.58–30.34	11.72–22.06	6.89–27.58
Elongation set, %	5–80			15–60
Compression set, %[a]	10–25	7–40	20–50	40–25[h]
Tear strength, kN/m				
Graves[b]	70–122.6	35–122.6	43.8–70	52.5–122.6
Split[c]	35–122.6			52.5–140.1
Resilience rebound, %	45–50	42–48	60–70	

[a]22 hr., ASTM D 395, Method B.
[b]ASTM D 624.
[c]FTMS 601–M-4221.
[d]Data cover Vulkollans (Farbenfabriken Bayer), Multrathanes (Mobay Chemical Co.), Vibrathanes (US Rubber) and Cyanaprene (American Cyanamid).
[e]Data cover Adiprenes (Du Pont) and Solithanes (Thiokol).
[f]Data cover Adiprene C (Du Pont), Genthane S (General Tire), Vibrathane (US Rubber), Elastothane (Thiokol), Chemigum SL (Goodyear), and Urepan (Farbenfabriken Bayer); includes black-reinforced polymers.
[g]Data cover Estane (BF Goodrich), Texin (Mobay Chemical Co.), Roylar (US Rubber), and Desmopan (Farbenfabriken Bayer).
[h]Data apply to Texin polymers, may not be typical of others,e.g., Estane gives (68–87% set at Shore A 85–88 hardness.
Source: J. H. Saunders, "Elastomers by Condensation Polymerization," in *Chemistry of Synthetic Elastomers*, Pt. 2, J. P. Kennedy and E. Tornqvist, eds. Copyright © 1969 John Wiley & Sons, Inc. Reprinted by permission of John Wiley & Sons, Inc.

moderate temperatures, and that at about 70–80°C polyurethane abrasion resistance is frequently little better than that of natural rubber.[59]

Good low-temperature flexibility is also a feature of many polyurethane elastomer compositions. Determining factors include polymer composition, polymer molecular weight (inverse relationship)[60a, 60b, 61a, 61b] and degree of chemical crosslinking (inverse relationship).[5a, 5b, 81] Gradual stiffening usually occurs as the temperature is reduced, with many types retaining toughness as low as −70°C.[59]

Relative to the diene-based hydrocarbon rubbers, the polar polyurethane elastomers show superior resistance to degradation by oxygen and by ozone, and to swelling by hydrocarbon fuels and oils. The upper temperature limit for continuous use in air may be about 90–100°C for polyurethanes, with hardness and a polyester base (vs. a polyether) being favorable factors.[59]

At this point, it is necessary to explain that the properties of the types of microcellular polyurethane elastomers obtained from reaction-injection molding (RIM), which is of growing commercial importance, were not included in Table 15.5 because of the "engineering-plastic" nature of RIM polyurethanes as opposed to the more "rubber-like" properties and products of cast, millable gum, and thermoplastic resin polyurethane elastomers. RIM has centered on polyurethane elastomer compositions of very high modulus that are frequently further stiffened by reinforcement with milled or chopped fibers (e.g., glass, carbon, aramid, and metal), mica flake, clay, and glass beads. RIM products include large, exterior automotive parts, business-machine housings, furniture, sports equipment, and so forth.[62] Therefore typical RIM polyurethane elastomer properties are shown separately in Table 15.6.

Table 15.6. Typical Properties of Unreinforced and Reinforced RIM Urethanes.[63]

Property	*Low Modulus*	*High Modulus*	*High Modulus with 20% Glass*[a]
Density, kgm/m^3	960–1040	990–1040	1120–1220
Hardness, Shore D	50–55	65–70	69–76
Tensile strength, MPa	17–24	29–35	>35
Elongation, %	175–340	75–105	17–20
Flexural modulus, MPa			
−29°C	410–550	1450–1800	2000–2800
22°–25°C	130–207	860–900	1400–2000
66°–70°C	55–124	275–520	620–650
Heat sag,[b] mm	7.5–13	7.5–13	2.5

[a]Milled glass (1.6 mm).
[b]For 1 hr at 120°C in 10-cm cantilever test.

POLYURETHANE COMPOUNDING

Polyurethane elastomers are compounded for several reasons. Their environmental resistance requires that in some applications they be protected from hydrolysis,[19a, 19b] UV-initiated autoxidation,[64a, 64b, 65a, 65b] thermooxidation,[66, 68] microbiological attack,[69, 70, 71, 72, 73, 74, 75, 76] and nitrogen dioxide,[77, 78] to name the more important degradation processes.

Reinforcing agents such as fine-particle blacks and silicas are effective in millable gum polyurethane elastomers, improving tensile strength, tear strength, and abrasion resistance. But polyurethane elastomer systems capable of developing enhanced VC are usually preferably reinforced by chemical (internal) compounding as discussed earlier in this chapter. These include the liquid castable and RIM types, and the thermoplastic resins.

In the castable system, where high fluidity is a requirement, all solid-additive levels must be restricted to a few parts of inert materials. On the other hand, RIM polyurethane fabricators, needing to exceed the modulus limits achievable by chemical compounding, have learned to incorporate high levels of reinforcing agents, such as milled glass fibers, into the polymer formulations, producing reinforced RIM (RRIM) parts.[62] This is obviously a consequence of the higher speed of the RIM mixing and mold-filling operations relative to those of the casting process. The use of polybutadiene glycol as the macroglycol component in polyurethane formulations (Table 15.3) produces unsaturated polymers with special affinity for, and reinforcement by, carbon black.[62]

Thermoplastic polyurethane resins can be compounded extensively; however, losses in the levels of some important properties, e.g., tensile strength, in this system and in any polyurethane system exhibiting strong VC, may result from compounding.[79, 80] Reasons for this vary with different polymers and may include polymer dilution, interference with VC network structure, and/or property tradeoffs, e.g., higher modulus for lower elongation and tensile strength.

Other polyurethane compounding ingredients include plasticizers, fillers, lubricants, and other polymers.[79, 80] Colorants and fire retardants are also commonly used.

POLYURETHANE PROCESSES

Laboratory procedures are provided to describe the preparation of the different polyurethane elastomer types, and reference to corresponding commercial manufacturing methods are then cited.

Castable Liquids.

Prepolymer Preparation (Eq. 8A).* The dry macroglycol ($\leq \sim 0.03\%$ water) is added to the excess diisocyanate with mechanical mixing keeping the

*If the isocyanate-terminated prepolymer is purchased, this step is omitted.

reaction mixture at ~85°C until it has reached the isocyanate level calculated for the specific diisocyanate-macroglycol charge.

Chain Extension-crosslinking Reaction (Eqs. 8B and 10). The liquid or melted chain extender component (Table 15.4) is rapidly added in one portion to the mechanically stirred prepolymer of step (1) at ~85°C, the mixture is briefly degassed under reduced pressure with stirring, and then is promptly poured into preheated, lubricated molds (~110°C) or onto a moving, heated, endless belt. The reactions of Eqs. 8B and 10, where applicable, occur here. Parts are demolded as soon as their "green strength" allows.

Post-cure (Eqs. 8B, 10). The demolded parts, still lacking ultimate strength, elasticity, and so forth, are heated in an oven at ~110°C for ~24 hours, or are allowed to stand at room temperature for one week for some systems. The post-cure completes the chemistry of Eqs. 8B and 10, where applicable. It also provides the time and conditions for VC formation to occur, where possible.

The detailed technology of castable polyurethane elastomer processes has been reviewed.[39,82,83,84,85,86]

Millable Gums

Preparation (Eq. 8A). The example cited here for polyurethane millable gum preparation follows Equation 8A but is the case where [OCN—R—NCO] $\lessapprox$ [HO—R′—OH] and no chain extender, HO—R″—OH, is charged.[24,91] To 0.23 mole of dry, molten poly(ethylene adipate) glycol ($\overline{M}_n$ 668, Acid No. < 1.0) in an internal mixer there is added 0.23 mole of *p*-phenylene diisocyanate, and the mixture is stirred in the closed reactor for 30 minutes at 105°C. The resulting poly(ester-urethane) is removed from the reactor and is allowed to cool at room temperature. The reaction product is a soft, snappy, rubbery gum that processes very well in standard rubber equipment.[91]

Chemical Crosslinking (Eq. 10A). An amount of 12.2 parts by weight of *p*-phenylene diisocyanate is mixed into 100.0 parts of the above poly(ester-urethane) gum on a hooded, two-roll rubber mill. This elastomer compound is then heated in a steam press in a ventilated hood for 60 minutes at 280°F. The resulting chemically/covalently cured (CC) rubber has excellent physical properties including 59.3 MPa tensile strength, 10.3 MPa 300% modulus, and 700% breaking elongation.[91] The CC in this vulcanizate are essentially all the allophanate-type crosslink of Eq. 10A. The peroxide cure of such gums to yield carbon-carbon crosslinks (Eq. 10B) is also possible[24] and is preferred. But, lacking ethylenic unsaturation-activated cure sites, such gums do not undergo sulflur-accelerator cure (Eq. 10C).

Thermoplastic Resins

Polymer Preparation (Eq. 9). The one-shot process for preparing TPU is described here. A mixture of 1.70 moles of poly(tetramethylene adipate) glycol ($\overline{M}_n$ 849, Acid No. 0.89) and 1.22 moles of dry 1,4-butanediol is mechanically stirred at 110°C with a spiral-ribbon stirrer in a 4-liter resin kettle, which is heated with a Glas-col mantle. After about 10 minutes of such mixing, 2.92 moles of diphenylmethane-*p*,*p*′-diisocyanate is added cleanly in one portion. Vigorous stirring is continued for 1 minute and the reaction mixture is then poured into a lubricated 1-gallon can which is promptly closed with a friction-fitting lid. The sealed can is placed in a 140°C oven for 3.5 hours to complete the polymerization, then cooled, and the snappy, elastomeric TPU product is removed. It has 85 degrees Shore A hardness, mills satisfactorily at 225°F on a plastics mill, dissolves in N,N-dimethylformamide, and has excellent physical properties.[4]

Other details on TPU preparation and commercial processes have been published.[87,88,89,90] The two latter references involve the use of a twin-screw extruder-reactor.

Reaction Injection Molding (Eqs. 9, 10A). Reaction-injection molding (RIM) is a process of growing importance in polyurethane elastomer fabrication. In principle, it is akin to polyurethane casting in that it is a liquid process and involves in situ polymerization to the finished product in the mold. But it differs from casting by the fact that the reaction components are rapidly impingement-mixed under pressure and then injected, rather than poured, into a mold, all in very rapid sequence. Formulations are adjusted to very fast reaction rates so mold residence time is short.

The example provided here is for a laboratory RIM polyurethane elastomer preparation that involves reaction components similar to those used in commercial RIM operations. It follows Eq. 9 with two exceptions: first, the polyol component is a polytriol rather than a macroglycol; and second, no chain extender is used in the formulation. Eq. 10 (urethane and possibly allophanate crosslinking) is involved in network formation.

An amount of 1.25 volumes of a prepolymer diisocyanate (B_2, 18.55 ± 0.2% NCO) made from excess hexamethylene diisocyanate and triethylene glycol is placed in a glass reactor and brought to temperature (32–50°C) in a water bath. Then 1.00 volume of a poly(ϵ-caprolactone) triol (A_3, $\overline{M}_n$ 939, 0.02% water, 99% primary hydroxyl groups) containing 2100 ppm by volume of dibutyl tin dilaurate catalyst D-22 (Union Carbide Corp., density 1.021 g/cc) and at the same temperature as B_2, is injected rapidly and cleanly into B_2 with intense and efficient stirring for 15 seconds. The ensuing thermoset polyurethane formation is allowed to proceed to completion in the reactor with the generation of considerable heat of reaction.[92]

In commercial practice, the above reactants would be intensely impingement-mixed (1-3 sec.) at pressures of ~7-20 MPa and instantly injected into a lubricated mold (fill time 2-3 sec., pressure ~350 kPa). Mold residence is no longer than required to develop polymer green strength adequate to allow demolding without damage to the part then, or distortion during later treatments. In any case, mold residence is quite short and must accommodate a total cycle time of <2 minutes.[62,92]

After demolding, post-curing (45-60 minutes at ≲96°C) is required to develop the temperture-sensitive (modulus ratio, heat sag) and strength properties of the molded part.[93]

REFERENCES

1. O. Bayer, Angew. Chem. **59,** (9), 257 (1947).
2. O. Bayer, E. Müller, S. Petersen, H-F. Piepenbrink and E. Windemuth, *Rubber Chem. Technol.* **23,** (4) 812, (1950).
3. C. S. Schollenberger, H. Scott, G. R. Moore: (a) *Rubber World* **137** (4), 549 (Jan. 1958); (b) *Rubber Chem. Technol.* **35** (3), 742, (1962).
4. C. S. Schollenberger (assignor to the B. F. Goodrich Co.), U.S. Pat. 2,871,218 (Jan. 27, 1959).
5. C. S. Schollenberger and K. Dinbergs: (a) *J. Elast. Plast.* **7,** 65, (1975); (b) *Advan. Ureth. Sci. Technol.* **6,** 60, K. C. Frisch and S. L. Reegen, eds., Technomic, Stamford, CT, 1978.
6. Z. H. Ophir and G. L. Wilkes, *Multiphase Polymers,* in *Advances in Chemistry* Series, No. 176, S. L. Cooper and G. M. Estes, eds., American Chemical Society, Washington DC, 1979, p. 53.
7. C. S. Schollenberger, K. Dinbergs, and F. D. Stewart, *Rubber Chem. Technol.* **55,** (1) 137 (1982).
8. M. Morton, A. Deisz, and M. Ohta, "Degradation Studies on Condensation Polymers," U.S. Dept. of Commerce Report PB-131795, Mar. 31, 1957.
9. E. P. Kohler, J. F. Stone, Jr., and R. C. Fuson, *J. Am. Chem. Soc.* **49,** 3181 (1927).
10. T. Mukaiyama and M. Iwanami, *J. Am. Chem. Soc.* **79,** 73 (1957).
11. I. C. Kogon, *J. Org. Chem.* **23,** 1594 (1958).
12. D. J. David and H. B. Staley, "Analytical Chemistry of the Polyurethanes," *High Polymers* Vol. 16, Part 3, John Wiley–Interscience, New York, 1969.
13. W. Siefken, *Annal. Chem.* **562,** 75 (1948).
14. A. A. R. Sayigh, H. Ulrich and W. J. Farrissey, *Condensation Monomers,* Chap. 5, J. K. Stille, ed., John Wiley & Sons, New York, 1972.
15. R. G. Arnold, J. A. Nelson and J. J. Verbanc, *Chem. Rev.* **57** (1), 47 (1957).
16. S. Ozaki, *Chem. Rev.* **72** (5), 457, (1972).
17. H. Ulrich, *J. Elastoplast.* **3,** 97 (1971).
18. R. Stolle, *Berichte der Deutsche Chemie Gesellschaft* **41,** 1125 (1908).
19. C. S. Schollenberger and F. D. Stewart: (a) *J. Elastoplast.* **3,** 28 (1971), (b) *Advan. Ureth. Sci. Technol.* **1,** 65, K. C. Frisch and S. L. Reegen, eds., Technomic, Stamford, CT, 1971.
20. H. G. Khorana, Chem. Rev. **53,** 145 (1953).
21. F. Kurzer and K. Douraghi-Zadeh, Chem. Rev. **67** (2), 107 (1967).

22. M. F. Brooks and V. Kerrigan (to Imperial Chemical Industries, Ltd.), British Patent 1,194,050 (June 10, 1970).
23. J. H. Saunders and K. C. Frisch, ''Polyurethanes: Chemistry and Technology: Part I, Chemistry,'' *High Polymers*, Vol. XVI, Part I, John Wiley-Interscience, New York, 1962.
24. E. E. Gruber and O. C. Keplinger, *Ind. Eng. Chem.* **51** (2), 151 (1959).
25. The General Tire and Rubber Co., Akron, Ohio, Technical Bulletins ''Genthane S'' (GT-S3) ''Genthane SR'' (GT-SRI).
26. Farbenfabriken Bayer, Leverkusen, F.R.G., Technical Bulletin, ''Urepan E, an Ester-Based Urethane Rubber for Crosslinking with Peroxides'' (July 1, 1961).
27. Naugatuck Chemical Co., Div. of U.S. Rubber Co. (Uniroyal), Naugatuck, CT, Technical Bulletins ''Vibrathane 5003,'' ''Vibrathane 5004''.
28. C. S. Schollenberger and K. Dinbergs (to B. F. Goodrich Co.), U.S. Patent 4,255,552 (Mar. 10, 1981).
29. Ş. V. Urs, *Ind. Eng. Prod. Res. and Devel.* **1** (3), 199 (1962).
30. L. B. Weisfeld, J. R. Little, and W. E. Wolstenholme, *J. Polym. Sci.* **56,** 455 (1962).
31. O. Bayer and E. Müller, *Angew. Chemie* **72** (24), 934 (1960).
32. E. F. Cluff and E. K. Gladding, *J. Appl. Polym. Sci.* **3,** 290 (1960).
33. Thiokol Chemical Corp., Trenton, NJ, Technical Bulletin, *Thiokol Facts* **5** (2) (1963).
34. M. M. Swaab, *Rubber Age* **92** (4), 567 (1963).
35. E. I. du Pont de Nemours & Co., Elastomer Chemicals Dept. Wilmington, DE, Technical Bulletins (a) ''Adiprene C, a Urethane Rubber,'' Development Products Report 4 (July 15, 1957); (b) ''Adiprene CM, a Sulfur-Curable Urethane Rubber,'' by S. M. Hirsty, Report A-66693 (1969).
36. D. B. Pattison, (to E. I. du Pont de Nemours & Co.), U.S. Pat. 2,808,391 (Oct. 1, 1957).
37. W. Kallert, *Kautsch. Gum. Kunst.* **19** (6), 363 (1966).
38. H. S. Kincaid, G. P. Sage and F. E. Critchfield, ''Polycaprolactone Millable Urethane Elastomers,'' American Chemical Society (Rubber Division) Meeting, Montreal, May 1967.
39. P. Wright and A. P. C. Cumming, *Solid Polyurethane Elastomers*, Gordon and Breach, New York, 1969.
40. Y. M. Boyarchuk, L. Y. Rappaport, V. N. Nikitin and N. P. Apukhtina, *Polym. Sci. U.S.S.R.* **7,** 859 (1965).
41. S. L. Cooper and A. V. Tobolsky, *J. Appl. Polym. Sci.* **10,** 1837 (1966).
42. G. L. Wilkes, S. Bagrodia, W. Humphries, and R. Wildnauer, *J. Polym. Sci.*, Polym. Letters Edition **13,** 321 (1975).
43. C. E. Wilkes and C. E. Yusek, *J. Macromol. Sci.-Phys.* **B7**(1), 157 (1973).
44. G. L. Wilkes and R. Wildnauer, *J. Appl. Phys.* **46,** 4148 (1975).
45. C. S. Schollenberger, Chap. 5, *Multiphase Polymers*, in *Advances in Chemistry Series* No. 176, S. L. Cooper and G. M. Estes, eds., American Chemical Society, Washington, DC, 1979.
46. R. W. Seymour, G. M. Estes, and S. L. Cooper, *Macromolecules* **3** (5), 579 (1970).
47. R. Bonart, *J. Macromol. Sci.-Phys.* **B2** (1), 115 (1968).
48. R. Bonart and E. H. Müller, *J. Macromol. Sci.-Phys.* **B10,** 345 (1974).
49. G. L. Wilkes and J. A. Emerson, *J. Appl. Phys.* **47** (10), 4261 (1976).
50. Z. Ophir and G. L. Wilkes, *J. Polym. Sci.-Phys. Ed.* **18,** 1469 (1980).
51. R. Hosemann and S. N. Bagchi, ''Direct Analysis of Diffraction by Matter,'' North-Holland, Amsterdam, 1962.
52. R. Hosemann, *J. Polym. Sci.*, Sympos. 50, 265 (1975).
53. J. Blackwell and C. D. Lee, *J. Polym. Sci.-Phys.* **21,** 2169 (1983).
54. L. Born, J. Crone, H. Hespe, E. H. Müller and K. H. Wolf, *J. Polym. Sci.-Polym. Phys. Ed.* **22,** 163 (1984).

55. L. L. Harrell, Jr., *Macromolecules*, **2** (6), 607 (1969).
56. S. L. Samuels and G. L. Wilkes, *Polym. Prepr.*, Div. Polym. Chem., Am. Chem. Soc., **12,** 2694 (1971), and *J. Polym. Sci.*, Part B, **9,** 761 (1971).
57. S. L. Samuels and G. L. Wilkes, *J. Polym. Sci.*, Sympos. 43, 149 (1973).
58. A. E. Allegrezza, Jr., R. W. Seymour, H. N. Ng, and S. L. Cooper, *Polymer* **15,** 433, (1974).
59. J. H. Saunders, Chap. 8, Part 2, *Polymer Chemistry of Synthetic Elastomers*, J. P. Kennedy and E. Tornqvist, eds., John Wiley-Interscience, New York, 1969, p. 727.
60. C. S. Schollenberger and K. Dinbergs: (a) *J. Elastoplast.* **5,** 222 (1973), (b) *Advan. Ureth. Sci. Technol.* **3,** 36, K. C. Frisch and S. L. Reegen, eds., Technomic, Westport, CT (1974).
61. C. S. Schollenberger and K. Dinbergs: (a) *J. Elast. Plast.* **11,** 58 (1979); (b) *Advan. Ureth. Sci. Technol.* **7,** 1, K. C. Frisch and S. L. Reegen, eds., Technomic, Westport, CT (1979).
62. L. J. Lee, *Rubber Chem. Technol.* **153** (3), 542 (1980).
63. J. A. Vaccari, *Product Engineering*, June 1978, p. 33.
64. C. S. Schollenberger and F. D. Stewart, (a) *J. Elastoplast* **4,** 294 (1972); (b) *Advan. Ureth. Sci. Technol.* **2,** 71, K. C. Frisch and S. L. Reegen, eds. Technomic, Westport, CT (1973).
65. C. S. Schollenberger and F. D. Stewart, (a) *J. Elast. Plast.* **8,** 11 (1976); (b) *Advan. Ureth. Sci. Technol.* **4,** 68, K. C. Frisch and S. L. Reegen, eds., Technomic, Westport, CT (1976).
66. H. J. Fabris, *Advan. Ureth. Sci. Technol.* **4,** 89, K. C. Frisch and S. L. Reegen, eds., Technomic, Westport, CT (1976).
67. L. Nicholas and G. T. Gmitter, *J. Cell. Plast.* **1,** 85 (1965).
68. A. Singh, L. Weissbein and J. C. Mollica, *Rubber Age*, Dec. 1966, p. 77.
69. British Pat. 797,576, assigned to Farbenfabriken Bayer (July 2, 1958).
70. J. G. DiPinto, ''Fungus Resistance of Adiprene L–100'', Technical Bulletin, E. I. du Pont de Nemours & Co. (Mar. 22, 1963).
71. G. A. Kanavel, P. A. Koons and R. E. Lauer, *Rubber Chem. Technol.* **39** (4), 1338 (1966).
72. R. T. Darby and A. M. Kaplan, *Appl. Microbiol.* **16,** 900 (1968).
73. A. M. Kaplan, R. T. Darby, M. Greenberger and M. R. Rogers, *Develop. in Ind. Microbiol.* **9,** 201 (1968).
74. O. C. Elmer, (to General Tire & Rubber Co.), U.S. Pat. 3,531,433 (Sept. 29, 1970).
75. British Patent 1,274,145, assigned to Mobay Chemical Co. (May 10, 1972).
76. S. J. Huang, ''Biodegradation of Polyurethanes,'' lecture, Polymer Conference Series, Univ. of Detroit, June, 1981.
77. H. H. G. Jellinek and T. J. Y. Wang, *J. Polym. Sci.-Polym. Chem. Ed.* **11,** 3227 (1973).
78. H. H. G. Jellinek, F. Martin and H. Wegner, *J. Appl. Polym. Sci.* **18** (6), 1773 (1974).
79. C. S. Schollenberger, in *Polyurethane Technology*, P. F. Bruins, ed., John Wiley–Interscience, New York, 1969, p. 197.
80. C. S. Schollenberger and D. Esarove, in *The Science and Technology of Polymer Films*, Vol. II, O. J. Sweeting, ed., John Wiley–Interscience, New York, 1971, p. 487.
81. J. H. Saunders, *Rubber Chem. Technol.* **3,** 1259 (1960).
82. *Polyurethane: Kunststoffe Handbuch*, Vol. VII, R. Viewig and A. Höchteln, eds., Carl Hanser Verlag, Munich, 1966.
83. D. P. T. Copcutt, *J. Inst. Rubber Ind.* **2** (1), 41 (1968).
84. P. A. Gianatasio, *Rubber Age* **101** (8), 57 (1969).
85. C. F. Blaich, Jr., in *Polyurethane Technology*, P. F. Bruins, ed., John Wiley–Interscience, New York, 1969, p. 85.
86. C. Hepburn, *Polyurethane Elastomers*, Applied Science Publishers, London and New York, 1982.
87. J. H. Saunders and K. A. Piggot (to Mobay Chemical Co.), U.S. Pat. 3,214,411 (Oct. 26, 1965).

88. G. F. Bartel, M. Klawitter and E. Denker, (to Die Kunststoffburo Orzabruck Dr. Rester), U.S. Patent 3,620,680 (Nov. 16, 1971).
89. K. W. Rausch, Jr., and T. R. McClellan, (to The Upjohn Co.), U.S. Pat. 3,642,964 (Feb. 15, 1972).
90. B. Quiring, G. Niederdellmann, W. Goyert, and H. Wagner (to Bayer A.G.), U.S. Pat. 4,245,081 (Jan. 13, 1981).
91. C. S. Schollenberger (to B. F. Goodrich Co.), U.S. Pat. 2,770,612 (Nov. 13, 1956).
92. S. D. Lipshitz and C. W. Macosko, *J. Appl. Polym. Sci.* **21,** 2029 (1977).
93. R. L. McBrayer and G. J. Griffin, *Plastics World* **35** (9), 62 (1977).
94. P. Dreyfuss, L. J. Fetters, and D. R. Hansen, *Rubber Chem. Technol.* **54**(1), 181 (1981)).

16

THERMOPLASTIC ELASTOMERS

GEOFFREY HOLDEN
Shell Development Company
Houston, Texas

Thermoplastic elastomers are materials that combine the properties of thermoplasticity and rubber-like behavior. That is, they are processed on conventional plastics equipment, such as injection molders, blow molders, sheet and profile extruders, and so forth, but develop their final rubber-like properties immediately on cooling. Conventional rubbers, on the other hand, must be vulcanized (see Chapter 2: Compounding and Vulcanization) to give useful properties. This is a slow, irreversible process and takes place on heating. In thermoplastic elastomers, however, the transition from a processible melt to a solid rubber-like object is rapid, reversible, and takes place on cooling. Some thermoplastic elastomers can be dissolved in common solvents and regain their properties when the solvent is evaporated. These transitions are shown in Fig. 16.1.

The advantages in processing flexibility, and so on, are obvious and have resulted in significant commercial applications. However, some end-use properties of thermoplastic elastomers (e.g., solvent resistance and upper service temperature) are usually not as good as those of the corresponding vulcanizates. Applications of thermoplastic elastomers are therefore in areas where the above properties are less important, including footwear, wire insulation, and adhesives. However they do not include other products such as automobile tires, radiator hose, drive belts, and so forth.

TYPES OF THERMOPLASTIC ELASTOMERS

Five types of commercially important thermoplastic elastomers are:

1. Polystyrene/elastomer block copolymers.
2. Polyester block copolymers.
3. Polyurethane block copolymers.

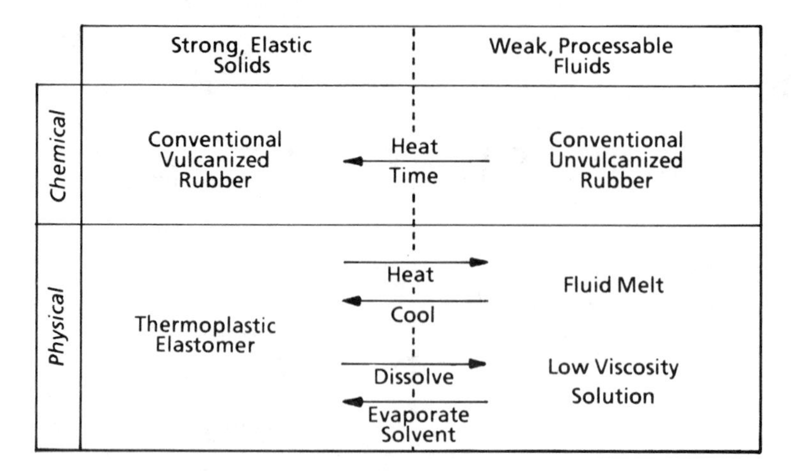

Fig. 16.1. Chemical and physical changes. (Reprinted from *Elastomerics*, Vol. 117, No. 10. © October 1966 by Communications Channels, Inc., Atlanta, Ga., U.S.A.)

4. Polyamide block copolymers.
5. Polypropylene/EP copolymer blends.

The polyurethane block copolymers are described in Chapter 15 and will not be covered in detail in this chapter.

All these thermoplastic elastomers are two-phase systems. One of the phases is a hard polymer that does not flow at room temperature but becomes fluid when heated. The other phase is a soft rubbery polymer. In the block copolymers, the two phases are formed from segments of the same chain molecule. The simplest arrangement is a three-block or A–B–A structure where A and B represent hard plastic and soft rubbery segments respectively. (Hard and soft, in this case, refer to the properties of the respective homopolymers.) This arrangement is typical of the polystyrene/elastomer block copolymers.

In these polysytrene/elastomer block copolymers, three types of elastomer segments (polyisoprene, polybutadiene, and poly(ethylene-cobutylene)) are used commercially. The products will be written as S–I–S, S–B–S, and S–EB–S respectively. In this system, the hard polystyrene end segments and the soft elastomeric center segments are incompatible and form a two-phase structure, similar to that shown diagramatically in Fig. 16.2[1,2]. Here the polystyrene end segments form separate regions ("domains") dispersed in the continuous elastomer phase.

At room temperature, these polystyrene domains are hard and join the elastomer chains together in a network. In some ways, this arrangement is similar to the network that gives strength to conventional vulcanized rubbers. The main difference is that in thermoplastic elastomers the crosslinks are physical rather than chemical. Thus the domains lose their strength when the polymer is heated

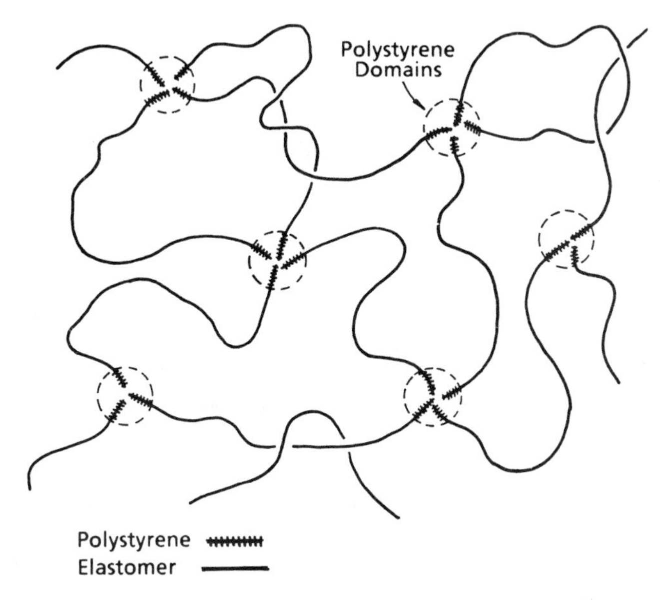

Fig. 16.2. Phase structure of themoplastic elastomers (diagrammatic). (Reprinted from *Elastomerics*, Vol. 117, No. 10. © October 1966 by Communications Channels, Inc., Atlanta, Ga., U.S.A.)

or dissolved in solvents, allowing the polymer (or its solution) to flow. When the polymer is cooled down or the solvent is evaporated, the domains become hard again and so the polymer regains its original properties.

Block copolymers with structures such as A–B or B–A–B are not thermoplastic elastomers, since for a continuous network to form, both ends of the elastomer (B) segment must be immobilized in the hard (A) domains. Instead, they are much weaker materials that resemble conventional unvulcanized synthetic rubbers.[1]

The polyester, polyurethane and polyamide block copolymers have a multiblock structure A–B–A–B–A . . . and so forth. In the polyesters, the hard segments are a crystalline copolymer, typically poly(butylene terephthalate) (PBT). The soft segments are copolymers of a higher-molecular-weight diol and terephthalic acid. The polyamide block copolymers are relatively new materials. The hard segments are a crystalline polyamide while the soft rubbery segments are polyethers or polyesters.

The polypropylene/EP copolymer blends are fine interdispersed multiphase systems and are often grafted or crosslinked. In addition to these commercial products, there is an almost unlimited number of possible block copolymers and blends having hard and soft phases. A list of some of these materials has been given in a recent article.[3] However, none of them is commercially important at this time.

MANUFACTURE

The polystyrene/elastomer block copolymers are made by block copolymerization.[3-7] Briefly, an alkyl lithium initiates the polymerization of a monomer (A). The resulting polymer $(A)_n$ can in turn initiate polymerization of a second monomer (B) to give a block copolymer $(A)_n - (B)_m$ and so forth.

Commercially, anionic polymerization is limited to three common monomers—styrene, butadiene, and isoprene. These give two useful A–B–A block copolymers, S–B–S and S–I–S. In both cases, the elastomer segments contain one double bond per molecule of original monomer. These bonds are quite reactive especially toward oxygen, and limit the stability of the product. To improve this property, the butadiene monomer may be polymerized in two structural forms, which on hydrogenation give essentially a copolymer of ethylene and butylene. The S–EB–S so produced is a fully saturated rubber and, therefore, much more stable.

The polyester thermoplastic elastomers are synthesized by condensation polymerization of three monomers. These are typically a long-chain glycol such as poly(tetramethylene ether glycol), which reacts with dimethyl terephthalate to give the elastomer segments. In the same polymerization, butanediol also reacts with dimethyl terephthalate to give the hard segments of poly(butylene terephthalate).[3-5] The polyurethane and polyamide block copolymers are believed to be manufactured by basically similar processes.

The production of the polypropylene/EP copolymer blends is relatively simple. The components, polypropylene and an elastomer (usually EPDM), are blended by intensive shear-mixing. In some cases, this process causes grafting. In another variation, the elastomer can be crosslinked while the mixing is taking place, a process described as "dynamic vulcanization."[8] This process forms a fine dispersion of the two phases, which is sufficient to give the properties of a thermoplastic elastomer.

EFFECT OF MOLECULAR STRUCTURE ON PROPERTIES

Changes in the molecular structure strongly affect the properties of the polymers just discussed. Four of the most important properties are stiffness, processibility, service temperature range, and environmental resistance. Their variations can be summarized as follows:

Stiffness

For all these multiphase systems, as the ratio of the hard to soft phase is increased, the polymers in turn become harder and stiffer. The effect has been well documented for the polystyrene/elastomer block copolymers.[1,2,9,10] As the styrene content is increased, they change from very weak, soft, rubberlike matter to strong elastomers, then become leathery. At high (>75%) styrene content, they are hard, glass-like products, which have been commerciallized as clear high-impact polystyrenes under the tradename K-Resin (Phillips Petroleum Co.).

As noted later, in many cases the stiffness of the final product is modified by compounding.

Processibility

Molecular weight usually controls polymer processibility. Low-molecular-weight materials are easier to process but have poorer service properties than higher-molecular-weight analogs. The polystyrene/elastomer block copolymer still remain a two-phase system in the melt, and their melt viscosity (and hence their processibility) is abnormally sensitive to variations in molecular weight.[1] As with stiffness, the processibility of many final products depends on compounding.

Service-temperature Range

Thermoplastic rubbers have two service temperatures which define the service-temperature range. Below the lower service temperature, the soft rubbery phase becomes hard and so the polymer is stiff rather than rubbery. Above the upper service temperature, the hard phase becomes soft and fluid, and so the polymer loses its strength. This effect is shown diagrammatically in Fig. 16.3. The lower service temperature depends on the glass transition temperature (T_g)[11] of the soft rubbery phase while the upper service temperature depends on the glass-transition temperature or crystal-melting temperature (T_m)[12] of the hard phase. Values of T_g and T_m for the various phases in commercially important thermoplastic elastomers are given in Table 16.1.*

Environmental Resistance

Environmental resistance can be divided into two types—the resistance of the polymer to chemical degradation and the resistance to swelling, dissolution, and so forth, by solvents and oils.

In the polystyrene/elastomer block copolymers, resistance to degradation is controlled by the nature of the soft elastomer segment. Polybutadiene and polyisoprene both have one double bond per original monomer unit, or more

*Since the hard phase begins to soften somewhat below T_g or T_m, the practical upper service temperatures are somewhat less than the values given in Table 16.1. They also depend on the stress applied. An unstressed part (e.g., one undergoing steam sterilization) will have a higher service temperature than a similar part that must support a load. Similary, the practical lower service temperatures are slightly above the T_g of the soft rubbery phase. See Fig. 16.3.

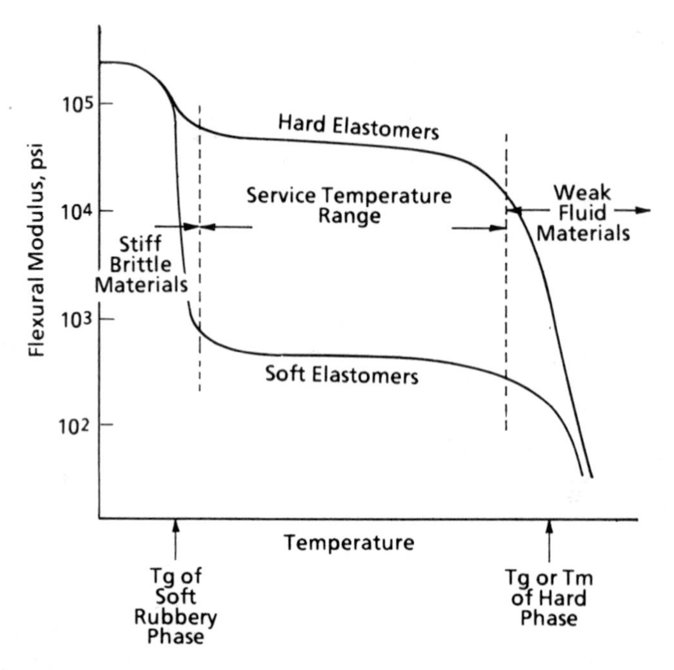

Fig. 16.3. Stiffness of typical thermoplastic elastomers at various temperatures. (Reprinted from *Elastomerics*, Vol. 117, No. 10. © October 1966 by Communications Channels, Inc., Atlanta, Ga., U.S.A.)

than 500 double bonds per molecule of S–B–S or S–I–S. These are an obvious source of chemical instability and so limit the thermal and oxidative stability of these block copolymers. In contrast, poly(ethylene-co-butylene) is completely saturated and so S–EB–S block copolymers are much more stable. Polyester, polyurethane, and polyamide block copolymers and polypropylene/EP are also

Table 16.1. Glass Transition and Crystal Melting Temperatures.[a]

Thermoplastic Elastomer Type	*Soft, Rubbery Phase* T_g (°C)	*Hard Phase* T_g or T_m (°C)
S–B–S	−90	95 (T_g)
S–I–S	−60	95 (T_g)
S–EB–S	−60	95 (T_g) 165 (T_m)[b]
Thermoplastic polyesters	−65 to −40	185 to 220 (T_m)
Thermoplastic polyurethanes	−40 (polyester) −60 (polyether)	190 (T_m)
Thermoplastic polyamides	−65 to −40	120 to 275 (T_m)
Polypropylene/EP blends	−60	165 (T_m)

[a]Measured by differential scanning calorimetry.
[b]In compounds containing polypropylene.

relatively stable to oxidative degradation, although the first three can have problems with hydrolytic stability. The latter condition is improved if polyethers rather than polyesters are used as the elastomer segments in the polymer molecule.

Polyisoprene, polybutadiene, and poly(ethylene-co-butylene) (especially the latter) are nonpolar elastomers. The corresponding block copolymers can thus be compounded with hydrocarbon-based extending oils, but do not have much oil resistance. Similarly, since both the polystyrene and the elastomer segments are soluble in many common solvents (e.g., toluene) these polymers do not have good solvent resistance. However, both oil and solvent resistance can be improved by compounding. The polar soft rubbery segments in the polyester, polyurethane, and polyamide block copolymers have little affinity for such hydrocarbon oils and the crystalline nature of the hard phases makes them insoluble. The elastomer network thus remains intact in the presence of oils and solvents. This gives these polymers excellent resistance to attack by oils and solvents.

In the polypropylene/EP copolymer blends, the EP copolymer phase can be swollen by oils and solvents but the crystalline polypropylene phase is insoluble except at elevated temperatures. Thus complete solubility of these materials is impossible at ambient temperatures, but they are swollen by hydrocarbons.

COMPOUNDING

The various classes of thermoplastic elastomers are quite different from each other in this respect. The polystyrene/elastomer block copolymers are always used as compounds; there are no known applications for the pure polymers. However, they are supplied both as compounds and pure polymers. The polyester, polyurethene, and polyamide block copolymers are most often used as pure materials, although filled and plasticized products have been described. The polypropylene/EP copolymer blends are supplied as precompounded materials and are normally used as supplied.

The types and amount of compounding ingredients added depends on the intended application of the final product. This is discussed in detail in the next section.

Compounding techniques vary with the type of polymer. There is one important generalization: the processing equipment should be heated to a temperature above the upper service temperature of the base polymer (see Table 16.1). The use of cold mills, and so on, can result in polymer breakdown that is not only unnecessary but also detrimental.

Unless large quantities of fillers are being added, a single-screw extruder having a mixing screw and with a length/diameter ratio of at least 24 should be adequate. The extruded product is chopped to give pellets. Both stand-cutting and underwater face-cutting have been used successfully. Unfilled or lightly filled compounds can also be made on twin-screw extruders, which give excellent mixing. Many of the higher-molecular-weight styrenic block copolymers are difficult to process as pure materials and so oiled versions are available. If desired, more oil can be added during compounding. If large amounts of fillers are used, these are best dispersed on a closed intensive mixer, such as a Farrel or a Banbury. The mixer discharges into a single-screw extruder that completes the mixing and gives steady feed for the pelletization.

Trial batches or even small-scale production runs can be made on a conventional Banbury. After several minutes mixing, the hot product is immediately transferred to a sheetout mill, then cut off in the usual way, allowed to cool, and finally granulated.

The end-use properties of compounded thermoplastic elastomers can be sensitive to both processing conditions and processing equipment. These elastomers often become harder and stiffer as the shear in processing is increased. Thus it is important to test samples on equipment and under conditions that will be used in production. Misleading results can be obtained if, for example, test or prototype parts are compression-molded when the final product is to be injection-molded.

Thermoplastic elastomers have relatively high friction characteristic of rubbers, and this can give ejection difficulties, especially with softer products. These difficulties can be reduced if the sides of the mold are tapered. Teflon-coating the mold surface and the use of mold release agents help, as does the use of air ejection and stripper rings in the mold design. Small-diameter injection pins should not be used with softer products, since the part will deform rather than eject.

Ground scrap from thermoplastic elastomers is reusable. Predrying is usually needed before scrap from the polyurethene, polyester, and polyamide block copolymers can be reused. Grinding of softer products can be difficult. Problems will be minimized if it is realized that rubbery materials must be cut rather than shattered. Thus the blades of the grinder should be sharp and the clearance between the rotating and fixed blades should be minimized.

APPLICATIONS

Polystyrene/Elastomer Block Copolymers

The applications of these materials have been described in detail.[13,14] They can be divided into three classes as indicated below.

Replacements for Conventional Vulcanized Rubbers. For this application, the products are intended to be processed by the fast machines originally developed for conventional thermoplastics, such as injection molders and blow molders. The S–B–S and S–EB–S polymers are preferred, and to achieve a satisfactory balance of properties they are compounded with oil, fillers, other polymers, and so forth. A list of some of the compounding ingredients and their effects on the compound properties is given in Table 16.2 and their effects may be quickly summarized.

Polystyrene hardens the products while oils soften them. Both improve processibility. Oils with high aromatic content should not be used; they plasticize the polystyrene domains. Instead of polystyrene, crystalline hydrocarbon polymers such as polypropylene, polyethylene, or ethylene-vinylacetate copolymer can be used. They improve ozone resistance and also give the compounds some solvent resistance. Recently, blends of S–EB–S block copolymers with silicone oils have been reported to be suitable for medical applications.[15]

Inert fillers such as whiting, talc and clays can be added to these compounds. They have only small effects on physical properties but reduce cost. Reinforcing fillers, such as carbon black, are not needed.

The major end use of S–B–S based compounds is in footwear. Canvas footwear (e.g., sneakers) and unitsoles can be made by injection molding. The shoes have good frictional properties, similar to those made from conventionally vulcanized rubbers and much better than those made from flexible thermoplastics such as plasticized polyvinylchloride. They also have excellent cold flexibility.

Compounds based on S–EB–S usually contain polypropylene. This added polymer gives improved solvent resistance, higher upper service temperatures and better processibility. Applications for which special compounds have been developed include materials intended for the automobile industry, and food-contact, wire-insulation, and pharmaceutical applications.

The S–B–S and S–EB–S compounds available commercially have hardnesses from 35 Shore to 45 Shore D and specific gravities from about 0.9 to about 1.95 (the latter are special products intended for sound deadening). Properties of representative grades are given in Table 16.3 and some typical formulations are given in Tables 16.4 and 16.5.

Molding and extrusion are relatively easy. Compounds based on S–B–S process similarly to polystyrene while S–EB–S-based analogs process similarly to polypropylene. Predrying is not needed and scrap is recyclable.

Adhesives, Coatings and Sealants. This is also an important end-use for these polymers.[16] Here the effective compounding ingredients are resins and oils. Resins and oils that are compatible with the elastomer segments soften the final product and so give tack. Resins that are compatible with the polystyrene segments harden the final product. Aromatic oils or similar resins with low softening points are not normally used since they plasticize the polystyrene domains. The products can be applied either from solvents or as hot melts.

Table 16.2. Compounding Styrenic Block Polymers.[a]

Component	*Hardness*	*Processibility*	*Effect on Ozone Resistance*	*Cost*	*Other*
Oils	Decreases	Increases	None	Decreases	Decreases U.V. resistance
Polystyrene	Increases	Increases	Some increase	Decreases	-
Polyethylene	Increases	Variable	Increases	Decreases	Often gives satin finish
Polypropylene	Increases	Variable	Increases	Decreases	Improves high-temperature properties
EVA	Small Increase	Variable	Increases	Decreases	-
Fillers	Some Increase	Variable	None	Decreases	Often improves surface appearance

[a]From N. R. Legge et al., in *Applied Polymer Science*, 2d Ed., R. W. Tess and G. W. Poehlein, eds., Ch. 9, Table III. Copyright 1985 American Chemical Society. Reprinted with permission of the American Chemical Society.

Table 16.3. Properties of Compounded Styrenic Block Copolymers.[a]

Product:	*KRATON® D2109*	*KRATON G2705*	*KRATON D3202*	*KRATON D5119*	*KRATON D5152*	*KRATON D5239*	*KRATON G7150*	*KRATON G7720*	*KRATON G7880*	*ELEXAR® 8313*	*ELEXAR 8451*	*ELEXAR 8614*
Applications: Properties	*Milk Tubing*	*Medical*	*General*	*Footwear Direct Molded*	*Footwear Unitsole*	*Footwear Crepe Look*	*Sound Deadening*	*Automotive Soft Parts*	*Automotive Bumper Cover*	*Automotive Primary Wire*	*Wire & Cable Flexible Cord*	*Wire & Cable Fire Retardant Control Cable*
Hardness (Shore A or D)	44A	55A	65A	47A	47A	36A	36D	60A	42D	54D	80A	95A
Tensile strength (psi)	950	1100	850	580	560	700	700	900	2100	3500	2000	2000
100% modulus (psi)	240	240	480	360	275	140	500	300	1300	1900	600	1400
Elongation, (%)	800	700	500	380	550	820	250	600	450	550	500	500
Melt Flow, (gm/10 min; Condition E or G) ASTM D 1238	15(G)	-	14(E)	27(E)	26(E)	20(E)	-	-	-	-	22(G)	3(G)
Specific gravity	0.94	0.9	1.0	1.09	1.0	0.98	1.94	1.20	1.05	1.28	1.0	1.2

[a]From G. Holden, in *Encyclopedia of Polymer Science and Engineering* 2d Ed., Vol. 5, J. I. Kroschwitz, ed., John Wiley & Sons, New York, p. 427. Copyright © 1966 by John Wiley & Sons, Inc. Reprinted by permission of John Wiley & Sons, Inc.

Table 16.4. Typical Formulations Based on KRATON® D Thermoplastic Rubbers.

	General Purpose	*Sneaker Sole*	*Crepe Sole*	*Booster Cable*
KRATON D rubber	100	100	100	100
Polystyrene or styrenic resin	60–80	80–100	15–25	–
Polyolefin	–	–	–	40–60
Filler	10–40	50–80	10–30	40–80
Oil	70–90	90–110	100–130	60–80
Stabilizer	0.8	0.8	0.8	1.5
Typical Shore A hardness	60–70	45–55	30–40	55–65

Table 16.5. Formulations Based on KRATON® G Thermoplastic Rubbers.

	Sound Deadening	*Medium Hardness*	*Wire & Cable*
KRATON G rubber	100	100	100
Polyolefin	80–120	20–60	100–125
Filler	500–1000	150–200	50–80
Oil	100–140	125–175	75–125
Stabilizer	0.5	0.5	1.0
Typical Shore A hardness	–	55–65	75–85
Typical Shore D hardness	30–40	–	–

Further details of these applications are given in the commercial literature.[17]

Blends with Thermoplastics. Styrenic block copolymers are technologically compatible with a surprisingly wide range of materials; that is, they can be blended to give useful products.[18] Blends of S–B–S with polystyrene, polyethylene, or polypropylene show improved impact resistance. Similarly, S–EB–S can be blended with many engineering thermoplastics, including poly(butyleneterephthalate), poly(phenylene oxide) and polycarbonate. Special grades of styrenic block copolymers have been developed as modifiers for sheet-molding compounds (SMC) based on thermoset polyesters.

Another application is as a flexibilizer for bitumens and asphalts.[19] S–B–S/asphalt blends have low viscosity at processing temperatures combined with remarkably elastic behavior at ambient temperatures. They can be used to make flexible built-up roofing membranes. S–EB–S block copolymers give similar improvements and in addition better weather resistance. Styrenic block copolymers are also used as modifiers for asphalt-based road crack sealants.

Polyester, Polyurethane and Polyamide Block Copolymers

Replacements for Conventional Vulcanized Rubber. This is by far the largest application for these thermoplastic elastomers.[20,21] They can be molded or extruded on equipment intended for processing thermoplastics at melt temperatures of about 175 to 225°C. Scrap is reusable and predrying is strongly recommended. Polyester block copolymers can be plasticized. Hardness of the unplasticized products varies from 60 Shore A to 70 Shore D and they have specific gravities of about 1.0 to 1.25. They are flexible, tough, and resistant to oils, solvents, and abrasion. Properties of representative grades are shown in Table 16.6. Applications include belting, hydraulic hose, tires, shoe soles, wire coatings and automobile parts.

Adhesives, Coatings, and Sealants. Some of these materials are used as hot-melt adhesives in shoe manufacture and as an adhesive interlayer in coextrusion.

Blends with Thermoplastics. Both polyurethane and polyester block copolymers can be blended with plasticized PVC. The blends are flexible with high abrasion resistance and are suitable for shoe soles.

Polypropylene/Ethylene-propylene Copolymer Blends

Replacements for Conventional Vulcanized Rubbers. This is the largest application for these polymers. They are available as precompounded materials designed for various end-uses.[22] Some are quite soft (60 Shore A) but harder products (up to 50 Shore D) are also produced. Specific gravities usually vary from about 0.98 to 1.05. They can be molded or extruded at conditions similar to those used for polypropylene and the scrap is reusable. The two most important applications are for wire insulation and as automobile exterior and interior parts (both painted and unpainted). Properties of some representative grades are shown in Table 16.7.

Blends with Thermoplastics and/or other Polymeric Materials. There is some use for these materials as impact modifiers for polyethylene and polypropylene. They are not compatible with more polar polymers such as polystyrene.

Trade names of representative thermoplastic elastomers are given in Table 16.8.

Table 16.6. Typical Properties of Thermoplastic Polyurethanes, Polyesters, and Polyamides.

Product	*ESTANE 55013*	*HYTREL 40xy*[a]	HYTREL 63*xy*[a]	HYTREL 72*xy*[a]	GAFLEX 547	PEBAX 2533	PEBAX 6333
Type	Polyurethane	←—	Polyester		—→	←Polyamide	→
Hardness (Shore A or D)	86A	40D 92A	63D	72D	47D	25D 75A	63D
Tensile strength (psi)	4300	3700	5700	5700	4500	4200	7100
100% modulus (psi)	650	925	2700	4100	1600	630	2700
Elongation (%)	630	450	350	350	650	350	680
Specific gravity	1.18	1.17	1.22	1.25	1.17	1.01	1.01

[a]Two-digit identifier for the individual grades denoted by *xy*.

Table 16.7. Properties of Polypropylene/Ethylene-propylene Copolymer Blends.

Product	*TPR 1600*	*TPR 1900*	*TPR 2800*	*TELCAR 340*	*SOMEL 301*	*SOMEL 601*	*SANTOPRENE 101-73 201-73*	*SANTOPRENE 103-50 203-50*
Hardness (Shore A or D)	67A	92A	87A	91A 41D	40D	50D	73A	50D
Tensile strength (psi)	650	1850	1800	1480	1660	4460	1100	3900
100% modulus, (psi)	500	1850	1250	1360	1550	2000	470	1450
Elongation (%)	230	230	150	400	230	495	375	600
Specific gravity	0.88	0.88	0.88	0.89	1.02	1.04	0.98	0.94

Table 16.8. Some Trade Names of Thermoplastic Rubbers.[c]

Trade Name (Mfgr)	*Type*	*Hard Segment*	*Soft Segment*	*Notes*
KRATON® D and CARIFLEX® TR (Shell)	3-block (S–B–S or S–I–S)	S	B or I	General purpose, soluble
SOLPRENE 400[a] (Phillips)	Branched $(S–B)_n$ $(S–I)_n$	S	B or I	General purpose, soluble
STEREON (Firestone)	3-block (S–B–S)	S	B	General purpose, soluble
TUFPRENE (Asahi)	3-block (S–B–S)	S	B	General purpose, soluble
EUROPRENE SOL T (ENICHEM)	3-block (S–B–S or S–I–S)	S	B or I	General purpose, soluble
KRATON G (Shell)	3-block (S–EB–S)	S	EB	Improved stability, soluble when uncompounded.
ELEXAR® (Shell)	3-block (S–EB–S and S–B–S)	S	EB or B	Wire and cable
C-FLEX (Concept)	3-block (S–EB–S and silicone oil)	S	EB	Medical applications

(*cont.*)

[c]From N. R. Legge et al., in *Applied Polymer Science*, 2d Ed., R. W. Tess and G. W. Poehlein, eds., Ch. 9, Table IV. Copyright 1985 American Chemical Society. Reprinted by permission of the American Chemical Society.

Table 16.8. (*Continued*)

Trade Name (Mfgr)	*Type*	*Hard Segment*	*Soft Segment*	*Notes*
ESTANE (B.F. Goodrich) TEXIN (Mobay) PELLETHANE[b] (Upjohn) CYANOPRENE (American Cyanamid) RUCOTHANE (Hooker)	Multiblock	Polyurethane	Polyether or polyester	Hard, abrasion- and oil-resistant, high cost.
HYTREL (Du Pont) GAFLEX (GAF)	Multiblock	Polyester	Polyester (amorphous)	Similar to polyurethane and better at low temp.
PEBAX (Rilsan) ESTAMID[b] (Upjohn)	Multiblock	Polyamide	Polyether	Similar to polyurethanes but can be softer. Very good at low temp.
TPR (Reichhold-Cook)	Blend	Polypropylene or polyethylene	EPDM (crosslinked)	Low density, hard, not highly filled.
SANTOPRENE (Monsanto)	Blend	Polypropylene	EPDM (crosslinked)	Low density, hard, not highly filled.
REN-FLEX (Research Polymers, Inc.) POLYTROPE (Schulman) SOMEL (Colonial) TELCAR (Teknor-Apex) VISTAFLEX (Reichhold-Cook) FERROFLEX (Ferro) ETA (Republic Plastics)	Blend	Polypropylene	EPDM	Low density, hard, not highly filled.

[a]Now withdrawn from the market. [b]Now sold to Dow.

REFERENCES

1. G. Holden, E. T. Bishop and N. R. Legge, *J. Poly. Sci.* **C,** (26), 37 (1969).
2. G. Kraus, C. W. Childers, and J. T. Gruver, *J. Applied Polym. Sci.* **11:** 1581 (1967).
3. G. Holden, "Elastomers, Thermoplastic," in *Encyclopedia of Polymer Science and Engineering*, 2d Ed., Vol. 5, J. I. Kroschwitz, ed., John Wiley & Sons, New York, 1966, pp. 416–430.
4. N. R. Legge, G. Holden, S. Davison, and H. E. De La Mare, in *Applied Polymer Science*, J. K. Craver and R. W. Tess, eds., American Chemical Society, Washington, DC, 1975, pp. 404–405. See also N. R. Legge, S. Davison, H. E. De La Mare, and M. K. Martin, in *Applied Polymer Science*, 2d. Ed., R. W. Tess and G. W. Poehlein, eds., American Chemical Society, Washington, DC, 1985, pp. 187–200.
5. P. Dreyfuss, L. J. Fetters and D. R. Hansen, *Rubber Chem. Technol.* **53,** 738 (1980).
6. M. Morton, *Anionic Polymerization: Principles and Practice*, Academic Press, New York, 1983.
7. M. Szwarc, M. Levy and R. Milkovich, *J. Am. Chem. Soc.* **78,** 2656 (1956).
8. A. M. Gessler (assignor to Esso Research & Engineering Co.), U.S. Pat. 3,037,954 (June 5, 1962).
9. M. Morton, *Rubber Chem. Technol.* **56,** 1096 (1983).
10. J. C. Saam, A. Howard and F. W. G. Fearon, *J. Inst. Rubber Ind.* **7,** 69 (1973).
11. R. F. Boyer, "Transitions and Relaxation," in *Encyclopedia of Polymer Science and Technology*, 1st Ed., Vol. 2, H. F. Mork, N. G. Gaylord, and N. M. Bikales, eds., John Wiley–Interscience, New York, 1977, p. 745.
12. R. L. Miller, "Crystallinity," Ibid., Vol. 4, p. 449.
13. G. Holden, *J. Elastomers and Plastics* **14,** 148 (1982).
14. G. Holden, in *Handbook of Plastics and Related Materials*, I. Rubin, ed., John Wiley and Sons, New York (in press), Ch. 67.
15. Anon., *Modern Plastics* **60** (12), 42 (1983).
16. D. J. St. Clair, *Rubber Chem. Technol.* **55,** 208 (1982).
17. Bulletin SC: 198-83, Shell Chemical Company, Houston, 1984.
18. A. L. Bull and G. Holden, *J. Elastomers and Plastics* **9,** 281 (1977).
19. E. J. van Beem and P. Brasser, *J. Inst. Petroleum* **59,** 91 (1973).
20. S. C. Wells in *Handbook of Thermoplastic Elastomers*, B. M. Walker, ed., Van Nostrand Reinhold, New York, 1979, Ch. 4, pp. 104–105.
21. S. Wolkenbreit, Ibid., Ch., 5.
22. H. L. Morris, Ibid., Ch. 2.

17
MISCELLANEOUS ELASTOMERS

R. C. KLINGENDER
Goldsmith & Eggleton, Inc.
Akron, Ohio

POLYSULFIDE ELASTOMERS

Polysulfide elastomers were first developed in the early 1920's, when J. C. Patrick discovered the condensation reaction of ethylene dichloride and sodium tetrasulfide. The first production of polysulfide was in 1929 to 1930 by Thiokol Corporation, who manufactured the product Thiokol Type A.

Polysulfides were the first elastomers produced that were vulcanizable and provided excellent solvent resistance. There were many drawbacks to these early elastomers—mainly poor compression set, difficult processibility, and a very strong sulfide odor.

The Type B polysulfide was developed circa 1934 with improved low-temperature characteristics and less gassing on milling. This type was found to be soluble in trichloroethane; hence it was useful for coating fabric, and was also resistant to mustard gas. Subsequently, Type FA polysulfide, introduced in 1939, was intended to replace both Type A and B polysulfides. The low-temperature characteristics of this polymer were again slightly improved over the Type B, and FA is one of the types that is produced and used today. In 1943, a Type ST polysulfide was developed that exhibited improved compression-set characteristics and better low-temperature properties. These improved characteristics were accompanied, however, by poor bin stability and poor solvent resistance because of the lower sulfur content. The cure system required for the ST type polysulfides was also different. The LP type polysulfides, which are liquids of various viscosities, were produced by the chain scission of higher-molecular-weight polymers. The liquid types are found to be extremely useful for sealants and for rocket binders.

The basic characteristics of polysulfides are their excellent solvent resistance

Table 17.1. Polysulfide Types and Raw Materials for Preparation.

Trade Name	*Base Monomer*	*Sulfur Content*	*Specific Gravity*
Thiokol A	Ethylene dichloride	84%	1.60
Thiokol B	Di-2-chloroethyl ether	64%	1.51
Thiokol FA	Di-2-chloroethyl formal/ ethylene dichloride	47%	1.34
Thiokol ST	Di-2-chloroethyl formal/ 1,2,3-trichloropropane	37%	1.27

(depending upon the sulfur content), very good low-temperature properties, excellent ozone resistance, low weathering, and impermeability. Some of the drawbacks of polysulfides are their difficulty in processing, the strong odor, poor compression set, and moderate physical properties.

FA type polysulfide is normally cured with a metal oxide, such as zinc oxide, but others, such as lead oxide, lead dioxide, cadmium oxide, and zinc hydroxide, and so forth, may be used. The basic viscosity of Type FA polysulfide is too high to be used as is; hence the polymer must be peptized prior to further compounding. To obtain the desired viscosity, it is peptized with MBTS or other disulfides such as TMTD, TETD, and so on. After the peptizing step, the reinforcing fillers are added, with stearic acid, followed by zinc oxide at the end—provided that the temperature is not too high. Reinforcement is best obtained with various grades of carbon black, depending upon the physical characteristics desired, and to a much lesser degree with light-colored fillers. Care must be taken not to use any acidic base materials, as this will affect the cure rate and lead to poor aging characteristics. Stearic acid is included only as a mill and mold release agent.

ST Type polysulfide is mercaptain-terminated, and the polymer is produced to a desirable viscosity; hence there is no need to peptize this polymer. Reinforcement is obtained with carbon black in a manner similar to that of FA Type polysulfide. The cure system used with ST Type polysulfide is slightly different from that for FA Types: in addition to zinc oxide or other metal oxides, metal peroxides, organic peroxide, quinoid compounds, or sulfur-bearing materials may be used. The major advantages of the ST Type polymer are its much improved low-temperature characteristics and better set characteristics. Because of the better state of cure, this type of polymer may be removed from the mold without cooling—a distinct advantage over the FA Types.

LP liquid polysulfides are also curable with metal oxides, the most typical being PbO_2 and MnO_2, but quinone dioxime with DPG may also be used. Being liquid, these polymers are handled in paste form. Depending upon the application, the liquid polymers are normally pumped or gunned into place and then cured at relatively low temperatures.

Table 17.2. Polysulfide Millable Types.

Grade	*FA (low)*	*FA (medium)*	*FA (high)*	*ST*
Color	Light brown	Light brown	Light brown	Light brown
Specific gravity	1.34	1.34	1.34	1.27
Odor	Distinct Polysulfide	Distinct Polysulfide	Distinct Polysulfide	Slight Polysulfide
Moisture content (% max)	0.5	0.5	0.5	-
Bin stability	Limited	Limited	Limited	Very limited
Mooney viscosity				
(ML 1 + 4 @ 121°C)	60–80	81–101	102–112	
(ML 1 + 3 @ 100°C)				30–40

Table 17.3. Properties of LP Liquid Polysulfide Polymers.

Specification Requirements	*LP-31*	*LP-2*	*LP-32*	*LP-12*	*LP-3*	*LP-33*
Color-MPQC-29-A	150 max.	100 max.	100 max.	70 max.	50 max.	30 max.
Viscosity-poises 25°C	950–1550	410–525	410–525	410–525	9.4–14.4	15–20
Moisture content, %	0.12–0.22	0.15–0.25	0.15–0.25	0.12–0.22	0.1 max.	0.1 max.
Mercaptan content, %	1.0–1.5	1.50–2.00	1.50–2.00	1.50–2.00	5.9–7.7	5.0–6.5
General Properties						
Average molecular weight	8000	4000	4000	4000	1000	1000
Refractive index$_D^n$	1.5728		1.5689		1.5649	
Pour point, °C	10	7	7	7	−26	−23
Flash point (PMCC), °C	225	208	212	208	174	186
% Crosslinking agent	0.5	2.0	0.5	0.2	2.0	0.5
Sp. gr. @ 25°C	1.31	1.29	1.29	1.29	1.27	1.27
Avg. viscosity poises, 4°C	7400	3800	3800	3800	90	165
Avg. viscosity poises, 65°C	140	65	65	65	1.5	2.1
*Low-temp. flex., °C (G 703 kg/cm^2)	−54	−54	−54	−54	−54	−54

*Cured compound.

Table 17.4. FA Polysulfide Formulation.

Ingredients	*phr*
FA polysulfide rubber	100.0
SRF black	60.0
Stearic acid	0.5
Benzothiazyl disulfide (MBTS)	0.3
Diphenylguanidine (DPG)	0.1
Zinc oxide	10.0
	170.9

Many applications have been developed for polysulfide elastomers, the first of which being hose tubes and gaskets, to take advantage of the excellent solvent resistance for the paint and lacquer spray industry. The excellent impermeability and good low-temperature characteristics made it a standard material as a diaphragm for gas meters. The development of rollers, based upon polysulfide, for the printing and lacquer coating industries was a very large boon to the rubber industry. The ability of polysulfide to be blended with other polymers, primarily polychloroprene, enabled some of the drawbacks in processing and physical properties to be overcome, while at the same time advantage was taken of polysulfide's excellent resistance to solvent, aging, ozone, and so forth. The advent of the liquid polymers with their inherently superior impermeability, and ozone, weathering, UV-light and aging resistance made them a standard material to be used in the sealant industry for installing glass in aircraft, automobiles, and curtain-wall-constructed buildings, particularly thermopane glass. The liquid polymers were also very useful as coatings and as modifiers for epoxies. In addition, they were instrumental in the development of a binder system for solid rocket propellants.

Although polysulfide has been displaced for quite a number of applications because of problems with odor, processing, and borderline physical characteristics, it is still used today in many of the aforementioned applications, and will continue to be used in the future.

Table 17.5. ST Polysulfide Formulation.

Ingredients	*phr*
ST polysulfide rubber	100.0
Carbon black	60.0
Stearic acid	1.0
Zinc peroxide	5.0
Calcium hydroxide	1.0
	167.0

Table 17.6. Typical Cured Properties of LP Formulations.

Stress-strain Properties	LP-31	LP-2	LP-32	LP-12	LP-3	LP-33
300% modulus, kg/cm^2	21.1	24.6	14.8	14.1	10.5	14.1
Tensile strength, kg/cm^2	26.4	28.8	21.1	21.1	21.1	26.4
Elongation, %	600	510	930	900	275	700
Hardness, Shore A	48	50	50	48	45	34

Recipe: LP-31, LP-32, LP-2 and LP-12

Base Compound	phr
Liquid polymer	100
N774 SRF	30
Curing paste	
PbO_2[a]	7.8
HB-40	4.8
Stearic acid	0.1
Alumina (Al_2O_3)	0.2

Cure: 2 hr at 70°C in closed mold (ASTM D15), post-cure 20 hr at 23°C, 50% RH after demolding.

Recipe: LP-33 and LP-3

Base Compound	phr
Liquid polymer	100
N990 MT	20
P-quinonedioxime (GMF)	6.67
Diphenylguanidine (DPG)	0.67
MgO	4.15
Sulfur	0.50

Cure: 20 hr at 77°C in mold (ASTM D15 with teflon coating), post-cure minimum 2 hr at 23°C and 50% RH after demolding.

[a] With LP-31, 5.0 parts of PbO_2 are used.

Table 17.7. Typical Black Printing-roller Compound.

Masterbatch	*phr*
FA polysulfide rubber	100.0
MBTS	0.35
DPG	0.15
Rubber substitute (aerated)	20.0
	120.5

Compound	*phr*	*Typical Physical Properties (Cure 40 min @ 148°C)*	
FA rubber masterbatch	120.5	Hardness, Shore A	20
Neoprene W	15.0	Tensile strength, kg/cm^2	25
Zinc oxide	10.0	Elongation, %	350
MgO	0.6		
Ethylene thiourea	0.075		
Stearic acid	0.5		
SRF N774	30		
Highly aromatic oil	40		
P-25 Cumar	10		
	226.275		

Table 17.8. FA Polysulfide Paint-can Gaskets.

Formulation	*phr*	*Typical Physical Properties (Cured 40 min @ 148°C)*	
FA polysulfide rubber	100.0		
MBTS	0.4		
DPG	0.1	Tensile strength, kg/cm^2	52.7
Stearic acid	1.0	Elongation, %	150
Zinc oxide	10.0	Hardness, Shore A	82
Sulfur	1.0	Specific gravity	1.615
MT N990	130.0		
	242.5		

HYDROGENATED NITRILE RUBBER

The hydrogenation of diene-containing polymers has been a well-known reaction for many years. Berthelot began the earliest experiments in 1869, with many other workers carrying out numerous experiments under different conditions and with different catalyst systems. Pummerer, et al., in the early 1920's

$$[CH_2-CH=CH-CH_2]_x[CH_2-\underset{CN}{CH}]_y \xrightarrow{H_2} [CH_2-CH=CH-CH_2]_z[CH_2-CH_2-CH_2-CH_2]_{x-z}[CH_2-\underset{CN}{CH}]_y$$

Fig. 17.1. Hydrogenated nitrile rubber.

and Standinger about 1930 are notable contributors to the hydrogenation of polymer systems. Catalyst systems employed are metals, noble metals, organometallics, and diimides. Variations in the catalyst system, coupled with temperature and pressure, allow for variations in the degree of hydrogenation of the diene component of a polymer.

Although butadiene and isoprene polymers, and also those containing styrene groups, have been both experimentally and commercially hydrogenated for many years, not until very recently has it been possible to hydrogenate those containing acrylonitrile. Nippon Zeon Co., Ltd., after eight years of development, succeeded in commercializing in March 1984 a range of hydrogenated nitrile elastomers, trade-named Zetpol. Late in 1984, Bayer also announced a commercial production of a single grade of hydrogenated NBR called Therban.

Hydrogenated NBR is produced by first making an emulsion-polymerized nitrile elastomer in the traditional manner. This polymer is then dissolved in an appropriate solvent and, with a noble metal catalyst at a designated temperature and pressure, undergoes selective hydrogenation to produce the so-called highly saturated polymer. Most, or nearly all of the butadiene in the nitrile elastomer is converted to ethylene to produce an ethylene butadiene acrylonitrile terpolymer. (See Fig. 17.1.)

The commercially available grades of hydrogenated nitrile rubber listed in Table 17.9 exhibit much-improved properties in comparison to a typical NBR elastomer, and advantages over other higher-temperature, oil-resistant elastomers, as indicated on the following page.

Table 17.9. Grades of Hydrogenated Nitrile.

Producer:	*Bayer*	*Nippon Zeon*			
Grade	*Therban*	*Zetpol 1010*	*Zetpol 1020*	*Zetpol 2010*	*Zetpol 2020*
Bound ACN, %	34	45	45	37	37
Mooney viscosity, ML 1 + 4 @ 100°C	75	80	80	85	80
Specific gravity	0.95	1.00	1.00	0.98	0.98
Saturation, %	100	95	90	95	90
Stabilizer	-	Nonstaining	Nonstaining	Nonstaining	Nonstaining

- Very good abrasion resistance.
- Excellent tensile strength.
- Higher-modulus retention at elevated temperatures versus other elastomers.
- High hot tear resistance.
- Good to very good long-term compression-set resistance.
- Good low-temperature resistance.
- Good oil and solvent resistance, depending upon ACN.
- Good heat resistance.
- Very good ozone and weathering resistance.
- Good resistance to hydrogen sulfide and amine corrosion inhibitors.
- Good resistance to grease and lubricating oil additives.
- Good resistance to ''sour'' gasoline.

The hydrogenated NBR polymers react to filler and plasticizer loadings in much the same way as standard nitrile elastomers, except for the higher physical properties obtained. Mixing, milling, calendering, extruding, and molding characteristics are similar to normal NBR. A good high-temperature anti-oxidant system, such as 1.5 phr Vanox ZMTI with 1.5 phr Naugard 445, is also advisable.

The cure system requirements of hydrogenated nitrile rubber are much more like those of an EPR polymer in the case of Therban; like a medium cure rate EPDM in the case of Zetpol 1010 and 2010; and like an ultrafast cure rate EPDM in the case of Zetpol 1020 and 2020. The level of peroxide required is 6 to 10 phr, depending upon the polymer used, and the sulfur or sulfur donor system requires very fast or ultrafast accelerators.

Hydrogenated NBR has found application not only in many oil-field products, but also in a multitude of automotive components, because of its unique balance of high-temperature and chemical resistance.

Oil-field applications, such as down-hole packers, well-head seals, valve seals, O-rings, ram and annular blow-out preventors, take advantage of the heat, hydrogen sulfide and amine resistance of hydrogenated nitrile, plus the retention of high-temperature modulus for improved pressure ratings. These properties, plus excellent abrasion and hot tear resistance, also make this elastomer ideally suited for mud-pump-piston cups and drill-stem protectors.

Automotive components that require resistance to heat, grease, oil, and the many oil and grease additives, are based upon hydrogenated NBR. These include rotating-shaft seals, lip seals, servo-piston seals, valve-stem seals, O-rings and gaskets. Fuel-system requirements, especially for resistance to ''sour'' gasoline, gasohol, weathering, and higher operating temperatures have led to the use of hydrogenated NBR in fuel-line hose, fuel-pump and fuel-injection components, and diaphragms, as well as in emission-control systems. In addi-

Table 17.10. Comparison of Hydrogenated NBR Formulations.

	1 *Zetpol 1020*	*2* *Zetpol 2020*	*3* *Zetpol 2020*	*4* *Zetpol 2010*	*5* *Therban*
Polymer (phr)	100	100	100	100	100
Zinc oxide	5	5	5	5	5
Stearic acid	0.5	0.5	0.5	0.5	0.5
SRF N774	60	60	60	60	60
Thiokol TP95	5	5	5	5	5
Sulfur	0.5	0.5	-	-	-
TMTD	2.0	2.0	-	-	-
MBT	0.5	0.5	-	-	-
Vulcup 40KE	-	-	7	7	7
	173.5	173.5	177.5	177.5	177.5
Compound Properties					
Mooney viscosity					
ML 1 + 4 @ 100°C	96	102	98	105	96
Rheometer					
Temperature, °C	160	160	170	170	170
T5 (min)	2.8	2.5	1.4	1.4	1.7
T95 (min)	19.8	16.1	29.1	23.6	24.5
V max., kgf. cm	62.0	67.3	76.2	64.8	60.5
Physical Properties: All Post-cured 4 hr @ 150°C					
Cure	*20' @ 160°C*	*20' @ 160°C*	*15' @ 170°C*	*15' @ 170°C*	*15' @ 170°C*
Shore A	74	74	73	74	73
Modulus @ 100%, MPa	6.2	4.7	8.3	5.6	4.5
Tensile, MPa	21.5	20	24	24.9	24.4
Elongation, %	340	310	210	270	330

(*cont.*)

Table 17.10. (*Continued*)

	1 Zetpol 1020	*2* Zetpol 2020	*3* Zetpol 2020	*4* Zetpol 2010	*5* Therban
Aged 72 hr @ 150°C in Air—Change					
Shore A, pt	+5	+4	+4	+4	+5
Modulus @ 100%, %	+65	+77	+54	+44	+65
Tensile, %	0	−8	−8	−1	0
Elongation, %	−38	−42	−24	0	−38
180° bend	Pass	Pass	Pass	Pass	Pass
Aged 72 hr @ 175°C in Air—Change					
Shore A, pt	+9	+7	+11	+7	+14
Modulus @ 100%, %				+121	
Tensile, %	−44	−49	−62	−19	−63
Elongation, %	−82	−81	−76	−41	−91
180° bend	Pass	Pass	Fail	Pass	Fail
Low Temperature, ASTM 1053 (°C)					
T-2	−19	−27	−26.1	−25.5	−21.5
T-5	−21.7	−29.3	−29.3	−28.5	−26.5
T-10	−23	−30.7	−30.8	−30	−29
T-100	−32	−39	−39.3	−38.5	−36
T-fp	−25.6	−33.3	−33.5	−32	−33.8
Compression Set B, 70 hr					
23°C	17	13	7	12	18
70°C	17	13	8	15	19
100°C	19	19	8	15	20
150°C	54	57	15	18	32
175°C	87	86	24	25	45
200°C	100	100	49	54	76

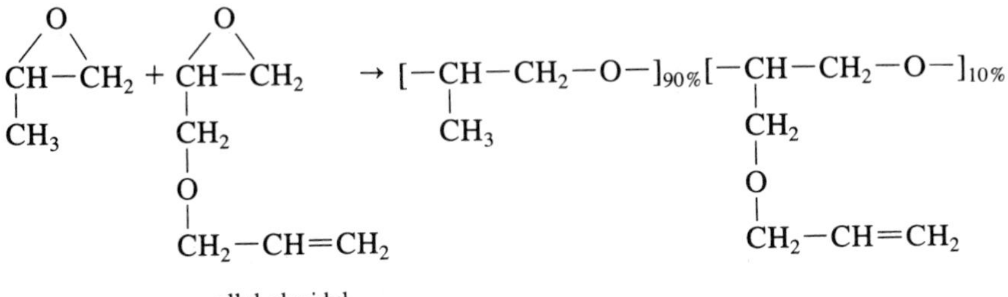

$$\underset{\underset{CH_3}{|}}{\overset{\overset{O}{/\ \backslash}}{CH-CH_2}} + \underset{\underset{CH_2-CH=CH_2}{\overset{|}{\underset{|}{O}}}}{\underset{CH_2}{|}}{\overset{\overset{O}{/\ \backslash}}{CH-CH_2}} \rightarrow [-\underset{\underset{CH_3}{|}}{CH}-CH_2-O-]_{90\%}[-\underset{\underset{CH_2-CH=CH_2}{\overset{|}{\underset{|}{O}}}}{\underset{CH_2}{|}}{CH}-CH_2-O-]_{10\%}$$

allyl glycidyl ether

Fig. 17.2. Propylene oxide polymers.

tion, its resistance to heat, weather, and ozone, and its modulus retention at elevated temperatures have led to the use of hydrogenated nitrile for timing belts.

Many other industrial uses have also been found for this elastomer.

PROPYLENE OXIDE POLYMERS

Propylene oxide elastomers are copolymers of propylene oxide and a cure-active monomer, allyl glycidyl ether. These were first developed in the early 1960's and originally marketed under the tradename Dynagen XP139 by The General Tire & Rubber Company, but are currently sold as Parel 58 by Hercules, Inc. (See Fig. 17.2 and Table 17.11.)

Propylene oxide polymers are noted primarily for their very good dynamic characteristics over a very broad temperature range, similar to but broader than that of natural rubber or polyisoprene. They do have the advantage of good ozone resistance and better high-temperature properties than polyisoprene, coupled with mild oil and aliphatic-solvent resistance.

The presence of approximately 10% allyl glycidyl ether units in the polymer allows for its capability of vulcanization with sulfur, as shown in Table 17.12. Moderate reinforcement can be obtained, depending upon the particle size of the filler used.

Applications for propylene oxide elastomers are engine mounts, body mounts, suspension bushings, dust seals, and boots, where higher-temperature, ozone

Table 17.11. Properties of Parel 58.

Physical form	Slab
Color	White or light amber
Specific gravity	0.92
Ash (%)	1.0
ODR viscosity—12 cpm, torque at 125°C	10–16

Table 17.12. Parel 58 Formulation.

Parel 58 (in phr)	100
Zinc oxide	5
Stearic acid	1
NBC	1
HAF N330	50
Sulfur	1.25
TMTM	1.5
MBT	1.5
	161.25
Compound Properties	
Mooney viscosity ML 1 + 4 @ 100°C	86
Mooney scorch	
Time to 3 pt. rise, min.	9
Time to 10 pt. rise, min.	12
Physical Properties Cured 30 min at 160°C	
Shore A	70
Modulus at 100%, kg/cm^2	37
Modulus at 200%, kg/cm^2	84
Tensile, kg/cm^2	136
Elongation, %	370
Low temperature, ASTM D1053 T 10,000	−58°C
Air Aged 70 hr at 150°C—Change	
Shore A, pt.	+3
Modulus at 100%, %	+20
Modulus at 200%, %	+21.8
Tensile, %	−18.1
Elongation, %	−43.2
Aged 70 hr at 23°C—Volume Change	
Water, %	+13
ASTM Fuel A, %	+69.7
ASTM Fuel B, %	+104
ASTM No. 1 Oil, %	+39.6*

*Aged at 150°C.

and weathering resistance are required in combination with excellent low-temperature, dynamic characteristics.

EPICHLOROHYDRIN POLYMERS

Epichlorohydrin elastomers were first produced in 1965 by both BFGoodrich and Hercules, Inc. The initial polymers were the homopolymers of epichloro-

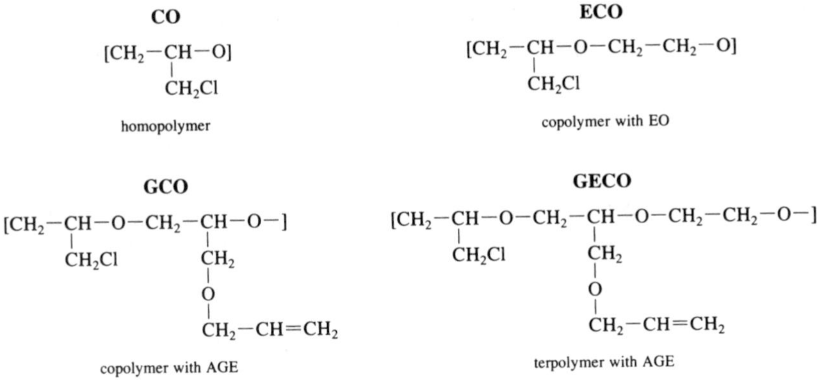

CO

$[CH_2-\underset{\displaystyle CH_2Cl}{\underset{|}{C}}H-O]$

homopolymer

ECO

$[CH_2-\underset{\displaystyle CH_2Cl}{\underset{|}{C}}H-O-CH_2-CH_2-O]$

copolymer with EO

GCO

$[CH_2-\underset{\displaystyle CH_2Cl}{\underset{|}{C}}H-O-CH_2-\underset{\displaystyle CH_2-O-CH_2-CH=CH_2}{\underset{|}{C}}H-O-]$

copolymer with AGE

GECO

$[CH_2-\underset{\displaystyle CH_2Cl}{\underset{|}{C}}H-O-CH_2-\underset{\displaystyle CH_2-O-CH_2-CH=CH_2}{\underset{|}{C}}H-O-CH_2-CH_2-O-]$

terpolymer with AGE

Fig. 17.3. Epichlorohydrin polymers.

hydrin and the copolymer of epichlorohydrin and ethylene oxide. Subsequently, BFGoodrich and Nippon Zeon produced epichlorohydrin elastomers with unsaturated side chains, which created much greater variations in possible cure systems.

In general, epichlorohydrin polymers have a good balance of low-temperature flexibility; fuel, oil, and solvent resistance; high-temperature resistance; low permeability; excellent ozone and weathering resistance; and very good dynamic properties. Although the original polymers did have problems with mold corrosion, poor resistance to oxidized fuel, and a tendency to revert upon long-term, high-temperature ageing, these drawbacks have been overcome to a large extent by compounding (addition of Stabilizer B) and improved polymer structures.

The homopolymer, CO, has excellent gas and solvent impermeability, but poor low-temperature characteristics. The copolymer, ECO, exhibits approximately 15°C better low-temperature resistance, but is poorer in resistance to solvent and gas permeation. The terpolymer, GECO, offers low-temperature flexibility and permeability equivalent to ECO, but has better resistance to sour gasoline and a greater flexibility in available cure systems. The terpolymer allows compounders to vary cure and scorch rate at will, and to overcome mold-fouling problems.

The major suppliers, BFGoodrich, Hercules, and Nippon Zeon, each produce a considerable range of homopolymers, copolymers, and terpolymers. These polymers have Mooney viscosities varying from 48 to 130, depending upon the process requirements; and ethylene oxide contents of zero for the homopolymer, 32 to 35% for terpolymers and up to 50% for the copolymers. The terpolymers contain a cure active third monomer such as allyl glycidyl ether (AGE).

Epichlorohydrin polymers, being somewhat more thermoplastic than other polymers, extrude and mold very well. During mixing, the compound temperature should be maintained between 49°C and 66°C to prevent roll sticking, above which range scorching may occur. Internal lubricants and metal oxides should be added early and curatives put in on a sheet-off mill or in a second pass. Normal practices should be employed for filler and plasticizer incorporation.

The original polymers used a red-lead/2-mercaptobenzimidazoline curing system, which was not only corrosive to molds, but also a probable health hazard. The use of dispersions, internal lubricants (metallic soaps), and anticorrosion agents such as Vanplast 201 or OTA (otho-toluic acid) provides some help in reducing mold fouling. Other cure systems such as the ECHO or the Zisnet F systems provide improvements in some properties and much-reduced mold fouling.

The epichlorohydrin terpolymers offer many advantages over the CO or ECO elastomers because they may be cured with sulfur, sulfur donor or peroxides, in addition to those previously mentioned. The sulfur/sulfur donor systems provide great flexibility in cure and scorch rates, but high-temperature resistance is reduced. The use of a peroxide helps to overcome the loss of heat-ageing with a sulfur cure system, and provides good compression set when combined with CaO, potassium stearate acid acceptors, and an acrylate monomer coagent. Higher curing temperatures must be employed with peroxides for optimum properties and reasonable cure times.

The excellent ozone, weathering, solvent, heat, and low-temperature resistance, combined with low solvent and gas permeability, make epichlorohydrin a natural selection for fuel and air-conditioning-system components, such as hose tube and/or cover, diaphragms, seals, gaskets, O-rings, and so forth. Very good dynamic characteristics coupled with the broad temperature range of these polymers make them well suited for vibration isolators, diaphragms, belts, rollers, and the like.

CHLORINATED POLYETHYLENE

Chlorinated polyethylene is produced by chlorinating a solution of polyethylene, in principle, in the same manner by which chlorobutyl is produced. The presence of chlorine provides not only a cure active site for vulcanization, but also reduces the crystallinity and conversely increases the rubberiness of the polymer. The nature of the original polyethylene (i.e., whether it is a more linear high-density type or a low-density, highly branched grade) also has a large bearing on the degree of crystallinity, or alternately, rubberiness of the CPE. This also greatly influences its high-temperature properties.

Originally, CPE was used exclusively for impact modification of plastic materials, but in the 1970's both Dow Chemical and Hoechst produced grades specifically for the rubber industry.

Table 17.13. Epichlorohydrin Cure Systems.

CO-homopolymer (phr)	60	60	100	100	-	-
ECO-copolymer	40	40	-	-	-	-
GECO-terpolymer	-	-	-	-	100	100
Stearic acid	-	-	2	1	0.8	-
Span No. 60	-	-	1	-	-	-
AgeRite White	-	-	-	-	3	-
MBI	1	1	-	-	-	-
NBC	1.25	1.25	1	1	-	1
Magnesium oxide	-	-	3	-	-	-
ZO-9	-	-	-	-	0.5	-
N330 carbon black	15	15	-	-	-	-
N326 carbon black	-	-	-	-	20	20
N550 carbon black	25	25	40	40	40	30
Paraplex G50	-	-	-	-	8	-
DOP	-	-	-	-	-	11
PETS	2	2	-	-	-	-
Chemlink 24	-	-	-	-	-	2.5
ETU, 70%	1.5	-	-	1.2	-	-
DPG	-	-	0.3	-	-	-
Cure-rite 18	-	-	-	-	2.0	-
BBTS	-	-	-	-	1.0	-
Sulfasan R	-	-	-	-	2.0	-
Zisnet F	-	-	0.9	-	-	-
Rhenofit UE	-	2	-	-	-	-
ECHO-MB	-	2.5	-	-	-	-
$BaCO_3$, 80%	-	4	-	-	-	-
Pb_3O_4, 80%	6	-	-	6.2	-	-
CaO	-	-	-	-	-	2.5
Potassium stearate	-	-	-	-	-	2.5
Zinc oxide	-	-	-	-	5.0	-
Di-Cup 40C	-	-	-	-	-	2.5
	151.75	152.75	148.2	149.4	182.3	172.0

Physical Properties

Cure (min @ °C):	*20 @170*	*20 @170*	*30 @160*	*30 @160*	*20 @160*	*20 @160*
Hardness, Shore A	68	71	69	69	65	58
Modulus @ 100%, MPa	3.3	4.3	5.7	5.7	2.3	2.6
Tensile strength, MPa	12.6	12.3	12.6	16.2	11.4	12.7
Elongation, %	390	330	400	400	820	350

CPE is supplied in a wide variety of grades with chlorine contents ranging from 25 to 42%, Mooney viscosities (ML 1 + 4 @ 121°C) varying from 35 to >110, and polymer morphology being either amorphous or with some degree of crystallinity.

The selection of the type of CPE depends upon the desired properties required for the end application. Increasing chlorine content increases oil, fuel, and solvent resistance, but decreases low-temperature flexibility. The amorphous grades are better for low-temperature flexibility, but exhibit increased swell in solvents. Low-residual sodium grades are required for lower water swell and better electrical properties. The Mooney viscosity of the polymer is selected to provide the processing characteristics desired in the final compound. Mixing is best

Table 17.14. CPE Applications.

	Hydraulic Hose Cover		*Cable Jacket*	
	4	*BX*	*XHD*	*MINING*
	Formulation			
CPE CM 0136 (in phr)	70	100	100	100
CPE XO 2243.45	30	–	–	–
EP202 (Eagle Picher)	–	–	10	–
Maglite D	10	10	–	2.5
DAP (Harwick)	–	–	10	–
Hi Sil EP	–	10	–	–
HAF N330	–	–	–	45
SRF N774	60	–	–	–
SRF N762	–	60	–	–
FEF N550	20	20	40	–
Catalpo clay	30	–	–	–
Chloro wax LV	–	–	20	–
Santicizer 79 TM	30	–	–	–
Sundex 790	–	–	–	30
DER 331 DLC (Harwick)	7	–	–	–
DCP 5275 (Burton)	10	–	–	–
Wax	–	–	3	–
Triallyl phosphate	2	–	–	–
SR 515	–	–	2	–
Percadox 17/40	9	–	–	–
Di-Cup 40 KE	–	–	8	–
Irgastab 2002	–	1	–	–
NDBC	–	1	–	–
ECHO-S	–	3	–	2.5
Vanax 808	–	1	–	6.8
	278	206	193	186.8
	Physical Properties—Cure			
Min. @ °C	*20 @ 160*	*20 @ 160*	*1 @ 193*	*2 @ 199*
Shore A	81	80		68
Modulus @ 100%, MPa	6.3	6.0	4.6	3.3
Modulus @ 200%, MPa	11.3	8.5	9.8	7.1
Tensile, MPa	12.4	10.1	19.9	19.9
Elongation, %	285	350	355	520

accomplished by the ''upside-down'' method, with the polymer loaded last into the internal mixer.

The initial method of vulcanization was by means of peroxides, with suitable acid acceptors and coagents. In 1980 and 1981, sulfur cure systems based on DETU (*NN'*diethylthiourea) and ECHO (a thiadiazole derivative) were introduced which provide much greater flexibility to the compounder. (*NOTE:* For all systems, *zinc oxide* and other *zinc derivatives must be avoided* because they catalytically decompose chlorinated polyethylene!)

As may be expected, the peroxide cure systems offer somewhat better heat resistance than the ECHO or DETU systems. The DETU system, in addition to causing bloom and being very-slow-curing, does produce much higher compression set.

The excellent ozone, very good heat and moderately good oil and aliphatic solvent resistance of chlorinated polyethylene have led to its use in a variety of applications, including hose, electrical-cable covers, automotive tubing, boots, dust covers, and so forth.

ETHYLENE/ACRYLIC ELASTOMER

An ethylene/acrylic elastomer with the trade name Vamac was introduced by Du Pont in 1975. The base polymer is a terpolymer of ethylene and methyl acrylate, with a small amount of a third monomer containing carboxylic acid to provide cure active groups in the polymer chain. The polymer is sold in the form of lightly loaded, either SRF or fumed silica masterbatches, which also contain stabilizers and process aids. Having a saturated backbone, the polymer has excellent ozone and weathering resistance, very good heat resistance, and fairly good low-temperature properties. Oil resistance of the ethylene/acrylic elastomer is fairly good and its flexing properties are excellent.

The very thermoplastic nature and the low viscosity of the ethylene/acrylic elastomer create problems in processing (mill-sticking), which must be overcome by the addition of release agents such as Armeen 18D, stearic acid, or Zelec UN. Mixing and milling should be performed at as low a temperature as possible. Much improved processing may also be achieved by the inclusion of 10 to 20% of cryogenically-ground cured ethylene/acrylic compound. Molds must be kept very clean to start with, and a mold release such as Frekote 33 should be applied before starting to cure.

$$[-CH_2-CH_2-]_x[-CH_2-\underset{\displaystyle COOCH_3}{\underset{|}{CH}}-]_y[-\underset{\displaystyle COOH}{\underset{|}{R}}-]_z$$

Fig. 17.4. Ethylene/acrylic elastomer.

Table 17.15. Ethylene/Acrylic Elastomers.

Vamac Masterbatch:	*B124*	*HGB124*	*N123*
Filler type	N774 SRF	N774 SRF	Fumed silica
Filler levels, phr	20	20	23
Stabilizer and process aids	4	4	
Specific gravity	1.12	1.12	1.08
Specific gravity, base polymer	1.04	1.04	1.04
Mooney viscosity, ML 1 + 4 @ 100°C	20	29	30

Although a peroxide cure is possible with the ethylene/acrylic elastomer, the properties achieved are not good. The best results are obtained with either methylene dianiline (MDA) for best scorch safety, or hexamethylene diamine carbamate (DIAK No. 1) for faster cure rate, both of which are combined with a guanidine accelerator. Care must be taken to exclude metal oxides, particularly zinc oxide, because of their reactivity with the carboxylic group, which creates nerve or a scorched appearance, slow cures, and poor physical and aged properties.

Ethylene/acrylic compounds have found use in automotive applications such as boots, grommets, seals, and so forth, where their very good flex, ozone resistance, and high-temperature properties are combined with fairly good low-temperature characteristics and oil resistance. The good dampening characteristics of ethylene/acrylic elastomers make it well-suited for vibration mounts, pads, isolators, and so forth.

PHOSPHONITRILIC FLUOROELASTOMERS

As long ago as 1895, H. N. Stokes prepared the first phosphazene polymer by causing phosphorus pentachloride to react with ammonium chloride to form hexachlorocyclotriphosphazene and subsequently polymerizing it to form a very-high-molecular-weight insoluble inorganic polymer; the latter, unfortunately, was degraded by moisture, which made it commercially unacceptable. In 1965, H. R. Allcock and R. L. Kugel were able to terminate the polymer chain at lower conversion to produce a soluble polymer, which allowed for the addition of side chains through a nucleophilic substitution reaction. S. H. Rose, by employing more than one nucleophile, was able to develop a rubbery elastomer, and through work on various nucleophiles and catalyst systems at Horizons Research and Firestone, there evolved a commercial phosphonitrilic fluoroelastomer, PNF®. PNF was originally produced and marketed by Firestone, but later the process was sold to Ethyl Corp. (See Fig. 17.5.)

The phosphonitrilic fluoroelastomer, with its saturated inorganic backbone, is resistant to oxygen, ozone, weathering, and heat up to 175°C. It should be

Table 17.16. Ethylene/Acrylic Curing Systems.

Formulation	*1A*	*1B*	*5B*
Vamac B124 MB (phr)	124	124	124
Armeen 18D	0.5	0.5	0.5
Stearic acid	2	2	2
Zelec UN	–	–	2
SRF N774	45	45	–
FEF N550	–	–	50
Dioctyl sebacate	10	10	10
Polyester plasticizer	10	10	10
DIAK No. 1	1.25	–	1.25
Methylene dianiline	–	1.25	–
DPG	4	4	4
Di-Cup 40C	–	–	7
	196.75	196.75	210.75
Compound Properties			
Mooney scorch—original, MS @ 121°C–10 pt rise (min)	14	>30	14
Rheometer, 3° arc, 3 cpm Tc 90 @ 177°C (min)	14	22	7.2
Physical Properties Cure, 20 min @ 177°C			
Shore A	56	59	57
Modulus @ 160%, MPa	1.6	1.8	2.8
Tensile, MPa	11.6	11.4	11.2
Elongation, %	570	530	280
Tear, Die C, kN/m	32.2	31.7	26.1
Air-aged 7 days at 177°C—Change			
Shore A, pt	+14	+17	+23
Tensile, %	+6	+9.1	+20
Elongation, %	−50.9	−54.7	−28.6
Compression Set B, 70 hr @ 149°C			
Post-cure 4 hr @ 177°C	24	19	29
Low Temperature			
Brittle point, °C	+4.7	+5.4	–
Fluid-aged 70 hr—Volume Change (%)			
Water, 100°C	+4.7	+5.4	–
ASTM No. 1 Oil, 149°C	+0.3	+1.2	–
ASTM No. 3 Oil, 149°C	+58	+53	–

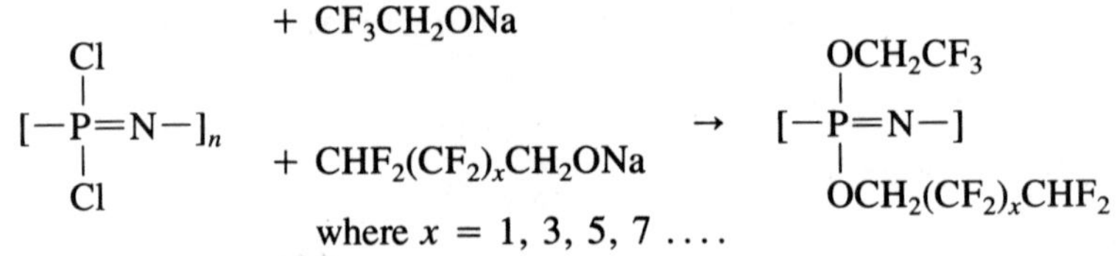

$$[-\underset{\underset{Cl}{|}}{\overset{\overset{Cl}{|}}{P}}=N-]_n \begin{array}{l} + CF_3CH_2ONa \\ \\ + CHF_2(CF_2)_xCH_2ONa \\ \text{where } x = 1, 3, 5, 7 \ldots. \end{array} \rightarrow [-\underset{\underset{OCH_2(CF_2)_xCHF_2}{|}}{\overset{\overset{OCH_2CF_3}{|}}{P}}=N-]$$

Fig. 17.5. Phosphonitrilic fluoroelastomer reactions.

Table 17.17. PNF® Elastomer.

Gum rubber

Mooney viscosity, ML 1 + 4 @ 100°C	15
Specific gravity	1.75
Low temp. T_g°C	−68
Fluorine content, %	55 approx.

Standard Compounds

PNF Compound	*Color*[a]	*Typical Applications*	*Shore A* ±5	*Tensile*[b] *MPa*	*Elongation*[b] %
240–001	B	Vibration mount	40	9.3	220
255–001	B	Diaphragm/vibration mount	55	8.3	150
260–004	G	Electrical/diaphragm	60	10.3	180
265–003	B	Diaphragm/vibration mount	65	10	150
265–004	B	High elongation	65	7.6	220
270–003	G	O-ring (general purpose)	70	10.3	120
275–009	G	T-seal (high tear resistance)	75	10.3	120
275–006	B	Fuel controls/vibration mount	75	9.7	150
280–001	G	O-ring (general purpose)	80	10.3	100
280–003	B	Vibration mount	80	9.5	120
280–008	B	Shaft seal	80	8.6	110
285–005	G	Shaft seal	85	8.7	110
290–003	G	T-seal (high tear resistance)	90	10.3	80

[a]B = black; G = green.
[b]Data obtained on microtensile rings, ASTM D3196.

noted that the physical properties of the polymer do not decrease nearly as rapidly at elevated temperatures as those of other elastomers. The high fluorine content of the polymer imparts extremely good resistance to solvents, fuels, oils, hydraulic fluids, and so forth. The basic flexibility of the polymer backbone and long pendant side chains containing fluorine have produced a polymer with excellent low-temperature characteristics, good damping, resistance to flex fatigue, and reasonably good physical properties. The dynamic properties are constant over the range 25°C to 150°C.

A small amount of unsaturation included in the polymer enables it to be sulfur-cured, but a peroxide system with a coagent, such as triallyl isocyanurate, is much preferred. Carbon black, clays, and silicas, and combinations thereof, can be used for reinforcement, but a silane coupling agent should be included with acidic fillers for optimum physical properties. A metallic oxide, preferably magnesium oxide, should be included for slightly higher modulus and good adhesion to metal. Mineral fillers give better heat, compression set, and fluid resistance, whereas carbon black gives better processing properties and flex characteristics.

Phosphonitrilic fluoroelastomer is invariably supplied in a compounded form, as shown in Table 17.18. As may be seen from that table, a broad range of compounds for various applications are commercially available. The properties of the polymer are shown in Table 17.19.

Table 17.18. Basic Phosphonitrilic Fluoroelastomer Compound.

PNF-200	100 parts
Filler	30–60 parts
Magnesium oxide	2.0–10.0 parts
Modifiers	Variable
Stabilizer	2.0 parts
Peroxide	0.4–2.0 parts

Table 17.19. Resistance of PNF® to Fluids.

	Temperature, °C	*Volume Swell, %*	*Micro-hardness Change*	*Tensile Change, %*	*Elongation Change, %*
ASTM Fuel A	93.3	13	−10	−29	−18
ASTM Fuel B	100	16	−7	−23	−22
ASTM Fuel C	100	19	−12	−24	−14
JP-4	110	9	−4	−12	−14
ASTM No. 1	149	0	+1	−15	−7
ASTM No. 3	149	2	−2	−16	−14

The very broad useful temperature range of −55 to +165°C for PNF, with its excellent resistance to solvents, oils, hydraulic fluids, etc., has created uses for it in many aerospace, military, petrochemical and other extremely critical applications. The use of this material is growing significantly.

REFERENCES

1. M. Morton, *Rubber Technology* 2nd ed., Van Nostrand Reinhold, New York, 1973.
2. J. A. Brydson, *Rubber Chemistry* Applied Science Publishers, London, 1978.
3. "FA Polysulfide Rubber," Morton Thiokol Inc., 1978.
4. "ST Polysulfide Rubber," Morton Thiokol Inc., 1977.
5. "LP Liquid Polysulfide Polymer," Morton Thiokol Inc., 1981.
6. K. Hashimoto, N. Watanabe, and A. Yoshioka, "Highly Saturated Nitrile Elastomer—A New High Temperature, Chemical Resistant Elastomer," Paper No. 43, 124th Meeting, Rubber Division, ACS, Houston, Oct. 1983.
7. K. Hashimoto, N. Watanabe, M. Oyama, and Y. Todani, "Highly Saturated Nitrile Elastomer—Zetpol," Annual Meeting, Swedish Institution of Rubber Technology, Gothenburg, Sweden, May 17-18, 1984.
8. J. Thoermer, J. Mirza, and H. Buding, "Therban A New Abrasion Resistant Special Elastomer with Good Heat and Swelling Resistance," Paper No. 87, 124th Meeting, Rubber Division, ACS, Houston, Oct. 1983.
9. "Parel, Propylene Oxide Elastomers," Hercules Inc., 1981.
10. "Herclor, Epichlorohydrin Elastomers," Hercules Inc.,
11. "Hydrin Elastomers," HM 13, B. F. Goodrich Company, 1978.
12. "Hydrin 400 Elastomer," HM 14, B. F. Goodrich Company, 1981.
13. E. N. Scheer, "Hydrin Elastomers Review," Nashville Seminar, Sept. 26, 1979.
14. H. Ehrend and W. Schuetta, "Crosslinking of Epichlorohydrin Elastomers," Paper No. 32, 125th Meeting, Rubber Division, ACS, Indianapolis, May 1984.
15. "Gechron Epichlorohydrin Rubber," Nippon Zeon Co., Ltd., 1984.
16. "Zisnet F Special Curative for Epichlorohydrin Rubber," Nippon Zeon Co., Ltd., 1984.
17. "CPE Technical Reports, Dow Chlorinated Polyethylene Elastomers," Dow Chemical Company.
18. R. I. Sylvest, C. Barnes, N. E. Warren, W. R. Worsley, Jr., Paper No. 15, "Thiadiazole Vulcanization of Chlorinated Polyethylene," 120th Meeting, Rubber Division, ACS, Cleveland, OH, Oct. 1981.
19. "Vamac, Ethylene/Acrylic Elastomer," E. I. du Pont de Nemours & Co., Inc., 1976.
20. "PNF Elastomer," Firestone Phosphazene Products, Firestone Tire & Rubber Co., Akron, OH, 1979.
21. D. F. Lohr, J. A. Beckman, "PNF Phosphonitrilic Fluoroelastomer: Properties and Applications." Paper 34, 120th Meeting, Rubber Division, ACS, Cleveland, Oct. 1981.
22. J. W. Fieldhouse and D. F. Graves, "Process for the Polymerization of Cyclic Polyhalophosphazenes Using a Catalyst Composition of Boron Trihalide and Oxygenated Phosphorus Compounds," U.S. Pat. 4,226,840, 1980.

18

RECLAIMED RUBBER

ROGER SCHAEFER AND R. A. ISRINGHAUS
Midwest Rubber Reclaiming Co.
East St. Louis, Illinois

HISTORY

Charles Goodyear invented the process of sulfur vulcanization in 1839 and as a result of that discovery the rubber industry was much expanded. The first products manufactured in the 1840's were footwear but soon many different types of proofed goods, sheet rubber, car springs, and hard rubber articles were being produced. In those days, before the establishment of india-rubber plantations, all raw rubber was collected from the jungles of Brazil. Demand for crude rubber soon outstripped supply, with prices tripling even before the Civil War. There was a great economic incentive to spur the development of methods of reusing scrap vulcanized rubber.

One method of recycling some scrap rubber is to grind it as fine as possible and work it into new rubber as an elastomeric filler. This was the first method of reclaiming and, in fact, a patent was granted to Charles Goodyear himself for this technique. This was suitable for compounding carriage springs, which were fairly large barrel-shaped molded articles. However, most of the demand and growth in those days was in the footwear business and simply using ground rubber was not acceptable.

Boots and shoes, then, were the most important rubber products in the nineteenth century. They were formulated with a fairly low level of sulfur in order to prevent unsightly sulfur bloom. As a result, they were lightly vulcanized and could be more easily reclaimed. Such material can be successfully devulcanized or partially replasticized by the action of steam heat alone at a higher temperature and for a substantially longer period of time than the original vulcanization cycle. This is known as the *heater process*, patented by Hiram L. Hall in 1858. The product of this reclaiming process was revulcanizable and could be blended with new crude natural rubber at a high percentage.

First Mechanization

In order to make a high-quality reclaim, the fiber must be removed from the rubber scrap. At one time this amounted to a cottage industry in which women and children stripped the rubber from the fabric of old boots "by soaking the rubber in water, and then taking a small knife and starting the rubber from the cloth and stripping it off." Even at the low wage these people must have received, this was not an effective way to produce large quantities of product. Eugene H. Clapp used an air-blast method of separating fiber from rubber after the scrap containing both had been ground. Clapp did not try to patent the technique but managed to keep it a secret for several years. Although Clapp's method was the ancestor of present-day defibering equipment, it was not developed to the point it could produce a raw material of more than mediocre quality. Better methods of producing large quantities of higher-quality reclaim were needed.

Development of Chemical Processes

It had been common knowledge for many years that acid has a degrading effect on fabric and that rubber is resistant to acid; so there is no single individual whom we can credit with the invention of the acid reclaiming process. N. Chapman Mitchell was a leader in the development of the acid process, and started the first company organized to make reclaimed rubber by that method. Mitchell filed for patents on the process in 1881, which were granted, but they were poorly drawn and were later voided. Mitchell was engaged in litigation with a number of competitors, who held similar patents, until 1887, when they all decided to be done with it and merged their companies.

In the acid process, the ground scrap was boiled for several hours in a fairly strong solution of either sulfuric or muriatic acid to destroy the fabric; the rubber was washed and then devulcanized with high-pressure steam in the pan process. It was a two-step operation.

Although the acid process was used quite successfully for a period of time, there were problems with the method. If the acid was not thoroughly washed, it could cause poor ageing characteristics. More important, there were coming on the market many types of higher-sulfur scraps, such as bicycle-tire scrap, that could not be processed by the acid method. If the high quantities of free sulfur in such types of rubber scraps are not removed, they tend to cause further vulcanization during the open steam heater process rather than devulcanization. Something better was needed.

A breakthrough came with the 1899 patent of Arthur H. Marks, which described an invention that came to be known as the *alkali digester process*. The ground rubber, fiber, and a dilute solution of caustic soda were cooked at high pressure for about 20 hours. This system caused the defibering, desulfuring, and devulcanization of the rubber scrap *all in one step*. This technique could be used to reclaim any type of rubber scrap available at that time. A year later, in 1900, Marks obtained another patent on an improvement which involved steam-jacketing the vessel and agitation of the mass. This is the cooking method used by most reclaimers for more than fifty years.

Marks had begun the development work at the Boston Woven Hose Co. under Raymond B. Price but completed it at the Diamond Rubber Co. of Akron, Ohio. He sold the domestic rights to the patents to Diamond for a block of stock. Diamond made a great amount of money manufacturing reclaimed rubber by the Marks method and from the royalties received from other U.S. companies; Marks soon became vice-president and general manager.

Under Marks' direction, Diamond pioneered many revolutionary developments in the rubber industry; he hired George Oenslager, whose discovery of organic accelerators was perhaps the most important contribution to rubber technology since Goodyear's discovery of vulcanization. Development work was also done on the use of carbon black in tire treads and on tire cords. Diamond Rubber merged with the B. F. Goodrich Company in 1912; Arthur Marks (who later founded and was president of several foreign reclaiming companies) became a multimillionaire as well as being recognized as one of the most important men in the history of the rubber industry.

So far we have discussed mostly the different methods for devulcanizing the rubber; two other important developments occurring at the turn of the century involved the introduction of new machines to greatly improve the quality of reclaimed rubber. One development was the *refiner*, which is a special type of mill used to produce a much more smooth and homogeneous product than would otherwise be possible. The other was the *strainer*, which is used to remove foreign matter from the rubber. With the exception of invention of the Reclaimator in 1953 (described below), developments in the reclaiming industry since the time of Marks have been mostly evolutionary rather than revolutionary.

DEFINITIONS

The following definition was adopted in 1981 by the Rubber Recycling Division of the National Association of Recycling Industries, Inc.:

> *Reclaimed Rubber* is the product resulting when waste vulcanized scrap rubber is treated to produce a plastic material which can be easily processed, compounded and vulcanized with or without the addition of either natural or synthetic rubbers. It is recognized that the vulcanization process is not truly reversible; however, an accepted definition for devulcanization is that it is a change in vulcanized rubber which results in a decreased resistance to deformation at ordinary temperatures.

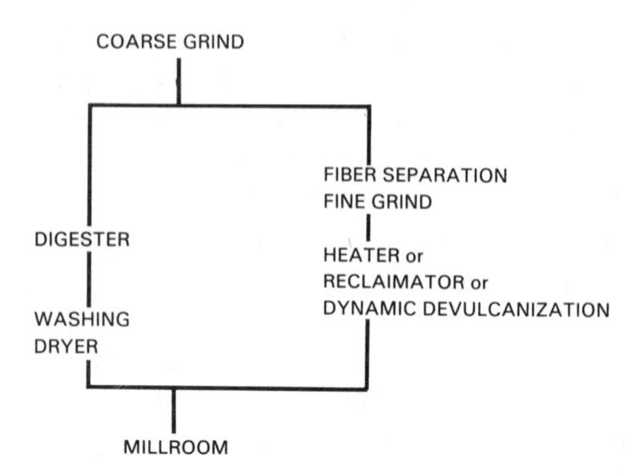

Fig. 18.1. Flow diagram of manufacturing process for whole-tire reclaim.

Recycled rubber can be more generally described as any sort of rubber waste that has been converted into an economically useful form such as reclaimed rubber, ground rubber, reprocessed synthetic rubber, and die-cut punched parts.

RECLAIMING PROCESSES

Scrap-rubber Preparation

The first stage of any reclaiming process is size reduction of the scrap; this is most generally accomplished using corrugated two-roll cracker mills. Other types of equipment have been employed such as a rotating cutter to obtain a rip, tear, or shear action, while others have rotating-knife slitter actions. The capital cost for these types of equipment is less than for a standard two-roll cracker installation; however, in most cases they have not been proven to be heavy-duty enough to withstand 24-hour operation.

The tire bead wire is either cut from the carcass before grinding or manually removed after the first pass through the cracker. At many points in the system, the ground rubber is passed over magnetic separation equipment to remove remaining bead wire and steel belt wire. Gyratory classifier screens return the oversize rubber particles to the cracker for further size reduction. There may be more than one stage of grinding before the desired size of approximately $\frac{3}{8}$-inch is obtained. At this point, the coarse-ground rubber could be conveyed to the digester or processed further for one of the other devulcanization methods.

The fiber from the coarse-ground tires is removed by some sort of fluidized-bed, specific-gravity table. The action of controlled streams of air moving through a bed of ground rubber on an inclined gyrating table of special design causes the fiber to separate from the clean rubber. The size of the rubber particles may now be further reduced to perhaps 20 or 30 mesh by fine grinding. Fine-corrugated, high-friction mills are employed in this operation.

Digester Process

At one time, most reclaim was made using the digester process. A digester is essentially a steam-jacketed, agitator-equipped autoclave mounted either horizontally or vertically. This is a wet process in which the coarsely ground scrap is submerged in a solution of water and reclaiming agents. These agents may include many types of light and/or heavy oils, naval stores, pine tar and coal tar pitches, and chemical peptizers.

Until the advent of synthetic rubber, the digesting solution also included caustic soda to remove free sulfur and to act as a defibering agent. In fact, the process was generally referred to as the *alkali digester method*. After World War II, it was found that scrap rubber containing SBR was totally incompatible with alkali cooks so the industry developed the *neutral process*, in which zinc or calcium chlorides are used as the defibering agents.

Weighed scrap, water, and reclaiming and defibering agents are dumped into the digester and the cook cycle is started. Steam, generally within a range of 150 to 250 psi, is injected into the jacket of the vessel for a digestion period that may be from 5 to 24 hours. During this time, the rubber becomes devulcanized and the fiber becomes hydrolyzed. After digestion, the charge is blown down and washed to remove the decomposed fiber and caustic soda or metallic chloride. The washed, devulcanized rubber particles are conveyed to a dewatering press and then to a dryer. The material is now ready for the final, refining stage of the reclaimed-rubber manufacturing process.

Heater or Pan Process

This is the simplest of all the reclaiming processes. A "heater" is a large, single-shell horizontal pressure vessel or autoclave. The ground rubber is mixed with reclaiming agents in an open ribbon mixer then placed into containers rolled into the vessel. The main consideration is to allow an even penetration of heat in the mass of rubber. To achieve this uniform steam penetration, shallow pans or boats equipped with hollow metal pipes or inverted "V" sections are used as the stock containers. Live steam at pressures of 100 to 250 psi with cycle times of 5 to 12 hours are typical.

This process yields fairly good results with some types of rubber scraps such as butyl inner tubes and marginal quality with other types such as fine-ground tires or low-specific-gravity natural-rubber scrap.

Reclaimator Process

The Reclaimator process is the only commercially successful continuous technique for devulcanizing tire scrap; all the others are batch processes. Tires are ground, the metal and fiber are mechanically separated, then the rubber is further ground to a fine particle size. This fine-ground rubber and the various reclaiming agents are all metered into a blending system and conveyed to the Reclaimator.

The Reclaimator is a special type of screw-extrusion machine. It is jacketed to provide for several zones of controlled temperature using either hot oil or cooling water; in addition, the clearances between the screw and the chamber wall are close and adjustable. The object is to subject the rubber to a controlled amount of high heat and pressure in a continuously moving environment. The residence time of the rubber in the machine is less than 5 minutes. During this period, the rubber undergoes devulcanization. After the softened rubber is discharged from the head of the machine, it is cooled and further processed in refining mills just as is done in other reclaiming methods.

Dynamic Devulcanization

The digester process has been used for decades, and millions of tons of excellent-quality reclaim have been produced. However, by the mid-1960's it became apparent that the many problems associated with this process doomed it to extinction. The digesting solution had to be disposed of after each cook; this caused water pollution. The rubber had to be washed after the cook; this caused a loss of some of the finer particles of rubber as well as produced more water pollution. The wet rubber crumb had to be dried with large amounts of warm air; this caused air pollution. The drying step also required large amounts of energy. The cycle times were long; the overhead costs were high. A better process was needed.

The heater or pan process solved many of the problems of the digester process but had its own problems. Although it is a satisfactory method for cooking some types of scrap rubber, it is highly unsatisfactory for whole-tire reclaim; the cooked mass is much too nonuniform.

The dynamic devulcanizer is essentially a marriage of the best features of the digester and heater processes. It is a large horizontal autoclave with a shaft down its axis on which are mounted a series of paddles arranged so when the shaft is rotated, the rubber inside receives the maximum amount of agitation. The mechanism must be built much more sturdily than the agitator on a digester because the rubber is to be cooked dry. The scrap rubber is agitated throughout the cook cycle so as to achieve total uniformity within each batch. The rubber is cooked with live steam and needs no washing or drying; there is no water pollution. The steam blowdown must be scrubbed or treated in some other way to avoid air pollution, but this is a much smaller problem than with the digester process. Since there is no loss of rubber in a wash cycle, the scrap tires can be ground much finer prior to the cook. This fine-ground rubber allows a much more uniform distribution of the reactive chemicals used in the process and results in a higher-quality product.

The dynamic devulcanizer enables a higher-quality product and also a more economical one. The finer particles of rubber and its continuous agitation allows higher steam pressures and shorter cycles. The greater uniformity of the cooked stock allows higher production rates in the millroom. Both of these factors increase productivity and reduce costs in a very capital-intensive industry—necessary if reclaim is to compete against synthetic rubber in the marketplace.

Millroom Operations

The rubber that has been softened by one of the methods described above is then conveyed to the millroom for the final stage of the reclaiming process. The stock is generally first mixed with weighed amounts of pigments, carbon black, or other ingredients in a blender. Next an extended barrel mixer strainer may be used to warm and plasticate the stock. It is then fed to a high-friction breaker mill and on to the strainer where tramp metal and other foreign particles are removed. The final milling step is the finishing refiner. A high friction ratio for the refiner mill is achieved with different-size rolls turning at considerably different speeds. The rolls are set quite tightly to obtain a thin sheet of rubber. The object is a product that is smooth, uniform, and free of grain or lumps.

In the past, the finished sheet from the refiner was pulled to a windup drum and allowed to build up to a thickness of perhaps an inch before being cut off by hand with a small knife. The resulting rubber slab was dusted and stacked. Today most reclaim manufacturers use balers to package their product.

THE ADVANTAGES OF USING RECLAIMED RUBBER

The primary, but not exclusive, advantage of using reclaimed rubber in the manufacture of tires and other rubber goods is one of cost. There are, of course, the direct cost savings resulting from the lower cost of the reclaimed product as compared to natural rubber and virgin synthetic rubbers, but there are equally important indirect cost benefits resulting from the proper use of reclaim.

Special Strengths through Reclaiming

Whole-tire reclaim is extremely uniform due to the extensive blending that occurs at every step of the manufacturing process. What is produced is best considered as a unique new raw material rather than some sort of modified SBR/natural blend. Just like any other ingredient chosen by the compounder to be incorporated into a formula, the raw material—reclaimed rubber—has its specific purposes. The purpose of reducing formula compound cost is certainly a legitimate one; however, there are other reasons to include reclaim in a formula that will result in processing advantages and indirect, but real, savings.

When the vulcanized tire rubber is reclaimed, it is softened through a reduction by scission of the molecular weight of the crosslinked elastomeric chains. That is, carbon-carbon bonds in the synthetic and natural-rubber polymers are broken, not the sulfur crosslinks that were formed when the tire was vulcanized. The result is a system of crosslinked fragments that is much more "three-dimensional" in nature than were the original polymers.

The three-dimensional character of the rubber fragments causes the reclaim to have properties different from the original tire compound. The shorter length of the polymer chains that make up the fragments give a lower tensile strength than the original compound when the reclaim is revulcanized. However, the three-dimensional nature of the fragmented elastomers in reclaim also produces some very desirable properties. The most important of these is the dimensional stability and reduced nerve imparted to compounds that include reclaim, improving the rate and the gauge stability of calendered and extruded stocks. The improved calendering production rate and lower reject rate result in significant cost savings that should be considered in the calculation of formula costs.

Further Advantages of Reclaiming; Applications

The intense mechanical work and chemical treatment given to the rubber in the reclaiming process result in further properties very beneficial to the end user. Reclaim can be expected to break down in mixing much more quickly than virgin polymers. This means that the mixing cycle can be shortened or, in the case of a two-pass mix, the reclaim can be added in the second pass. The reduced mixing time equals more indirect savings in the form of greater productivity—more product using less labor and overhead cost per unit.

In addition to the reduced mixing time when using reclaim, studies have shown that there is a sizable reduction in power consumption during mixing. This is significant at today's power cost, and will become even more important as the cost of electric power continues to escalate.

Reclaim can be used in a great many different applications. Where cost is a primary consideration and tensile strength requirements are not particularly demanding, reclaim can be the sole source of polymer. Examples of such products are mats and semipneumatic tires. In other cases, reclaim is blended with virgin synthetic and natural rubber. The virgin polymer adds strength and the reclaim contributes to improved processing, cooler mixing, and less sensitive, less scorchy cure systems. Such applications include most of the constituent parts of both passenger and truck tires, bias and radial-ply.

Formulating with reclaim is not difficult, but the compounder must remember that reclaim, unlike most processing additives that do little to the cure except for dilution, enters into the cure system. The reclaim elastomers become part of the overall crosslinked system. However, it is unrealistic to expect that the optimum cure system used with synthetic and/or natural rubber polymers will remain the same when reclaim is added. The optimum cure system must be reestablished when reclaim is added, just as it should be whenever changes are made to the types or proportions of virgin rubber.

The material, mixing, manpower, and energy-consumption savings associated with reclaimed rubber are but a part of the total benefit of its use. These savings are substantial and significant, but there are further positive aspects to the use of reclaim, which accrue to our society as a whole. They include the employment of hundreds of people in the reclaim industry and the alleviation of solid-waste-disposal problems. They also include the conservation of petrochemical raw materials, conservation of the energy required to convert the petrochemicals into synthetic rubber, and conservation of the foreign exchange needed to pay for imported oil, a large proportion of the petrochemicals' source.

MAJOR USES OF RECLAIMED RUBBER

1. Carcass, sidewall and undertread of passenger, light-truck and off-road tires.
2. Tubeless passenger-tire innerliner.
3. Inner tubes.
4. Semipneumatic tires.
5. Automotive floor mats.
6. Mechanical goods.
7. Adhesive, sealing, and tape compounds.
8. Rubberized asphalt.

TYPICAL FORMULATIONS

Automobile Floor Mat

Whole tire premix	250
SBR 1712	70
Zinc oxide	2
Stearic acid	2
Benzothiazyl disulfide	2.3
Methyl zimate	0.56
Sulfur	5
Hard clay	70
Calcium carbonate	85
Austin Black	95
Process oil	20
Cure	316°F—6 min

Semi-pneumatic Tire

Whole-tire premix	170
SBR 1849	98
Hard clay	150
Whiting	50
Mineral rubber	20
Tackifier	15
Zinc oxide	3
Stearic acid	3
Antioxidant	0.5
Sulfur	6
N-Tert-butyl-2-benzothiazole sulfenamide	2
Tetramethylthiuram disulfide	0.4
30-mesh ground rubber	10
Aromatic process oil	20
Cure	350°F—7 min

Butyl Inner Tube

Butyl 218	81
Butyl reclaim	25
Paraffinic plasticizer	20
Zinc oxide	5
Tetramethylthiuram disulfide	1
Mercaptobenzothiazole	0.5
Sulfur	1.75
N-660 carbon black	60
Cure	340°F—5 min

Innerliner

Halobutyl	50
SBR 1500	35
Butyl reclaim	30
Stearic acid	1.5
N-660 carbon black	30
Dixie clay	60

Piccopale 100	5
Austin Black	10
Paraffinic process oil	16
Zinc oxide	3
Mercaptobenzothiazole	1.25
Tetramethylthiuram disulfide	0.25
Sulfur	0.25
Cure	350°F—15 min

Carcass

No. 1 ribbed smoked sheet	40
SBR 1500	40
Whole-tire reclaim	46
N-660 carbon black	45
Stearic acid	1
Naphthenic-process oil	14
Petroleum resin	3
Antioxidant	1
OBTS accelerator	1
Insoluble sulfur	3
Cure	350°F—15 min

RUBBERIZED ASPHALT

The concept of using rubber to modify asphalt can be traced back to the very early days of paved-highway construction, when asphalt became a basic construction material. Many different types and forms of rubber have been tried. In the earliest work, latices of natural rubber, SBR, and neoprene were evaluated along with dry forms of these polymers. In the early 1960's, the concept of using recycled rubber began to emerge as an economical and technically effective method for enhancing asphalt properties. Many combinations of asphalts and rubbers have been found to have utility in various paving applications.

The incentive for asphalt modification derives from the extreme temperature sensitivity of asphalt. The viscosity of asphalt will change greatly between 10° and 50°C. Moreover, at room temperature and above, most asphalts are only weakly viscoelastic. On the other hand, their brittle points lie fairly high, typically between 40 and 45°F (4 and 7°C). With these asphalt properties in mind, the objective of polymer modification of asphalt is to inhibit flow at elevated use temperatures while simultaneously improving low-temperature flexibility. In practical terms, addition of rubber to asphalt will typically increase its softening point and lower the brittle point.

During the past 20 years, as the use of recycled rubber in paving applications has increased, terminology defining the various types of recycled materials has also evolved. The majority of usage has concerned ground whole-tire rubber and ground tread rubber. Use has also been made of devulcanized or reclaimed rubber.

Applications

Asphalt rubber in paving applications takes many forms, but the principal ones are as follows:

1. Asphaltic concrete hot mixes.
2. Stress-absorbing membrane innerlayer.
3. Stress-absorbing membrane.
4. Joint and crack sealers.

In the preparation of asphaltic concrete hot mixes, the rubber may be added directly to the pug mill used for mixing the asphalt and aggregate. For example, a powdered devulcanized reclaimed rubber can be added in an amount ranging from 0.7 to 1.0 weight-percent based on total aggregate. The resulting asphalt-concrete mix results in a pavement that exhibits less reflection cracking and reduced rutting, shoving, and bleeding.

In stress-absorbing membranes (SAM) or seal coat, the asphalt-rubber blend is prepared by preheating the asphalt, adding a diluent, if required, and then blending in the rubber components at 350° to 425°F (175° to 220°C.) The proportion of rubber is usually in the range of 20–25% by weight of the total blend. The resulting asphalt-rubber mixture is then sprayed on the road surface using a pressure-type bituminous-distributor truck. The application rate is typically 0.60 gallons per square yard. After the asphalt rubber binder has been applied, rock chips are immediately laid into the asphalt rubber layer at a rate of 25–40 pounds per square yard.

In stress-absorbing membrane inner layers (SAMI) applications, the preparation and installation of the asphalt-rubber system is identical to the system used for stress-absorbing membranes (SAM). However, the application rate may be somewhat higher, e.g., up to 1.0 gallons per square yard. After the asphalt-rubber membrane has been applied, 1 to $1\frac{1}{2}$ inches of asphalt concrete is applied over the asphalt-rubber layer. Field-prepared joint sealers may be produced by dissolving powdered devulcanized rubber into asphalt cement at approximately 20 weight-percent. Joint sealers may also be prepared at a factory and prepackaged for use in the field.

Economics

It is beyond the scope of this discussion to consider in detail the economics of rubberized asphalt in paving applications, but a few generalizations can be made. At current prices, the cost of an asphalt-rubber binder is about 1.7 times that of a straight asphalt-cement binder. Obviously, with a first-cost differential of this magnitude, total life-cycle costs must be considered. Available data indicate that lower maintenance costs of asphalt-rubber systems lead to a 25% less total life cost as compared to a conventional 3-inch concrete overlay. In summary, the use of recycled rubber in paving applications provides a means for utilizing a material that is otherwise discarded and, in addition, provides the following advantages:

1. Reduction in reflective cracking.
2. Waterproofing the paving structure and thus preventing water infiltration into the base of the paving structure.
3. Less tendency for rutting, shoving, and bleeding.
4. Reduction in spalling of asphaltic concrete around pot holes and large cracks.
5. Lower paving-maintenance costs.
6. Lower total life-cycle costs.

ACKNOWLEDGEMENTS

The authors wish to thank Mr. Bobby LaGrone, President of GSX Polymers, Inc., for his contribution of the section on rubberized asphalt. We also wish to acknowledge Ace Products, Inc.; BFGoodrich Co.; Cupples Co.; and Uniroyal, Inc., for the contribution of typical formulations using reclaimed rubber.

REFERENCES

J. M. Ball, *Reclaimed Rubber*, Rubber Reclaimers Association, Inc., New York, 1947.

Reclaimed Rubber: Its Development, Applications and Future, A. Nourry, ed. Maclaren & Sons, London, for the Association of British Reclaimed Rubber Manufacturers, 1962.

19

LATEX AND FOAM RUBBER

Robert F. Mausser
R. T. Vanderbilt Company
Norwalk, Connecticut

EARLY BACKGROUND

At the time the preparation of this chapter was begun, we received a poster from Grupo Hulero Mexicano, A.C. It informed us of the Mexican Rubber Group's sponsorship of the First International Symposium and Exhibition of the Rubber Industry—Mexico City, September 1984. The poster also contained a drawing under which was inscribed the following:

> A solar god, possibly Tonatiuh or maybe Xiuhtecuhtli offering rubber, while on the high altar a rubber ball is burning.

The drawing and inscription served to symbolize the fact that the natives of South and Central America were the first persons to produce and use rubber goods. These goods were derived from natural latex. Excavations indicate that the natives used rubber in religious ceremonies as early as the sixth century A.D.

The recorded history of Western Civilization contains no mention of rubber or the plants that produce it prior to the discovery of the Americas by Christopher Columbus. It was he, who, during the course of his second voyage to the Americas, saw Haitian natives playing games with balls produced from the latex of the rubber (Hevea brasiliensis) tree. Later, early explorers of South America found that natives, living in areas where the rubber tree grew, had the capability of fabricating rubber goods such as water-proof clothing, water bottles, and shoes from latex.

Today it is known that the rubber tree was not indigenous to South America alone. Rubber trees grew wild in many regions of Africa and the Far East at the time Columbus unknowingly set sail for the New World. They grew then and grow today in many places in a band ranging roughly 10° north and south of the equator, around the world. Surprisingly, no other culture developed the capability of latex-goods fabrication; if they did, history did not record the event.

Although the fabrication of rubber goods from latex was the first-known rubber process, the latex industry did not develop to any extent until the second decade of the twentieth century. Till that time, latex-manufacturing technology was hampered by the lack of adequate supplies of properly preserved concentrated latex in the manufacturing and research centers of Europe and America.

This fact, too, is surprising. A Frenchman, A. Fourcroy, discovered in 1791 that natural latex could be preserved by adding to it a small amount of ammonia. Fourcroy's important contribution lay dormant approximately 125 years until it was rediscovered.

GROWTH OF LATEX USAGE

The year 1921 saw considerable progress in the development of organic accelerators in the United States. C. W. Bedford and L. B. Sebrell discovered that mercaptobenzothiazole (MBT), its homologues, and its metal salts were particularly effective accelerators. The development of organic accelerators that permit the fast curing of latex films, and the widespread dissemination of knowledge pertaining to colloidal chemistry, both gave considerable impetus to the manufacturing technology of latex goods. The development of synthetic latices prior to World War II and their continued development to this day have also contributed greatly to the growth of the latex industry. The phenomenal growth of latex goods manufacture has not been at the expense of milled-rubber goods. Growth has occurred primarily in the manufacture of those products that cannot be fabricated by milled rubber technology.

''Latex'' originally referred to the milky-white substance occurring in certain trees and plants; natural-rubber latex occurs in the Hevea brasiliensis tree and guayule plant. Prior to the advent of the synthetic latices, the foregoing definition may have been adequate; it is no longer so today. A more suitable definition is that of Blackley:

> Latex is a stable dispersion of a polymeric substance in an essentially aqueous medium.

Latex is a commodity that by itself has little practical use. As with dry rubber, its properties must be altered. Ingredients must be added to the latex that will, at some point during latex-goods manufacture, modify the characteristics of the latex film, thus permitting the manufacture of saleable goods. The process of incorporating ingredients into either latex or dry rubber to modify its characteristics is called *compounding*. Latex technology is a specialized field. The handling and processing of latex is still considered by many persons to be both an art and science.

WET VS. DRY

The following formula (Table 19.1) is representative of those formulas used by the Latex Laboratory, R & D Division, R. T. Vanderbilt Company, Inc. With slight modification it is suitable for quality assurance testing and for evaluation of either new materials or competitive chemically identical compounding ingredients. The formula is presented at this point to use it as a basis for comparing the similarities and differences of the compounding of natural-rubber latex and dry natural rubber. For the purposes of illustration and comparison, the dry-rubber formula shown in Table 19.2 is also offerred.

Comparison of the formulas in Tables 19.1 and 19.2 will reveal that both compounds are based on 100 parts (phr) dry rubber. Both formulas are similar in that they contain a cure system based on sulfur, zinc oxide, and organic accelerator; however, the actual phr of each individual ingredient present is different in the formulations. Dry natural rubber is derived from NR latex. Thus, both compounds are again similar in that they contain an antioxidant for ageing protection.

The first step in the Banbury mixing of dry rubber involves its mastication. Processing aids such as REOGEN, a plasticizer, are mixed into the elastomer to facilitate its breakdown. Fatty acid is also mixed into the rubber. Its presence in the nonhydrocarbon constituents of natural rubber contribute to the efficiency of the accelerators of vulcanization. Stearic acid is the fatty acid most commonly used to perform this function. Neither plasticizer nor fatty acid is re-

Table 19.1. Base Natural-rubber (NR) Latex Compound.

Ingredient	Weight		Purpose
	Dry	*Wet*	
60% centrifuged NR latex	100	167	Film former
10% aqueous potassium hydroxide solution	0.5	5	Stabilizer
15% ammonium caseinate solution	0.5	3.33	Protective colloid
68% sulfur dispersion	1	1.47	Crosslinking agent
60% zinc oxide dispersion	3	5	Cure activator
50% BUTYL ZIMATE slurry	1	2	Cure accelerator
65% AGERITE SUPERLITE emulsion	1	1.54	Antioxidant for aging protection
Totals	107.0	185.34	

Cure: air-dried film (from compound) for 15 minutes at 93°C in circulating warm air.

Table 19.2. Base Dry NR Compound.

Ingredient	*Dry Weight*	*Purpose*
NR (SMR-5)	100	Film former
REOGEN	2	Plasticizer; enhances breakdown of rubber
Stearic acid	2	Cure activator
Zinc oxide	5	Cure activator
AGERITE STALITE S	1.5	Antioxidant; aging protection
THERMAX (N990)	35	Reinforcement
HAF Black (N330)	30	Reinforcement
Sulfur	2.75	Crosslinking agent
ALTAX	1	Primary cure accelerator
METHYL LEDATE	0.1	Secondary cure accelerator

Press Cure: 20 minutes at 143°C

quired for the compounding of NR latex. Latex does not utilize and does not require mastication as a processing step in compounding.

The first step in the compounding of latex involves the addition of stabilizer and protective colloid to the latex. These ingredients protect the colloidal integrity of the latex. Latex not containing these ingredients "breaks" (separates into water and rubber) when subjected to mechanical agitation used to add ingredients to it during compounding.

The dry-rubber compound shown contains HAF and THERMAX carbon blacks. Carbon black reinforces dry rubber; it does not reinforce cured latex film. In latex compounds, carbon black, at $1\frac{1}{2}$–2 phr, serves primarily as a colorant.

Both compounds contain organic accelerator. BUTYL ZIMATE, as well as METHYL LEDATE, is a dithiocarbamate ultra-accelerator. The latex compound contains ultra-accelerator at 1 phr; it is the sole accelerator for the compound. The dry-rubber compound utilizes an accelerator blend. ALTAX is used as the primary accelerator and METHYL LEDATE at 0.1 phr as the secondary accelerator. From the latex compounder's viewpoint, ALTAX is a slow-curing accelerator. During compounding, dry rubber experiences heat build-up, hence mixing of high levels of ultra-accelerator into the dry rubber would cause the rubber to scorch. Little if any heat is generated during the compounding of latex. Therefore, higher levels of ultra-accelerator can safely be added to latex compound; neither incipient cure nor scorch will occur. The difference in type and level of accelerator present in the compounds ultimately reflects itself partially in the time and temperature conditions of cure. As a general rule-of-thumb, latex compounds can be cured at lower temperatures than can dry rubber compounds.

Another reason latex goods can be cured more quickly is that they are usually much-thinner-walled than are dry-rubber articles. As a consequence, the problem of heat transfer through the wall of the rubber article is not a factor.

Perhaps the greatest difference existing between the formulas is the manner in which they are written. The latex formulation contains two columns pertaining to weight of ingredient. The dry-weight column is similar to that used by the dry-rubber compounder. All ingredients are added to the compound on a phr usage basis for purposes of calculations and convention. However, in most instances, the latex compounder must first prepare his or her ingredients to render them suitable for compounding prior to their addition to the latex compound. Water-insoluble powders must first be either dispersed or slurried and water-immiscible liquids must be emulsified prior to their addition to the compound. Thus, the need for a wet-weight column.

The capital cost of machinery for production of latex articles is considerably less than that for production of goods from dry rubber. The power requirements of machinery for production of latex products is also lower than that for fabrication of dry-rubber goods. The major disadvantage of latex-goods processing is that water must be removed from the compound at some point in the process. This generally limits applications to products having either thin-wall or cellular structure.

SELECTING A LATEX FOR PRODUCT MANUFACTURE

Once the decision has been made to make a specific product, the compounder must choose a latex that is to be used to fabricate the item. The latex selected must be capable, after compounding and processing, of imparting those properties and performance features that are required to the finished product.

Selection of a latex may be predicated on:

1. Physical properties.
2. Chemical resistance properties of the latex film.
3. Cost.

After a specific polymer latex has been selected, the compounder must choose between the various latex varieties within that type. Selection may be based on:

1. Ratio of monomers.
2. Particle size.
3. Particle-size distribution.

The final choice of a specific latex and the plant equipment that is available to produce the goods and the service requirements of the finished goods determine what compounding materials are to be added to the latex.

Table 19.3. Elastomeric Latices.

ASTM D1418 Nomenclature	*Common Name*	*Obsolete Name*
NR	Natural latex	
IR	Polyisoprene	
SBR	Styrene-butadiene	GR-S, Buna S
BR	Polybutadiene	
CR	Polychloroprene	GR-M, Duprene
NBR	Nitrile	GR-A, Buna N
IIR	Butyl	GR-I
EPDM	Ethylene-propylene	
NCR	Nitrile-chloroprene	
NIR	Nitrile-isoprene	
PSBR	Pyridine-styrene-butadiene	
SCR	Styrene-chloroprene	
SIR	Styrene-isoprene	
XNBR	Carboxylic nitrile	
XSBR	Carboxylic SBR	
XCR	Carboxylic neoprene	

Table 19.3 comprises of a listing of latices. Except for isoprene latex, which is no longer produced, all other latices are currently available commercially.

It can be seen from Table 19.3 that many different chemical types of latices are available to the compounder. Even a brief discussion of individual latices within each chemical type is beyond the scope of this chapter, but each *type* of latex will be discussed.

NATURAL-RUBBER LATICES

Natural Latex—NR Ref. ASTM D 1076

Natural-rubber latex is sometimes referred to as Hevea latex. It is a stereoregular polymer and consists essentially of *cis*-1,4 polyisoprene.

Natural latex is obtained directly from the *Hevea brasiliensis* rubber tree, of the family *Euphorbiaceae*, by tapping and collecting the fluid that flows from the spiral cut made almost through the bark, but just short of the cambium or growing layer. This fluid, natural latex, is in no way related to or derived from the tree's sap. The individual latex particles may vary in size between .1 and 2 microns, and are composed of rubber hydrocarbon with a surface layer of complex composition. This coating of nonhydrocarbon constituents amounts to about 2% of the total particle weight, consisting of proteins, fatty acids, sugars and traces of other organic materials. Natural-rubber latex is anionic,

Table 19.4. Characteristics of NR Latices.

Type	*% Total Solids*	*Specific Latex*	*Gravity Solids*	*pH Initial*	*Weight Lb/Gal*
Normal	38–41	.97	.92	10.5	8.1
Centrifuged	61.5 min	.96	.92	10.2	8.0
Creamed	68 min	.95	.92	10.5	7.9
Heat-concentrated	72–75	.97	.92	>10.5	8.1
Low-ammonia	61.5 min	.96	.92	9.5	8.0

i.e., the particles carry a negative charge. The isoelectric point is in the 4–5 pH range.

Natural latex as collected is a highly perishable material that will putrefy and coagulate in a few hours unless a preservative is added. Ammonia is added to the calabash, during tapping, to prevent bacterial decomposition of the latex. The rubber content of natural latex as it is obtained from the tree is 30 to 40%. Selected lots of higher concentration, 35 to 40%, are marketed as "normal" latex. "Normal" latex, which contains all the serum substances, finds limited uses because of its dark color, strong odor, and other objectionable features.

Present commercial practice calls for concentrating by one of several established methods to a total solid content of 60% or higher. Centrifuging (Utermark BP 219, 635; 1953) and creaming are the most widely used methods of concentration for producing latices of 62–70% total solids. Heat-concentrated latices (by evaporation) having 70–75% total solids content are also produced (Revertex BP 243, 016; 1926).

NR latex preserved only by ammonia is known as high-ammonia (HA) latex. The trend in the U.S., since 1955, has been to use NR latex containing a reduced level of ammonia and a secondary latex preservative; it is called low-ammonia (LA) latex. Currently most latex plantations are using a blend of tetramethyl thiuram disulfide and zinc oxide as secondary preservatives of latex. Many plantations use the ingredients in a 1/1 blend ratio. This secondary preservation system has given added impetus to the change from (HA) to (LA) NR Latex.

Modified NR Latex. Graft copolymers of methyl methacrylate and natural rubber latex are available under the generic name of Heveaplus MG. This latex may be blended with other latices to increase modulus.

Synthetic latices are produced by either one of the following processes:

1. Emulsion polymerization.
2. Solution polymerization.

To a degree, emulsion polymerization is similar to nature's production of natural rubber. The first product obtained is latex, from which water must be removed to produce dry synthetic rubber. Both natural and synthetic latex are coagulated with various acids to produce dry rubber. On the other hand, such latices as IIR and EPDM are produced from dry polymer. The polymer is first dissolved in solvent. The solution is then emulsified to produce the latex. The term "artificial latex" is sometimes used to identify latices produced by this method.

SBR and XSBR Latices (Ref. ASTM D 1417)

SBR Latex contains a copolymer of styrene and butadiene. When the styrene content is 50% or less, the copolymer is classed as a rubber latex. SBR resin latices contain more than 50% styrene. The two extremes are polybutadiene and polystyrene latices. SBR latices are prepared by the emulsion polymerization of the desired ratios of butadiene and styrene monomers together with the required modifiers and catalysts. Many SBR latices are shortstopped with a dithiocarbamate and will stain in the presence of copper. However, SBR latices are available that are not so shortstopped.

Carboxylic-modified styrene-butadiene latices (XSBR) find use where adhesion to textile fibers is important.

"Hot" latices are those polymerized at 49–66°C, while "cold" latices are polymerized at 5–10°C. The cold SBR latices are superior in elastomeric properties. The hot SBR latices are usually superior in colloidal properties. Commercial SBR latices are anionic, i.e., carry a negative charge. The stabilization systems vary considerably. In most cases, the polymer producer lists the type of emulsifier used.

SBR latices are used in conjunction with fibrous products as coatings, binders and adhesives. They are of particular interest to rug, textile, and paper manufacturers. They are also used as blends with natural latex in large-volume applications such as foam rubber. SBR latices are produced with a wide range of particle sizes, solids concentration and film properties. Table 19.5 shows the four broad categories.

Resin Latices

The styrene-butadiene copolymers in which the styrene is the predominant part, generally 60% or more of the monomer ratio, are termed resin latices. When used in combination with other latices, they increase modulus and hardness of the vulcanizates and decrease elongation. Resin latices are used as reinforcing agents in foam and other latex compounds and as individual materials for paper and textile coatings, adhesives, pigment binders, and so forth.

Table 19.5. Types of SBR Latex.

Type	*Primary Application*	*Approximate Total Solids*	*Approximate Styrene/Butadiene Ratio*
Large particle size High solids	Foam	70	30/70
General purpose High solids	Foam blends Fabric backing	60	30/70 to 50/70
General purpose Medium solids	Adhesives* Textile application*	40–50	30/70 to 50/50
Small particle size Low solids	Impregnation	25	23/77

*XSBR as well as SBR

Pyridine-styrene-butadiene—PSBR

Terpolymers of pyridine/styrene/butadiene are used to provide adhesion to synthetic tire cords.

Nitrile Latices—NBR and XNBR

Nitrile latex contains a copolymer of acrylonitrile and butadiene produced by emulsion polymerization (a copolymer of acrylonitrile and isoprene is also available). The oil resistance varies directly with the acrylonitrile content. The higher the acrylonitrile content, the stiffer or more ''leathery'' the vulcanizate. The approximate monomer ratios are shown in Table 19.6.

Carboxylic-modified nitrile latices are available for use where adhesion to fibers is desired. Nitrile latices have been used for many years to upgrade products made from paper, textile and other fibers. Recently, several producers have developed latex systems for making unsupported coagulated dip films with excellent physical properties. Nitrile latices are available with an extremely wide variety of emulsification systems.

Table 19.6. Acrylonitrile Ratios.

Acrylonitrile Content	*Acrylonitrile/ Butadiene Monomer Ratio*
Low	22/78
Medium	33/67
High	45/55

Polychloroprene Latex—CR

Polychloroprene latices are aqueous colloidal dispersions of polychloroprene or its copolymers. If the latex contains methacrylic acid as a comonomer, it is designated XCR. Du Pont, the major American producer of polychloroprene latex offers about 12 types of the latex to the trade under its Neoprene brand name. Ten of the latices are anionic. One is cationic and the last is carboxylic (XCR). Recently Denka has begun marketing CR Latex in the U.S.A. The total solids content of commercially available chloroprene latices is in the range of 31–60%.

LATEX COMPOUNDING

The ingredients used in latex compounding may be divided into three general classifications. The categories and their major constituents are shown in Table 19.7.

Water-insoluble solid and water-immiscible liquid compounding ingredients used as elastomer-phase modifiers must be converted to water-compatible systems that can be mixed uniformly into the latex without upsetting the latex colloidal stability. The task is accomplished by use of surfactants for preparations of dispersions, slurries, and emulsions. Most elastomeric latices are anionic, thus most surfactants used are either anionic or nonionic. Materials that function as dispersing agents have very slight, if any, tendency to reduce the surface tension of water. Those acting as either wetting agents or emulsifiers, or capable of functioning as both, appreciably reduce the surface tension of water (i.e., serve as surfactants).

Liquid-phase modifiers are used to protect the colloidal properties of latex

Table 19.7. Classification of Compounding Ingredients.

1. Surfactants
   a. Dispersing agents
   b. Wetting agents
   c. Emulsifying agents

2. Liquid Phase Modifiers
   a. Stabilizers
   b. Thickeners
   c. Wetting agents
   d. Coagulants
   e. Heat sensitizers
   f. Gelling agents
      (1) Primary
      (2) Secondary

3. Elastomer Phase Modifiers
   a. Sulfur
   b. Zinc oxide
   c. Organic accelerators
   d. Clay and other loading materials
   e. Softeners and plasticizers
   f. Dyes and pigments
   g. Reodorants

particles. They prevent premature coagulation and formation of prefloc during compounding and processing. Certain members of this group modify the liquid properties of compounded latex. They help control compound flow and wettability for processing. Coagulants, gelling agents, and heat sensitizers permit the transition of the compound from liquid phase to solid.

Elastomer-phase modifiers can be defined as materials that impart desired properties to finished rubber goods after transition of the latex compound from the liquid to the solid phase. Often sulfur, zinc oxide, and organic accelerator comprise the cure system for vulcanization of latex (solid) film. Antidegradents, i.e., antioxidants and antiozonants, provide age-resisting properties to latex articles. Latex compounds used for foam rugbacking, coatings, mastics and adhesives contain filler loading. Clays, talc, whiting and hydrated alumina are examples of loading materials used to modify physical properties and reduce the cost of the product.

DISPERSIONS

Mention has been made that water-insoluble powders must be rendered suitable for latex compounding. Frequently this is accomplished by preparing a dispersion of the material. During preparation of the dispersion, the particle size of the material being dispersed is reduced. Most dispersions used for latex compounding contain particles ranging in size from 0.2 to 5 microns. The finer the particle size of the dispersion, the more uniform the distribution of the material in the finished product.

The following types of equipment are used to prepare dispersions; each supplies the energy required to reduce the particle size of the material being dispersed:

1. Horizontal pebble mill.
2. Vibratory pebble mill.
3. Attrition mill.
4. Colloid mill.
5. Ultrasonic dispersion unit.

Each piece of equipment has its individual strengths and weaknesses. For example, colloid mills can be used to prepare both dispersions and emulsions. However, dispersions of materials which become pressure-sensitive under shear cannot be produced using a colloid mill.

The formula shown in Table 19.8 is representative of those used to produce dispersions in a horizontal ball mill.

Table 19.8. 55% Active METHYL ZIMATE Dispersion.

Ingredient	Weight Dry	Weight Wet	Function
METHYL ZIMATE	55	55	Active ingredient
DARVAN No. 1	2.2	2.2	Dispersing agent
10% Igepal CO-630 solution	0.1	1	Wetting agent
VAN GEL B	0.2	0.2	Thickening agent, viscosity control
15% ammonium caseinate solution	1.65	11	Colloidal stabilizer
Water	-	30.6	Grind vehicle
Totals	59.15	100.0	

Add all ingredients to ball mill. Grind 24 hours. Discharge mill.

SLURRIES

Compounders prepare clays, talcs, or other mineral fillers for compounding by slurrying the dry powder into water containing a dispersing agent. The particle size of the material being processed is not reduced during preparation of the slurry. Table 19.9 displays a typical formula for preparation of a slurry.

EMULSIONS

To prevent destabilization and improve compatability, water-immiscible liquids and low-melting-point, water-insoluble solids are emulsified prior to addition to latex.

An *emulsion* is a two-phase system consisting of fine droplets of one liquid suspended in another liquid with which the first liquid is incompletely miscible. For latex compounding, oil-in-water emulsions are used. The oil is the disperse phase; the water is the continuous phase. By definition, the oil is the material that is a liquid at the temperature of emulsification. The oil might be an antiox-

Table 19.9. 60% NYTAL Talc Slurry.

Ingredient	Weight Dry	Weight Wet	Function
NYTAL 200	60	60	Active ingredient
DARVAN No. 7	0.3	1.2	Dispersing agent
Water	-	38.8	Slurry vehicle
Totals	60.3	100.0	

Stir dispersing agent into water. Add talc to the solution, using high speed mechanical agitation.

idant, such as AGERITE STALITE, heated to reduce viscosity for easy emulsification. Or again, the oil might be either a tackifying resin or a wax dissolved in a solvent to reduce viscosity.

Emulsions used in latex applications consist of:

1. The insoluble material.
2. Water (the quantity used controls viscosity).
3. The emulsifying agent, which can be either a soap or (synthetic) surfactant. Soaps used as emulsifiers may be formed in situ by reaction of a base with a fatty acid.
4. Stabilizing agent as required. Stabilizing agents may be of several classifications.
   a. Insoluble solids:
      gelantious aluminum hydroxide
      zinc oxide
   b. Protective colloids:
      casein
      gelatin
      methyl cellulose
      sodium carboxy methyl cellulose
      carboxy vinyl polymers (high molecular weight)
      hydroxy propyl methyl cellulose
   c. Other:
      water swelling clays
      DARVAN No. 7

Emulsification methods follow.

1. In situ method—most generally used.
   a. Mix acid and oil.
   b. Dissolve base in water.
   c. Add (a) to (b) with vigorous stirring. With some viscous liquids, the combining order may be reversed to advantage.
2. Water method—best for waxes.
   a. Mix soap in water.
   b. Melt wax and add slowly with rapid agitation.
   c. Continue stirring until below melting point of wax.

Emulsion formulas should specify the order of addition of the ingredients and the temperature.

The following observations may prove of value:

1. Mechanical stability improvement: the addition of 2% to 10% of a petro-

leum lubricating oil or a vegetable oil (for example, castor oil) to the oil phase will generally improve the mechanical stability of an emulsion; the two oils should be miscible.
2. Storage stability: a slightly viscous emulsion is more stable during shelf storage than a water-thin emulsion; homogenization improves storage stability.
3. Stabilizer—inorganic: the use of fine particle size, insoluble material will stabilize some emulsions; zinc oxide can be used to increase emulsion stability.

The quality of an emulsion depends upon the following, in order of decreasing importance:

1. Technique.
2. Equipment.
3. Formula.

LATEX STABILIZATION

The first step in compounding involves the addition of a stabilizer to the latex. Latex is a colloidal system; it is critical to both the compound- and goods-manufacturing processes that the correct amount of and proper chemical type surfactant be added to the latex.

NR Latex, as tapped from the tree, is stabilized with proteins and fatty acids. On plantations, ammonia is added to tapped latex. The ammonia acts as a preservative; that is, it protects the protein present in the latex against bacterial attack. The ammonia also reacts with the fatty acid present in the latex to produce an anionic soap. Therefore, colloidal or anionic stabilizers are generally used to modify the stability of the latex, thus permitting the addition of other prepared additives to the compound.

The first commercially available alkaline synthetic latices produced contained either rosin acid or fatty-acid soaps, or blends of these soaps that functioned as stabilizers for the latices. Later, latices were polymerized utilizing either nonionic surfactants or blends of nonionic and anionic surfactants for stabilization. Cationic latices require the use of cationic and amphoteric surfactants for their stabilization during polymerization.

The compounder's selection of a stabilizer for addition to latices is sometimes made difficult by the latex producers. They are reluctant to divulge to the compounder the specific surfactant used to produce the latex. However, they will disclose the ionic type surfactant utilized for polymerization. Thus, the compounder must engage in some trial-and-error testing to determine the stabilizer addition best suited for the compound.

Latex compounding, of necessity, involves compromise, especially true when pertaining to stabilization of latices. The compounder wants the compound being processed to be stable during compounding, storage, and up to a certain point in manufacturing. Then, at some precise point in the process, the compounded latex is to be destabilized or coagulated; examples of this sequence are the manufacture of latex thread, household gloves and latex foams.

Compounders use soaps and surfactants (synthetic soaps) for stabilization of their compounds. Compound pH influences the choice of stabilizer. Potassium oleate soaps lose their stabilizing ability and revert to fatty acids if the pH of the compound is lowered sufficiently. To overcome this problem, many compounders prefer to use anionic, nonionic, or a blend of both types of surfactants in alkaline latex compounds.

Soaps and surfactants lower the interfacial tension of latices to which they are added. There exists an ideal interfacial tension that a specific compound must possess for a specific application or process.

If the compounder wishes to modify a nonloaded adhesive by addition of a filler he or she would have to increase the amount of stabilizer present in the compound. The extra amount of stabilizer required would be predicated not only on the amount of filler added to the compound but also on the particle size of the filler. Individual latex particles are surrounded by surfactant. Filler added to the latex adsorbs some of the surfactant surrounding the latex particle. The smaller the particle size of the filler, the greater is the amount of surfactant taken. If a sufficient amount of stabilizer is taken from the latex particles, the latex will destabilize and the compound will become unusable.

A similar situation often exists when the compounder blends latices of different particle size, since the smaller-particle-size latex may remove the stabilizer from the large-particle-size latex and cause thickening or coagulation of the compound. Again, it is required that additional stabilizer be added to the compound.

VULCANIZATION

ISO 1382 Rubber Vocabulary (First Edition, 1972), defines *vulcanization* as a process in which rubber, through a change in its chemical structure (e.g., cross-linking) is converted to a condition in which the elastic properties are conferred, reestablished, improved, or extended over a greater range of temperatures. In some cases, the process is carried to the point the substance becomes rigid.

In 1839, the American Charles Goodyear discovered that rubber, after having been baked with sulfur and white lead, exhibited little change after its exposure to heat and cold. Goodyear thereby discovered, and named, the process of vulcanization. In 1841, Steven Moulton brought samples of vulcanized rubber to England. Eventually the samples found their way to Thomas Hancock. Based on his experiments, Hancock became convinced that sulfur had a specific action on rubber, which he termed "change." He found that "change" could be brought about by immersion of rubber in molten sulfur. Hancock filed for a provisional patent in Britain for the process in 1843; Goodyear applied for a U.S. patent for his process several weeks after Hancock made application for a patent. In Britain, Alexander Parks found, in 1846, that vulcanization could be achieved by treatment of rubber with sulfur chloride.

The first accelerators used to shorten uneconomically long vulcanization times were oxides of calcium, magnesium, and lead. By current standards these accelerators are very slow. Hoffman and Gottlob of Bayer were the first to use dithiocarbamates for acceleration of rubber vulcanization. However, they failed to observe the influence of zinc oxide on vulcanization of rubber compounds. Today, hardly any elastomeric latex compounds do not contain zinc oxide. The pH of a latex compound has direct influences on the degree of solubility of the zinc oxide present in it. The amount and type of zinc ion present in the compound also have a direct bearing on the properties of cured film derived from the compound.

Colloidal free sulfur, in the absence of both zinc oxide and organic accelerator, is still capable of crosslinking latex film during vulcanization, but only very slowly. Cured film containing only sulfur, in the range of 1–10 phr, possesses extremely low unaged properties; products prepared in this manner would be unsuitable for use. The addition of zinc oxide to NR latex compound containing colloidal free sulfur helps to improve the tensile properties of vulcanized film. However, the properties are still inadequate for commercial goods. Organic accelerator must also be present in the compound.

Increasing the level of zinc oxide, over the range 0.5–6.0 phr, present in NR Latex compound containing 1 phr sulfur and 1 phr accelerator improves the state of cure of vulcanized film. It does not appreciably increase the rate of vulcanization. However, increased levels of zinc oxide render films more resistant to overcure.

Some compounders may not realize that the pH of a latex compound has direct influence on the degree of solubility of zinc oxide present in the compound. The type and amount of zinc ion present in a compound have a direct bearing on the cured film properties of the compound. If the zinc oxide has been solubilized, less is available for activation of cure. The pH of the latex compound should be adjusted to levels indicated in Fig. 19.1.

Mention should be made that not all latex compounds require sulfur for crosslinking during film vulcanization. Certain polychloroprene latices fall into this category; zinc oxide acts as the crosslinking agent. In polychloroprene compounds, zinc oxide also functions as an acid acceptor. Cured polychloroprene film liberates chloride ion during its normal ageing process, and this can combine with hydrogen ion to form hydrochloric acid, the presence of which ac-

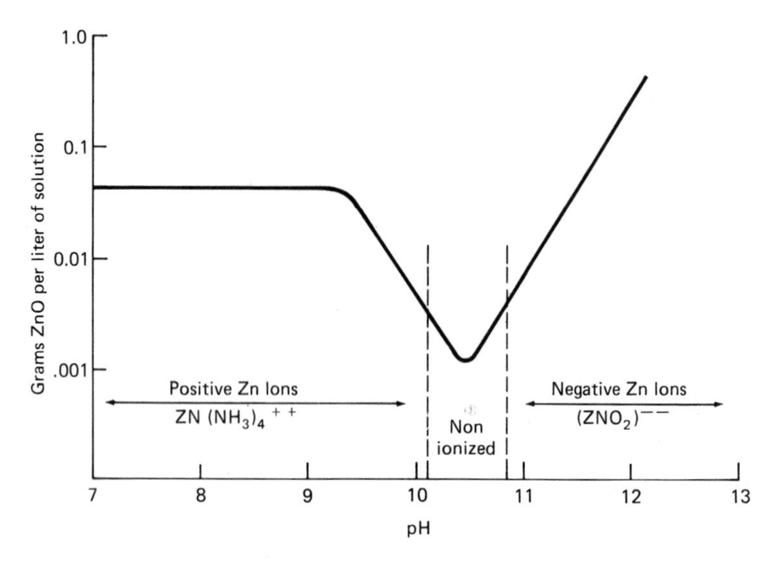

Fig. 19.1. Effect of pH on ZnO solubility.

celerates degradation of the film. The presence of zinc oxide in the film retards degradation of this type.

Many latex accelerators are active at temperatures below 100°C. If used for dry-rubber compounding, they cause the rubber to scorch on the mill; this precludes their use for dry-rubber compounding. Little, if any, heat is generated during latex compounding. Consequently, no incipient cure or scorch occurs during compounding. The chemical types and structures of materials shown in Tables 19.10 and 19.11 are commonly used as accelerators for latex compounding.

The dithiocarbamates are the fastest curing and most widely used group of accelerators for latex compounding. The xanthates are also very reactive. Zinc

Table 19.10. Types of Latex Accelerators.

Type	*Example*
Thiuram	Tetramethylthiuramdisulfide
Dithiocarbamate	Zinc dibutyldithiocarbamate
Xanthate	Zinc isopropyl xanthate
Thiazole:	
Mercapto	MBT and ZMBT
Sulfenamide	N-cyclohexyl-2-benzothiazolesulfenamide
Thiourea	1,3-dibutyl thiourea
Guanidine	Diphenylguanidine
Aniline condensation product	Butyraldehyde-aniline condensation product

Table 19.11. Empirical or Structural Formulas of Latex Accelerators.

Type Accelerators	*Empirical or Structural Formulas*	*Radical*	*Chemical Name*
1. Thiuram: Monosulfide	$R_2N{-}C({=}S){-}S{-}C({=}S){-}NR_2$	CH_3	Tetramethylthiuram monosulfide
Disulfide	$R_2N{-}C({=}S){-}S{-}S{-}C({=}S){-}NR_2$	CH_3	Tetramethylthiuram disulfide
		C_2H_5	Tetraethylthiuram disulfide
2. Dithiocarbamate: Copper Salt	$R_2N{-}C({=}S){-}S{-}Cu{-}S{-}C({=}S){-}NR_2$	CH_3	Copper dimethyldithiocarbamate
Zinc Salt	$R_2N{-}C({=}S){-}S{-}Zn{-}S{-}C({=}S){-}NR_2$	CH_3	Zinc dimethyldithiocarbamate
		C_2H_5	Zinc diethyldithiocarbamate
		C_4H_9	Zinc ci-n-butyl dithiocarbamate
3. Xanthate	$R{-}O{-}C({=}S){-}S{-}Zn{-}S{-}C({=}S){-}O{-}R$	C_3H_7	Zinc isopropyl xanthate
4. Thiazole: Mercapto	$C_6H_4(N{=})(S{-})C{-}SH$ (benzothiazole ring)		MBT 2-mercaptobenzothiazole
	$C_6H_4NSC{-}S{-}Zn{-}S{-}CSNC_6H_4$ (two benzothiazole rings)		ZMBT Zinc-2-mercaptobenzothiazole
Sulfenamide	$C_6H_4NSC{-}S{-}N(H){-}C_6H_{11}$ (benzothiazole ring, cyclohexyl ring)		N-cyclohexyl-2-benzothiazole sulfenamide
5. Thiourea Accelerator	$C_4H_9{-}N(H){-}C({=}S){-}N(H){-}C_4H_9$		1,3-dibutyl thiourea

isopropyl xanthate, the most stable of commercial xanthates, in combination with either zinc 2-mercaptobenzothiazole or 4-morpholinyl-2-benzothiazyl disulfide can be used to accelerate compounds whose films must display excellent resistance to copper staining. To obtain maximum resistance to copper staining, dithiocarbamates should not be used.

The thiazoles are too slow-curing to be used on their own for latex compounding. Generally, they are used as secondary accelerators in combination with the dithiocarbamates; the resultant vulcanizates are distinguished by their high modulus. The use of a thiazole in combination with a dithiocarbamate yields latex foam having higher compression modulus and load-bearing capacity.

Thiurams, also, are too slow-curing to be used alone for acceleration of latex compounds. Their cure rates can be increased by their use with either the dithiocarbamates or thioureas. The thiurams are generally used in low- or non-colloidal, free/sulfur-bearing latex compounds.

Also used, but in considerably smaller volume, are the guanidines, thioureas, and some aniline condensation products.

Some factors which influence the time and temperature of vulcanization are:

1. Cure system employed, especially
   a. Chemical type of accelerator or accelerator blend.
   b. Total accelerator level.
2. Type latex or latex blend employed.
3. Amount of sulfur.
4. Amount of zinc oxide.
5. Film gauge of product produced.

A study performed by the Rubber Research Institute of Malaysia, using a rheometer, shows the influence of accelerator structure on the vulcanization rate at 120°C. This is illustrated in Fig. 19.2.

The latex-foam rug-backing industry has been aware for many years of the synergistic effect obtained by use of a 1/1 ZDEC/ZMBT, blend ratio as shown in Fig. 19.3.

Similar results can be obtained with a 1/1 ZIX/ZMBT blend ratio. This accelerator blend performs better in low-ammonia NR-latex compounds than it does in high-ammonia NR-latex compounds.

It has been noted that dithiocarbamates are the most appropriate for latex compounding. At least 22 dithiocarbamate accelerators are available commercially. All have the basic chemical structure shown on p. 538.

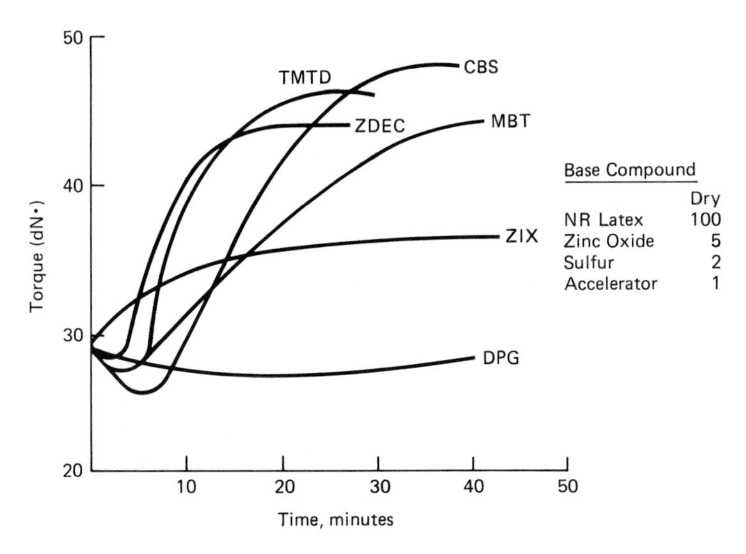

Fig. 19.2. Influence of accelerator structure on vulcanization rate at 120°C.

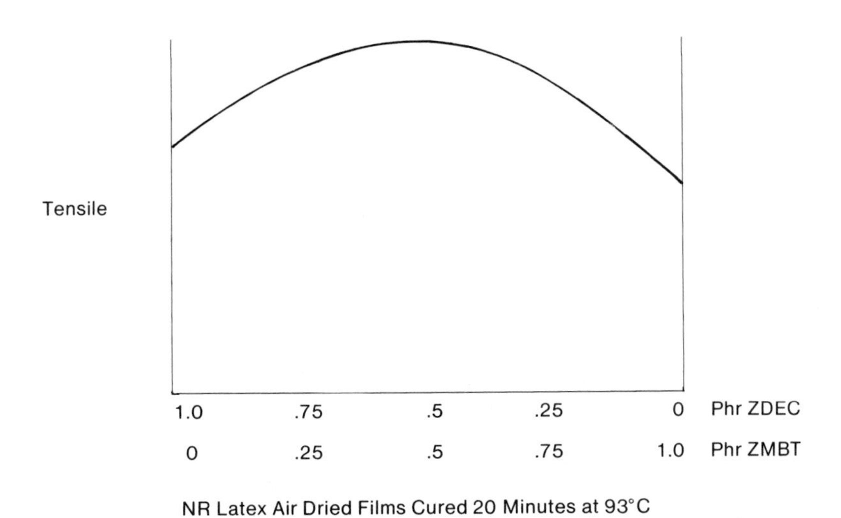

Fig. 19.3. Synergistic effect resulting from use of SDEC and ZMBT blend.

$$\left[\begin{matrix} R \\ & \diagdown \\ & & N-\underset{\underset{\displaystyle S}{\|}}{C}-S \\ & \diagup \\ R^1 \end{matrix}\right]_n M$$

Dithiocarbamate accelerators can be divided the four types as shown in Table 19.12.

Typical dialkyl dithiocarbamate accelerators used for latex compounding are shown in Table 19.13.

AGING

Aging is a general term used to describe the change in film properties of latex articles that occur over a long period of time. The more resistant a latex article to aging, the longer the useful service life of the article.

Exposure of latex articles to

1. Heat,
2. Humidity,
3. Light: ultraviolet and high energy radiation,
4. Oxygen,
5. Ozone,
6. Chemicals: strong acids and bases, solvents, oils, oxidizing agents and heavy metals, and
7. Stress

decreases their resistance to aging. Most often the article is subjected to several types of exposure simultaneously, which causes a more rapid deterioration of its film properties.

Latex-film Aging Symptoms

Depending upon the base elastomer employed to produce the film and the type of "aging" exposure, one or more of the symptoms on p. 540 will be noted.

Table 19.12. Four types of Dithiocarbamate Accelerators.

Type	*Example*
1. Zinc salt	Zinc diethyldithiocarbamate
2. Other metal salt	Copper dimethyldithiocarbamate
3. Water-soluble salt	Sodium dibutyldithiocarbamate
4. Activated type	SETSIT 5, 9, 51 and 104

Table 19.13. Typical Dialkyl Dithiocarbamate Accelerators.

Name	*Mol. Wt.*	*M Metal*	*n (Valence)*	*R*	*R*[1]	*Water-soluble or Insoluble*
Zinc dimethyl-dithiocarbamate	306	Zn	2	CH_3	CH_3	I
Zinc diethyl-dithiocarbamate	362	Zn	2	C_2H_5	C_2H_5	I
Zinc dibutyl-dithiocarbamate	474	Zn	2	C_4H_9	C_4H_9	I
Zinc dibenzyl-dithiocarbamate	612	Zn	2	$C_6H_5-CH_2$	$H_2C-C_6H_5$	I
Copper dimethyl-dithiocarbamate[a]	304	Cu	2	CH_3	CH_3	I
Potassium dibutyl-dithiocarbamate	243	K	1	C_4H_9	C_4H_9	S
Sodium dibutyl-dithiocarbamate	227	Na	1	C_4H_9	C_4H_9	S
Activated liquid dithiocarbamates			composition not disclosed			S

[a]For use only in certain SBR and NBR latex compounds. The presence of 20 pphm or more copper in cured NR latex films will cause degradation of the films.

1. Film hardening or embrittlement.
2. Film softening.
3. Film tackiness.
4. Cracking of film surface: orientated.
5. Cracking of film surface: non-orientated.
6. Loss of tensile and tear strength.
7. Loss of elasticity.
8. Tendency to "take permanent set."

OXIDATION

The forms of oxidation are:

1. Chain scission, resulting in a reduction in chain length and average molecular weight.
2. Crosslinking, resulting in a three-dimensional structure and/or higher molecular weight.
3. Chemical alteration of the molecule by introduction of new chemical groups.

Polymer degradation by oxygen and by the oxides of nitrogen causes natural rubber to soften; SBR, neoprene, and nitrile to stiffen. Softening is predominantly cutting of the elastomer chain, or chain scission. Stiffening is crosslinking of the elastomer chain and is similar to vulcanization. Oxidation may also bring about alteration in the side chains. Chain scission causes the tensile to drop very rapidly and the surface becomes soft and sticky. Stickiness is especially objectionable in fabric coatings. Crosslinking causes stiffening, embrittlement, and loss of elastomeric properties. Crosslinking will alter the "hand" of latex-coated fabrics. Alterations in the side chains may alter solubility. A change in solubility affects the dry-cleaning resistance of "solvent-resistant" coatings.

In general, the three reactions occur simultaneously. The nature of the polymer and the temperature, together with oxygen concentration, will determine whether chain scission, crosslinkage, or side-chain alteration predominates.

Oxygen degradation is autocatalytic. Exposure to light tends to speed up the reaction. The fatty-acid salts of copper, cobalt, manganese, and iron also catalyze the reaction. Fatty-acid soaps are used as emulsifiers for synthetic latex production, and are frequently used as compound stabilizers. Because of this, the use of antioxidants that inhibit metal poisoning is required. Antioxidants protect the polymer by interfering with the free-radical chain reaction of oxidation. The antioxidant is consumed in the process of protection. It is therefore desirable to keep the antioxidant level high to afford the needed protection. Natural latex, as tapped from the tree, contains antioxidants. The concentration

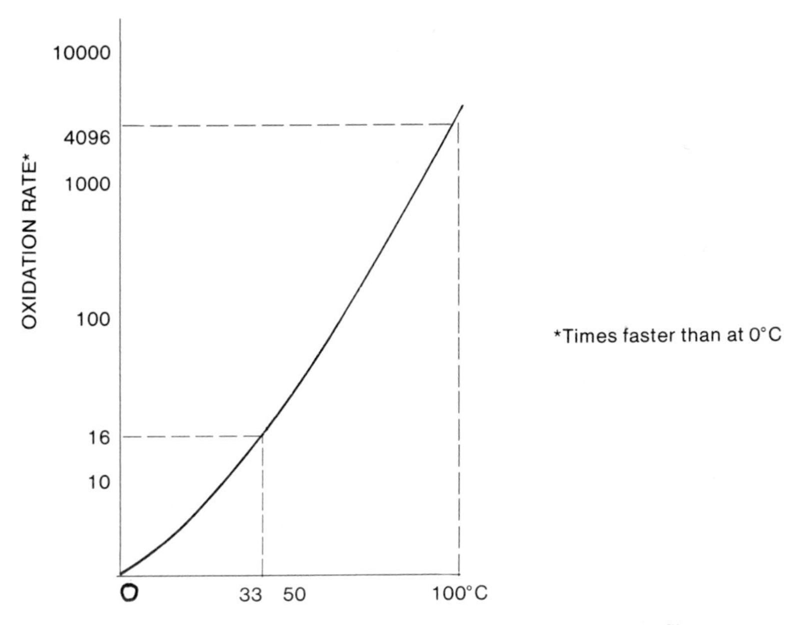

Fig. 19.4. Temperature effect on oxidation rate of NR latex films.

procedure tends to reduce the antioxidant materials originally present. Synthetic latices usually contain antioxidant. Experience indicates that additional antioxidant protection is required and should be added during compounding.

Effect of Heat on Oxidation Rate of Latex Films

The oxidation rate of latex films doubles for each 8.33°C rise in temperature, an exponential increase as shown in Figure 19.4.

Effect of Ultra-violet Light on Latex Films

UV light attacks the surface of the latex film, which often becomes brittle and discolored. Stretching or bending films exposed to UV light causes the film to craze (random order of crack formation). This type of crack formation differs from that caused by ozone. Ozone cracks form perpendicularly to the direction of stress during exposure.

Effect of Strain on the Oxidation of Latex Films

Strain applied to a film stretches the polymer chains causing some of them to break. In addition to creating a flaw in the network, the ends of the broken chains are free radicals that contribute to the oxidative mechanism. If the strain is dynamic, as in the case of flexing, the temperature is also increased, which results in a further acceleration of the oxidative process. Fatigue failure from flexing usually takes the form of what is called "flex cracking." Flex cracking is generally limited to the area that receives the greatest strain and in this way differs from ozone cracking, which will cover a larger region.

Effect of Heavy Metals on the Oxidation of Latex Films

Certain chemicals increase the rate of oxidation of elastomers. Probably the most common materials of this type are compounds containing metals such as copper, iron, and manganese. Common sources of contamination are processing equipment and compounding materials. P-phenylendediamines, commonly used as antioxidants, are also effective as metal deactivators. Reaction to form an insoluble product will also inactivate the metal catalyst. Thus, iron is converted to insoluble sulfides during the cure. The deleterious effects of metals are largely alleviated by the vulcanization process. Conversely, heat accelerates elastomer degradation caused by film contact with copper or copper salts.

Preventing Ozone Attack

The most economical way of stopping ozone attack is to add a microcrystalline wax to the compound that will bloom to the surface of the latex film and create an ozone barrier. This works until the film is broken by flexing or stressing. Immediately, ozone goes on the attack. The attack is concentrated at those points where the wax surface is flawed. Microcrystallize waxes are useful only for static applications. However, for dynamic applications, an antiozonant is added to the compound. Like waxes, the antiozonants must be present at the interface between the ozone atmosphere and the rubber compound. The compounder usually adds the antiozonants at high concentrations, such at 2.0 to 5.0 phr, to form a reservoir of antiozonant to replace the material at the surface as it is consumed, destroying the attacking ozone.

Blending of latices is also being used to combat the destruction of elastomers caused by ozone. The ozone attack is limited by the addition of a polymer that is not attacked by ozone. Examples are CR 400 Latex or EPDM Latex blended with NR Latex. CR 400 Latex or EPDM Latex is added at rates to comprise 15–30 phr of the total dry rubber content of the compounded latex blend.

LATEX PROCESSES

It should be remembered that the latex must process satisfactorily in the liquid state or a high percentage of defects and/or rejects will result. The importance of the proper type of colloidal stabilizer cannot be overemphasized. All equipment used for compounding and processing of latex articles should be free of copper, brass, or galvanized iron.

Dipped Goods

The basic requirements for dipping compounds are as follows:

1. *Good mechanical stability:* this will permit the use of stir-strain dip tanks for the elimination of surface-skimming without build-up of coagulum on the leading edge of the propeller blades.
2. *Predictable precure rate:* the compound level in the dip tank must be maintained by frequent additions of fresh compound; since there is always a residue, the precure rate of a dip compound is of considerable importance. Too little precure and drying will become a problem; too much precure and cracks will occur at stress points.
3. *High wet gel strength:* the wet gel strength must be sufficiently high so that cracking or rupture does not occur during the drying process.
4. *Film forming; deposition rate:* all latex dip compounds must produce excellent films over the desired portion of the dipping form; in coagulant dipping, the deposition rate should be easily controllable so that the film thickness will meet specifications.
5. *Entrapped air:* the dipping compound must be free of entrapped air.

Dipping Techniques

Straight Dipping (Dip and Dry). This is the oldest and still a desirable method of producing a large variety of thin-wall articles. In most cases, two or more dips are made with an air-dry between dips. Fingercots, prophylactics and disposable surgeons' gloves are generally made by this process.

Coagulant Dipping. Most dipped articles having a dry film thickness greater than 0.008 inches are made using a coagulant. The reason for this is economics; faster film build-up is obtained when a coagulant is used.

The two methods employed for coagulant dipping of latex goods are:

1. *Anode process:* the base former is first dipped into coagulant and then into the latex compound.
2. *Teague process:* the base former is first dipped into the latex compound and then it is dipped into the coagulant.

The anode process is generally used to produce latex goods having a dry film gauge in the range of 12–25 mils. Articles having a dry film gauge greater than 25 mils are generally produced by the Teague process.

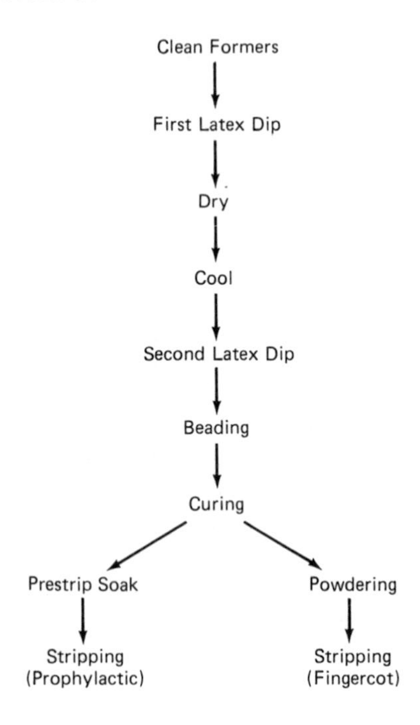

Fig. 19.5. Prophylactic and fingercot dipping processes (straight dipping).

The coagulant generally used in both processes is Norwegian-produced calcium nitrate. The solvent for the coagulant may be alcohol, water, or a blend of both.

A word concerning (dipping) formers is appropriate at this point. Dipping formers are usually made from:

1. Porcelain (gloves, boots).
2. Aluminum (gloves, girdles).
3. Glass (balloons).
4. Hard rubber (girdles).

Porcelain formers are generally used for latex dipped gloves and boots:

1. The surface is resistant to coagulants, latex, and most cleaning solutions.
2. Porcelain withstands thermal shock of wet stripping operation.
3. Close size tolerance can be maintained.

The surface of the porcelain former affects the appearance of the finished product; with a glaze surface for example, the product has a shiny appearance and colors are vibrant.

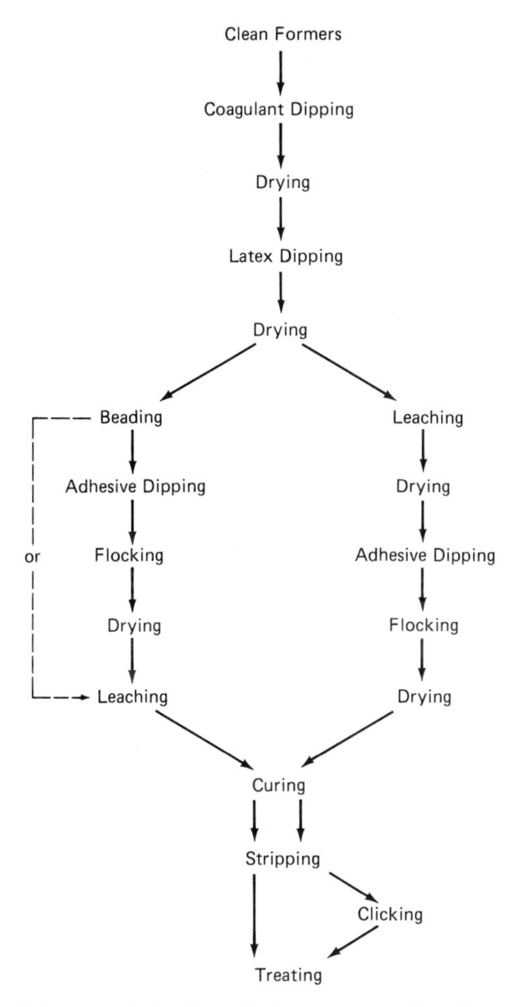

Fig. 19.6. Household-glove dipping processes (anode process).

It is important that formers be mounted so that they do not entrap air while being immersed in the latex compound. Formers should be so designed that they do not contribute to entrapment of air. Formers should be rotated immediately upon their withdrawal from coagulant and latex dip tanks to avoid runs and droplet formation.

Prevulcanization of Latices

All of the products discussed up to now can be produced by the post-vulcanization technique. It is an intriguing fact that latex can be vulcanized in the liquid state and then used to form a deposit of vulcanized rubber without a need to resort to further heating for cure. Only drying is necessary. When latex is vulcanized, it retains its original fluidity and general appearance. Cure takes place in the individual latex articles without altering their state of dispersion.

Vulcanized latex is made by adding sulfur, zinc oxide, and an accelerator to a stabilized latex, and then heating the mixture under conditions whereby a loss of moisture is prevented. Originally, this process was carried out by heating the compounded latex to temperatures above the boiling point of water in a pressure vessel. However, with the introduction of ultra-accelerators, it has become possible to vulcanize latex under atmospheric pressure at temperatures considerably below 100°C. Satisfactory vulcanization can be obtained easily in about an hour at 80°C.

In practice, the vulcanization of latex in liquid form must be rather carefully controlled. If it is carried beyond a certain rather low combined sulfur figure, the resulting rubber deposit will be short and weak. On the other hand, if the latex is properly vulcanized, physical properties almost equal to those obtained by conventional cure after drying are obtained. To prevent further cure from taking place during storage, the compound may be centrifugally clarified to remove excess sulfur and zinc oxide. In some cases, if the curing ingredients are not too finely ground, decantation of the compound after the materials have settled will prevent further curing during compound storage.

Extruding

Natural-latex thread is made by extruding matured compound through a glass orifice. The manufacturing process is shown in Figure 19.7.

LATEX FOAM

The two major methods for latex-foam production are:

1. Talalay Process
2. Dunlop Process
   a. Heat-gel Foam
   b. No-gel Foam

1. Talalay Process. Foam rubber of good structure can be produced without the use of mechanical means for air introduction into the compound. By this method, known as the Talalay process (U.S. Patent 2,432,353), the compounded latex is foamed by the catalytic decomposition of hydrogen peroxide, and then rapidly frozen. The frozen structure is then coagulated by permeation with carbon dioxide.

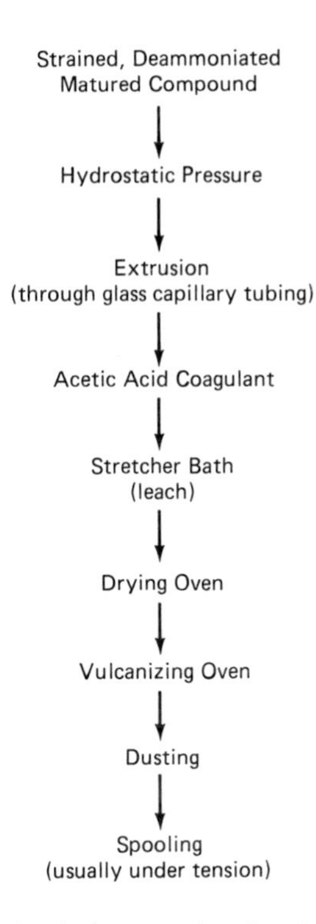

Fig. 19.7. Strained, deammoniated mature compound.

Latex to be used for foam manufacture by this process must be compounded. The compounding ingredients used are identical to those used to compound latex for manufacture of foam by the Dunlop process. The exception is that primary and secondary gelling agents are not used in Talalay-process latex foam.

2. Dunlop Process. This process involves the mechanical whipping (or mixing) of air into compounded latex to froth it.

a. *Heat-gel foam:* gelling agent added to compound at whip.
b. *No-gel foam:* gelling agent is not used to produce the foam. The whipped foam gells during drying and vulcanization (heat-gel foam gels prior to this point). High levels of surfactant are added to the latex; those surfactants used must ensure foam stability during the drying operation.

Regardless of the process used to produce latex foam, the latex must be compounded prior to foam manufacture. The following section describes in a general way the ingredients required to compound latex for use in foam manufacture. It should be remembered that not every compounding ingredient discussed will be required; those required will be determined by the process employed to produce the foam.

Foam-compound Ingredients

Soaps. These materials are present to promote the frothing action during whipping, which is the basis of foam formation. Synthetic latices incorporate soaps in the polymerization reaction, but in some cases additional soap or surfactant is required for frothing.

Sulfur. Requirements vary from 2.5 phr for natural latex to 2 phr for 100% SBR. Blends should be compounded accordingly. Nitrile foam requires sulfur at a 2-part level.

Zinc Oxide. Requirements vary between 3 and 5 phr in all types of foam compounds, except neoprene specialties for which higher quantities are used as the vulcanizing agent. For general activation, 5 phr is a good recommendation. Contrary to most latex practices, zinc oxide is added to a foam compound near the end of the frothing cycle, rather than early in the mix with the rest of the curing system.

Accelerators. Combinations of a thiazole (ZETAX) and ultra-accelerators are used to obtain fast low-temperature cures. Among the ultras, SETSIT 9 and ETHYL ZIMATE are effective in quantities of .25 to 1 phr, depending on the slab thickness and vulcanizing method. If the foam is molded, or run as a continuous slab, sufficient cure must be obtained in the first heating cycle to permit handling, washing, and squeezing without deformation. Accelerator persistency must continue through the drying cycle to complete the cure in foam on fabric applications, for which dry-heat vulcanization is a single operation and the foam is a thin sheet, ZETAX-ETHYL ZIMATE-SETSIT 9 acceleration produces good results.

Antioxidant. The use of a good antioxidant or antioxidant blend at 2 phr is suggested for latex-foam applications.

Gelling Agents. Latex foam before gelling has been accurately described as a two-phase system, in which soap stabilizes both the dispersed air bubbles and rubber globules suspended in the latex serum. To obtain the required results, the rubber serum system must coagulate before the air-serum system breaks. Timing must be such that the coagulum is weak enough to permit coalescence and escape of the air bubbles, yet strong enough to prevent collapse of the foam structure.

Sodium silicofluoride, prepared as a 50% dispersion and diluted to 20% concentration just prior to use, generally functions as the primary gelling agent.

In processes in which heat is applied to the ungelled foam, as in continuous-slab units, potassium silicofluoride or blends of the sodium potassium salts may be used to reduce premature coagulation on doctor blades or other leveling devices. As the gelling agent is added, decomposition with the release of hydrofluoric acid begins. The acid in turn reacts with existing zinc ammonium complex ions, releasing zinc that in turn insolubilizes the fatty-acid soap. As this reaction proceeds and the pH becomes lower, the destabilized rubber particles coalesce and form a gel. With further pH decrease, the discrete air bubbles break to form the desired interconnecting cell structure in the coagulant foam.

Sodium silicofluoride is widely used because its use at elevated temperature shortens the gelation time. Potassium silicofluoride is active at temperatures above 32°C. Ammonium salts such as the chloride, nitrate, sulfate, or nitrile can be used in thin cross-sections permitting uniform heat transfer. These require temperatures of 43°C or higher.

Foam Stabilizers. Materials of this classification are sometimes referred to as *secondary gelling agents*. Their primary purpose is to permit the latex to gel without collapse of the foam structure. Trimene Base (an amine compound), diphenylguanidine, and certain quaternary amines are widely used in this capacity. These materials function by partially destabilizing the latex, thus permitting the primary gelling agent to become effective before the soap bubbles start to coalesce. Trimene Base diluted to 50% with water is added to the latex near the end of the compounding cycle. Normally 0.75 to 1.25 phr are sufficient to prevent collapse. Diphenylguanidine is prepared as dispersion, frequently as a mixture with zinc oxide, and added to the compound near the end of the foaming operation; 0.5 to 1.0 phr DPG dispersion will act effectively as a foam stabilizer.

Miscellaneous. For special foam products, it is sometimes desirable to add an internal lubricant to provide better resilience or "snap" during cutting. This may be obtained by adding 1.5 phr castor oil prepared as an emulsion. The odor of foam products may be improved by adding up to 0.25 phr of reodorant.

Foam formulations of the types listed may be converted into consumer products by molding into various individual shapes, forming into slab stock, or applying to fabrics.

LATEX FOAM FABRICATION

Molded Goods

In this field, the trend is away from maintaining a large number of individual molds. It is now customary to mold large slabs of pin-cored foam and cut the required shapes to customer specifications. Slabs up to 6 inches in thickness can be produced in these cord molds.

Slab Stock

Foam up to 2 inches in thickness is produced continuously with equipment now in use in modern plants. This is the most fully automated operation in the foam industry.

Table 19.14. Natural-latex Foam.

	Dry	*Wet*
60% deammoniated natural latex—total alkalinity as ammonia 0.10–0.12% (wet basis)	100	167
20% potassium oleate soap	1	5
50% ETHYL ZIMATE slurry	0.75	1.50
68% sulfur dispersion	0.5	0.735
Mature the above for 16–20 hours at 24°C, then add the following:		
55% AGERITE WHITE dispersion	1	1.82
or		
50% AGERITE STALITE S dispersion	2	4
50% ETHYL ZIMATE slurry	0.25	0.50
65% ZETAX dispersion	1	1.54
68% sulfur dispersion	1.5	2.21
50% Trimene Base solution	1	2
20% potassium oleate soap:		
for medium density	0.6	3
for light density	1	5
For dry-heat cures only, use the following in place of accelerators listed directly above.		
SETSIT 9 (mix)	–	0.5
Water of dilution (mix)	–	1
65% castor oil emulsion	2	3.08
Add the following to line which brings compounded latex to head of Oakes mixer.		
60% zinc oxide dispersion	5	8.33
20% sodium silicofluoride dispersion	1.5	7.5

Notes:

1. Ammonia level must be adjusted to the above limits for optimum processing.
2. Maturing step required.
3. Castor oil emulsion is added just before foaming.

Foam on Fabrics

Foam is spread and cured continuously on carpeting materials. Resulting products provide durable low-cost floor coverings that have a deep-pile feel and are easy to install. Fabrics of various types are foam-coated to improve dimensional stability and increase tear resistance. When dry cleaning is frequent, the new nitrile latex foams are used.

FOAM FORMULATIONS

The foam formulations listed in Tables 19.14–19.20 are representative of the types that are currently being used. We have listed a formula calling for 100%

Table 19.15. 60 Natural/40 SBR Latex-foam Rug Backing.

	Dry	*Wet*
60% natural latex—deammoniated to 0.18% ammonia (wet basis)	60	100
20% potassium oleate soap	1	5
70% SBR latex (large particle size)	40	57.2
68% sulfur dispersion	0.5	0.74
50% ETHYL ZIMATE slurry	0.6	1.2
Mature the above for 16–20 hours at 24°C, then add the following:		
20% potassium oleate soap	0.7	3.5
68% sulfur dispersion	1.5	2.21
65% ZETAX dispersion	1	1.54
50% ETHYL ZIMATE slurry	0.4	0.8
50% AGERITE STALITE S dispersion	2	4
or		
55% AGERITE WHITE dispersion	1	1.82
50% Trimene Base solution	1	2
For dry-heat cures only, use the following in place of accelerators listed directly above.		
SETSIT 9 } mix	–	0.5
Water of dilution } mix	–	1
65% castor oil emulsion	2	3.08
65% MCNAMEE CLAY slurry	10–30	15.4–46.2
Add the following to line which brings compounded latex to head of Oakes mixer.		
60% zinc oxide dispersion	5	8.33
20% sodium silicofluoride dispersion	3	15

Note: Castor oil emulsion should be added to compound on day of use.

Table 19.16. 25 Natural/75 SBR Latex High-density-foam Rug Backing.

	Dry	*Wet*
60% centrifuged natural latex	25	41.7
20% potassium oleate soap	1.5	7.5
71.7% Pliolite 5352 (SBR) latex	75	104.6
50% ETHYL ZIMATE slurry	1.5	3
68% sulfur dispersion	2.5	3.7
65% ZETAX dispersion	1.5	2.3
50% SM2064 (silicone emulsion)	0.25	0.5
Barytes (dry)	30	30
Alumina C-31 (dry)	50	50
PYRAX A (dry)	60	60
65% P-33 dispersion	1.5	2.3
50% AGERITE NEPA emulsion	3	6
or		
55% AGERITE WHITE dispersion	3	5.45
Add the following to line which brings compounded latex to head of Oakes mixer.		
20% ammonium acetate solution	0.6	3
60% zinc oxide dispersion	5	8.33

Table 19.17. 100% SBR Latex Foam.

	Dry	*Wet*
70% SBR latex	100	143
68% sulfur dispersion	2	2.94
65% ZETAX dispersion	1	1.54
50% ETHYL ZIMATE slurry	1	2
65% AGERITE SUPERLITE emulsion	1	1.5
50% Trimene Base solution	1	2
65% MCNAMEE CLAY slurry	10	15.4
Add the following to line which brings compounded latex to head of Oakes mixer.		
25% zinc oxide dispersion	5	20
20% sodium silicofluoride dispersion	2–2.5	10–12.5

Notes:

1. Soap may be required with some latices.
2. Deammoniation and maturing not required.
3. Compound is designed for low shrinkage slab stock or rug underlay. For molded foam, loading may be omitted.

Table 19.18. High Density Latex Foam for Carpet Backing.

	Heat Gel		No Gel	
	Dry	Wet	Dry	Wet
60% centrifuged natural latex	25.0	41.7	20.0	33.0
20% K Oleate soap solution	1.5	7.5	–	–
35% Cyansol 18	–	–	2.0	5.7
73% FRS 230 (SBR) latex	75.0	102	80.0	110
68% sulfur dispersion	2.5	3.7	1.0	1.47
50% ETHYL ZIMATE slurry	1.5	3.0	1.5	3.0
65% ZETAX dispersion	1.5	2.3	1.5	2.3
50% SM 2064 Silicone emulsion	.25	.5	.25	.5
65% P-33 (carbon black) dispersion	1.5	2.3	1.5	2.3
50% AGERITE NEPA emulsion	3.0	6.0	3.0	6.0
Feldspar	30.0	30.0	30.0	30.0
Barytes	30.0	30.0	30.0	30.0
Alumina C-31 (hydrated)	80.0	80.0	80.0	80.0
60% zinc oxide dispersion	–	–	4.0	6.67
At whip (mechanical frothing), add:				
20% ammonium acetate solution	.6	3.0	–	–
60% zinc oxide dispersion	5.0	8.3	–	–

Table 19.19. No-gel 100% SBR Foam Rug Underlay.

	Dry	Wet
70% SBR Latex	100	143
35% surfactant solution	2	5.7
50% silicone emulsion	0.25	0.5
50% antioxidant emulsion	0.5	1
55% antioxidant dispersion	0.5	0.91
68% sulfur dispersion	1.7	2.5
60% zinc oxide dispersion	1.5	2.5
65% carbon black dispersion	1	1.54
Calcium carbonate filler	150	150

	A		B	
	Dry	Wet	Dry	Wet
50% ZDEC slurry	1.25	2.5	-	-
65% ZMBT dispersion	1.0	1.54	-	-
SETSIT 104	-	-	-	5 } premixed
Water of dilution	-	5	-	5 } premixed
Total froth time	6 minutes		6 minutes	
Compression set: 50% compression for 22 hours at 70°C	14%		9.1%	

Accelerator(s) added to compound just prior to whip. Whipped foam spread onto cotton duck. The spread foam air-dried and cured in a circulating-warm-air oven at 135°C for (combined time) 24 minutes.

Table 19.20. Nitrile Latex Foam.

	Dry	*Wet*
66% NBR Latex	100	152
50% ETHYL ZIMATE slurry	1	2
65% ZETAX dispersion	1	1.54
68% sulfur dispersion	2	2.94
55% AGERITE WHITE dispersion	1	1.82
or		
55% AGERITE STALITE S dispersion	1	1.82
50% Trimene Base solution	1.25	2.5
Whip the above volume, then add:		
60% zinc oxide dispersion	5	8.33
30% sodium silicofluoride dispersion	4	13.33

natural latex and also one for 100% SBR latex. Usually a blend of 90 natural/10 SBR will be used rather than 100% natural in order to provide better ageing characteristics. A 10 natural/90 SBR blend will likewise give better ageing characteristics than 100% SBR. There would be no change in the compounding ingredients listed.

Major compound adjustment in partial or complete replacement of natural latex with SBR is reduction of soap content to compensate for that introduced with the synthetic. Maturing is not required in processing 100% SBR foam, and in some compounds of this type, zinc oxide may be added to the batch before foaming. Curing-rate comparisons indicate that properly compounded 100% SBR foam may be handled at increased production speeds.

PROCESSING NOTES

Mixing. Latices and prepared compounding materials are combined by mixing in the order listed in typical formulas. Propeller speeds are 1200–2000 rpm and propeller positioning should be such that vortexing is avoided. Mixing time is generally 15–20 minutes. Compounds in maturing tanks should be kept in motion with slow-moving action and a minimum agitation.* Mixing and maturing temperature may be 24–38°C. After maturing to the required extent, compounds are stored at 4–10°C until used.

Foaming. Compounds are metered continuously with a positive-displacement pump into the Oakes mixer. Initial whipping temperature of the compound

*Maturing is required for natural latex only.

is 15–21°C. This operation should be conducted in a room maintained at 21–25°C and initial mold temperature should be 29–38°C.

In order to produce foam which will conform to RMA compression values, it is necessary to know the approximate dry density and relate this figure to wet density or "blow."

FLEXIBLE POLYURETHANE FOAM

Prior to World War II, the only commercially available foam was that produced from latex. After the war, foam producers looked at other materials from which to produce foam. Their objective was to produce a foam having properties superior to those of latex foam.

One such foam that was produced was polyurethane foam. In the early days of its development, the ingredients were placed in 55 gallon containers, mixed, and allowed to react. The foam was removed from the containers upon completion of the reaction. Then the cylindrically shaped foam was cut, in a fashion (translated to two dimensions) similar to cutting a master record, to yield a continuous sheet.

The name *polyurethane* pertains to the chemical product obtained from the interaction of a diisocyanate and a glycol. The glycol, in this case a polyol, is a dihydroxy-terminated resin.

$$\underset{\text{polypol}}{HO{-}R{-}OH} + \underset{\text{diisocyanate}}{O{=}C{=}N{-}R'{-}N{=}C{=}O} \rightarrow$$

$$\underset{\text{polyurethane}}{\left[{-}O{-}R{-}O{-}\overset{\overset{\displaystyle O}{\|}}{C}{-}NH{-}R'{-}NH{-}\overset{\overset{\displaystyle O}{\|}}{C}\right]_n}$$

Polyols can be of either the polyester or polyether types. Polyesters generally have hydroxyl groups that are free to react with isocyanate. The polyethers are low-molecular-weight macroglycols; frequently they are based on polyalkylene oxides.

The production of polyurethane foam involves yet another reaction. The reaction of isocyanate and water produces carbon dioxide gas. The carbon dioxide gas expands the liquid urethane into a foam and it also enters into chain propagation functions and crosslinking.

$$R{-}N{=}C{=}O + HOH \rightarrow \left[R{-}NH{-}\overset{\overset{\displaystyle O}{\|}}{C}{-}OH\right] \rightarrow R{-}NH_2 + CO_2\uparrow$$

The amine then reacts with additional isocyanate to form a substituted urea.

$$R{-}NH_2 + R{-}N{=}C{=}O \rightarrow R{-}NH{-}\overset{\overset{\displaystyle O}{\|}}{C}{-}NH{-}R$$

Most often, in this type reaction, the diisocyanate also reacts with urea and the urethane to produce biuret and allophonate cross-linkages.

As mentioned, water is used in most polyurethane foam processes to produce carbon dioxide gas for foaming. However, most commercial processes also utilize low-boiling halocarbons such as either methylene dichloride or monofluorotrichloromethane to blow the foam. Use of an inert gas to blow the foam offers the following advantages:

1. Evaporation of the inert gas moderates the foam-processing temperature.
2. Affords the compounder greater control of the reaction.
3. Permits production of flexible low-density foam.
4. Provides the opportunity to produce foam at lower cost than does the reaction of water and isocyanate for manufacture of CO_2 (for blowing the foam).

Otto Bayer and his associates at Farbenfabriken Bayer developed the chemistry of urethane foam in 1937. That year they developed a method for the synthesis of high-molecular-weight compounds involving the reaction of glycols and diisocyanates.

Today polyurethane foam manufacture involves many other ingredients in addition to the basic three ingredients (polyol, diisocyanate and water):

1. Auxiliary blowing agents (discussed previously).
2. Organosilicone surfactants.
3. Antioxidants.
4. Fillers.
5. Dyes.
6. Plasticizers.
7. Catalysts (frequently tin compounds).
8. Flame retardants.
9. Smoke suppresants.

Polyurethane foams have been divided into two general classifications by the American Society for Testing and Materials (ASTM):

1. Flexible.
2. Rigid.

Flexible slab urethane foams are broken down into two major types:

1. Slab urethane foam (Test Method ASTM D 1564).
2. Molded urethane foam (Test Method ASTM D 2406).

Depending upon foam density, cell structure, size and hardness, flexible foam is further subdivided into many different product types. Likewise, molded flexible urethane foam is subdivided into different categories:

1. Hot-molded foam.
2. Cold-molded foam or highly resilient foam.
3. Integral-skin foam.

POLYETHER POLYURETHANE FOAM

Most polyurethane foam produced today is of the flexible polyether variety. Polyethers are lower in cost and are more hydrolysis-resistant than are the polyesters. Additionally, they provide greater ease of processing. Polyether polyols used are either diols of about 2000 molecular weight, as are polypropylene glycols; or 3000–6000-molecular-weight triols. Frequently, the triol, if used, has a molecular weight of approximately 3000; it is an adduct of propylene oxide and ethylene oxide with glycol. High-resilient (HR) foam requires the use of highly reactive polyether triols in the molecular-weight range of 4500–6000. Special polyethers containing acrylonitrile/styrene copolymers (grafted organic polymers) are used in HR foams because they impart improved load-bearing characteristics to the foams.

In *Rubber Technology*, Second Edition, T. Rogers and K. Hecker presented a formula similar to that shown on p. 558. Many firms in Third World countries are currently producing urethane foam using this basic formula with only slight modifications.

The ingredients may be blended in either a tank or a mixing head. Regardless of method of manufacture, the order in which ingredients are added is important. If a mixing head is used, then blends of some ingredients are added to the head as streams as listed on p. 558.

Table 19.21. Polyether/Urethane Ratios.

Type Polyether Foam	*Isocyanate(s) Used*
Flexible	80/20 TDI (isomer ratio 2,4- and 2,6-tolylene diisocyanates
HR	75–80% 80/20 TDI and 20–25% PMDI
Semiflexible	PMDI

Table 19.22. Formula for Base Flexible Polyether Urethane Foam.

Polyol (3,000 MW; OH number 45)		100.0
Silicone (surfactant)		1.0
N-ethylmorpholine	activator	0.4
Triethylene diamine		0.5
Stannous octoate (polymerization activator)		0.45
Water		3.5
TDI 80/20		42.86
Freon 11		Variable
Methylene chloride		Variable

1. First Stream: water is mixed with the amine and silicone prior to the addition of the polyol.
2. Second Stream: the tin catalyst, often diluted in a 1 : 2 ratio with polyol.
3. Third Stream: water free TDI.
4. Fourth Stream: depending upon equipment setup, the halocarbons are either sent to the mixing head as the fourth stream or they are mixed with the polyol stream.

The individual streams are metered, proportionately, into the mixing head.

The amount of TDI required can be calculated with the following formula:

$$[(.155 \times \text{Polyol OH number}) + (9.67 \times H_2O \text{ content})] \times \text{TDI index desired}$$
$$= \text{Amount TDI required}$$

Example: Substituting information from our formula,

$$[(.155 \times 45) + (9.67 \times 3.5)] \times 1.05 = 40.85 \times 1.05 = 42.86 \text{ parts TDI}$$

Flexible foam generally has a TDI index in the range of 1.05–1.12. As the index increases, the foam tends to become:

1. Harder.
2. Less flexible.
3. Poorer-ageing.
4. More prone to hysteresis.

Silicone is used as a system stabilizer to prevent foam collapse. Foam density is determined by the quantity of water and organic blowing agent present in the formulation. Under ideal conditions, gas formation progresses at a rate approximately equal to the rate of polymer crosslinking. The amount of tin catalyst present in the formulation is generally higher when Freon is used for blowing than when it is absent. Cell size may be varied by either increasing or decreasing pressure in the mixing head. Foam density is determined by the amount of carbon dioxide produced and by the amount of halocarbon used.

Only amine catalyst is used for manufacture of either sponge or fabric foam in formulations based on polyester polyol.

Once intimately mixed, the ingredients are discharged from the mixer head and deposited onto a conveyor, where the complex chemical reactions take place. The reactions are primarily exothermic; they convert the liquid to foam. The original method of continuously pouring slab stock foam was to discharge material from the mixer head onto either polyethylene film or paper carried by the conveyor. Friction between the rising foam and side walls of the conveyor restricted the rise at the sides and resulted in manufacture of crowned foam; its height was greater at the center than at the sides. The Draka process was the first successful commercial method developed to overcome crowning. In this process, the sides of the foam were pulled upward at a rate equal to that at which the foam was rising.

Another method developed to overcome crowning is known as "flat-top" or Planiblock. In this method, platens are placed on paper covering the top of the foam mass. The weight of the platens forces the foam to flow to the sides of the conveyor, constricting the foam into a rectangular configuration. A newer method for preventing crowning is known as Maxform. In this method, either high or low mixhead pressure can be used. The intimately mixed ingredients are discharged from the mixhead, through tubing, to a trough. There the ingredients cream and begin to rise. In about a half-minute, the creamed mixture flows from the trough onto moving paper passing over an adjustable fall plate. The plate is adjusted at an angle such that the drop in slope coincides with the increase in height of the rising foam. Thus the top of the foam remains flat as the foam expands downward.

Recently, a new process has been developed for manufacture of polyurethane foam. The Vertifoam process was designed, initially, to produce rectangular-block foam. However, it became apparent that the process would lend itself to round-block product thereby providing savings in chemical cost and reduction in scrap. The Vertifoam process feeds small amounts of premixed chemicals into a pre-expansion chamber. The chemicals expand in the chamber to a predetermined configuration as they are moved vertically. After expansion, the blocks are moved vertically through a cut-off machine, where they are cut to length. The cut blocks are then sent to the curing area.

SUGGESTED REFERENCES

R. J. Bender, *Handbook of Foamed Plastics*, Lake Publishing Corp., Libertyville, IL, 1965.
D. C. Blackley, *High Polymer Latices*, Palmerton, New York, 1966.
B. Golding, *Polymers and Resins*, D. Van Nostrand, New York, 1959.
W. K. Lewis, L. Squires, and G. Broughton, *Industrial Chemistry of Colloidal and Amorphous Materials*, Macmillan, New York, 1978.
E. W. Madge, *Latex Foam Rubber*, Interscience Publishers, New York, 1962.
M. L. Miller, *The Structure of Polymers*, Reinhold, New York, 1966.
Neoprene Latex, E. I. du Pont de Nemours and Co. Inc., Wilmington, DE, 1962.
H. J. Stern, *Rubber Natural and Synthetic*, 2d Ed., Palmerton, New York, 1967.
The Vanderbilt Latex Handbook, 3d Ed., R. T. Vanderbilt Co., Inc., Norwalk, CT, 1987.
The Vanderbilt News, Vol. 34, No. 2 (R. T. Vanderbilt Co.), 1972.

20

RUBBER-RELATED POLYMERS

Part I

Poly(vinyl chloride)

C. A. DANIELS AND K. L. GARDNER
BFGoodrich Co.
Elastomers & Latex Division
Avon Lake, Ohio

INTRODUCTION

Poly(vinyl chloride) [C.A.9002-86-2], commonly referred to as PVC, has probably the most diverse and widespread use of any man-made synthetic polymer material.

Applications of PVC include bloodbags, wire insulation, rigid and flexible tubing, potable water and drainage pipe, house siding, bottles, meat wrap, injection molded articles, floor coverings, upholstery, and so forth.

Polyethylene, PVC, and polystyrene are the three major plastic materials produced in the United States, with annual sales volumes of 13.7 [HDPE 5.7, LDPE 8.0], 6.2, and 3.6 billion pounds, respectively. PVC, however, is shown to be unique when its properties and low cost are compared to those of other thermoplastic polymers. These differentials have led to a significant sales growth over the last 20 years and replacement in traditional applications of many natural materials such as glass, leather, metal, rubber, and wood. Finished products either can be rigid, or, if compounded with plasticizers and additives, they can be flexible. Generally speaking, PVC plastics exhibit good mechanical toughness, resistance to weathering, and electrical insulating properties. Although they have a high-melt viscosity and require heat stabilizers, they are fairly easily processed by extrusion, calendering, milling, or injection-molding techniques. PVC materials have excellent resistance to inorganic acids, alkalis, water, and very good resistance to oxygen and ozone degradation.

PVC is produced by the free radical polymerization of vinyl chloride monomer:

$$n\ CH_2{=}\underset{\displaystyle Cl}{\underset{|}{CH}} \xrightarrow[40\text{–}70°C]{\textit{free-radical initiation}} \text{-}\!\!\left[CH_2{-}\underset{\displaystyle Cl}{\underset{|}{CH}} \right]\!\!\text{-}_n$$

Polymerization temperature primarily controls molecular weight. Commercially, products of various molecular weights are available, ranging from 20,000–120,000 g/mole (n, corresponding degree of polymerization is 320–1920). Molecular weight controls the performance and processing characteristics, such as glass transition, melting range, and resulting melt rheology of PVC.

The low cost and excellent range of properties briefly summarized in this introduction thus account for the diverse and widespread use of PVC. This chapter will review the monomer production process and the different commercial polymerization methods now in common usage that have led to the growth of PVC into a multibillion-pound-per-year industry. The physical and chemical properties together with compounding methods will then follow and finally the chapter will describe the fabrication processes and end-use market applications.

VINYL CHLORIDE MONOMER [C.A.75-01-04] $CH_2{=}CHCl$]

Vinyl chloride monomer is a colorless, slightly sweet-smelling gas at room temperature with a boiling point of −13.4°C. Since 1974, vinyl chloride has been an OSHA-regulated material with established worker-exposure standards. For polymerization purposes, it is used under pressure in liquid form and no human contact is allowed. The physical properties of vinyl chloride are listed in Table 20.1 (adapted from Reference 1).

The original commercial process for manufacturing vinyl chloride developed in the 1930's was based on the reaction of hydrogen chloride with acetylene. Most production today is carried out using what is termed the "balanced process" with ethylene and chlorine as the feedstock raw materials. The process involves the chlorination of ethylene to 1,2 dichloroethane (EDC) followed by the pyrolysis of EDC to vinyl chloride. The EDC pyrolysis creates HCl as a by-product. Oxychlorination then uses this HCl along with oxygen to react with ethylene to form more EDC and water. In the balanced process all of the HCl is used up in oxychlorination and the EDC production is divided equally between direct chlorination and oxychlorination. The chemical reaction scheme for vinyl chloride production is as follows:

direct chlorination $CH_2{=}CH_2 + Cl_2 \rightarrow ClCH_2\ CH_2Cl$

EDC pyrolysis $2ClCH_2\ CH_2Cl \rightarrow 2CH_2{=}CHCl + 2HCl$

oxychlorination $CH_2{=}CH_2 + 2HCl + \frac{1}{2}O_2 \rightarrow ClCH_2\ CH_2Cl + H_2O$

Table 20.1. Physical Properties of Vinyl Chloride[1].

Property	*Value*
Molecular weight	62.5
Melting point, °C	−153.8
Boiling point, °C	−13.4
Specific heat, J/[kg.K]	
Vapor at 20°C	858
Liquid at 20°C	1352
Critical temperature, °C	156.6
Critical pressure, MPa	5.60
Critical volume, cm^3/mol	169
Vapor pressure, kPa	
0°C	175
20°C	348
40°C	624
60°C	1031
Explosive limits in air, vol, %	3.6–22
Self-ignition temperature, °C	472
Flash point (open cup), °C	−77.75
Liquid density (20°C), g/cm^3	0.9098

Overall reaction $$2CH_2{=}CH_2 + Cl_2 + \tfrac{1}{2}O_2 \rightarrow 2CH{=}CHCl + H_2O$$

The annual U.S. production of vinyl chloride was rated at a capacity of 4,014,000 metric tons in 1983.[2] Major producers in thousands of metric tons are Borden, 277; Dow Chemical, 907; Vista Chemical, 318; Formosa Plastics, 363; Georgia-Gulf, 454; BFGoodrich, 908; PPG Industries, 408; and Shell Chemical, 381.

POLY(VINYL CHLORIDE) MANUFACTURING METHODS

There are four polymerization methods used to manufacture PVC commercially: suspension, mass, emulsion, and solution. Industry capacity in the United States in thousands of metric tons is approximately as follows: suspension, 3,195; mass, 200; emulsion, 250; and solution, 50. Table 20.2 lists the principal manufacturers of PVC, and descriptions of the suspension, mass, and emulsion processes follow in this section.

Suspension Polymerization

The suspension polymerization process was developed in the early 1950's, has largely superseded emulsion polymerization for producing many different PVC resins, and now accounts for approximately 82% of all PVC made in the United States. The process simply consists of polymerizing monomer droplets in water.

Table 20.2. 1983 United States Poly(Vinyl Chloride) Industrial Capacity.[2]

Producer	*Capacity, Metric Tons (1000's)*
Air Products & Chemicals Inc.	182
Borden Inc.	354
Certain Teed Corporation	86
Vista Chemicals	322
Formosa Plastics Corporation	367
Georgia-Gulf Corporation	377
BFGoodrich Chemical Group	779
Goodyear Tire & Rubber Company	54
Occidental Chemical Corporation	295
Shintech	308
Tenneco Inc.	406
Union Carbide Corporation	57
Total	3587

Water acts as the heat-transfer medium to remove the large heat of polymerization (111 kJ/mol) and to transport the suspending agents to the monomer droplet surface.[3] The reactor is a pressure vessel equipped with an agitator and cooling jacket. Reactor sizes vary from about 7.57 to 190 m^3 (2,000–50,000 gal). Vinyl chloride is charged to the reactor along with water and by vigorous agitation is finely dispersed into small droplets about 50–80 micrometers (μm) in diameter. Suspending agents such as hydrolysed polyvinylacetate, hydroxypropyl methyl cellulose, gelatin, and so forth, are added as suspension stabilizers and also to control the subsequent coalescence and morphology of the growing grains. Buffers are frequently added to react with HCl liberated during the polymerization reaction from vinyl chloride peroxide. Finally, the initiator is charged to the reactor and the mixture is heated to reaction temperature, usually 40–70°C (624–1291 kPa). Initiators are most times of the oil-soluble type such as a peroxydicarbonate, e.g., 2-ethylhexyl peroxydicarbonate, or a peroxyester such as tertiary butylperoxyneodecanoate.

The monomer droplet polymerizes and coalesces with other droplets resulting in a final white-resin grain about 100–200 microns in diameter. As previously mentioned, the polymerization temperature controls the PVC molecular weight because the predominant chain termination reaction is by chain transfer to monomer, a temperature-dependent reaction. The chain-transfer constant to the monomer, defined as the ratio of the rate coefficient of transfer to monomer to that of chain propagation, is 6.25×10^{-4} at 30°C and 2.38×10^{-3} at 70°C. Porosity of the resin particle is controlled by conversion and the use of porosifiers such as low-hydrolysis polyvinylacetates. Polymerization is normally terminated, after approximately 4–16 hours, when the pressure in the reactor falls to some predetermined value (30–200 kPa), which corresponds to a monomer conversion in the 80–90% range.

At the end of polymerization the PVC, resin–water slurry is transferred to a tank where most of the unreacted monomer is recovered with compressors. The remaining monomer is removed from the resin by passing the resin through a stripping column. The column usually consists of a series of stainless-steel perforated trays or plates through which steam flows counter-current to the resin slurry flow. The slurry is then centrifuged, dried, and screened (see Fig. 20.1). (For reviews of PVC resin polymerization processes see references 4 and 5.)

Vinyl chloride became an OSHA-regulated material in 1974, following the finding of cases of angiosarcoma (a form of liver cancer) among workers exposed to it. In 1976, EPA issued emission standards for production processes using vinyl chloride; and the PVC industry set a standard of 10 ppm vinyl chloride concentration in dried resin. Developments in the last 10 years have thus concentrated towards processes to reduce worker exposure to the monomer. The development of steam-stripping columns drastically reducing residual monomer levels in the resin to just trace quantities, and the development of reactor-coating treatments to prevent polymer build-up in the reactor with the result of a lower frequency of reactor openings, were major advances in the industry.[6]

Recent Developments. More recently BFGoodrich has announced the development of a spherical suspension resin, Geon Advantage, in contrast to existing resins, which are irregularly shaped due to agglomeration of the grains.[7] This new spherical resin is reported to have significant economic benefits for pipe extrusion, with 20–25% higher output rates due to improved bulk density and improved packing in the extruder-screw flights.

Mass Polymerization

Mass polymerization, or, as it is sometimes called, bulk polymerization, was originally commercially developed as a single-stage process by Saint Gobain (France) in the 1940's. Polymerization of the monomer alone is normally difficult to control because of the heterogeneous reaction product. The reaction product becomes extremely viscous with poor heat transfer and at high conversion results in a low-molecular-weight polymer with a broad molecular-weight distribution. The two-stage process developed by Saint Gobain and now licensed by its successors Ato-Chem has overcome these problems. The first stage of the process is carried out in a prepolymerizer, a reactor equipped with a flat-blade turbine agitator and baffles, where the monomer is polymerized to about 7–10% conversion, forming a seed. In the second stage of the process, the seed mixture is transferred to an autoclave along with more monomer and initiator (see Fig. 20.2). Like suspension resin, initiators are either peroxydi-

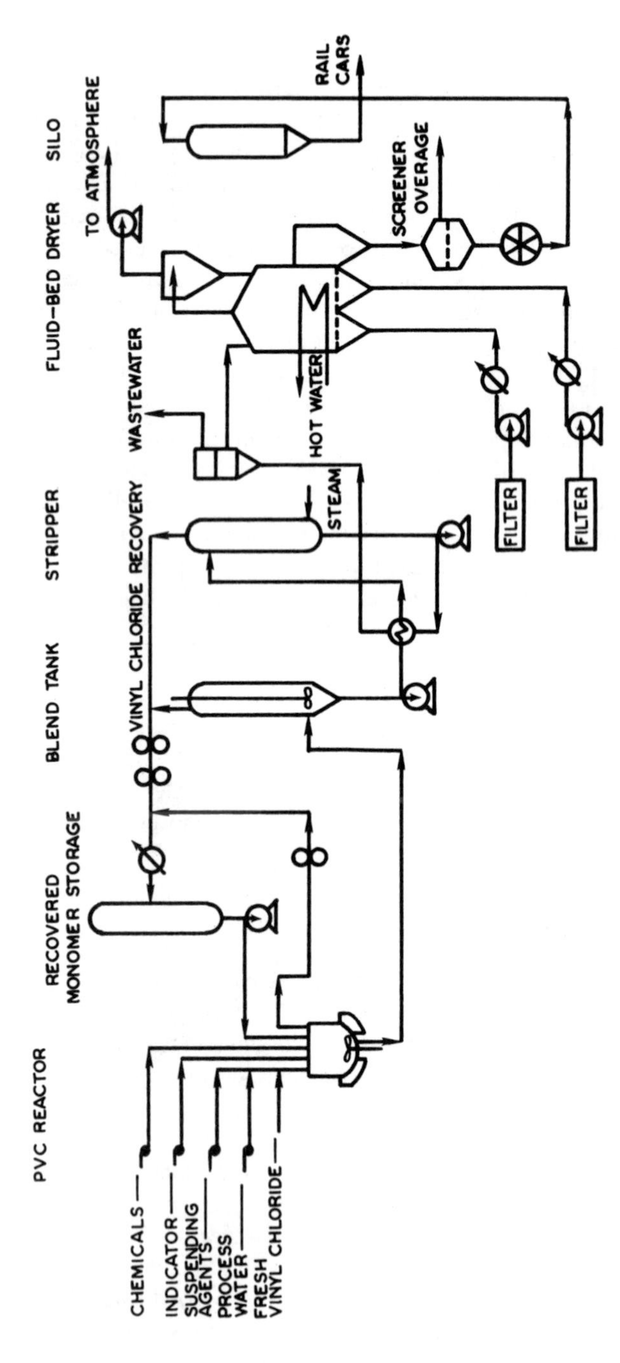

Fig. 20.1. Suspension polymerization plant. (Courtesy of Hydrocarbon Processing[8])

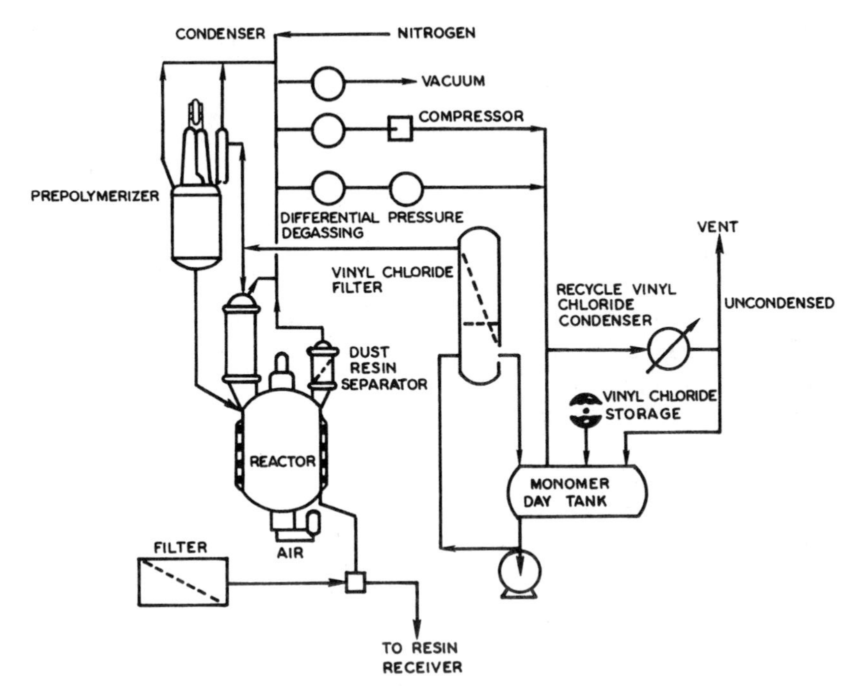

Fig. 20.2. Two-step mass PVC plant with prepolymerizer and vertical autoclave. (Courtesy of Hydrocarbon Processing[8])

carbonates or peroxyesters and the rate of polymerization is dependent upon the initiator half-life, concentration, and polymerization temperature.

The autoclave may be either horizontal or vertical. Inside the horizontal autoclave is a slowly rotating cagelike agitator that sweeps very close to the reactor wall and inside the vertical autoclave is a screw that circulates the product. Autoclaves range in size from 15–47 m^3 (4,000–12,500 gal) and, to remove the large heat of polymerization, are cooled through a hollow agitator shaft, condenser, and jacket. The reaction starts out in the liquid phase, but at approximately 20–25% conversion becomes a dry powder; after 3–9 hours (65–75% conversion) depending upon the molecular weight being produced, the reaction is terminated and the unreacted monomer is removed.

Mass PVC resins have narrow particle-size distributions with slightly smaller particles than the average suspension resin. These resins are fairly spherical in shape, possess good bulk density, high porosity, and good fusion characteristics. The process is attractive in that only monomer and initiator are required with no need for dispersants or water. Also, because no water is used in the process, no drying stage is required. The disadvantages are the limited autoclave size, due to heat-removal design; also the process tends to have less flexibility than the suspension process in producing PVC resins with varying requirements.

Emulsion Polymerization

The PVC emulsion polymerization process in the United States is mainly used to make what is known as *dispersion* or *paste resin*. This resin has an average (and fairly uniform) particle size of about 1 micron, whereas a true emulsion or latex has a much smaller average particle size, about 0.1 micron. The vast majority of PVC resin made by the emulsion process is thus homopolymer dispersion resin and the remainder comprises PVC copolymer latexes using such comonomers as vinyl acetate, acrylates, olefins, and so forth, with only a minor amount of true PVC homopolymer latex being produced. Microsuspension polymerization is also included in this section because it produces dispersion resins with a particle-size distribution of 0.1–1.5 micron and uses the same ingredients and equipment as dispersion resins. The major differences are the use of monomer-soluble initiators and the monomer is mechanically homogenized into very small droplets using the microsuspension method.

The emulsion process emulsifies vinyl chloride in water, using anionic or nonionic surface-active agents. Most of the monomer is in the form of emulsion droplets and a small amount is dissolved in surfactant micelles. Initiation takes place in the droplets with oil-soluble initiators, and the rate of polymerization is dependent upon the initiator half-life and temperature. With water-soluble initiators, polymerization takes place in the micelles and the rate of polymerization is a function of the micelle concentration and formation rate in addition to temperature and initiator. Most dispersion resins are steam-stripped and spray-dried. The drying process is expensive and leaves emulsifier on the polymer surface, which can affect resin color, clarity, heat stability, and so forth. Also, the drying process must be carefully controlled because it causes agglomeration and particle-size changes that subsequently affect the paste-dispersion properties.

Dispersion resins when mixed with plasticizers form suspensions or pastes known as *plastisols*. When hydrocarbon diluents are added to reduce the paste viscosity, these are termed *organosols*. Viscosity of plastisols is extremely important in terms of application properties. Blending resins (particle size about 25–75 μm nonporous resins) are often incorporated for controlling plastisol viscosity.

POLYMER PROPERTIES

Poly(vinyl chloride) is characterized on several structural levels, from molecular to particulate, because each has an important effect on how the polymer performs in applications[10]. Typical properties that are routinely monitored in-

Table 20.3. Selected ASTM Tests for PVC Resins.[11]

Polymer Property	*ASTM Test*	*Title of the Test*
Molecular weight	D 1243-66 (1972)	Test for dilute solution viscosity of vinyl chloride polymers
Specific gravity and density of plastics	D 792-66 (1975)	Test for specific-gravity displacement
Homopolymer PVC resin classification	D 1755-78	Specifications for poly(vinyl chloride) resins
Volatile matter	D 3030-72	Test for volatile matter (including water) of vinyl chloride resins
Density and resin (powder) flow	D 1895-69 (1975)	Test for apparent bulk density and pourability of plastics materials
Porosity	D 2873-70 (1976)	Test for internal porosity of poly(vinyl chloride) resins by mercury intrusion porosimetry
Electrical-volume resistivity	D 257-78	Test for DC resistance or conductance of insulating materials
Particle size	D 1921-63 (1975)	Test for particle size by sieve analysis of plastic materials
Plasticizer sorption	D 3367-75	Test for plasticizer sorption of poly(vinyl chloride) resins under applied centrifugal force
Powder mix	D 2396-69 (1976)	Recommended practice for powder-mix test of poly(vinyl chloride) using a torque rheometer.

Source: American Society for Testing and Materials, 1916 Race St., Philadelphia, PA 19103

clude molecular weight and distribution, particle size and distribution, porosity, rheology of PVC compounds and particle structure. ASTM Tests for PVC are listed in Table 20.3.[11]

Molecular Weight

The most commonly employed method for monitoring PVC molecular weight is the *inherent* or *intrinsic viscosity* (I.V.) measurement (ASTMD 1243). PVC for various applications are graded by I.V. value, ranging from 0.5 for injection

Table 20.4. Molecular Weights of PVC Required for Various Processes and Applications.

I.V. Range	*Process Used*	*Application Examples*
0.50–0.65	Injection molding	Video disc, appliance cabinets
0.65–0.85	Injection molding and Extrusion	Pipe-fittings, sheet, film
0.85–0.98	Extrusion	Pipe, sheet, siding, film, flexible, rigid profile
0.98–1.16	Extrusion	Flexible, rigid profile, wire and cable

molding to about 1.0 for extrusion resins. Table 20.4 lists the ranges, the process, and the end use of resins of varying molecular weights expressed as I.V.

Note that the true molecular weight, i.e., weight average, can be calculated from the intrinsic viscosity-molecular weight relationship.[12] For cyclohexanone solutions at 25°C.:

$$[\eta] = 2.69 \times 10^{-2} M_w^{0.72}$$

Porosity

The porosity of PVC resins can be controlled by the choice of type of polymerization, temperature of polymerization, and polymerization chemistry; and is, of course, designed for the end use. Plastisol resins are virtually nonporous, whereas suspension resins designed for film application are much higher in porosity. The porosity of PVC resins is routinely determined by mercury intrusion porosimetry,[13] although recently methods employing liquid absorption (e.g., plasticizer or oil) have been adopted.[14,15] The rate at which plasticizer is absorbed is important to the mixing and processing of the resin, as is the total absorption capacity of the resin grains. The higher the porosity, the more plasticizer can be absorbed into the resin while still producing a dry-powder mix.

Particle-size Distribution and Shape

Particle size of resins of PVC is also an important physical property controlled to ensure proper processing use. General-purpose PVC resins produced by mass or suspension processes have an average size in the 80–200 μm range[5]. Dispersion PVC consists of particles averaging about 0.7–2.0 μm. Each type has a fairly broad particle-size distribution, generally determined by the agitation, dispersant, or emulsifiers employed.

Suspension resins are analyzed for particle size distribution by dry sieving methods, such as the Sonic Sifter.[16] Newer methods employ light blockage, and can rapidly determine size and distribution.[17]

Dispersion resins are designed to be much smaller in diameter, to produce nonsettling pastes in plasticizers for coating operations.

Particle shapes of PVC typcially are spherical in dispersion resins, but irregular and puckered from a suspension process. These differences are due to the level of protection afforded by the choice of dispersant, agitation, shrinkage due to the densification of polymer (ρ_{PVC} = 1.40 g/cc; ρ_{PVC} = 0.91 g/cc), and agglomeration during polymerization. Particle shape and porosity strongly determine bulk density, an important property for feed rates to extruders, handling, and shipping. Mass resin tends to be smoother and more spherical than suspension PVC.

Rheology

The flow properties of PVC compounds are controlled by a number of variables, all interrelated. The molecular weight of the base resin and the compound ingredients are chosen to match the shear rate of the process, the temperature of the finished-good formation, and the desired mechanical properties.

PVC, like many other polymers, displays non-Newtonian melt rheology, so that the shear rate of the process must be known to match the flow properties desired. Since several decades of shear rate encompass the processes in which PVC is used, laboratory rheometers such as the Instron[10] and Sieglaff-McKelvey[11] can be used to obtain flow curves for selected materials.

Plastisol viscosities are also measured to determine the proper resin/compound formulation for a given application. Although Brookfield rotational viscometers[12] are used to monitor low-shear flow properties and Severs efflux rheometers[13] are used for high-shear, the best method is to employ cone/plate or parallel-plate rheometers to cover the wide shear-rate ranges encountered in applications. And, since viscosities vary with time (aging) in plastisols, these measurements become all the more important.

Plastisols can display shear thinning, shear thickening, dilatent, and time-dependent (thixotropic) flow behavior, making it important to understand whether the process to form an end-use product is done at high shear (e.g., strand coating), low shear (e.g., casting) or somewhere in-between.

COMPOUNDING OF POLY(VINYL CHLORIDE)

It is well-known that PVC, like other polymers, does not have wide processing latitude without being compounded with other ingredients. The mechanical properties and processing windows can be optimized only by proper choice of additives, which in turn must be matched to the final application. Typical PVC compounds can contain plasticizers, thermal stabilizers, lubricants, pigments, flow modifiers, impact modifiers, UV stabilizers, filler, and so forth. It is extremely important that the PVC compounder recognize the environment in which the final product will function, because the choice of compounding ingredients, as well as the method of mixing can affect the acceptance and price/performance balance so important in today's marketplace.

Plasticization of PVC

Plasticizers are chemicals employed to enhance the flexibility of compounds by lowering the glass-transition temperature of the matrix polymer. These materials are most commonly liquids of high boiling point, such as phthalate esters, but can also be polymeric. The physical properties and chemical resistance of PVC compounds are affected by the type and amount of plasticizer employed, and so the choice is driven primarily by end-use needs, but also by price. Plasticizers, in general, reduce the modulus of a PVC compound, decrease hardness, decrease mechanical strength, but increase elongation, creep, and friction.

Earliest plasticizers for PVC were generally phthalate esters, such as di-2-ethylhexyl phthalate, commonly called DOP. Other phthalate esters such as butyl, butyl-benzyl, and diisodecyl are also used, where the choice depends upon compatibility with the PVC (i.e., efficiency of lowering the T_g, volatility, permanence and compatibility with the final compounded matrix. More efficient plasticizers, such as adipates and sebacates, yield lower T_g compounds on a per-part basis, but are more volatile and more expensive than phthalates. Phosphates are sometimes used to enhance flame retardance.

Secondary plasticizers, commonly employed with the esters cited above, include epoxidized soybean oil and epoxidized linseed oils. These are not as efficient as the primary ester plasticizers and are subject to migration because of limited compatibility. These oils are, however, secondary stabilizers against thermal degradation, and so, are popular because of this added property.

Higher-molecular-weight plasticizers are also used. They can be polymeric esters of glycols, sebacates, and other difunctional acids. Ranges of molecular weights are available, commonly ranging from 800 to 4000, and these are sometimes employed with lower-molecular-weight esters to produce compounds of acceptable cost.

Plasticizer choice is often significantly affected by extraction, bleeding, ease of fusion, and other phenomena that reflect the compatibility of these lower-molecular-weight substances with the PVC. Thus resistance to staining, oxidation, soapy-water extraction, and hexane extraction as well as environmental factors all influence the design of the plasticizer system.

Nitrile rubber, actually a copolymer of butadiene and a designed amount of acrylonitrile, has been used as a polymeric plasticizer/impact modifier because of its high molecular weight, but is difficult to process and thoroughly mix because of its high-melt viscosity compared to liquids. Moreover, weathering of such compounds may not be as good as other choices because of unsaturated sites in the nitrile rubber.[18]

Thermal Stabilization of PVC

One of the most serious limitations to the latitude of processing PVC is its strong tendency to discolor and dehydrochlorinate at temperatures above 100°C. Structures within the PVC chain that have been shown to be weak sites are tertiary chlorine, chloromethyl branches, and unsaturation. Although the frequency of these structures in PVC is extremely low, nevertheless at processing temperatures, the reaction will occur when no stabilizers are employed.

Typical thermal stabilizers for PVC include organo-metallic compounds, inorganic metal salts, metal soaps, phosphite esters, and epoxy compounds. Examples include calcium stearate, lead stearate, epoxidized soybean oil, octyl-tin-thioglycolate, and others. Synergistic mixtures of these stabilizers are commonly marketed, yielding packaged blends that can increase processing latitude through lubrication as well as giving thermal stabilization. Such packages might contain phosphites as chelators for metal stabilizers.

The choice of stabilizer is another compounding step governed not only by efficiency and cost but also by end use. For example, food-contact applications such as bottles or packaging film require the use of FDA-approved stabilizers, such as oxtyl-tin types. More choices are available if the use is less stringent. For example, in the United Kingdom, lead-based stabilizers are approved for potable water pipe, but not in the United States, where tin-based materials are used. Barium-lead salts can be employed in drain-waste. Barium-cadmium combinations are less favored in applications because of toxicity, and are being replaced by others such as strontium-zinc.[19]

The mechanism by which stabilizers function in PVC is an area studied and reported upon at length in the technical literature. Some stabilizers merely function as HCl acceptors, while others are believed to react with unstable sites along the polymer chain, preventing or delaying unzipping of the chain through autocatalysis. For example, some metal soaps used as PVC stabilizers are claimed to exchange carboxylate moieties with the more labile tertiary chlorine sites in the PVC chain. This is the mechanism specifically cited for soaps of zinc and cadmium. Stabilizers utilizing magnesium, calcium, barium and strontium, on the other hand, are reported to react with chlorides initially formed, preventing further dehydrochlorination.

Organo-tin compounds seem to function via reaction of substituent groups, such as *n*-octyl or *n*-butyl, with double bonds or free radicals that form early in thermal decomposition. Despite numerous in-depth studies of stabilizer mechanisms, no universally accepted mechanism for all stabilizer functions has yet been established.

Fillers

Fillers are employed to extend and modify the properties and the processibility of rigid, flexible and plastisol compounds of PVC.

Poly(vinyl chloride) formulations can contain both reinforcing and nonreinforcing fillers. Typical reinforcing fillers are glass fibers, graphite, asbestos, and other mineral microfibers. Of the nonreinforcing fillers, calcium carbonate, clays, silica, and silicates are employed. $CaCO_3$ is the most commonly used filler in PVC. Pipe, wire, and cable siding, and plastisol compounds can contain $CaCO_3$. Generally, the finer the particle size of the $CaCO_3$, the greater the impact strength. $CaCO_3$ is used normally at a level of about 3 to 4 parts per 100 parts of resin. Excessive levels of $CaCO_3$ can cause extruder wear in the screw and barrel areas.

Clay (kaolin) is also widely used, for example, in flooring compounds. In PVC plastisols and organosols, the choice of kaolin particle size can control the flow properties during the application, again depending on particle size and loading. Asbestos has been extensively employed in vinyl flooring compounds to improve hardness, and thermal and impact resistances, but is being replaced by other fillers.

Colorants

Although colorants for plastics are generally divided into three classes: dyes, and inorganic and organic pigments, the latter two are the most commonly employed in PVC. The poor chemical resistance and limited stability of dyes has led processors to specify the other types.

The choice of pigments is based on the color imparted, the cost, ease of dispersing, and uniformity of tone. Inorganic pigments include the commonly employed titanium dioxide, as well as oxides, sulfates, sulfides, and other salts of metals such as iron, lead, chromium, cadmium, and so forth. Specialty pigments such as copper or aluminum flakes give a metallic tone, whereas bismuth and other salts can give pearlescence. Organic pigments, such as phthalocyanines and azo compounds, are essentially insoluble dyes.

Ease of handling has promoted the use of color concentrates and pastes, some of which are made using plasticizers as the vehicle. These are especially useful in coloring plastisols. Matching of colors from batch to batch has become a specialized science, employing in some cases spectrophotometric analyzers to ensure reproducibility.

Lubricants

Lubrication of PVC encompasses materials essentially incompatible with the polymer matrix, an external lubricant, or if compatible, an internal lubricant. External lubricants function by being active at the interface between the PVC stock shape and the metal mold or roll surface. Compounds such as silicones or paraffin-based waxes can migrate to the surface, and prevent adhesion at processing temperatures. Internal lubricants such as metal stearates ease the processing behavior of PVC compounds, improving the melt flow characteristics.

Recently, polyfunctional, that is, combined internal and external lubricant packages, have been developed and employed. They are comprised of hydrocarbon chains with amide, alcohol, or ester groups. A commonly employed external/internal lubricant for PVC is oxidized polyethylene.

The amount of lubricant, generally in the 1%-or-less range, is important, because overlubrication can result in slippage, poor fusion, and subsequent surface blooming, whereas too little lubrication can result in degraded compounds. Extruder surging, lumping, and poor impact are common manifestations of over lubrication.[20]

Modifiers of Flow and Impact

Polymeric additives such as nitrile rubbers (acrylonitrile-co-butadiene copolymers), ABS (acrylonitrile-co-butadiene-co-styrene) and MBS (methyl methacrylate-co-butadiene-co-styrene) are examples of PVC impact modifiers. They contain chemical structures that are partially compatible with the PVC, forming good interfacial bonds; and incompatible segments (e.g., butadiene runs) that serve as rubber-impact-modifying domains. These materials toughen the PVC matrix, making brittle failure of the compound less likely.

More compatible modifiers, such as those based on acrylic copolymers, function similarly to plasticizers, which decrease the melt viscosity. Chlorinated polyethylene is another type of processing aid that has a high degree of compatibility with PVC, and is nonmigrating.

All impact modifiers reduce chemical resistance, tensile strength and stress rupture of PVC compounds. Pipe generally has about 0–2 phr whereas siding might contain 0–10 phr. Chlorinated polyethylene, ethylene-vinyl acetate, and acrylic copolymers are chosen for weatherable compounds whereas MBS and ABS are good for low-temperature properties.[18]

Flame and Smoke Retardants

Although PVC itself is extremely flame-retardant due to its high chlorine content, burning of PVC can result in dehydrochlorination. This is extremely important in plasticized compounds, where some plasticizers have been found to enhance smoke generation.

Compounds based on antimony trioxide, aluminum, molybdenum, and zinc oxides and halides are used to suppress HCl formation in PVC. These additives are important where low smoke and low flammability are requirements for the application: upholstery, wire and cable jacketing, carpet backing, and other building materials. New regulations in aircraft and housing codes are being met by proper combination of these additives without sacrifice of mechanical and aesthetic properties.[18]

Compounding and Fabrication Rigid and Flexible PVC

The blending of PVC with other additives is an extremely important step in the fabrication of a finished product. The degree of mixing uniformity, heat, and shear history imparted to the compound can have as much influence over the quality of the product as does the choice of resin or additive.

PVC is normally blended in ribbon-type blenders, high-speed internal-batch mixers, or continuous internal mixers, yielding powder compounds, cubes, or pellets for thermoforming operations (for example, extrusion, blown film, injection molding). Some operations are run in series, such as Banbury mixing, followed by calendering, followed by pelletizing before extrusion. Reference 22 lists some of the suppliers of mixing equipment and their products.

Forming of PVC compounds into finished articles takes place via a variety of methods including extrusion, calendering, injection molding, extrusion- and injection-blow molding, film blowing, and sheet molding.

A typical mixing procedure for a PVC compound is as follows:[21]

1. Turn the high-intensity mixer to the slow speed and add resin.
2. Shift speed to the high setting, then add a stabilizer.
3. Add remaining dry ingredients (TiO_2, lubricant, wax, $CaCO_3$, etc.).
4. Mix until temperature reaches 88°C (190°F), and drop into the cooler.
5. Cool at temperature at or below 43°C (110°F).
6. Transfer to extruder.

Some molders add a pelletizing step to the mixing cycle. This procedure gives extra work to the compound, but requires a pelletizer at the end of a multiscrew extruder. Finally, quality control labs often run a laboratory mixing simulation employing a Brabender Plasticorder, which measures the time necessary commonly called *dry-up time*, for a resin to completely incorporate the compounding ingredients.

SHAPING PVC COMPOUNDS

Extrusion

Single-screw extrusion of rigid PVC compounds is done at high RPM (25-80) and at high compression ratios (3.0/1–3.5/1). According to manufacturers, a length-to-diameter ratio of 24/1 is a minimum to achieve good fusion and at-

Table 20.5. Temperature Settings for a Single-screw Extrusion of Pipe.

Barrel Zones	*Die*	*Screw Oil*	*Spider*	*Adapter*
1 350–360°F	370–380°F	300–320°F	350–360°F	340°F
2 340–350°F				
3 320–340°F				
4 300–320°F				

tainment of mechanical properties. The PVC compound is melted, transported, and forced through the die as a function of the helical flight pattern of the screw.

Twin- and multiscrew extruders are extremely useful in PVC siding and profile markets. Design characteristics vary from staged twin-screw to two-stage parallel twin-screw extruders, which contain four helical screws. Twin screws normally run at 10–40 rpm.[23]

Comparison of single-screw and twin-screw geometries show higher shear, higher RPM, and higher melt temperatures for single screws. This forces the extruded compound to meet more stringent stability criteria. Single screws do not present as uniform a temperature profile as are seen in multiscrew machines. Lubrication internally is more important in the compound for single-screw extruders, whereas the balance shifts to external in twin-screw extruders, primarily due to the shear differences between the two geometries. The setting of temperatures on the extruder is critical to achieve properties. For example, Cincinnati Milacron recommends for their CM-111 twin-screw extruder for pipe the following settings to control zone temperatures:[24]

Calendering

Production of PVC film and sheets often involves successively pressing a thick film into the dimension desired on a series of roll mills, the total process being labeled *calendering*. A compound is prepared and fluxed in a batch or continuous mixer, and delivered to the nip of the first rollers in the bank, where the film is eventually fed continuously through the rolls and taken up at the end of the process. The tolerances in the calender are obviously critical, as is temperature control. This is especially true in the take-up area, because uneven temperature profiles can cause sheet or film to lose dimensional stability; shrinkage or stretching of the film after calendering can cause loss of mechanical properties and material waste. The size of the roll can be as large as 3 feet in diameter by 10 feet long, running at speeds processing up to 18,000 pounds of film per hour, using pressures measuring thousands of pounds per linear inch of surface. Calendering equipment is very expensive, but is an excellent method for producing rigid or flexible PVC for packaging applications.[23]

Molding

There are several kinds of molding operations useful for PVC compounds. Basic designs employ a process to melt the compound and inject it into a mold under high pressure, cool the article, release and remove the article, and begin over again in a continuous operation.

Reciprocating-screw molding operations use a design similar to extrusion to melt the PVC compound before delivery to the molding section of the process. Hydraulic mold-closing mechanisms are designed to clamp the article under high pressures to produce precision-part designs. Control of viscosity and elasticity of the fluxed PVC compound are important, because complete mold filling is required during the forming operation, and minimum mold shrinkage is also required.

The cycle for reciprocating-screw-injection molding is complex. The molten plastic is pumped down the screw to the nozzle area, where back pressure builds due to the restriction. At this point, the mold is closed, and the screw, moved backward by the restriction pressure, is driven forward as a plunger, forcing the melt into the mold. Holding pressure on the mold is set to maintain the melt within the confines of the design, and the part begins to cool. Meanwhile, the reciprocating screw is preparing the next "shot" of fluxed compound. As the mold cools, an ejector mechanism is activated at the proper temperature, opening the mold and ejecting the parts.

Compression molding is another method of forming parts using pressure and temperature to cause the compound to flux and form a desired shape. Here, powder or pellets are pressed and heated between two halves of the mold. These operations are obviously simple and less expensive than injection molding, but are not as automated.

Blow molding is a method used to produce film or bottles of PVC, by several processes such as extrusion-, injection-, and stretch-blow molding. Designs for extrusion-blow molding incorporate basic extruder-die combinations which feed a mold, into which air can be forced to produce the hollow object. Injection-blow molding incorporates the injection molding process to feed the mold, whereas stretch-blow molding produces biaxially oriented products, such as bottles, by mechanically stretching the article linearly, and then blowing the melt to the final axial dimension. Bottles prepared by this method have enhanced mechanical properties due to the orientation process.[23]

Thermoforming

Recently, PVC sheet compounds have been used to thermoform large articles, such as bathtubs and shower stalls, producing lightweight, tough articles of excellent quality, replacing ABS. Extrusion of the sheet is followed by softening and forming, either by vacuum or compression, into the desired shape.

DISPERSION RESIN—COMPOUNDING AND FABRICATION

Compounding

PVC plastisol, or "paste," resins are most frequently used as dispersions in plastisols, which contain not only the resin and plasticizer, but also stabilizers, pigments, diluents, and fillers.

Compounding of plastisol or organosols is achieved by mixing the ingredients in either high- or low-shear equipment, the recipe for which will vary to control the amount of shear history. Shear heating occurs, but should be minimized to prevent premature gelation. Resultant compounds are usually very viscous, stable suspensions of PVC dispersion resin (particle size 0.30–2.0 μm in diameter) in the continuous plasticizer phase. These pastes exhibit unusual rheology, varying from shear-thinning to shear-thickening, depending on resin, plasticizer, concentration, and shear rate. Lowering the plastisol viscosity by addition of a volatile diluent such as mineral spirits produces a more easily applied organosol, and can allow for reduced levels of plasticizer in the compound and a product of increased hardness and lower flexibility.

Fabrication

Molding. Organosols and plastisols of PVC are applied by dip, slush, and rotational molding. In dip molding, an article such as a glove, wire rack, or tool handle is coated by dipping a heated form into a plastisol, gelling the required amount of compound, and then transferring this preformed article to a finishing oven set at a higher temperature to complete fusion.

In *slush molding*, organosols or plastisols are transferred to open molds, and then heated to complete the plasticization of the PVC. The molds are then cooled, and the articles removed. Products such as toys, shoes, vinyl boots, and traffic cones are formed by filling the mold with the proper compound, heating the mold to gel the outer layer, emptying the mold of excess compound, and fusion of the finished object.

Rotocasting is a method of forming large, hollow parts or toys by rotation of the mold during the loading and fusing of the PVC plastisol. Machines with a number of arms containing molds spin the plastisol to form the desired shape while the mold is programmed through its heating and cooling cycle. Boat bumpers, dolls, and vinyl balls are all produced by this method. Table 20.6 shows a representative recipe for a rotocasting operation. This should be compared to a rigid extrusion recipe for pipe, shown in Table 20.7.

Knife coating is a method by which the plastisol is applied to a substrate fabric, such as cotton or pretreated nylon, to form an outer layer of water-resistant plasticized PVC. The coated fabric is passed through an oven on a conveyor line, where the plastisol is fused.

Table 20.6. Rotocast Doll Formulation Using Dispersion Resin.

Ingredient	*Level (parts per 100 parts Resin)*
Geon 121	70
Geon 212	30
Di-2-ethylhexyl phthalate	45
Butyl benzyl phthalate	15
Liquid Ba/Cd/Zn stabilizer	3

Table 20.7. A PVC Compound for Rigid Pipe.[26]

Ingredient	*Parts per 100 parts Resin*
Geon 103EPF-76	100.0
MBS impact modifier	0.0–10.0
Acrylic processing aid	0.0–3.0
Sn stabilizer	1.0–2.5
Ca stearate	1.0–2.0
Oxidized polyethylene	0.0–0.5
Paraffin wax	0.4–1.2
TiO_2	0.5–3.0
$CaCO_3$—stearate coated	0.0–5.0

Roll coating is the process for forming vinyl-flooring sheet goods, representing about one-fourth of all the consumption of PVC dispersion resin. Vinyl flooring is a unique composite layered in structure: a reinforced backing, a foam PVC cushion layer, and a wear layer of a hard plastisol compound to resist staining and abrasion. Unique and sometimes proprietary processes are used to form the embossed and color-printed designs on modern vinyl flooring.

MAJOR MARKETS FOR PVC PRODUCTS

In the early years of the growth of the PVC industry, the three largest markets were calendered film and sheeting, calendered flooring, and extruded wire and cable. The development of a rigid-PVC-processing technology and the modernization of local building codes to permit plastic pipe has resulted in the pipe segment of the PVC industry becoming the single largest end use. The pipe market is divided into different applications: pipe for water supply, 38% of the end use; sewer, 22%; electrical and telephone-conduit, 20%; drain-waste-vent, 10%; and agricultural irrigation, 5%. Further market developments include the growth of rigid PVC house siding into a very large segment of the business. Window profile is also a relatively new and expanding end use following on its great success in this application in Europe. Extrusion of film for packaging uses has grown into a large application. Improvements in formulation technology to provide transparent PVC compounds with good sparkle has led to a high rate of growth of PVC bottles for general-purpose applications. Food-grade PVC bottles now successfully compete with glass and other plastics and significant growth is forecast in this area with the development of ultra-low residual vinylchloride compounds. The development of low-molecular-weight resins has led to rigid custom injection molding of appliance and business-machine housings, and the like. A comprehensive breakdown of market applications for PVC can be found in reference 25.

Over the past 5 years, however, PVC manufacturers have experienced intense competition because of excess industry capacity. This has driven resin prices to such low levels that several manufacturers have withdrawn from the marketplace. But other major manufacturers have begun to develop specialized applications for PVC in order to obtain greater profit margins. As a result of this ongoing activity we can expect to see many new and diverse applications of PVC during the next 10 years.

REFERENCES

1. J. A. Cowfer and A. J. Magistro in *Encyclopedia of Chemical Technology*, 3d Ed., Vol. 23, I. Kirk and D. F. Othmer, eds., John Wiley & Sons, New York, 1983, p. 865.
2. *Chemical Economics Handbook*, SRI International, Menlo Park, CA, 1984.
3. K. J. Ivin in *Polymer Handbook*, 2nd Ed., J. Brandrup and E. H. Immergut, eds., John Wiley & Sons, New York, 1975, Sect. 2, p. 421.
4. D. E. M. Evans in *Manufacture and Processing of PVC*, R. N. Burgess, ed., Macmillan, New York, 1982, Chapt. 3.
5. J. A. Davidson and K. L. Gardner in *Encyclopedia of Chemical Technology*, 3d Ed., Vol. 23, I. Kirk and D. F. Othmer, eds., John Wiley & Sons, New York, 1983, p. 886.
6. M. G. Morningstar and H. J. Kehe (to BFGoodrich Company) U.S. Pat. 4,024,330 (May 17, 1977).
7. L. G. Shaw, A. R. DiLuciano, P. O. Hong, K. Dinbergs, and T. H. Forsyth, SPE Preprints, 42d Antech, pp. 820–822, 1984.
8. J. B. Cameron, A. J. Lundeen, and J. H. McCulley *Hydrocarbon Processing* **59** (3), 39 (1980).
9. N. Fischer and L. Gioran, *Hydrocarbon Processing* **60** (5), 143 (1981).
10. C. A. Daniels, *J. Vinyl Technology* **1** (4), 212 (1979).
11. T. J. Kraus and J. E. Copus, *Plast. Comp.* **3** (5), 23 (1980).
12. G. Schroeter, in *Polyvinylchloride*, Vol. 2, K. Krekeier and G. Wick, eds., Carl Hauser Verlay, Munich, FRG, 1963, Chapt. 2.
13. ASTM Test, D 2873.
14. J. A. Davidson, *J. Powder Technol.*, **2** (2), 67 (1985).
15. R. Tregan, *J. Macromol. Sci.-Phys.*, **B14,** 7 (1977).
16. ATM Corporation, Sonic Sifter Division, P.O. Box 2405, Milwaukee, WI 53214.
17. Pacific Scientific, HIAC Instruments Division, P.O. Box 3007, 4719 W. Brooks St., Montclair, CA 91763.
18. K. Eise and H. Harris, "Additives," in *Plastics Polymer Science and Technology*, M. D. Baijal, ed., Wiley-Interscience, New York, 1982, pp. 571–612.

19. D. A. Tester, "The Processing of Plasticized PVC," in *Manufacturing and Processing of PVC*, R. H. Burgess, ed., Macmillan, New York, 1982, pp. 215–245.
20. W. V. Titow, "PVC Additives," in *Developments in PVC Production and Processing-1*, A. Whelan and J. L. Craft, eds., Applied Science Publishers, London, 1977, pp. 63–90.
21. Technical Service Bulletin No. 8, "Mixing Rigid PVC," BFGoodrich Chemical Co., 6100 Oak Tree Blvd., Cleveland, OH 44131.
22. "Manufacturing Handbook and Buyer's Guide," *Plastics Technology*, **30,** 7, 1984.
23. E. O. Allen, "Processing" in *Plastics Polymer Science and Techology*, M. D. Baijal, ed., Wiley-Interscience, New York, 1980, pp. 571–611.
24. G. A. Thacker, Jr., *Rigid PVC Extrusion: Problems and Remedies*, Carstab Corp., West St., Reading, OH 45215.
25. *Mod. Plast.* **61** (1), 53 (1984).
26. T. J. Kraus and J. E. Copus, *Plast. Comp.* **3** (6), 23 (1980).

Part II

Polyethylene

JAMES E. PRITCHARD (*Ret.*)
Phillips Petroleum Company
Bartlesville, Oklahoma

HISTORY

Polyethylene became an important commercial product as a result of high-pressure-reaction studies that were undertaken in the laboratories of Imperial Chemical Industries (I.C.I.) in England during the 1930's. However, polyethylene was prepared prior to 1900 by the decomposition of diazomethane:

$$n\ H_2C\langle{N \atop N}\!\| \rightarrow (CH_2)_n + n\ N_2$$

This reaction was described by Bamberger and Tschirner, who stored diazomethane in ether solution over pieces of unglazed china. The white powder that separated out was found to be polymer containing repeating methylene units melting at 128°C.[1] Apparently this linear polyethylene or polymethylene was observed earlier by Pechman,[2] who did not determine its composition. Subsequently, other workers have confirmed this reaction and have studied the properties of polyethylene derived from diazomethane.[3]

The I.C.I. polyethylene, which has become known as "high-pressure polyethylene" or "low-density polyethylene," was produced from ethylene monomer in a high pressure reaction with trace amounts of oxygen as catalyst. According to Swallow,[4] the success of the initial work at I.C.I. depended on the fact that a leaky reactor was used for the high-pressure studies. Fresh ethylene, which was required to maintain pressure, contained traces of oxygen in proper catalytic amounts to ensure continuous polymerization. The subsequent development of reliable compressors capable of maintaining pressures to 3,000 atmospheres made large-scale commercial production of high-pressure polyethylene feasible.

Twenty years after the I.C.I. work, ethylene polymers of very different properties were developed by two separate research groups. These materials, which have come to be known as high-density or linear polyethylenes, also were prepared from ethylene monomer but are much closer in physical properties to the original polyethylenes of the 1900 era than to the more recent high-pressure-process polymers. The leading process for the production of high-density polyethylene is that of Phillips Petroleum Company, which uses a chromium oxide catalyst on a silica or silica-alumina support for polymerization of ethylene in hydrocarbon solution or in hydrocarbon slurry. Polymer produced in this process may be completely linear or unbranched materials exhibiting a nominal density of at least 0.96. In addition, a wide rage of polymers including low-density materials (0.925–0.939) have been produced.

A second low-pressure polyethylene process is based upon the work of Karl Ziegler of Germany.[5] The Ziegler process uses an organometal catalyst comprised of Group IV to Group VI metal halides such as titanium tetrachloride ($TiCl_4$) and an aluminum alkyl such as triethyl aluminum [$(C_2H_5)_3Al$] for polymerization of ethylene monomer. The original polymers produced by the various Ziegler licensees were somewhat lower in density (0.94 to 0.95) than the Phillips process material, but subsequently polymers of 0.96 density have been produced.

In contrast to synthetic rubber, polyethylene has no counterpart in nature and technology is not available by direct transfer from known materials. However, some processing and fabrication techniques have been borrowed from rubber and plastics technology and modified to meet the requirements of the new highly crystalline materials. This has contributed to the rapid growth in annual consumption of polyethylene, which now exceeds any other plastic material. Domestic use of polyethylene of all types in 1984 was more than 14 billion pounds. This is well above the volume of other polymers such as poly(vinyl chloride), 6.8 billion pounds, and styrene polymers, 5.9 billion pounds.[6]

MANUFACTURE

High-pressure Process

The high-pressure process for the polymerization of ethylene has undergone many improvements and modifications since the original work of I.C.I. However, all major producers depend upon the bulk polymerization of ethylene in molten polyethylene at pressures of 1,000 to 3,000 atmospheres and temperatures of 80 to 300°C.[8] Reactors may be stirred autoclaves or continuous tubes. At one time, the products from the two types of reactors differed in physical properties and applications, but in recent years the versatility of the processes has been improved and a broad range of products may be obtained from either system. The catalyst may be trace amounts of oxygen (0.01%), organic peroxides, or other free radical sources. Certain reaction modifiers also may be used to permit production of somewhat higher-density resins at the same reaction pressure.

The effluent from the reactor is comprised of molten polymer containing unreacted monomer that is removed from the polymer stream and recycled to the reactor after purification. The degassed polymer is fed to extruders for processing into pellets. In some instances, strands of polymer are extruded into water baths and subsequently chopped into small pellets. In other cases, molten polymer exudes from a die face and is cut by rotating knives under water to give a spherical rather than a pillow-shaped pellet. The pelleted material may be used in the natural state or it may be homogenized in Banbury-type mixers and re-extruded to yield polymers exhibiting improved properties for film or other special applications. Antioxidants, slip agents, pigments, or other additives may be added in the various processing and extrusion steps.

The original high-pressure-process polyethylene exhibited a density of about 0.92, but more recently the scope of the high-pressure process has been broadened to include polymers having densities as high as 0.93 to 0.94. One of the most direct techniques for increasing density of free-radical polymerized ethylene polymers is to increase reaction pressure. In laboratory work, polymers of 0.95 density have been obtained at 7000 atmospheres pressure.[9] However, production of medium-density or high-density polyethylene by the high-pressure process involves sacrifice in production capacities, which severely limits the feasibility of this approach. In recent years, physical blends of high-density and low-density polyethylene have been developed to fulfill many of the demands for medium-density polyethylene.

A wide variety of copolymers may be derived from the high-pressure process by copolymerization of ethylene with other 1-olefins[10] or polar monomers such as vinyl acetate[11] or ethyl acrylate.[12] The 1-olefins are said to function as chain transfer agents and to yield polymers exhibiting lower melt elasticity than the ethylene homopolymer. In general, the effectiveness of the 1-olefin increases with increasing chain length. Copolymerization of ethylene with vinyl acetate leads to polymers that are rubbery at 20 to 30 mole % comonomer; very tacky and soft at 50 to 60% comonmer, and hard and glassy at 80 to 100% vinyl acetate. The copolymers containing minor amounts of vinyl acetate are used as wax modifiers to improve ductility. Ethylene-ethyl acrylate copolymers also are more rubber-like than the ethylene homopolymer. They may be used as tough, ductile injection-molding resins and, in some instances, are excellent replacements for plasticized polyvinyl chloride.

Phillips Process

The Phillips low-pressure process uses a chromium oxide catalyst on a silica or silica-alumina support for polymerization of ethylene at low pressure in paraffinic or cycloparaffinic solvent.[13,14] In one variation of the process, the polymer is prepared in solution. In another modification the polymer is formed as discrete particles in hydrocarbon slurry. The catalyst is prepared by impregnation of the support with a solution of chromium trioxide or other chromium compound and then is activated by calcining in air at 400 to 850°C.

In the solution process, solvent, monomer, and catalyst slurry are charged continuously to the reactor which operates at 125 to 175°C and 20 to 30 atmospheres pressure. In typical commercial practice, 1000 pounds or more of polymer are obtained from 1 pound of catalyst. The reactor effluent is a solution of polymer containing small amounts of catalyst that may be removed by filters or centrifuges. The solvent is removed by flashing and steam stripping or by cooling to precipitate polymer followed by filtration. The polymer is pelletized and compounded with antioxidants, pigments, or other additives as desired.

In the slurry process, the reaction media, polymerization temperature, and other environmental conditions are designed to avoid dissolving of the polymer. The product is obtained in the form of a slurry of polymer particles in hydrocarbon diluent and is recovered from the reactor effluent by a simple flashing step. The productivity of the slurry process in terms of pounds of polymer per pound of catalyst is so high that the catalyst removal step may be eliminated.

A wide range of polymers and copolymers of various densities and molecular weights is available from the Phillips process. Commercial resins include density ranges of about 0.925 to 0.962 and molecular weights from perhaps 25,000 to more than 1,000,000.

Ziegler Process

Although a number of producers throughout the world have obtained licenses[14,15] for the use of the Ziegler polymerization catalyst, they have not followed identical routes to the finished polymer. In a typical Ziegler process, the catalyst is prepared by combining triethyl aluminum and titanium tetrachloride in a diluent. The tetrachloride is reduced, at least in part, to titanium trichloride, which may be isolated and combined with more triethyl aluminum to yield a catalytic mixture. Alternatively, the mixture of aluminum alkyl and tetrachloride may be charged to the reactor without isolation of the trichloride. Polymerizations are carried out in batch or continuous systems and the polymer is obtained as a slurry in hydrocarbon diluent.

The hydrocarbon diluent is recovered from the reactor effluent and recycled. The catalyst is removed from the finely divided polymer by washing with an alcohol, such as isopropyl alchol, or with other reagents. After the polymer cake is recovered and dried, it is pelletized and compounded with various additives by standard techniques.

The very early Ziegler-type polyethylenes were of high molecular weight and relatively difficult to process. However, various techniques have been developed for controlling molecular weight, including increasing the proportion of titanium tetrachloride in the reaction media[16] or introduction of hydrogen.[17]

Union Carbide Process

Union Carbide was a pioneer in low-density polyethylene and a licensee of both the Ziegler and Phillips technology. The Carbide "Unipol" process, which combines features of the Ziegler and Phillips processes, is a major source of a low-density polyethylene dubbed "linear low-density polyethylene," or LLDPE. In the "Unipol" process, a Ziegler catalyst on a silica support is used to copolymerize ethylene and 1-butene.[18] The reaction is carried out in a gas-phase, fluid-bed reactor. Granular polymer is removed continuously from the system and unreacted monomer is recycled.

In comparison with low-density polyethylene from the high-pressure process, LLDPE exhibits superior tensile strength, higher stress-cracking resistance, and greater overall toughness. Optical clarity is lower. In 1984 1,160,000 metric tons of LLDPE was consumed.[19] Film and injection molding accounted for 764,000 and 102,000 metric tons, respectively.

PHYSICAL PROPERTIES

Information on physical properties is used to characterize polyethylenes and to predict their performance in end-use applications. In addition, some physical measurements are sensitive to structural features of the polyethylene molecule and are of value in fundamental polymer research. Because polyethylene, for the most part, is used in the natural state, free of additives and reinforcing fillers, structural features follow through to end-product performance to a greater degree than is the case with an elastomer. Polymer deficiencies cannot be masked or compensated for by subsequent compounding and curing steps. Consequently, it is of greatest importance to recognize and control the composition of the polymer.

Density

The properties of polyethylene are largely dependent upon crystallinity and molecular weight. The level of crystallinity may be determined by X-ray studies or by nuclear magnetic resonance techniques.[20] Because there is a direct correlation between the degree of crystallinity and polymer density, it is convenient to measure density using a gradient column[21] or another technique.[22] As indicated in Table 20.8, a high-density polyethylene, of 95% crystallinity and 0.96 density, exhibits a higher melting point, higher tensile strength, higher flexural modulus, and greater hardness than does a low-density polyethylene, of 65% crystallinity and 0.92 density in the table. Chemical properties of polyethylenes

Table 20.8. Properties of Polyethylenes.

Polyethylene	*Crystallinity, %*	*Melting Point °C*	*Density*	*Ultimate Tensile, psi*	*Elongation at Break, %*	*Flexural Modulus, psi*	*Hardness Shore D*
High density	95	134–137	0.96	4400	25	230,000	68
Low density	65	114–116	0.92	2000	500	45,000	52

also are dependent upon crystallinity and density, with high-density polymers exhibiting superior resistance to oils, solvents, and permeation by water vapor and gases.[23]

Melt Index

The molecular weight of polyethylene, like other high polymers, may be determined by solution viscosity, melt viscosity, light scattering or ebulliometric techniques. A common method for estimating molecular weight involves measurement of melt viscosity in a simple extrusion plastometer, or melt indexer. In this device, molten polymer is extruded through a 0.0825-inch orifice at a temperature of 190°C with a dead-load piston providing a force of 2160 grams. The weight of polymer in grams extruded in 10 minutes is termed the *melt index*.[24] Typical melt-index values for commercial polyethylenes are in the range of 0.1 to 20. The polymers of low-melt index or high molecular weight are preferred for heavy-duty applications such as plastic pipe, while those of high-melt index or low molecular weight may be preferred for high-speed molding or extrusion operations. It has been recognized that a simple instrument that provides a single determination of melt viscosity under an arbitrary set of conditions is inadequate to characterize fully the melt-flow properties of polyethylene. In many respects, the melt indexer occupies the same position in polyethylene technology as does the Mooney viscometer in rubber technology. Both instruments are of value in characterizing polymers providing the limitations of a single-point determination of viscosity are recognized. An additional problem with the melt indexer resides in several sources of error inherent in the device, which further limits its utility.[25] Nevertheless, melt-index determinations are universally used in the polyethylene industry and, providing the limitations of the equipment are recognized, considerable use can be made of them.

Stress-strain

Stress-strain properties of polyethylenes are understandable to rubber technologists although it is recognized that polyethylene is relatively inelastic beyond a few-percent elongation. As indicated in Fig. 20.3, stress increases sharply

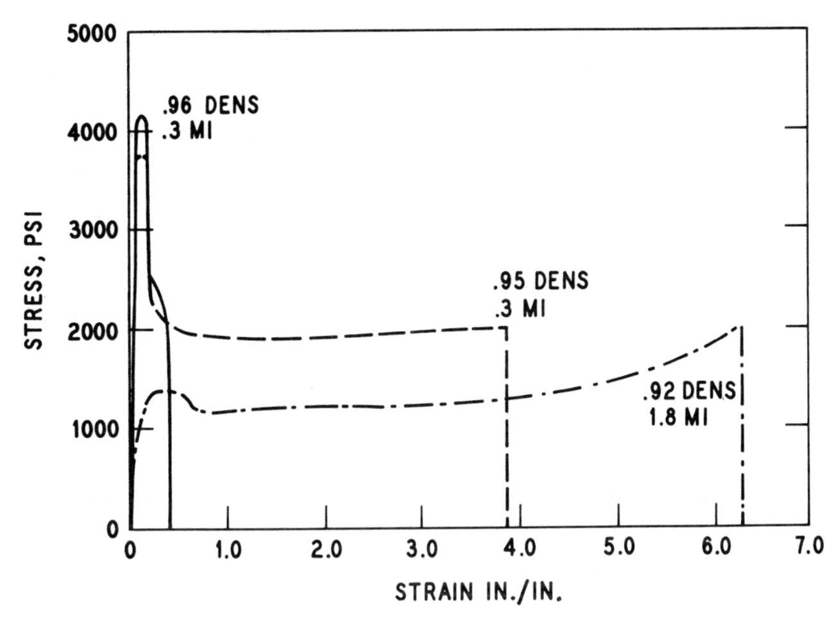

Fig. 20.3. Stress-strain curves for ethylene polymers.

with small increments of strain up to the point of yield, after which it drops precipitously and remains more or less constant during the cold drawing step that precedes final rupture. Polyethylenes of 0.96 density are free of branching and exhibit relatively little chain entanglement. This may account for the fact that stress levels decrease regularly with increasing strain and linear polymers do not exhibit the reinforcing and orientation effect immediately preceding the break point as do lower-density polyethylenes and other polymeric materials. The decrease in stress or stress decay exhibited by unbranched polyethylene is particularly noticeable at low rates of strain or at temperatures approaching the melting point of the polymer.

Stress-strain properties of polyethylenes, like other thermoplastic materials, are time-dependent. For example, the tensile strength of 0.96-density polyethylene at a strain rate of 20 in./in./min is above 4000 psi. However, if this polymer is subjected to a static load of 500 psi, it will break within a relatively few hours and at very low elongation. This type of failure is sometimes called "glassy" or "brittle" and is unexpected unless one is familiar with the load-bearing properties of the polymer. If molecular weight is increased substantially, some improvement in load-bearing properties is obtained and the failures become more ductile in nature. A more effective way of improving load bearing properties is to incorporate branches in the polymer molecule. It can be shown that three or four short chain branches per molecule will have a beneficial effect upon load-bearing properties and will reduce the time dependence of the stress-strain properties quite substantially. The effect of branching on polymer performance and application will be considered in more detail in the section on structural features.

Flexural Modulus

Flexural modulus values reflect the relative stiffness or rigidity of the polymer. In the determination of flexural modulus, a rectangular beam of polymer supported at both ends on a tensile testing machine is subjected to a deflecting force at midpoint.[26] Values for low-density polyethylenes are relatively low and range from 15,000 to 45,000 psi, while those for high-density polyethylenes fall in the range of 150,000 to 250,000 psi. Other tests may be used to determine the rigidity of polyethylenes; one of the most common is the stiffness-in-flexure test.[27] In this test, the force required to deflect a cantilever beam or strip of plastic is measured. Stiffness values for polyethylene are 10,000 to 30,000 psi for low-density materials and 100,000 to 175,000 for high-density polymers.

Impact Strength

Many laboratory tests have been devised for the determination of impact strength, but none of these is universally useful in predicting end-product performance. The most common impact value is derived from the Izod test,[28] which uses a pendulum machine to break a cantilever beam of the polyethylene. The beam may be notched or unnotched, and notches of different radii may be used to provide an indication of notch sensitivity. High-density polyethylenes yield Izod values of 1 to 5 ft lb/in. of notch, while low-density polyethylenes, particularly of high-molecular-weight and ductile materials, may give values above 10. Flexible, tough polymers yield meaningless values because the plastic beam simply deflects and does not break. More recently, techniques for determining impact in tension have been studied.[29] In these tests, the energy to break a tension impact specimen is determined by means of a pendulum machine that, in some instances, is a simple modification of the standard Izod device.

Because of the inadequacies of laboratory impact tests, it is common practice to determine impact strength directly on fabricated items such as injection-molded containers, polymer films, or plastic bags. Such direct testing is necessary because mold designs, extrusion rates, molding temperature, or other fabrication variables may affect product properties. A common technique for determining impact strength of injection-molded containers involves dropping a steel ball on the area of the item most susceptible to failure. Impact strength of films may be determined by a similar technique that requires dropping a missile or dart from various heights on a stretched sample of film.[30] Blow-molded items such as bottles and plastic bags often are filled with water, sand, polymer, or other material and subjected to drop tests.

Environmental Stress-cracking Resistance

Environmental stress-cracking (ESC) is the term applied to the premature failure of polyethylene in the presence of soap, detergent, water, or other active environment, usually under conditions of relatively high strain.[31] This phenomenon first was recognized in polyethylene-coated wire that often was lubricated with surface-active materials to facilitate installation in conduits. Under these conditions, polyethylenes that appeared to perform satisfactorily in the laboratory very rapidly developed severe cracks that propagated completely through to the conductor. A laboratory test, commonly known as the Bell ESC test, or bent-strip test, was devised by the Bell Telephone Laboratories and has been extensively studied and widely used throughout the world. In this test, a polymer specimen 1.5 in. × 0.5 in. × 0.125 in. is provided with a razor slit 0.75 inch long, 0.020 inch deep in the center, and parallel to the longest side. The specimen is bent in a "U" shape, held in a metal channel, and exposed to detergent or other environment at elevated temperature, commonly 50°C. The time for 5 out of 10 specimens to fail is reported as the F_{50} value. Stress-cracking resistance of polyethylenes is improved by increasing the molecular weight and by including short chain branching in the molecule. Very-low-molecular-weight materials may fail in this test in a matter of minutes, while high-molecular-weight materials or polymers of moderate molecular weight and moderate degrees of branching will survive for several thousand hours. This test is used widely for evaluating polyethylenes and predicting performance in wire coating applications and other uses.

Electrical Properties

Polyethylene has been of interest as an insulation material almost from the beginning of the development of the high-pressure process. It offers a very low dielectric constant as might be expected from the relatively low density of the polymer. High pressure polyethylene of 0.92 density exhibits a dielectric constant of 2.28, while the value for 0.96-density materials is about 2.35. Standard techniques for determination of dielectric constant, power factor, and other electrical properties are applicable to polyethylene.[32] In general, electrical properties deteriorate very rapidly if the polymer is oxidized,[33] and every effort must be made to provide efficient antioxidant systems if the original electrical and mechanical properties are to be maintained for an extended period of time.

Effect of Additives

Antioxidants must be incorporated into polyethylene prior to fabrication of test specimens, determination of melt viscosity, or investigation of structural features if useful data are to be obtained. This is important because virgin polyethylene, particularly in the molten condition, is rapidly oxidized to yield a variety of chain-scission products, network structures, and carbonyl or other oxygen-containing groups. In addition, anioxidants, and in some instances ultraviolet light stabilizers, are required if reasonable service life of end products is expected.

Antioxidants for polyethylene differ from those used in many rubber applications in that virtually all polyethylene antioxidants are food-grade materials that are generally colorless and odorless. Antioxidant levels in polyethylene are relatively low, usually in the range of 0.001 to 0.1%. One of the most widely used materials is 2,6-ditertiarybutyl-4-methyl phenol:

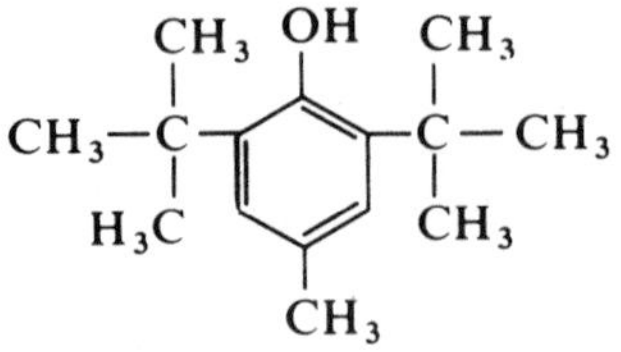

CH_3 OH CH_3
CH_3—C— —C—CH_3
H_3C CH_3
CH_3

For heavy-duty applications 4,4-thiobis(6-tert-butyl)-*m*-cresol often is used:

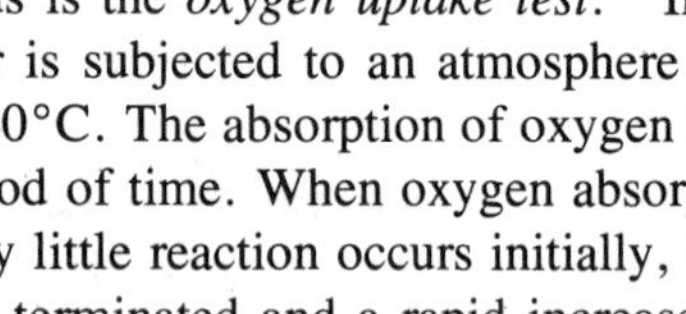

CH_3 CH_3
S
HO OH
$C(CH_3)_3$ $C(CH_3)_3$

One of the most common methods for determining oxidation resistance of polyethylene compounds is the *oxygen uptake test*.[34] In this determination, a sample molten polymer is subjected to an atmosphere of oxygen at elevated temperatures such as 140°C. The absorption of oxygen by the polymer sample is measured over a period of time. When oxygen absorption is plotted against time, it is seen that very little reaction occurs initially, but within a few hours this induction period is terminated and a rapid increase in oxygen absorption occurs. Antioxidants extend the induction period and reduce the subsequent rate of oxygen absorption. In the absence of oxygen, polyethylene is relatively stable to heat, and, in fact, is said to be more stable than polyisobutylene or SBR.[35] At temperatures above about 300°C, chain scission predominates, and low-molecular-weight polymers may be formed. However, polyethylene does not "unzip," and very little ethylene monomer is obtained by thermal degradation of the polymer.[36]

Polyethylene does not absorb ultraviolet light and one would expect it to be resistant to ultraviolet degradation. However, traces of carbonyl, resulting from oxidation, are always present and the carbonyl groups absorb light in the ultraviolet region below 3300 Å. It has been shown that carbonyl groups exercise a very important effect on the resistance of polyethylene to ultraviolet light; and, in fact, wavelengths outside of the carbonyl absorption band (2200 to 3200 Å) do not cause polymer degradation by photo-oxidation.[37]

Unprotected polyethylene degrades on outdoor exposure within a few months to a year depending upon the hours of exposure and the intensity of the sunlight. In applications such as pipe, wire coating, and cable sheathing for which a black color is acceptable, it is common practice to add 2 to 3% carbon black for protection against ultraviolet-light. Preferred carbon blacks are fine- to medium-particle blacks, although soft blacks at relatively high loadings will give some protection. In commercial practice, carbon black is dispersed in polyethylene as a concentrate that may contain 50% carbon black in low-density polyethylene or 25% black in high-density polymer. The concentrate may be "let down" with virgin polymer to give the required black loading. Experience has shown that a good dispersion of carbon black in the concentrate virtually ensures a good dispersion of black in the let down product. Because stability under ultraviolet light depends upon the quality of the dispersion as well as the particle size of the black, particular attention is paid to the composition and preparation of the concentrate when maximum resistance to outdoor exposure is required.

For nonblack polyethylene applications such as rope, fabric, or molded items designed for outdoor exposure, various types of UV-absorbing organic compounds are incorporated into polyethylene. Among the most common of these are benzophenone derivatives and salicylates, which are used in conjunction with antioxidants to provide outdoor life of perhaps 2 to 5 years. Although this extension represents a substantial improvement over additive-free polyethylene, it is substantially below the life expectancy of 25 years or more that may be obtained from the best carbon-black stocks. Typical Weather-Ometer data for UV stabilizers in high-density polyethylene at a level of 0.25% are presented in Table 20.9.

The condition of the polymer after various periods of exposure is indicated by a measurement of tensile strength and elongation from which the number of hours required to provide a 50% loss in properties may be determined. As in-

Table 20.9. Weather-Ometer Data for UV Stabilizers

Exposure of High Density Polyethylene in a Weather-Ometer	*Hours to 50% Loss in*	
	Elongation	*Tensile*
2,2′-Dihydroxy-4-octoxybenzophenone	1000	>1600
tert-Butylphenyl salicylate	400	800
Control–no UV stabilizer	125	300

dicated in the table, the most effective stabilizer is the benzophenone derivative. However, in some instances the lower-cost salicylate material may be preferred.

STRUCTURAL FEATURES

Branching

One of the most important structural features of the polyethylene molecule is the number and type of branches. Highly branched polymers such as low-density polyethylene containing 15 to 20 branches per 1000 carbon atoms are low in crystallinity and density as well as stiffness and hardness because the branches reduce chain regularity. Completely linear polyethylenes, on the other hand, exhibit maximum crystallinity, density, hardness, and rigidity. The branches on low-density polyethylene chains are mainly ethyl and butyl.[38] Similarly, some high-density polyethylenes contain short-chain branches which, in the case of Phillips process polymers are determined by the type of 1-olefin comonomer that is used. Propylene and 1-butene have been used extensively in this process to give methyl and ethyl branches, respectively. The effect of branches on the density of Phillips-type ethylene polymers is illustrated in Fig. 20.4. It can be seen that ethyl branches are more effective in reducing density than are methyl branches. Additional studies of this type have shown that longer branches are even more effective than ethyl, but the differences become less pronounced as the length of the branch increases.

With the advent of the Phillips process, completely unbranched polyolefins became available in quantity. This provided a base point for the study of branching effects that was previously unavailable. In recent years, it has become quite apparent that the first two or three branches on a polymer molecule exert a very marked effect on physical properties, particularly long-term load-bearing properties, stress-crack resistance, and cold-drawing performance. For example, an ethylene homopolymer of 0.96 density may exhibit an F_{50} value in the Bell ESC test of 60 hours, while an ethylene-1-butene copolymer of 0.95 density containing four or five ethyl branches per 1000 carbons will have an F_{50} value of 400 hours.[39] In the form of monofilament fiber, the copolymer will withstand a given static load, 20,000 psi for example, for 10 times as long as the homopolymer. When these materials are cold-drawn at temperatures below the crystalline melting point, the unbranched homopolymer necks down to smaller and smaller cross-section and reaches no limiting dimensions until it eventually breaks as a fine thread. On the other hand, the copolymer, when cold-drawn, necks down to a limiting dimension that is a predictable characteristic of the polymer under the conditions of test. At one time, it was believed that most polymers would cold-draw to a limiting dimension and would give a draw ratio of the order of 4/1. It is now known that this draw ratio is characteristic only

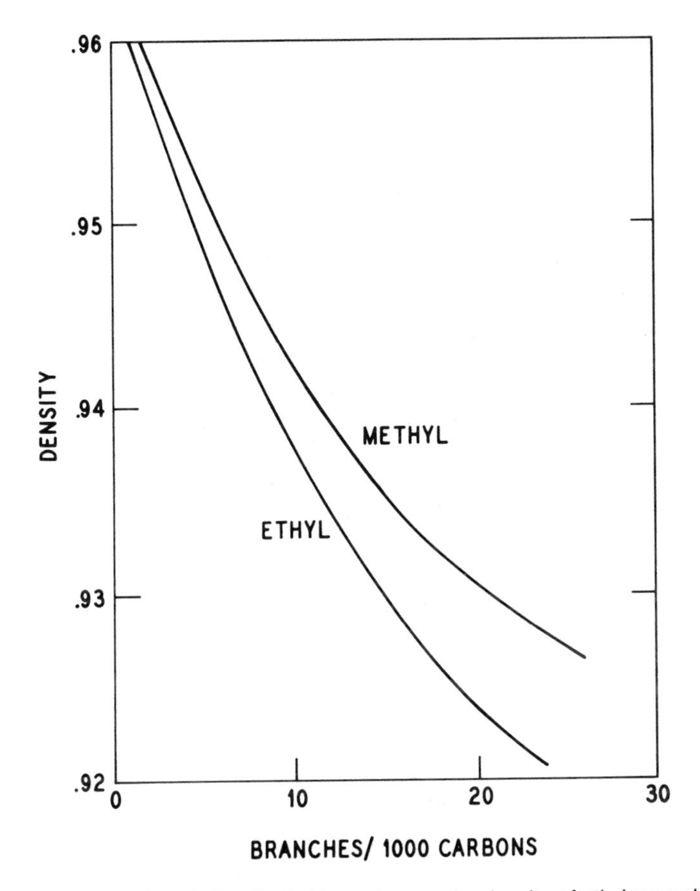

Fig. 20.4. Effect of methyl and ethyl branches on the density of ethylene polymers.

of highly branched materials of the type that were available prior to the development of truly linear polyethylenes.

Crystallinity

Single crystals of polyethylene when grown in dilute solution are found to be flat plates in which the chain direction of the molecule is perpendicular to the plate.[40] The crystal has distinct growth layers about 100 Å thick indicating that any one molecule is confined largely to a single layer and probably is folded. However, it is suggested that some molecules may participate in more than one layer, in which case the resistance to shear failure of the crystal under stress may be increased.

Although polyethylenes, particularly high-density, highly crystalline materials, are relatively opaque as a result of an effective scattering of light, this does not involve polymer crystals because they are smaller than the wavelength of visible light. Light-scattering and opacity result from larger structural units such as spherulites, which are formed by outward growth of crystals from nucleation points. It has been noted that spherulitic growth continues until the spherulite meets an obstruction such as another spherulite or a solid surface. Spherulites may be 0.05 mm in diameter and can be seen very clearly with a polarizing microscope under low power. In addition to internal diffraction of light, spherulitic structures often protrude from the surface of blown film in such a way that surface roughness from this source accounts for much of the observed opacity. The maximum growth of a spherulite requires a finite time; consequently, it is possible to limit spherulites' size and number by quenching, orienting, or applying other external or internal forces which will partially immobilize the polymer until it is completely chilled. In the quenched condition, spherulitic growth is nonexistent or so slow that it may be regarded as permanently inhibited.

As has been noted previously, there is a good correlation between crystallinity, density, and other physical properties of polyethylenes. For this reason, there is definite interest in not only providing polymers of inherently different levels of crystallinity but in modifying and controlling the crystallinity of molded and extruded products from a given type of polyethylene. Very little information has been published in this area, but it is known, for example, that crystallinity is reduced by quenching of polymer melts, and advantage of this fact is taken in the production of film by chill-roll and water-quench techniques. Nucleation or seeding techniques will produce many growth sites and yield a finer crystal structure. In other applications such as injection molding or blow molding, it is desirable to increase the rate at which polymer freezes in the mold, thereby reducing cycle time and increasing production rates.

Molecular Weight

Determinations of molecular weight and interpretations of data for polyethylene are similar to those for rubber and other polymers. One of the most common techniques for molecular weight determination is the measurement of dilute solution viscosity, usually in tetralin at 130°C. Dilute solution viscosities may be correlated with weight average molecular weight (M_w) as determined by light scattering[41] as follows:[42]

$$n = 3.78 \times 10^{-4} M_w^{0.72}$$

Commercial products for the most part exhibit dilute solution viscosities of 1 to 3 corresponding to weight-average molecular weights of 50,000 to 250,000.

Table 20.10. Dependence of Pipe Performance on Molecular Weight.

Polyethylene	*Predicted 20-year Maximum Hoop Stress at 88°C, psi*
High MW (500,000)	700
Med. MW (150,000)	10

$$\text{Hoop stress, psi} = \frac{\text{Pressure (psi)} \times \text{outside diameter (in.)}}{2 \times \text{wall thickness (in.)}}$$

Another commonly used molecular weight average is the number average (M_n), which may be determined by osmometry, or in the case of polyethylene, more satisfactorily by boiling-point elevation or ebulliometry.[43] Number-average molecular weights for commercial polyethylenes are in the range of 5000 to 20,000, which gives weight-average to number-average ratios of 10/1 to 15/1. This ratio has been used as an indication of the breadth of the molecular-weight distribution and indications are that most general-purpose polyethylenes exhibit relatively broad distributions. In recent years, however, numerous polymers of relatively narrow molecular-weight distribution having weight-to-number average ratios of perhaps 3/1 to 5/1 have become available for special uses, particularly for film-paper coating and injection molding.

In general, the higher the molecular weight, the better the physical properties of a polyethylene and the more reliable the performance of the end product. However, as molecular weight increases, processibility, as measured by extrusion rates, surface quality, or other fabrication limitations, decreases. It is therefore important to optimize molecular weight and fabrication rates in each instance. High-molecular-weight polymers are particularly useful for structural and load-bearing applications requiring long-term reliability. Pipe, wire coating, and monofilaments are applications in which these factors are important. In Table 20.10, for example, the effect of molecular weight on the performance of high-density polyethylene pipe is indicated. It may be seen that the predicted 20-year hoop stress, as indicated by long-term stress-life curves, is substantially better for the high-molecular-weight material.

Molecular-weight Distribution

Although polymers of very narrow molecular-weight distribution may be prepared by special fractionation or synthesis techniques, commercial polymers without exception contain molecules of widely varying size. In recent years, it has been shown that a single average-molecular-weight value such as M_w or M_n or even a combination of these is inadequate to describe fully the molecular-weight distribution of the polymer and to predict its performance in fabrication equipment and in end-use applications. One technique for studying molecular-weight distribution is *fractionation*, which may be done by methods developed for other polymers except that crystalline polyethylenes require the use of hot solvents. One type of apparatus used is the packed column containing glass beads, clay, or similar material on which polymer is deposited.[44] Elution of the polymer from the column is achieved by adding solvent/nonsolvent mixtures with a gradual increase in the ratio of solvent to remove higher-molecular-weight materials as the fractionation progresses. One important deficiency of this technique is that 5 or 10% of highest-molecular-weight material is not recovered, or at best is poorly fractionated. In many instances, this is the most important component, and lack of information on the high-molecular-weight fraction severely limits the value of the data.

The most widely used technique for fractionation of polymers is gel permeation chromatography.[45] Polymer solution is passed through a packed column containing a crosslinked microporous polystyrene. Large molecules are least readily absorbed, and therefore appear in the first fraction of effluent; the smallest molecules appear last. A correspondence between fraction number and molecular weight can be obtained by comparison with an established standard. Fig. 20.5 indicates molecular-weight distribution of two polyethylenes by gel-permeation chromatography.

An important aspect of molecular weight distributions that applies directly to fabrication problems is the dependence of the apparent melt viscosity upon shear rate. As indicated in Fig. 20.6, polymers of broad molecular-weight distribution exhibit the greatest reduction in apparent viscosity with increasing rate of shear. These materials consequently are of interest for fabrication steps such as filament extrusion, in which high shear rates are involved. As a corollary to this observation, it may be seen that broad-distribution polymers of relatively high molecular weight may be fabricated satisfactorily if high-shear-rate fabrication steps are involved. On the other hand, polymers of relatively narrow molecular-weight distribution exhibit low-melt viscosity at low shear and are of primary interest for an operation such as paper coating, in which low-shear flow and spontaneous ''knitting'' or ''healing'' of the melt on the substrate are required.

Crosslinking

In general, any uncontrolled crosslinking of polyethylene resulting from inadvertent oxidation during the production of fabrication steps is undesirable. It may be shown that low-order crosslinking leads to very substantial increases in melt viscosity without the concomitant improvement in physical properties normally expected from high-melt-viscosity (high-molecular-weight) material. In some instances, crosslinking may be done intentionally by addition of peroxide

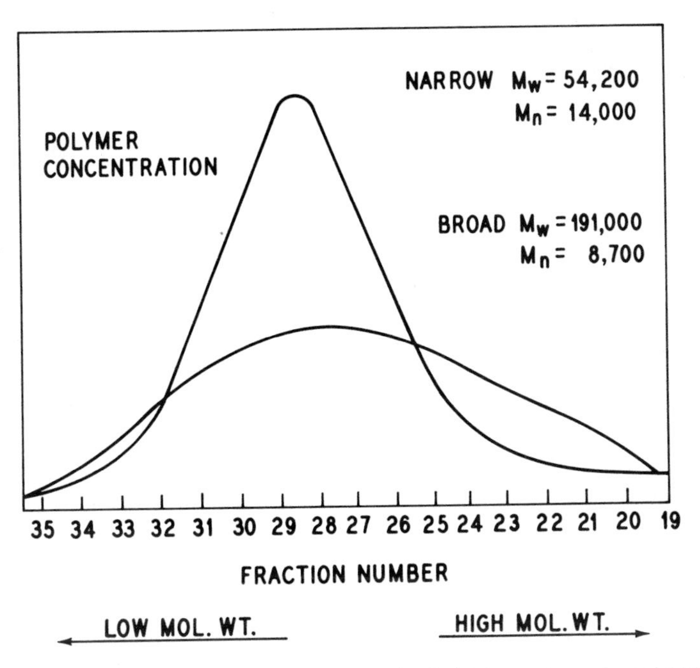

Fig. 20.5. Molecular-weight distribution of two polyethylenes as shown by gel permeation chromatography.

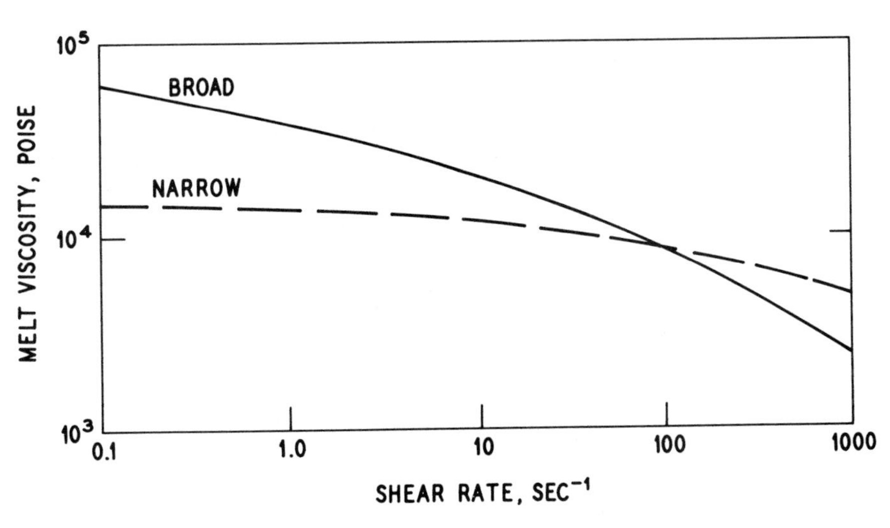

Fig. 20.6. Dependence of melt viscosity on shear rate for two polyethylenes of different molecular-weight distributions.

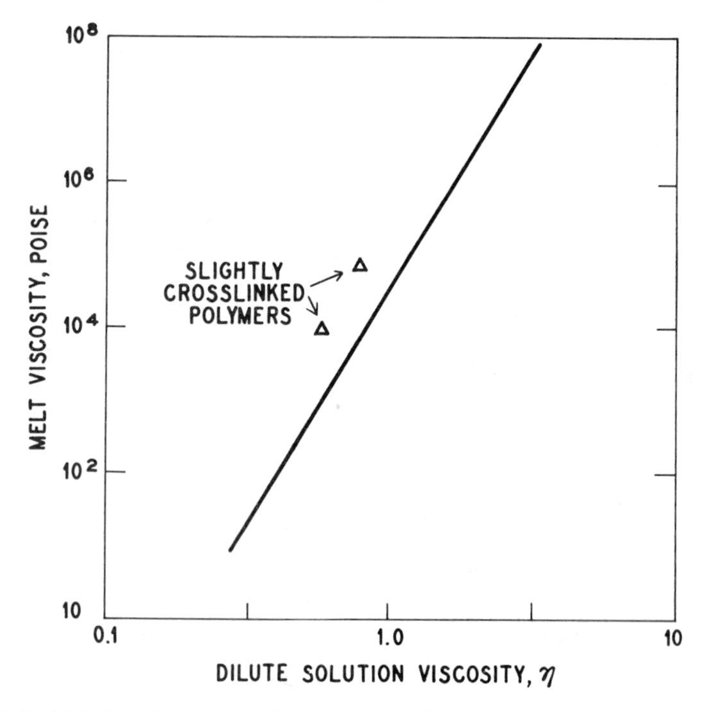

Fig. 20.7. Melt viscosity-solution viscosity relation for linear polyethylenes of variable molecular weight showing the effect of crosslinking.

or other curing agents to provide resistance to creep at high temperatures or to improve environmental stress-cracking resistance for special applications. Deliberate crosslinking to provide a vulcanized polyethylene compound will be considered in the section on applications. Because crosslinking exerts such a marked effect on melt viscosity, particularly when the viscosity is measured under conditions of low shear rate, it is possible to use rheological tools to detect the presence of crosslinks. One of the most effective devices for this purpose is a low-shear capillary melt viscometer that operates at shear rates in the range of 0.1 reciprocal seconds.[46] The detection of crosslinking requires development of data on low-shear melt viscosity and dilute-solution viscosity for uncrosslinked polyethylenes of various molecular weights (Fig. 20.7). When similar data are obtained on slightly crosslinked polyethylenes, it is seen that the low-shear melt viscosity is unusually high for a given solution viscosity. This technique also may be sensitive to the presence of long-chain branching, and, in fact, in the limiting case, low-order crosslinking and long-chain branching may be indistinguishable.

APPLICATIONS

Fabrication Equipment

In the polyethylene industry, extruders, Banbury mixers, and injection-molding machines are basic tools although roll mills and calenders are used in some instances. Extruders are used in polyethylene manufacture for pelletizing operations and in fabrication plants for wire and cable coating, production of pipe, sheet, and film, and for feeding blow-molding units in the production of bottles and other containers. Because polyethylene must be extruded at an elevated temperature and because of the inherent high-heat requirements for a crystalline material, polyethylene extruders must be capable of high-temperature operation and high-heat input rates. This is apparent from a study of a specific-heat-versus-temperature plot.[47]

The majority of polyethylene extruders are electrically heated although some oil-heated machines are in use.[48] In general, they offer length-to-diameter ratios of 20/1 or 30/1 compared to 8/1 or 10/1 for machines designed for rubber and noncrystalline plastics. The long-barrel design ensures adequate melting, mastication, and metering of the polymer, and is particularly important for high-speed extrusion of high-density polyethylenes. Screen packs are used to develop back pressure and for removing charred polymer and foreign particles. Stock temperatures range from 150 to 250°C or higher, depending upon the application. Screw diameters of 1 to 12 inches are common and outputs range from a few pounds per hour to several tons per hour. There is a wide variety of machine designs including twin-screw machines and single-screw types containing high-shear sections or other modifications for special blending, compounding, or homogenizing operations.

Injection-molding machines for polyethylene are similar to those used throughout the plastics industry. In the injection-molding process (Fig. 20.8), polymer is melted in the barrel of the machine and forced into a mold under high pressure. Most machines have hydraulically operated rams and mold clamps although some very small laboratory machines may have partially or totally mechanical systems. Ram pressures commonly are in the range of 10,000 to 30,000 psi with clamping forces of 30 to 2500 tons. Injection-molding machines are the most common devices for making housewares such as dishes, wastebaskets, and garbage cans as well as automobile and machine parts. In recent years, reciprocating-screw injection-molding machines have become quite common in the plastics industry.[49] These machines offer advantages in reduced cycle time, high plasticizing capacity, improved color dispersion, low stock temperatures, and less warpage or internal stress in the molded part.

Banbury mixers, roll mills, and calenders are similar to their counterparts in the rubber industry. Banbury mixers often are used for homogenizing polyeth-

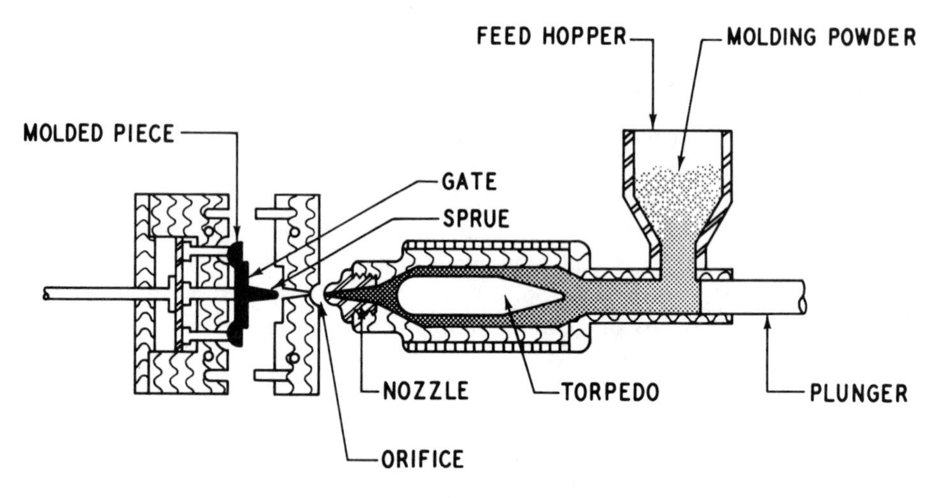

Fig. 20.8. Injection molding.

ylene and for incorporation of pigments or additives. Very frequently intensive mixers of the Banbury type are mounted above a pelletizing extruder that receives molten polymer from the mixer and converts it into finished resin. Roll mills are similar to rubber mills with provision for high-temperature operation. Small mills are very common in laboratories, but massive units of the type found in the rubber industry are rare. Calenders in the polyethylene industry are roughly similar to rubber calenders and are used for lamination and sheeting of polyethylene. In some instances, special types of film or tape are derived from calender operations. However, most laminating film or sheet production in the polyethylene industry involves extruder-fed units rather than calenders.

Polyethylene Consumption

Domestic polyethylene consumption in 1984 was 14.9 billion pounds, of which 5.2 billion was high-density, 7.3 billion was low-density and 2.4 billion was linear-low-density.[50]

Film and Sheeting

The term *film* is applied to material of less than 0.010-inch thickness, while *sheeting* is material above this dimension. There are three basic processes for the production of polyethylene film: tubular, chill-roll, and water-quench. In the tubular-film process, an extruder feeds molten polymer to a heated die designed to give a tubular extrudate. Air pressure is maintained inside the tube to provide a bubble of the desired size. The melt is cooled in air or by cooling rings, which provide an even distribution of forced air. The tubing is pulled from the die, collapsed by guide systems, and wound up on rolls. It may be used directly for making bags, or it may be slit to provide flat films up to 20 feet or more in width.

In the chill-roll process, the extruder feeds molten polymer to a slit-type die from which the melt extrudes as a thin sheet or web. The die is placed within a few inches of the polished chill roll and the polymers pulled under and over a series of rolls onto a wind-up roll. Speeds of up to 500 to 1000 feet per minute are common and film clarity is outstanding. In recent years, the advantage for this process in film clarity has been reduced somewhat by the development of new resins that provide excellent optical properties to the product of the tubular film process.

In general, fresh polyethylene-film surfaces are nonreceptive to printing ink and must be treated by flame or electrical discharge to promote a mild oxidation of the surface. This operation may be done in-line while the film is being produced, or it may be part of the subsequent printing step. Very often, antiblock and slip additives are incorporated in polyethylene-film resins to reduce adhesion of film surfaces and permit high-speed operation in packaging equipment. Fatty-acid amides and finely divided silica are among the most common additives for this purpose.[51]

Low-density tubular films offer good elongation and tear strength as well as relatively low haze. Materials of this type have been the utility films of the construction and agricultural industries. Medium-density polyethylene films derived from chill-roll techniques are used widely in the packaging of dry goods, bakery products, and other consumer items because of their excellent balance of properties and low level of haze. High-density water-quench film also offers good optical properties and is used in overwrap packaging where stiffness, low-moisture transmission, and the tear-tape feature are desired. High-density tubular film exhibits high haze and relatively low tear strength, but recent rubber-modified films of this type offer outstanding puncture and tear resistance and are used for heavy-duty industrial packaging.

In the production of polyethylene sheet, the most common practice involves extrusion of a molten web on a polished chill roll. This system differs from chill-roll film operations in that the chill roll is heated to a temperature slightly below the freezing point of the polymer and linear speeds are very much lower. Polyethylene sheet is used for production of trays, boats, chairs, containers, and many other items via thermoforming techniques.[52,53,54]

In a thermoforming operation, the polymer sheet is heated above the melting point, drawn into a mold, or draped over a mold that contains numerous small holes. A vacuum is applied to propel the molten sheet against the sides of the mold. Many variations of the basic vacuum-forming idea have been developed to provide low-cost production of large or complex items. Thermoforming also is of interest when the demand for an item is too low to justify investment in the more expensive injection-molding or blow-molding equipment.

Injection Molding

Polyethylene injection-molding technology originally was based on that developed for polystyrene, which continues to be an important injection-molding resin. Injection-molding machines commonly are rated for capacity on the basis of the number of ounces of polystyrene that can be injected per shot. Machines of 1 to 400 ounces capacity or greater are available. In general, equipment of this type is relatively expensive with machine prices falling in the range of $5000 to $100,000 or more, and molds costing $500 to $10,000 each. For this reason, injection molding is economical only for long continuous runs in which thousands of identical parts are produced. The choice of polyethylene for injection-molding operations depends upon performance and price considerations. High-density resins are gradually gaining acceptance in this area because they provide greater stiffness or permit the use of thinner walls. However, low-density polyethylenes continue to be used in great volume, particularly for houseware and for general-purpose containers. Colored injection-molded items may be obtained by adding pigment as a dry blend on virgin pellets or by including concentrates of pigment and polymer along with the natural pellets fed to the machine. If the ultimate in color dispersion is demanded, a color compound prepared by "letting-down" a concentrate with natural polymer in a Banbury mixer or extruder is used.

Injection-molded items are identified by the presence of a tiny button of polymer in the "gate" area of the part. This is the point at which the polymer enters the mold cavity and very often is located at the midpoint of the bottom area in items such as dishes, pails, or other containers; vacuum-formed and blow-molded items do not have this characteristic. Gates also may be located at the edge of a part and occasionally more than one gate is used. Mold design is a sophisticated art. Some designs that require close temperature control at different levels for specific areas in the mold or contain several cavities equipped for independent, sequential, or simultaneous operation can be quite complex.

Blow Molding

Blow molding is a major application for high-density polyethylene. The early technology was based on that of glass,[55,56] and some terms, such as "parison," which denotes the molten tube from which the item is blown, are used in both the glass and plastics industries. In recent years, most major bottle, can, and container manufacturers have entered the blow-molding field. In addition, some

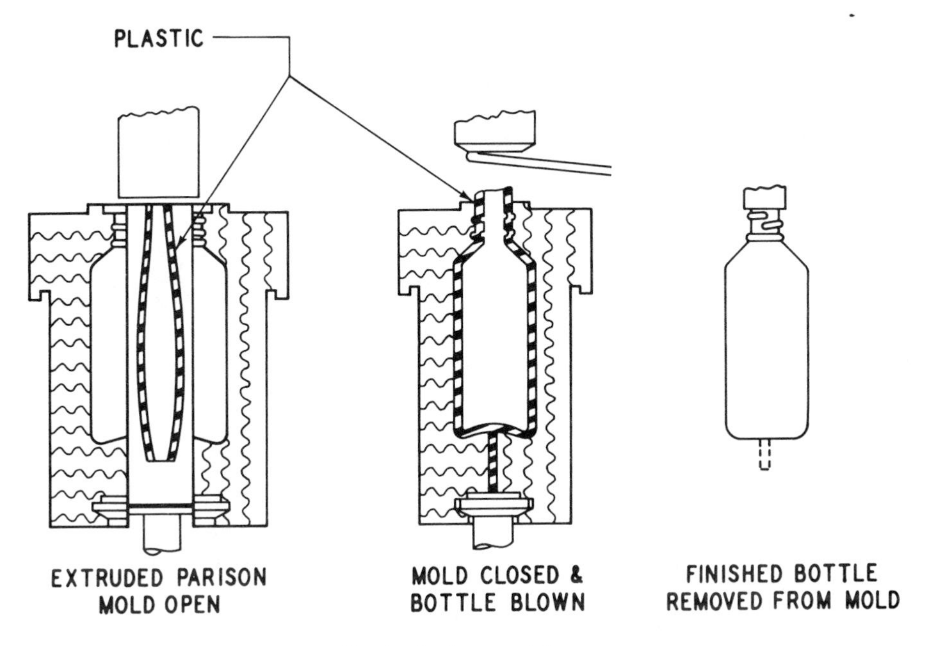

Fig. 20.9. Blow molding.

equipment manufacturers, resin suppliers, and others are interested in special fields of blow molding such as the production of large tanks, bag and box containers, large toy cars and animals, beverage cases, and the like.

In principle, all blow-molding systems require that a molten tube or parison of polymer be delivered to an open mold. The mold closes on the parison and air pressure is applied to force it against the inside surfaces. The solidified part is ejected from the mold and trimmed if necessary (Fig. 20.9). Although injection-molding machines may be used to supply the molten parison, most producers utilize extruders for this purpose. In some designs, the extruder feeds an accumulator which in turn forms the parison. This permits continuous operation of the extruder even though the parison extrusion is intermittent. In other designs, several molds are placed on a rotary table and are automatically presented under the extrusion head to receive the parison at the proper time. Continuous extrusion of tubing to be received by molds on a belt or wheel is another technique that is used widely in the United States. Blow molding is one of the fastest growing segments of the polyethylene industry. Important products derived from blow-molding operations include bottles for detergent and bleach packaging, 5- to 50-gallon containers for home and industry, automobile ducts, and machine parts.

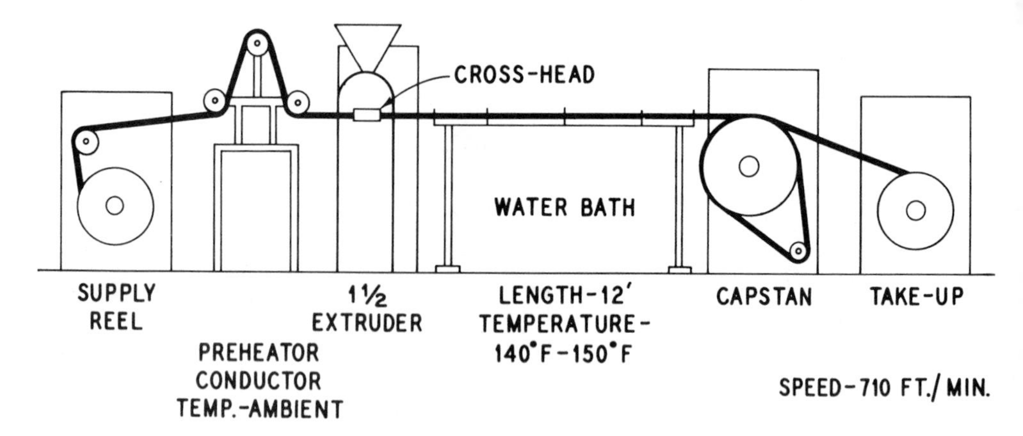

Fig. 20.10. Wire coating.

Wire and Cable

One of the first applications for polyethylene involved the use of low-density pilot-plant material for a 1-mile length of submarine cable in England in July 1939.[54] Polyethylene continues to be a material of choice for this and similar applications because of its low water absorption and low power factor. Power cable is the most rapidly growing insulation application.[57] Cables handling up to 15 KV have been in service for 15 years and polyethylene is the only proved material for buried cathodic protection cable.

Polyethylene wire and cable coatings are applied by means of crosshead dies that are fed by extruders usually placed at right angles to the path of the wire[58] (Fig. 20.10). The wire may be cleaned, preheated, and straightened before it enters the die. It is continuously encapsulated by molten polymer at linear speeds up to 2000 feet per minute. As it emits from the die, the wire is cooled and tested electrically for coating integrity. Cable sheathing involves the same principle as wire coating, but diameter and thickness are greater and linear rates are lower.

In recent years, considerable interest has developed in the use of crosslinked polyethylene for wire-coating applications.[59,60] Crosslinked polymers offer advantages over noncrosslinked materials in resistance to stress-cracking and to creep particularly at temperatures above the crystalline melting point. In many instances, high levels of carbon black are included in the formulation to provide weather resistance and reduce costs. A typical formulation for a carbon-black-loaded polyethylene stock with organic peroxide crosslinking agent follows:

Polyethylene	100
Carbon black	25–300
Peroxide	2

Cure:	
Temperature, °C	175–200
Time, min	10–0.5

According to Dannenberg et al.,[61] crosslinking occurs between the polymer and carbon black as well as between polymer molecules.

Pipe

Low-density polyethylene continues to be an important raw material for the production of plastic pipe, but in recent years the trend has been toward the use of medium- and high-density polyethylenes for this application. The basis of choice is largely an economic one in which the lower cost of the low-density polymer is weighed against the higher performance of the high-density resin that permits use of thinner walls in equivalent grades of pipe. In any case, the top grades of pipe exhibiting the highest pressure ratings and greatest reliability are produced from high-density polymers that for this purpose must be of very high molecular weight. The major outlet for polyethylene pipe is in farm and home outdoor applications such as sprinkler systems or other cold-water service. Polyethylene pipe of 1- to 2-inch diameter is the most popular and is sold in coils of 100 feet or more in length. Pipe of diameters above about 2 inches is too rigid to coil easily and is sold in 20- to 40-foot lengths for use in industrial applications, drain lines, and low-pressure service.

There are five important techniques for producing polyethylene pipe: external-sizing tube, sizing rings, internal sizing, positive-pressure dies, and biaxial orientation. In each case, the molten polymer is forced through a suitable die by means of an extruder.

1. In the external-sizing technique[62] the molten polymer is forced against the cold walls of a sizing tube by internal air pressure or by the application of a vacuum. The pipe is pulled from the sizing tube into a water bath and onto a reel. This is one of the preferred methods for extrusion of high-density polyethylene.
2. In the sizing-ring technique, pipe is extruded into a water bath and pulled through one or more sizing rings to obtain the desired dimensions.
3. The internal-sizing technique is used extensively for medium- and low-density polyethylene. The extruded pipe is pulled over a mandrel fitted with a sizing knob on the end to provide proper internal dimensions. External dimensions and wall thickness are controlled by extrusion rate and take-up speed.
4. High-quality pipe is prepared from high-molecular-weight, high-density polyethylene by a positive-pressure technique in which the polymer is

pushed through the die by the extruder rather than pulled through dies or sizing equipment by the take-up unit.[63] In this process, both the external and internal walls of the pipe are in contact with cold surfaces and the finished pipe exhibits a very glossy appearance. In one of the preferred methods of operation, a small amount of water is used as a lubricant.

5. High-performance pipe may be prepared, preferably from high-density polyethylene, by orienting the material under carefully controlled conditions.[64] In this process, a thick-walled, small-diameter pipe is extruded in the conventional way and brought to the required orientation temperature, which is usually slightly below the melting point. It is drawn down, then blown to several times its original circumference. The oriented pipe prepared in this way offers working pressures up to twice that of nonoriented pipe from the same material.

Because plastic pipe is expected to offer predictable performance over many years of service, it is necessary to develop adequate accelerated testing techniques. In general, this is accomplished by subjecting short lengths of pipe to a wide range of pressures, usually in water baths at temperatures of 20 to 100°C. The time to fail at a given pressure and temperature is recorded in the form of stress-life curves.[65,66,67] By using various established techniques, one may estimate the pressure rating for a life expectancy of 20 years or more with reasonable certainty.

Coatings

The extrusion coating of polyethylene on paper or other substrate is an important segment of the polyethylene industry. In commercial practice, an extruder is used to feed polymer through a slit die somewhat similar to that required for chill-roll film (Fig. 20.11).[68] The polymer melt is laid down continuously on the substrate and then chilled by contact with a polished roll. Linear speeds of 300 to 1000 feet per minute are common. Coatings may be 0.25 to 2 mils in thickness. Important substrates for polyethylene coating include kraft paper for multiwalled bags and paper board for milk cartons and other food containers.[69] When applied to cellophane, polyethylene provides heat sealability, tear resistance, moisture-barrier properties, and improved resistance to shelf-ageing. Polyethylenes of very low molecular weight may be applied with roll-coating machines of the type used for making wax paper. Other techniques involve the use of polyethylene emulsions or micro-pulverized powders. Automobile carpet backing that can be molded to fit the contours of the floor is an important new use for micro-pulverized, low-density polyethylene.

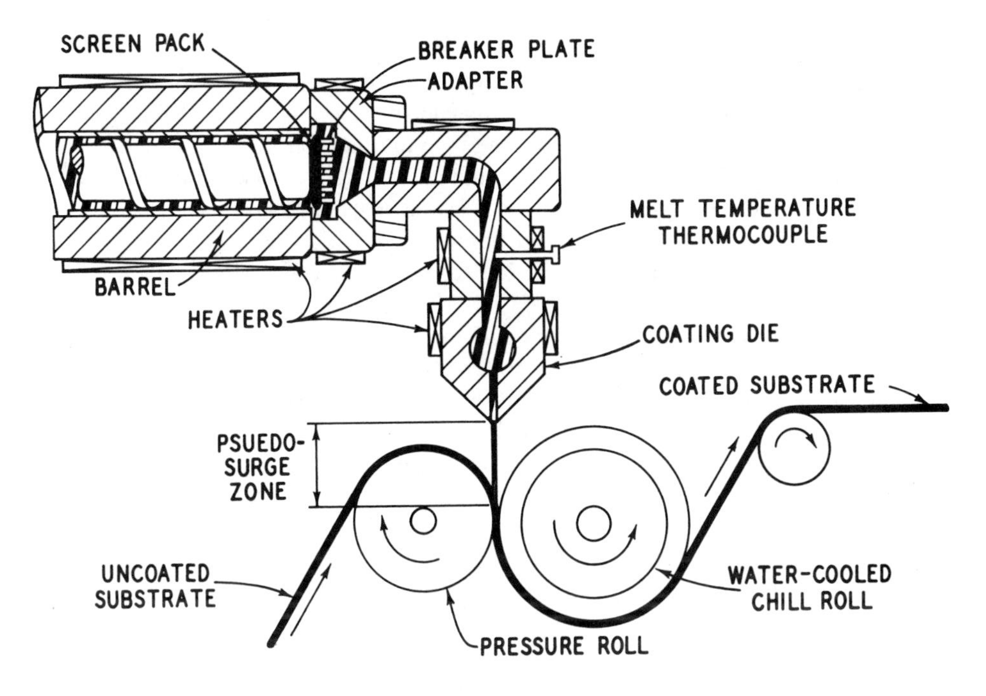

Fig. 20.11. Extrusion coating.

Filaments

One of the first markets for high-density polyethylene was the production of monofilaments for rope, automobile seat covers, and similar applications. At one time this represented a major market for high-density polyethylene, but growth has been slow and the volume for monofilaments has been surpassed many times over by demands for blow-molding and injection-molding resins. Polypropylene has displaced polyethylene for polyolefin fibers and filaments. In the production of monofilament, several strands of molten polymer are extruded through an orifice plate and quenched in a water bath. At this point, the filament has a tensile strength of only 2000 to 4000 psi. In the next step, the strands are wrapped on godet rolls, heated in a steam bath, and subjected to draw ratios of 10/1 to 12/1. The drawing operation increases tensile strength to 50,000 to 100,000 psi.

Powder Molding

Powder molding of polyethylene consumed 265 million pounds in 1984.[50] The early techniques for powder molding of polyethylene included the Engel process,[70] the centrifugal processes of Heisler,[71] and rotational molding. In the Engel process, powder is placed in a cold mold that then is heated to 300 to 400°C. A portion of the polyethylene powder fuses on the walls of the mold. The unfused polymer is dumped out and the mold is reheated to fuse all of the remaining resin. When the mold cools, the polyethylene shrinks from the walls and is removed with ease. Large items such as boats and tanks have been prepared by this process, particularly in Europe.

In the centrifugal process, the molds are preheated and partially filled with polyethylene. The mold is rotated on its long axis to give an even distribution of polymer on the walls. This process is not widely used, but it is convenient for making objects that are open at both ends such as pipe.

Rotational molding techniques have been used for years for molding vinyl plastisols. In this system, the mold is loaded with the exact amount of polymer to be used. It is then rotated simultaneously around two axes at right angles to each other while it is being heated in an oven or by infrared lamps. After all of the polymer is fused to the walls of the mold, the mold is cooled and then opened to release the polyethylene product. In general, powder molding is of interest for (1) short runs or prototypes because of the very low investment involved, and (2) large items such as 30-gallon tanks, or large tubs or vats, which are impractical to mold by any other technique. Commercial polyethylene molding powders are either low- or high-density materials that are mechanically ground to give particles in the 100- to 300-micron range.

Modified Polyethylenes and Polyethylene Compounds

Very early in the development of low-density polyethylene, it was found that substantial improvements in low-temperature flexibility and environmental stress-cracking resistance could be obtained by the addition of 10 to 15% polyisobutylene.[73] Currently these requirements are met by ethylene-1-butene copolymers and high-molecular-weight polymers. Polyethylene is blended with ethylene-propylene rubber, with or without crosslinking, for use in automotive gaskets, weatherstripping, and other injection-molded or extruded products.[74] High-density polyethylenes containing relatively high loadings of special elastomers are being used for industrial bags and other heavy-duty film applications.

Polyethylene also has been suggested as an additive for rubber where it functions as a reinforcing agent to reduce cold-flow and increase hardness.[75] Low-molecular-weight polyethylene has been suggested as a lubricant and processing aid for butyl, natural rubber, SBR, neoprene, and chlorosulfonated polyethylene. In natural-rubber tread stocks, it is said to improve tear resistance and processibility and increase resistance to crack growth and abrasion. When incorporated in neoprene and chlorosulfonated polyethylene, it reduces the tendency to stick to mill rolls.[76]

Polyethylene may be chlorinated in the dry state or in solutions of carbon tetrachloride or other chlorinated solvents. At levels of 20 to 35% chlorine, the product from high-density polyethylene resembles plasticized polyvinyl chloride. At 50 to 65% chlorine, it offers many of the properties of rigid polyvinyl chloride. High-density polyethylenes containing 40 to 50% chlorine have been suggested as additives for improving the impact strength of rigid polyvinyl chloride.[77]

Low-density polyethylenes are chlorinated in the presence of sulfur dioxide to yield chlorosulfonated products that typically contain 27.5% chlorine and 1.5% sulfur.[78,79,80] The chlorosulfonated polymer may be cured in systems containing magnesium oxide or other metal oxides, litharge, or cribasic lead maleate. The vulcanizates, which are resistant to ozone and oxygen, provide good abrasion resistance, good flex life, and resistance to crack growth. Compounds of this type are also resistant to most chemicals with the exception of aromatic and chlorinated solvents.

REFERENCES

1. Bamberger and Tschirner, *Berichte* **33,** 955–959 (1900).
2. Pechman, *Berichte* **31,** 2643 (1898).
3. Meerewein and Burneleit, *Berichte* **61,** 1840 (1928).
4. Swallow, *Polythene*, Renfrew and Morgan, eds. Interscience Publishers, New York, 1960, p. 3.
5. Ziegler et al., *Angew. Chem.* **67,** 541 (1955).
6. *Chemical and Engineering News*, June 10, 1985.
7. Dobson, *Polythene*, Renfrew and Morgan, eds., Interscience Publishers, New York, 1960, p. 13.
8. Hines, Bryant, Larchar, and Pease, *Ind. Eng. Chem.* **49,** 1071 (1957).
9. Davison and Erdmon, U.S. Pat. 2,839,515.
10. Perrin, Fawcett, Paton, and Williams, U.S. Pat. 2,200,429.
11. Brubaker, Coffman, and Hoehn, *J. Am. Chem. Soc.* **74,** 1509 (1952).
12. Hogan and Banks, U.S. Pat. 2,825,721.
13. Clark, Hogan, Banks, and Lanning, *Ind. Eng. Chem.* **48,** 1152–5 (1956).
14. Ziegler, Belgian Pat. 533,632.
15. Goppel and Howard, *Polythene*, Renfrew and Morgan, eds., Interscience Publishers, New York, 1960, pp. 17–27.
16. British Pat. 779,540 (1957).
17. Belgian Pat. 549,910 (1958).
18. Catalysis Reviews, *Science and Engineering* **26,** No. 3, 4 (1984).
19. *Modern Plastics*, Jan. 1985.
20. Smith, *Ind. Eng. Chem.* **48,** 1161 (1956).
21. Bayer, Wiley, and Spencer, *J. Polymer Sci.* **1,** 249 (1946).
22. ASTM D-792-60T.
23. Jones and Boeke, *Ind. Eng. Chem.* **48,** 1155–60 (1956).
24. ASTM D-1238-57T.
25. Harban and McGlamery, *Materials Research & Standards* **3,** (11) (1963).
26. ASTM D-790-59T.

27. ASTM D-747-58T.
28. ASTM D-256-56.
29. ASTM D-1822-61T.
30. ASTM 1709-59T.
31. DeCoste, Malm, and Wallder, *Ind. Eng. Chem.* **43,** 117 (1951).
32. ASTM D-150-59T.
33. Biggs and Hawkins, *Modern Plastics* **31,** 121 (1953).
34. Hawkins, Lanza, Loeffler, Matreyek, and Winslow, *J. Appl. Polym. Sci.* **I,** 43–9 (1959).
35. Madorsky, Straus, Thompson, and Williamson, *J. Polym. Sci.* **4,** 639 (1949).
36. Oakes and Richards, *J. Chem. Soc.*, 2929 (1949).
37. Haywood, *Polythene*, Renfrew and Morgan, eds., Interscience Publishers, New York, 1960, pp. 132–133.
38. Dole, Keeling, and Rose, *J. Am. Chem. Soc.* **76,** 4304 (1954); Miller and Willis reported by Willbourn, *J. Polymer Sci.* **34,** 569 (1959); Boyd, Voter, and Bryant, *Abstracts*, 132nd National ACS meeting, Sept. 1957, p. 8T.
39. Pritchard, McGlamery, and Boeke, *Modern Plastics* **37,** 132 (Oct. 1959).
40. Till, *J. Polym. Sci.* **24,** 301 (1957); Keller, *Phil. Mag.* **2,** 1171 (1957); Fischer, *Z. Naturforsch* **12a,** 753 (1957).
41. Debye, *J. Appl. Phys.* **15,** 388 (1944); *J. Phys. Chem.* **51,** 18 (1947).
42. Stacy and Arnett, *J. Poly. Sci.*, Part A **2,** pp. (1964) 167–179.
43. Arnett, Smith, and Buell, *J. Poly. Sci.*, Part A **1,** 2753 (1963).
44. Guillet, Combs, Colner, and Slonaker, *J. Polym. Sci.* **47,** 307–320 (1960).
45. Moore, *J. Poly. Sci.* Part A **2,** 835 (1964).
46. McGlamery and Harban, *SPE Eighteenth Annual Technical Conference*, Vol. 8, 1962.
47. Wash, *Modern Plastics Encyclopedia* **40,** 228 (1963).
48. Kennaway and Weeks, *Polythene*, Renfrew and Morgan, eds., Interscience Publishers, New York, 1960, pp. 437–462.
49. Elliot, *Modern Plastics Encyclopedia*, 419–430 (1969–70).
50. *Modern Plastics*, **62,** No. 1 (Jan. 1985).
51. Barker, Lewis, and Happoldt, U.S. Patent 2,770,608.
52. Santer, *Polythene*, Renfrew and Morgan, eds., Interscience Publishers, New York, 1960, pp. 615–622.
53. *Modern Plastics* **47,** 67–69 (June 1970).
54. Doyle and Allison, *SPE Journal* **16,** No. 3 (March 1960).
55. Wood, *Polythene*, Renfrew and Morgan, eds., Interscience Publishers, New York, 1960, pp. 571–579.
56. Ferngren, U.S. Pat. 2,128,239 (1948); 2,175,053 (1939); 2,175,054 (1939).
57. Wood, *Modern Plastics Encyclopedia* **40,** 223 (1963).
58. *Modern Plastics Encyclopedia* **40,** 765 (1963).
59. Precopio and Gilbert, U.S. Pat. 2,888,424.
60. Ivett, U.S. Pat. 2,826,570.
61. Dannenberg, Jordan, and Cole, *J. Polym. Sci.* **31,** 127 (1958).
62. Croley and Doyle, *Plastics Technology* **4,** 717 (Aug. 1958).
63. U.S. Pat. 3,066,356.
64. Gloor, *Modern Plastics*, **38,** 111 (Nov. 1960).
65. Richard, Diedrich, and Gaube, Paper, ACS Meeting, New York, 1957.
66. Gloor, *Modern Plastics* **36,** 144 (1958).
67. Richard and Ewald, *Plastics* **36,** 153 (1959).
68. Wilbert and Grant, *Polythene*, Renfrew and Morgan, eds., Interscience Publishers, New York, 1960, pp. 585–597.
69. *Modern Plastics Encyclopedia*, 181 (1969–70).

70. Engel, U.S. Pat. 2,915,788.
71. Heisler, U.S. Pat. 2,736,925.
72. Zimmerman and Johnson, *Modern Plastics Encyclopedia* **40,** 717–720 (1963); Zimmerman, *British Plastics* **36,** 84 (1963).
73. Williams, British Pat. 514,687 (1939).
74. *Modern Plastics Encyclopedia*, **61,** No. 10A (1984–85).
75. Railsback and Wheat, *Rubber Age* **82,** 664–671 (Jan. 1958).
76. Bulifant, *Rubber Age* **82,** 89 (1957).
77. Frey, *Kunststoffe* **49,** 50 (1959).
78. McQueen, U.S. Pat. 2,212,786.
79. AcAlvey, U.S. Pat. 2,586,363.
80. Warner, *Rubber Age* **71,** 205–221 (1952).

Appendix

RUBBER INFORMATION RESOURCES

RUTH MURRAY
Rubber Division ACS Library
The University of Akron
Akron, Ohio

The body of rubber literature continues to increase as new materials enter the picture and as old and new ones find new applications. It is difficult for the expert in the field to keep track of its many developments, while to one not so skilled in the art as the engineer or designer, the rubber literature can be obscure.

The references listed below are intended only to highlight the major sources of information. Books review the known and basic information; handbooks, dictionaries and directories point out the specific facts and data; periodicals keep one up-to-date on the relatively late information. The newest information is usually found in conference proceedings.

Forming a little classification of their own are the books which cannot be included in the technical listing per se but which will appeal to the reader interested in the history of the rubber industry and of rubber technology.

GENERAL BOOKS

K. W. Allen, ed., *Adhesion 1–9*, Applied Science Publishers, London, 1977–1985.

The books contain papers that were given at the Annual Conferences on Adhesion and Adhesives at the City University in London.

G. Alliger and I. J. Sjothun, *Vulcanization of Elastomers*, Reinhold, New York, 1964.

A series of lectures presented by the Akron Rubber Group, edited and published in book form.

J. C. Arthur, Jr., ed., *Polymers for Fibers and Elastomers*, American Chemical Society, Washington, DC 20036, 1984.

A state-of-the-art volume in fiber technology, based on a symposium sponsored by the Macromolecular Secretariat at the 186th Meeting of the ACS, 1983.

R. O. Babbit, ed., *The Vanderbilt Rubber Handbook*, 12th ed., R. T. Vanderbilt Co., Norwalk, CT 06855, 1978.

A practical manual describing available elastomers, explaining basic principles of compounding, and giving frequently used physical test methods.

L. Bateman, ed., *The Chemistry and Physics of Rubber-like Substances*, Maclaren & Sons, London, 1963.

A compilation of the main studies undertaken by the Natural Rubber Producers' Research Association. A book of classic scientific information on natural rubber.

F. W. Billmeyer, Jr., *Textbook of Polymer Science*, 3d Ed., John Wiley & Sons, New York, 1984.

An up-to-date text dealing with basic principles of polymer science and including descriptions of new polymer materials and composite materials.

D. C. Blackley, *High Polymer Latices: Their Science and Technology*, Maclaren & Sons, London, 1966.

A thorough coverage of polymer latices, what they are, and the uses to which they may be put.

D. C. Blackley, *Synthetic Rubbers: Their Chemistry and Technology*, Applied Science Publishers, London, 1983.

An up-to-date survey of the principal types of synthetic rubber that have been and are currently available.

C. M. Blow and C. Hepburn, *Rubber Technology and Manufacture*, 2d Ed., Butterworth Scientific, London, 1982.

An up-to-date guide for students, beginners in the rubber manufacturing and associated supplying industries, and users of rubber products in other industries.

R. P. Brown, *Physical Testing of Rubbers*, Applied Science Publishers, London, 1979.

A handbook of testing methods for workers in the field and for engineers and others who specify or use rubber products.

J. A. Brydson, *Plastics Materials*, 4th Ed., Butterworth Scientific, London, 1982.

Preparation, structure, and properties of many classes of plastics, including diene rubbers.

J. A. Brydson, *Rubber Chemistry*, Applied Science Publishers, London, 1978.

The chemistry underlying rubbery polymeric materials with particular emphasis on rubber technology.

K. O. Calvert, *Polymer Latices and Their Applications*, Applied Science Publishers, London, 1982.

An up-to-date look at latex technology compiled by authors from industry.

S. K. Clark, ed., *Mechanics of Pneumatic Tires*, 2d Ed., Superintendent of Documents, U.S. Government Printing Office, Washington, DC 20402, 1981.

A complete treatise on tire mechanics, including rubber and textile properties, friction, material properties, tire design and construction, skid and handling.

C. C. Davis and J. T. Blake, eds., *The Chemistry and Technology of Rubber*, Reinhold, New York, 1937.

Originally the basic, comprehensive reference volume on rubber; now of historical interest.

J.-B. Donnet and A. Voet, *Carbon Black: Physics, Chemistry, and Elastomer Reinforcement*, Marcel Dekker, New York, 1976.

In one volume, the most important contributions to the physics and chemistry of carbon black and its manufacture.

F. R. Eirich, ed., *Science and Technology of Rubber*, Academic Press, New York, 1978.

A postgraduate text, covering the most important aspects of rubber science and some technology.

C. W. Evans, ed., *Developments in Rubber and Rubber Composites 1–3*, Applied Science Publishers, London, 1980–1983.

Chapters on topics relating to rubber and rubber compounding and processing by different authors.

C. W. Evans, *Powdered and Particulate Rubber Technology*, Applied Science Publishers, London, 1978.

A guide and introduction to the subject.

C. W. Evans, *Practical Rubber Compounding and Processing*, Applied Science Publishers, London, 1981.

Discussion of all phases of rubber manufacture, from raw materials to typical recipes, mixing and equipment, processing and vulcanization steps, to the finished product.

R. A. Fleming and D. I. Livingston, eds., *Tire Reinforcement and Tire Performance*, American Society for Testing and Materials, Philadelphia, PA 19103, 1979.

Proceedings of a symposium sponsored by ASTM Committees D13 and F9, Montrose, OH, October 1978.

P. K. Freakley and A. R. Payne, *Theory and Practice of Engineering with Rubber*, Applied Science Publishers, London, 1978.

Of interest to technologists and to scientists, covering properties of elastomeric materials, theory and practices of design and many applications.

P. K. Freakley, *Rubber Processing and Production Organization*, Plenum, New York, 1985.

A guide to the analysis and synthesis of manufacturing systems in rubber product companies.

K. C. Frisch and J. H. Saunders, eds., *Plastic Foams* (2 parts), Marcel Dekker, New York, 1972.

The fundamental principles of foam formation in general as well as coverage of the specific varieties of the flexible and rigid types.

J. M. Funt, *Mixing of Rubbers*, RAPRA Technology Ltd., Shawbury, Shrewsbury, Shropshire, SY4 4NR, England, 1977.

Aimed at the production engineer or machine designer, dealing with the mechanical operations of rubber in the processing plant.

J. B. Gomez, *Anatomy of Hevea and its Influence on Latex Production*, Malaysian Rubber Research and Development Board (MRRDB Monograph No. 7), Kuala Lumpur, 1982.

A summary of the present state of knowledge of the anatomy and physiology of Hevea.

J. B. Gomez, *Physiology of Latex (Rubber) Production*, Malaysian Rubber Research and Development Board (MRRDB Monograph No. 8), Kuala Lumpur, 1983.

Provides a summary of several aspects of latex production and relates the background research to the practical problems of the rubber producers.

C. A. Harper, ed., *Handbook of Plastics and Elastomers*, McGraw-Hill Book Co., New York, 1976.

Handbook featuring the important plastic and elastomer materials, and a survey of testing, standards, and specifications within industry.

D. F. Hays and A. L. Browne, eds., *The Physics of Tire Traction, Theory and Experiment*, Plenum, New York, 1974.

Contains the papers and discussions of a symposium held at General Motors Research Laboratories, October 1973, and details fundamental aspects of rubber friction and tire traction.

C. Hepburn and R. J. W. Reynolds, eds., *Elastomers: Criteria for Engineering Design*, Applied Science Publishers, London, 1979.

The published papers of a symposium held at Loughborough University of Technology as a tribute to Dr. Bob Payne. Its subjects provide information on the academic, theoretical, and practical aspects of rubber physics and engineering.

C. Hepburn, *Polyurethane Elastomers*, Applied Science Publishers, New York, 1982.

The essentials of the industrially important and established materials and processes described for people with both technical and semitechnical backgrounds.

W. Hofmann, *Vulcanization and Vulcanizing Agents*, Applied Science Publishers, London, 1967.

Comprehensive survey of the methods of crosslinking and the necessary implementation systems, for the rubber technician.

D. I. James, ed., *Abrasion of Rubber*, trans. by M. E. Jolley, Maclaren & Sons, London, 1967.

Consists of papers presented at the conference on abrasion in Moscow, December 1961, and reviews the problems of wear mechanism of rubber and rubber products, mainly tires.

J. P. Kennedy and E. G. M. Tornqvist, eds., *Polymer Chemistry of Synthetic Elastomers*, 2 vols., Interscience Publishers, New York, 1968.

A comprehensive treatise on synthetic elastomers. Chemical aspects of polymer formation are emphasized rather than polymer physics.

F. J. Kovac, *Tire Technology*, Goodyear Tire & Rubber Co., Akron, OH 44316, 1973.

A small basic tire manual, going into types, components, materials, reinforcing systems, engineering features and building.

G. Kraus, ed., *Reinforcement of Elastomers*, Interscience Publishers, New York, 1965.

A collection of chapters covering the basic phenomena of reinforcement, properties of filled rubbers, and processing operations.

A. Krause, A. Lange, and M. Ezren, *Chemical Analysis of Plastics and Elastomers: A Guide to Fundamental Qualitative and Quantitative Chemical Analysis*, Macmillan, Riverside, NJ, 1982.

Brings together in one volume the classical methods of chemical analysis and modern instrumental analytical methods. An appendix lists service companies and reference spectra.

H. L. Long, ed., *Basic Compounding and Processing of Rubber*, Rubber Division, American Chemical Society, Akron, OH 44325, 1985.

A textbook emphasizing practical rather than theoretical aspects of compounding and designed for Rubber Group use.

W. Lynch, *Handbook of Silicone Rubber Fabrication*, Van Nostrand Reinhold, New York, 1978.

Describes all of the main methods of fabricating silicone rubber, covering the electrical, appliance, automotive, aerospace and medical applications. It is a guide to industry processes, materials and applications.

E. W. Madge, *Latex Foam Rubber*, Maclaren & Sons, London, 1962.

Surveys and analyzes the technical factors that influence the process and product of foam rubbers based on both natural rubber latex and on synthetic rubber latices.

J. E. Mark and J. Lal, eds., *Elastomers and Rubber Elasticity*, ACS Symposium Series No. 193, American Chemical Society, Washington, DC 20036, 1982.

Based on a symposium sponsored by the Division of Polymer Chemistry, American Chemical Society, New York, August 1981, the papers discuss the important advances in the molecular theories of rubberlike elasticity.

D. F. Moore, *Friction and Lubrication of Elastomers*, Pergamon Press, New York, 1972.

Theories for the slide of elastomers over a surface.

D. F. Moore, *Friction of Pneumatic Tires*, Elsevier Publishing Co., New York, 1975.

Deals with the pneumatic tire as a complete system and discusses the tread/pavement interaction under varying conditions.

S. H. Morrell, *Progress of Rubber Technology*, Applied Science Publishers, London.

Annual review of most topics in the industry. Begun in 1937, its last volume was published in 1984. It has been succeeded by the quarterly journal *Progress in Rubber and Plastics Technology*.

M. Morton, ed., *Introduction to Rubber Technology*, Reinhold, New York, 1959.

Covers the compounding of natural and synthetic rubber, as well as testing and properties. Though now out of print, it is still a good introduction to rubber technology.

W. J. S. Naunton, *The Applied Science of Rubber*, Edward Arnold (Publishers), London, 1961.

A comprehensive text-book of rubber technology which combines theory and practice.

R. H. Norman, *Conductive Rubber and Plastics: Their Production, Applications and Test Methods*, Applied Science Publishers, 1970.

This book covers the properties, processing of, and uses for conductive rubbers and plastics.

A. R. Nutt, *Toxic Hazards of Rubber Chemicals*, Elsevier, New York, 1984.

The toxicities and effects of all the chemicals commonly used in rubber processing are noted, and toxicological testing methods and atmospheric monitoring methods used by the industry are reviewed.

A. R. Payne and J. R. Scott, *Engineering Design with Rubber*, Maclaren & Sons, London, 1960.

Discussion of the properties, testing, and design of rubber as an engineering material.

C. K. Riew and J. K. Gillham, eds., *Rubber-Modified Thermoset Resins*, American Chemical Society, Washington, DC 20036, 1984.

Covers chemistry and physics of thermoset polymerization.

W. J. Roff and J. R. Scott, *Handbook of Common Polymers, Fibers, Films, Plastics and Rubbers*, CRC Press, Cleveland, 1971.

Information on high polymers and natural rubber and all types of synthetic rubbers: data, trade names, synonyms, applications, structural characteristics, properties.

W. M. Saltman, ed., *The Stereo Rubbers*, John Wiley & Sons, New York, 1977.

A treatise on the preparation, processing, basic physical, mechanical, and technological properties, and major uses of the elastomers made with solution and stereospecific catalysts.

D. J. Schuring, ed., *Tire Rolling Resistance*, Rubber Division, American Chemical Society, Akron, OH 44325, 1982.

Papers and discussion of a symposium, Chicago, Fall 1982, dealing with the measurement of rolling resistance, conditions that affect rolling loss, and mathematical modelling.

M. Sittig, *Stereo-Rubber and Other Elastomer Processes*, Noyes Development Corp., Park Ridge, NJ, 1979.

A chemical engineering study based primarily on the patent literature supplemented by other commercial information and data.

H. J. Stern, *Rubber: Natural and Synthetic*, 2d Ed., Maclaren & Sons, London, 1967.

An account of all aspects of the subject from polymers synthesis to reclaiming, suitable for anyone with basic scientific training but not necessarily any prior knowledge of rubber technology.

J. A. Szilard, *Reclaiming Rubber and Other Polymers*, Noyes Data Corp., Park Ridge, NJ 07656, 1973.

First book on the subject in many years, supplying detailed technical information. It can also be used as a guide to the U.S. Patent literature in the field.

L. R. G. Treloar, *The Physics of Rubber Elasticity*, 3rd Ed., Clarendon Press, Oxford (U.K.), 1975.

A detailed presentation of the main developments in the field of the equilibrium elastic properties of rubber and the associated theoretical background.

C. M. van Turnhout, *Rubber Chemicals*, D. Reidel Publishing Co., Dordrecht, Holland, 1973.

A revised and enlarged edition of J. van Alphen's book of the same title. It gives names, trade names, and suppliers of many chemicals.

W. C. Wake, *Adhesion and the Formulation of Adhesives*, 2d Ed., Applied Science Publishers, London, 1982.

Emphasizes applications and formulations of adhesives.

W. C. Wake, B. K. Tidd, and M. J. R. Loadman, *The Analysis of Rubber and Rubber-like Polymers*, 3d Ed., Applied Science Publishers, London, 1983.

Coverage of a variety of techniques and applications from the ''burn test'' to those using instrumentation costing many thousands of dollars.

W. C. Wake and D. B. Wootton, *Textile Reinforcement of Elastomers*, Applied Science Publishers, London, 1982.

Discusses from the textile viewpoint the various reinforced rubber structures used industrially with the exception of the tire.

B. M. Walker, *Handbook of Thermoplastic Elastomers*, Van Nostrand Reinhold, New York, 1979.

Providing information on commercially available TPE's, is organized by major elastomer groups, manufacturers, and brand names.

M. A. Wheelans, *Injection Moulding of Rubbers*, Halsted Press, New York, 1974.

Shows how natural rubber may be injection-molded. This book is designed for rubber manufacturers, chemists, technologists, and engineers.

A. Whelan and K. S. Lee, eds., *Developments in Rubber Technology*, 3 vols.: 1. *Improving Product Performance*; 2. *Synthetic Rubbers*; 3. *Thermoplastic Rubbers*, Applied Science Publishers, London, 1979–1981.

Many subjects covered in this series, each chapter written by one who has extensive experience in topic or field.

A. Whelan, *Injection Moulding Machines*, Elsevier Science Publishers, London, 1984.

Detailing of the principles of injection molding.

G. S. Whitby, C. C. Davis, and R. F. Dunbrook, eds., *Synthetic Rubber*, John Wiley & Sons, New York, 1954.

A comprehensive post-World War II publication covering all that was known about synthetic rubber at that time; of great historical interest.

G. G. Winspear, *The Vanderbilt Latex Handbook*, R. T. Vanderbilt Co., New York, 1954.

A source of information for those engaged in the latex branch of the rubber industry.

P. Wright and A. P. C. Cumming, *Solid Polyurethane Elastomers*, Maclaren & Sons, London, 1969.

Emphasizes process and applications, but also considers chemistry, chemical reactions, and analytical techniques.

PERIODICALS

Caoutchoucs & Plastiques, 5, rue Jules-Lefebre, 75009 Paris, France. Monthly.

Practical and technical information. Includes summaries of its papers in English.

Elastomerics, Communication Channels, Inc., 6255 Barfield Rd., Atlanta, GA 30328. Monthly.

A journal covering new developments, business conditions, production, sales, personalities, and forthcoming meetings and events.

European Rubber Journal, Crain Communications Ltd., 20–22 Bedford Row, London WC1R 4EW England. Monthly.

Reviews European technical developments, standards, testing, patents, economic trends, and forecasts.

GAK Gummi Fasern Kunststoffe, A. W. Gentner Verlag GmbH & Co. KG, Postfact 688, Stuttgart 1, FR Germany. Monthly.

Contains articles of practical information, review of patent literature and short abstracts of journal articles.

International Polymer Science and Technology, RAPRA Technology Ltd., Shawbury, Shrewsbury, Shopshire SY4 4NR, England. Monthly.

Includes abstracts of papers from leading Soviet, Eastern European and Japanese periodicals in polymer science and technology, and translation of papers recommended by subscribers.

NR Technology, The Malaysian Rubber Producers' Research Association, England. Quarterly.

Provides scientific and technical information on many aspects of natural-rubber use.

Natural Rubber News, Malaysian Rubber Bureau, 1925 K St., Washington, DC 20005. Monthly.

Provides news and information to consumers of natural rubber in North and South America.

Plastics and Rubber International, The Plastics and Rubber Institute, London. Six issues per year.

A trade publication for the professionals in the plastics and rubber industries.

Plastics & Rubber News, Thomson Publications SA, Johannesburg, SA. Monthly.

The news and official publications of the South African section of the Plastics and Rubber Institute.

Plastics and Rubber Processing and Applications, Elsevier Applied Science Publishers, London. Four issues per year.

The journal publishes original papers at a specialist level, in particular in three areas: processing, the effect of processing on properties, and applications.

Progress in Rubber and Plastics Technology, Plastics and Rubber Institute and RAPRA Technology Ltd., Shawbury, Shrewsbury, Shopshire SY4 4NR England. Quarterly.

Critical and authoritative reviews of rubber and plastics topics.

Rubber & Plastics News and *Rubber & Plastics News II*, Craine Automotive Group Inc., 34 N. Hawkins Ave., Akron, OH 44313. Weekly.

Newspapers, published alternate weeks, containing technical articles and events and developments of interest to the rubber industry.

Rubber Chemistry and Technology, Rubber Division, ACS, The University of Akron, Akron, OH 44325. Five times a year.

Contains original manuscripts and papers of theoretical and technical interest presented at Rubber Division meetings or published elsewhere.

Rubber Developments, The Malaysian Rubber Producers' Research Association, England. Quarterly.

Reviews developments in natural rubber research, technology and use.

Rubber World, Lippincott & Peto Inc., 1867 W. Market St., Akron, OH 44313. Monthly.

A technical service and news magazine for the rubber industry.

Tire Science and Technology, The Tire Society, Inc., Munroe Falls, OH. Quarterly.

Authoritative articles and reviews on the subject.

ABSTRACTS AND INDEXES

Researchers and practitioners concerned with rubbers and polymers have been well-served by two important indexes: *RAPRA Abstracts* and *Chemical Abstracts*. The *Bibliography of Rubber Literature* is also a useful reference for publications prior to 1972.

RAPRA Abstracts.

Rubber and Plastics Research Association of Great Britain began publication of its monthly *Rubber Abstracts* in 1922, and in 1972 a computer-readable form of the printed abstracts also became available. This data base offers worldwide access to commercial and technical aspects of these industries, including machinery, raw materials, compounding, synthesis and polymerization, properties, testing, and applications. The printed fortnightly journal *RAPRA Abstracts* is available from RAPRA Technology Ltd., Shawbury, Shrewsbury, Shropshire, SY4 4NR, England; the online data base is available from Pergamon International Information Corporation, 1340 Old Chain Bridge Road, McLean, VA 22101.

Chemical Abstracts: Section 39—Synthetic Elastomers and Natural Rubber.

Chemical Abstracts (*CA*) has indexed the world's chemical literature since it began publication in 1907. Its online data base contains records for documents covered in printed *CA* for 1967–present.

Section 39 includes the analysis, preparation, manufacture, testing, processing, and composition of synthetic elastomers and natural rubber, and the chemicals used in their manufacture. Also included is chemical engineering related to the production of synthetic rubbers and rubber chemicals and to the fabrication of tires and preparation of the components.

CA is published weekly by Chemical Abstracts, a Division of the American Chemical Society, P.O. Box 3012, Columbus, OH 43210. *CAS Online* is available from The Scientific & Technical Information Network, c/o Chemical Abstracts Service, P.O. Box 02228, Columbus, OH 43202; Dialog Information Services, 3460 Hillview Avenue, Palo Alto, CA 94304; BRS, 1200 Route 7, Latham, NY 12110; and Pergamon International Information Corporation, 1340 Old Chain Bridge Road, McLean, VA 22101.

Bibliography of Rubber Literature.

An annual volume of the Rubber Division of the American Chemical Society that provided coverage of the periodical and patent literature from 1935 until 1972, when it ceased publication. The entries were grouped by subject classes and included a short abstract. Each volume also included an author index.

DICTIONARIES, DIRECTORIES, REFERENCES

Blue Book—Materials, Compounding Ingredients and Machinery for Rubber, by *Rubber World Magazine*, Lippincott & Peto, Akron, OH 44313.

An annual directory, its intention being to furnish the compounder, chemist, researcher, purchasing agent, or market agent with a short sketch of the properties of these materials, and the how and why they are used in rubber-making processes.

A. S. Craig, *Dictionary of Rubber Technology*, Philosophical Library, New York, 1969.

Some of the more important entries expanded into short articles and sources for further reading given for many entries.

Compilation of ASTM Standard Definitions, American Society for Testing and Materials, 3d ed., 1916 Race St., Philadelphia, PA, 1976.

A compilation of terms to promote and encourage preparation and use of standard definitions. The terms are listed alphabetically, each definition followed by the designation of the ASTM standard containing it together with the ASTM committee having jurisdiction.

A. F. Dorian, ed., *Six-Language Dictionary of Plastics and Rubber Technology*, Iliffe Books, London, 1965.

A comprehensive dictionary in English, German, French, Italian, Spanish, and Dutch.

Elastomeric Materials: A Desk Top Data Bank, International Plastics Selector, San Diego, CA, 1977.

Compilation of specific technical data of more than 1400 individual commercially available elastomers.

Encyclopedia of Polymer Science and Engineering, 2d Ed., editorial board H. F. Mark, N. M. Bikales, C. Overberger, G. Menges; J. I. Kroschwitz, ed.-in-ch. (Vols. 1–4); additional volumes to a total of 20 will be available as published. John Wiley & Sons, New York, 1985.

Articles in these volumes include polymer descriptions, processes, and uses. Bibliographies supply supporting references to articles of review or particularly good subject treatments.

"Glossary of Terms Relating to Rubber and Rubber-Like Materials," Special Technical Publication 184A, American Society for Testing and Materials, 1916 Race St., Philadelphia PA, 1972.

Authoritative definitions that reflect the state-of-the-art of the rubber industry.

K. F. Heinish, *Dictionary of Rubber*, John Wiley & Sons, New York, 1974 (English edition of *Kautschuk-Lexikon*, Gentner Verlag, Stuttgart, 1966).

Provides compositions and properties of various commercial products, equivalent materials, concepts, jargon, abbreviations, and so forth. This dictionary also includes obsolete products, outdated concepts, and processes.

W. Hofmann, *Kautschuk-Technologie*, Gentner Verlag, Stuttgart, 1980.

In German, a detailed book, encyclopedia in scope, on rubber technology.

H. E. Horton, *Plastics and Rubber Machinery in Four Languages*, Elsevier, London, 1970.

Detailed manufacturers' descriptions, with drawings, of processing machinery in English, German, French, and Spanish. International symbols, conversion tables, and a four-language dictionary are included.

K. P. Jones, *Thesaurus of Rubber Technology*, Malayan Rubber Fund Board, The Natural Rubber Producers' Research Association, Welwyn Garden City, U.K. (England) 1972.

Slanted toward natural rubber, consisting of main heads under which words appear and coded to indicate relation to head word.

R. E. Kirk and D. F. Othmer, eds., *Concise Encyclopedia of Chemical Technology*, Wiley-Interscience, New York, 1985.

A single-volume condensation of all subjects covered in the 25-volume *Kirk-Othmer Encyclopedia of Chemical Technology*, 3d ed.

R. E. Kirk and D. F. Othmer, eds., *Encyclopedia of Chemical Technology*, 3d Ed., Wiley-Interscience, New York, 1979.

Gives properties and describes manufacture of many substances. The rubber articles summarize many aspects of rubber technology, giving bibliographies to journal and patent literature.

L. R. Mernaugh, ed., *Rubbers Handbook*, Morgan-Grampian, London, 1969.

Covers aspects of rubber pertinent to engineers. Part 1 gives basic characteristics and properties, Part 2 shows particular applications, and Part 3 covers specific rubbers.

New Trade Names in the Rubber and Plastics Industries, Pergamon Press Inc., Maxwell House, Fairview Park, Elmsford, NY 10523.

An annual volume of new trade names applicable to the rubber and plastics industries, as noted by RAPRA from trade literature, journals, and books. The trade names are listed in alphabetical order, each followed by a description, company, and reference.

Rubber Red Book, Communication Channels, Inc., 6285 Barfield Rd., Atlanta, GA 30328.

An annual directory of manufacturers and suppliers to the rubber industry. Product listings for machinery and equipment, chemicals, fabrics, natural, synthetic, and reclaimed rubber are also covered. Information on courses in rubber chemistry and technology is provided. Trade and technical organizations throughout the world that deal with rubber are also listed.

Rubbicana, by *Rubber & Plastics News*, Akron, OH 44313.

An annual directory of rubber product manufacturers and rubber industry suppliers in North America. Tradename information and information about industry associations and societies is also included.

Rubbicana Europe, by *European Rubber Journal*, 25 Bedford Sq., London WC 1B 4HG. England.

Annual directory, started in 1985, this volume provides names of suppliers of raw materials, equipment and services to the rubber and polyurethane industries in Europe.

The Semperit Tyre Dictionary, 3d ed., Semperit Aktiengesellschaft, Austria, 1973.

A trilingual tire dictionary.

Synthetic Rubber, International Institute of Synthetic Rubber Producers, Inc., 2077 S. Gessner St., Houston, TX 77063.

Includes world rubber consumption and production statistics for synthetic rubber and the capacities of world production facilities for general and specialty rubbers. The producers of these rubbers are also chronicled. An annual compilation.

STATISTICS (REPORTS)

International Rubber Digest, International Rubber Study Group, Brettenham House, 5-6 Lancaster Place London, England. Monthly.

Report on the natural rubber market and the main influences underlying rubber price movements.

Natural Rubber News, Malaysian Rubber Bureau, 1925 K. St., N.W., Washington, DC 20006. Monthly.

Newsletter presenting topical developments in rubber, and industry statistics.

Rubber Manufacturers Association Industry Rubber Report, Rubber Manufacturers Association, 1400 K. St. Washington DC 20005. Monthly.

Statistics covering U.S. production, imports, exports, stocks, and apparent consumption of natural and synthetic rubber.

Rubber Statistical Bulletin, International Rubber Study Group, Brettenham House, 5-6 Lancaster Place London, England. Monthly.

Statistical tables dealing with production, consumption, imports, exports, stocks and prices of natural and synthetic rubber in most of the major producing and consuming countries.

Rubber Trends, Economist Intelligence Unit, Spencer House, 27 St. James Pl., London, England. Quarterly.

A review of production, markets, prices, and so forth.

The Synthetic Rubber Manual, International Institute of Synthetic Rubber Producers Inc., 2077 S. Gessner Rd., Houston, TX 77063. Annual.

Tabulates the synthetic rubbers according to their method of manufacture and general physical and chemical properties.

STANDARDS

American National Standards Institute, 1430 Broadway, NY 10018.

Serves as a clearinghouse for nationally coordinated voluntary safety, engineering, and industrial standards. It publishes an annual catalog of all approved American National Standards and periodic supplements.

Annual Book of Standards, Part 09.01: ''Rubber, Natural and Synthetic—General Test Methods, Carbon Black''; Part 09.02: ''Rubber Products, Industrial Specifications and Related Test Methods; Gaskets; Tires,'' American Society for Testing and Materials, 1916 Race St., Philadelphia, PA 19103. Annual.

Geared particularly toward development of standards.

CONFERENCES

Rubber Division, American Chemical Society, The University of Akron, Akron, OH 44325.

This Division was created in 1919 and is the oldest and largest body devoted to the chemistry and technology of rubber. National meetings are held twice a year at various locations in the U.S. and Canada; the technical presentations are avilable for purchase. Every two years, a rubber trade show is sponsored in association with the technical program.

International Rubber Conferences (RUBBERCON)

Annual conferences held in a different country each year and sponsored by a leading trade organization of that country. Its programs are international in scope and proceedings are published.

HISTORIES

T. R. Dawson, *Rubber Industry in Germany During the Period 1939–1945* (British Intelligence Objectives Sub-Committee, Overall Report No. 7), HMSO, London, 1948.

Summary of the technical information collected at governmental level after World War II concerning wartime activities in Germany and Japan.

J. D. D'Ianni, *Alfin Rubber History*, American History Research Center, The University of Akron, Akron, OH 44325, 1984.

Summary of the program on research in Alfin polymerization, 1945–1947.

J. H. Drabble, *Rubber in Malaya, 1876–1922: The Genesis of the Industry*, Oxford University Press, London, 1973.

The establishment of rubber cultivation in Malaya is traced through combination of capital, labor, entrepreneurial and government activity.

H. Fry, *In Tribute to the Chemists Who Tame Rubber: Celebrating the 75th Anniversary of the ACS Rubber Division*, Suppl., *Rubber & Plastics News*, Akron, OH 44325, 1984.

The story of the chemists who have played, and are playing, a major role in rubber chemistry's history.

W. C. Geer, *The Reign of Rubber*, The Century Co., New York, 1922.

A record of the early successes, failures, and hopes of lives spent in creating rubber products.

C. Goodyear, *Gum-Elastic*, facsimile of 1855 edition by *The India-Rubber Journal* Vol. I, *Gum-Elastic and its Varieties, with a Detailed Account of its Applications and Uses and of the Discovery of Vulcanization*; Vol. II, *The Applications and Uses of Vulcanized Gum-Elastic; with Descriptions and Directions for Manufacturing Purposes*, Maclaren and Sons, London, 1937.

Charles Goodyear's own account of the early history of rubber and his discovery of vulcanization, as well as descriptions of its applications.

V. Herbert and A. Bisio, *Synthetic Rubber: A Project That Had to Succeed*, Greenwood Press, Westport, CT, 1985.

The development of synthetic rubber from its inception, through its beginnings, up to its current relatively mature status is encapsulated.

F. A. Howard, *Buna Rubber: The Birth of an Industry*, Van Nostrand, New York, 1947.

The story of synthetic rubber from its beginnings in foreign patents and research, through World War II.

P. E. Hurley, *History of Natural Rubber*, publ. by *J. Macromol. Sci.* (Chem.) **A15,** 1279 (1981); also in *History of Polymer Science and Technology*, R. B. Seymour, ed., Marcel Dekker, New York, 1982, p. 215.

The development of natural rubber into a key industrial material is chronicled.

P. Mason, *Cauchu The Weeping Wood: A History of Rubber*, The Australian Broadcasting Commission, Sydney, 1979.

A light-readable and illustrated history of rubber, going back to its origins in 1500 up through World War II and on briefly to the present.

F. M. McMillan, *The Chain Straighteners*, Macmillan, London, 1979.

The book is an interesting account of the discoveries and developments that have been the basis for the growth in production of stereoregular high polymers. Ziegler and Natta are the principle characters.

M. Morton, "History of Synthetic Rubber," *J. Macromol. Sci. (Chem.)*, **A15,** 1289 (1981); also in *History of Polymer Science and Technology*, R. B. Seymour, ed., Marcel Dekker, New York, 1982, p. 225.

An account of the synthetic-rubber project, including the post-World War II period.

R. A. Solo, *Across the High Technology Threshold, the Case of Synthetic Rubber*, Norwood Editions, Norwood, PA, 1980.

A case in the technological development of synthetic rubber under governmental direction.

E. Tompkins, *History of the Pneumatic Tire*, Eastland Press, London, 1981.

Recounts the development of automobile tire and rim construction, improvements in tread patterns and tread compounds, and the development of tire science, with emphasis on the contributions by Dunlop over the years.

R. F. Wolf, *India Rubber Man; The Story of Charles Goodyear*, The Claxton Printers, Caldwell, ID, 1939.

The first biography of Charles Goodyear.

H. Wolf and R. Wolf, *Rubber: A Story of Glory and Greed*, Covici Friede, New York, 1936.

A readable history of rubber from the discovery of the raw material, to its promotion by invention and research, then on to rubber as a big business.

INDEX